Birkhäuser

Classic Texts in the Sciences

Classic Texts in the Sciences offers essential readings for anyone interested in the origin and roots of our present-day culture. Considering the fact that the sciences have significantly shaped our contemporary world view, this series not only provides the original texts but also extensive historical as well as scientific commentary, linking the classical texts to current developments. Classic Texts in the Sciences presents classic texts and their authors not only for specialists but for anyone interested in the background and the various facets of our civilization.

More information about this series at https://link.springer.com/bookseries/11828

Michel Janssen • Jürgen Renn

How Einstein Found His Field Equations

Sources and Interpretation

Michel Janssen
School of Physics and Astronomy
University of Minnesota
Minneapolis, MN, USA

Jürgen Renn
Max Planck Institute
for the History of Science
Berlin, Germany

ISSN 2365-9963 ISSN 2365-9971 (electronic)
Classic Texts in the Sciences
ISBN 978-3-030-97957-7 ISBN 978-3-030-97955-3 (eBook)
https://doi.org/10.1007/978-3-030-97955-3

This book is published under the imprint Birkhäuser, www.birkhauser-science.com by the registered company Springer Nature Switzerland AG
The registered company address is: Gewerbestrasse 11, 6330 Cham, Switzerland

Dedicated to John Stachel,
Einstein Studies Nestor
and Mentor Extraordinaire

Preface

"Today you should finally be happy with me," Albert Einstein wrote to his friend and colleague Paul Ehrenfest in Leiden in early 1916, "I will not rely on the papers at all but will go through all the calculations for you" (see Ch. 8 of Pt. II of this volume). It is in this spirit that we assembled this volume. In Pt. I, we offer a concise and largely non-technical account of the tortuous path that took Einstein away from field equations extracted from the Riemann curvature tensor (see the 1912–13 Zurich notebook documenting his early search for gravitational field equations in collaboration with Marcel Grossmann) only to lead him back to them two and a half years later (see his four short communications to the Berlin academy of science in November 1915). In Pt. II, we present, both in the original German and in English translation, these four papers of November 1915, the most important pages of the Zurich notebook, and (the relevant parts of) eight other documents that together, we hope, will allow the reader to experience vicariously what arguably was the most thrilling intellectual ride of Einstein's life. As Einstein did for Ehrenfest in the letter quoted above, we cover the key calculations in these sources in great detail—using, where helpful, modernized notation—so that readers familiar with general relativity at the level of an undergraduate course on the subject should be able to follow along without having to take out pen and paper. Readers who want to keep a textbook on general relativity at hand as they work their way through some of the derivations in this volume have many options. At the risk of dating ourselves, we mostly consulted Weinberg (1972).

Physicists we talk to are often puzzled that historians working on Einstein and general relativity have spent so much time and effort on Einstein's formulation of the field equations. Modern textbooks tend to devote only a handful of pages to the introduction of the field equations. Once one accepts that the potential of Newtonian gravitational theory should be replaced by the metric tensor, it is easy enough to make it plausible that the Poisson equation for the gravitational field produced by a given mass distribution should be replaced by an equation that sets the Einstein tensor equal to the energy-momentum tensor of matter (for an elegant example of this type of plausibility argument, see Carroll 2004, sec. 4.2, pp. 155–159). That we want a generally-covariant tensor of the same rank as the energy-momentum tensor with second-order derivatives of the metric directs us to the Ricci tensor. If we want field equations that guarantee energy-momentum conservation, we need to add another term, thus replacing the Ricci tensor by the Einstein tensor, which has a vanishing covariant divergence courtesy of the contracted Bianchi identities. Another way to proceed is to note that the Riemann curvature scalar is the natural candidate for the Lagrangian for the metric field, to show that the Euler-Lagrange equations for this Lagrangian set the Einstein tensor equal to zero and to generalize these vacuum field equations to ones in the presence of matter by setting the Einstein tensor equal to the energy-momentum tensor of matter. This is essentially the way in which Einstein introduces the field equations in the last paper presented in this volume (Einstein 1916d, Pt. II, Ch. 10). Yet, even in that paper, published in November 1916, Einstein does not appeal to the contracted Bianchi identities to prove that the equations imply energy-momentum conservation. He also does not explicitly evaluate the Euler-Lagrange equations. Instead he follows step-by-step the argument in a review article published two years earlier, in November 1914, for a version of the theory with field equations of severely limited covariance (Einstein 1914c, Pt. II, Ch. 3)—a review article that, in March 1916, he had already replaced with one for the November 1915 version of the theory (Einstein 1916b, Pt. II, Ch. 9).

As these preliminary remarks indicate, Einstein's path to the Einstein field equations was considerably more complicated than their introduction in modern textbooks would lead us to expect. It is also strikingly different from the way Einstein later claimed he found these equations. This makes the convoluted story of how Einstein actually arrived at them all the more irresistible to us as historians. The story serves as a forceful reminder that the way Einstein thought about his theory as he was developing it was markedly different in several respects from the way he himself later thought about it and even more so from the way modern relativists think about it. Both in the essay in Pt. I and in our commentary on the sources in Pt. II, we have tried to recapture as best as we can how Einstein originally conceived of his theory. This will also help the reader understand better, we hope, the infamous "competition" between Einstein and David Hilbert in the "race" toward the final version of the field equations. We will argue that Einstein's work in November 1915 owed nothing to Hilbert's while Hilbert's owed much to Einstein's work of 1913–14. Drawing attention to features of the theory that have since fallen by the wayside (such as energy-momentum conservation giving rise to conditions restricting the covariance of the theory), we will argue that Hilbert did not arrive at the Einstein field equations before Einstein, even though he almost certainly had identified the Riemann scalar as the Lagrangian for the metric field before Einstein had hit upon the field equations that, as we now recognize, follow immediately from this Lagrangian.

General relativity, however, was by no means the work of "a lone genius" (Janssen and Renn 2015b). First of all, one of the central claims we will argue for in this volume is that Einstein in developing his theory for the gravitational field was guided predominantly not by considerations of mathematical elegance, as he would maintain in his later years, but by an elaborate analogy with the Maxwell-Lorentz theory for the electromagnetic field reformulated in the four-dimensional formalism of special relativity by Hermann Minkowski, Arnold Sommerfeld, Max Laue and others. Moreover, although he did not always acknowledge this in print, Einstein, as he was developing his theory, received substantial help and feedback from a number of individuals, some well-known, others long forgotten, such as Paul Bernays, Michele Besso, Paul Ehrenfest, Adriaan D. Fokker, Erwin Freundlich, Marcel Grossmann, Paul Hertz, Friedrich Kottler, Tullio Levi-Civita, Hendrik A. Lorentz, Gunnar Nordström and Moritz Schlick.

More so than general relativity itself, our account of its genesis—and that goes both for the introductory essay in Pt. I and for the transcriptions, translations and annotations in Pt. II—is the result of elaborate team efforts. In fact, the main goal of this volume is to bring together in one place what we see as some of the most important results of these collaborations. As things stand, the story of how Einstein found his field equations has to be pieced together from sources scattered over several volumes. The most important collection of sources we mined for this book are Vols. 1 and 2 of *The Genesis of General Relativity* (Renn 2007a), which we co-authored with John Norton, Tilman Sauer and John Stachel. The four volumes of *Genesis*, together taking up about 2,000 pages, were the final product of a long-term research project of *Abteilung I* of the *Max-Planck-Institut für Wissenschaftsgeschichte* in Berlin, which actually predated the founding of the institute (see Castagnetti et al. 1993, and the preface in Vol. 1 of *Genesis*). The centerpieces of Vols. 1 and 2 are a facsimile and a transcription of the Zurich notebook and a detailed commentary on the 57 pages of the notebook that deal with gravity. This commentary and the accompanying essays in these volumes built on the annotation of the notebook and other documents related to the development of general relativity in Vols. 3–8 of *The Collected Papers of Albert Einstein* and on papers in various volumes in the series *Einstein Studies*, many of them based on talks given at conferences dedicated to the history of general relativity held between 1986 and 2005.[1] Two ground-breaking papers in Vol. 1 of *Einstein Studies*, written by two of our co-authors and mentors deserve to be singled out: John Norton's (1984) "How Einstein Found his Field Equations, 1912–1915" (the title of which we recycled for this volume) and John Stachel's (1989) "Einstein's Search for General Covariance, 1912–1915," based on a talk he gave in Jena in 1980.

1. *Einstein Studies* Vols. 1, 3, 5, 7, 11 and 12 are based on proceedings of these conferences (Howard and Stachel 1989; Eisenstaedt and Kox 1992; Earman, Janssen, and Norton 1993; Goenner et al. 1999; Kox and Eisenstaedt 2005; Lehner, Renn, and Schemmel 2012). Also relevant are Vol. 6, based on a conference on Mach's principle (Barbour and Pfister 1995), and Vol. 14, based on a conference to celebrate the centenary of general relativity (Rowe, Sauer, and Walter 2018).

For the introductory essay, we drew on several earlier efforts of both of us, some of them in collaboration with Hanoch Gutfreund, to give a concise summary of our findings accessible to a wider audience (Janssen 2004, 2005, 2014; Renn 2006, 2020a, 2020b; Gutfreund and Renn 2015, 2017). We drew on an early draft of this essay for an article in *Physics Today* to commemorate the centenary of the Einstein field equations (Janssen and Renn 2015a) as well as for similar contributions to German and French periodicals (Janssen and Renn 2015c, 2015d, 2015e). A general discussion of the metaphor of "arches and scaffolds" used in our *Physics Today* article and in the introductory essay in this volume can be found in Janssen (2019). Similar ideas are presented in Renn (2006, 2020a). The essay also informed and was informed by several talks we gave on the occasion of the centenary of the Einstein field equations. Renn spoke, for instance, at a conference at the Perimeter Institute in Waterloo in June 2015;[2] Janssen at a conference at the Institute for Advanced Study in Princeton in November 2015.[3]

Given that we are drawing on work that began three decades ago, we owe a debt of gratitude to a great many people. We will only list the most important ones. First of all, we thank our co-authors of Vols. 1 and 2 of *Genesis*, John Norton, Tilman Sauer and John Stachel. We dedicate this volume to John Stachel, the Nestor of this quintet and the founding editor of both *The Collected Papers of Albert Einstein* and (together with Don Howard) the *Einstein Studies* series. But the *Genesis* project involved many more people than the five of us. We would be remiss if we did not at least mention four of them. Heinz Reddner led the group responsible for a new transcription of the Zurich notebook that we reused for this volume. Matthias Schemmel not only co-edited Vols. 3 and 4 but was a valued participant in the discussions out of which the commentary on the notebook and the accompanying essays in Vols. 1 and 2 grew. The late Peter Damerow was closely involved with the project from its inception and made sure we did not lose sight of the forest for the trees. Lindy Divarci was the assistant editor for all four volumes.

Lindy also coordinated the work on this volume, supervising work-study students, keeping them (and us!) on track, and oversaw its production. This volume would never have seen the light of day without her. Her trusted lieutenant in this endeavor was Bendix Düker, who prepared the LaTeX code of most transcriptions and translations in Pt. II of this volume.

We are also grateful to editors and staff (past and present) of the Einstein Papers Project, especially its current director Diana Kormos Buchwald (who kindly shared the electronic files for transcriptions and translations used in Pt. II), its former director Robert Schulmann, editors Dan Kennefick, Anne Kox, Christoph Lehner and Jeroen van Dongen, translators Alfred Engel and Ann Hentschel (whose work we used as the basis for the translations

---

[2] See <`https://www.youtube.com/watch?v=bj8rZnOUjWU`> for a video of this talk.

[3] See <`https://www.youtube.com/watch?v=-n6-9IgiqHo`> for a video of this talk.

in this volume) and editorial assistant Rita Fountain (responsible for the transcription of the Einstein-Besso manuscript reused for this volume).

We have also been inspired by the work of some younger colleagues, who have taken over the baton from the *Genesis* crew, such as Alex Blum and Roberto Lalli (see, e.g., Blum, Lalli, and Renn 2015, 2020) and Dennis Lehmkuhl (2014, 2022).

That still does not exhaust the list of people who helped us with this volume in one way or another. With apologies to those we forgot to include here, we thank Chaya Becker, Laurent Besso, Steven Blau, Kaća Bradonjić, Ernie Brandt, Jed Buchwald, Leo Corry, Jeff Crelinsten, Siska De Baerdemaeker, Tony Duncan, Suzanne Durkacs, John Earman, Johanna Ege, Ole Engler, Sam Fletcher, Allan Franklin, Jan Guichelaar, Ted Jacobson, Michael Janas, Christian Joas, Shadiye Leather-Barrow, Ad Maas, Charles Midwinter, the late Ted Newman, Jim Peebles, Serge Rudaz, Rob "Ryno" Rynasiewicz, Urs Schoepflin, Bernard Schutz, Chris Smeenk, George Smith, Laurent Taudin, Roberto Torretti and Alex Wheeler.

We are grateful to Sarah Kempf, Leonie Kunz and Dorothy Mazlum at Springer for making the production process of this volume run as smoothly as we could have hoped for.

We thank Jürgen Jost for the invitation to publish this volume in the series *Klassische Texte der Wissenschaft*. We thank the *Alexander von Humboldt Stiftung* for granting one of us (MJ) a Humboldt Research Award, which made our transatlantic cooperation possible. Finally, we thank the Albert Einstein Archives at the Hebrew University of Jerusalem for their permission to reproduce the generous selection of Einstein documents presented in this volume and to quote liberally from many more.

Minneapolis, Minnesota, USA *Michel Janssen*
Kleinmachnow, Brandenburg, Germany *Jürgen Renn*

July 2021

Contents

Part I
Essay: How Einstein Found His Field Equations

Chapter 1
Overview of the contents of this volume

November 25, 2015 marked the centenary of Albert Einstein's submission of a short paper, "The Field Equations of Gravity," to the Prussian Academy of Sciences in Berlin. In this paper, Einstein first proposed what are now known as the *Einstein field equations*, the jewel in the crown of his scientific career, the general theory of relativity.[1]

The November 25 paper was the last in a series of short communications submitted to the Berlin Academy on four consecutive Thursdays that month (Einstein 1915a, 1915b, 1915c, 1915d).[2] In the first, Einstein replaced the field equations he had published two years earlier with his friend and colleague Marcel Grossmann by equations within hailing distance of the Einstein field equations (Fig. 1.1). The old equations—known as the *Entwurf* ('outline' or 'draft') field equations after the title of Einstein and Grossmann's (1913) paper—are only covariant (i.e., retain their form) under a severely restricted class of coordinate transformations. The Einstein field equations are generally covariant: they retain their form under arbitrary coordinate transformations. The field equations of Einstein's first November communication, while not yet generally covariant, are covariant under a much broader class of coordinate transformations than the *Entwurf* field equations. In his second communication of November 1915, Einstein adopted a highly speculative hypothesis about the nature of matter to make these equations generally covariant. In the fourth, he achieved the same goal by tweaking the field equations of the first paper in a different and more convincing way (Fig. 1.2).

Throughout the period under consideration here (i.e., from late 1912 to late 1916), Einstein conflated general covariance and general relativity of motion.

---

[1] To commemorate this event, we published an article based on a draft of this introductory essay in the November 2015 issue of *Physics Today* (Janssen and Renn 2015a).

[2] These four papers along with all other Einstein documents (papers, manuscripts, letters) cited in this volume can be found in Vols. 3–8 of *The Collected Papers of Albert Einstein*. These volumes, referred to hereafter as CPAEX (with X the volume number), are prepared by the *Einstein Papers Project*, housed, since 2000, at Caltech, under the directorship of Diana Kormos Buchwald. The 15 volumes that have appeared so far (all with companion volumes with English translations) cover writings and correspondence up to May 1927. They are freely available on-line at www.einstein.caltech.edu. The *Einstein Archive* at Hebrew University in Jerusalem has also made most of its holdings freely available at *Einstein Archives Online* (EAO) at www.alberteinstein.info.

Throughout this volume we will refer to Einstein's correspondence and manuscripts via the document numbers in CPAE3–8. We generally follow the translations in the companion volumes to the Einstein edition but sometimes (silently) deviate from them.

M. Janssen, J. Renn, *How Einstein Found His Field Equations*, Classic Texts in the Sciences, https://doi.org/10.1007/978-3-030-97955-3_1

Gesamtsitzung vom 4. November 1915

$$G_{im} = \{il, lm\} = R_{im} + S_{im}$$

$$R_{im} = -\frac{\partial \left\{ {im \atop l} \right\}}{\partial x_l} + \sum_{\rho} \left\{ {il \atop \rho} \right\} \left\{ {\rho m \atop l} \right\}$$

$$S_{im} = \frac{\partial \left\{ {il \atop l} \right\}}{\partial x_m} - \left\{ {im \atop \rho} \right\} \left\{ {\rho l \atop l} \right\}.$$

Figure 1.1 In Einstein's communication to the Berlin Academy of November 4, 1915, the Ricci tensor, G_{im}, is split into two parts (the indices take on the values 1 through 4). The quantities $\{\ldots\}$ are the Christoffel symbols of the second kind (for a definition, see Ch. 3, note 7 below). Both parts transform as tensors under what today are called unimodular transformations, for which the transformation determinant equals 1 and under which g, the determinant of the metric field, transforms as a scalar (see Ch. 3, note 7). As his new field equations, Einstein proposed $R_{im} = -\kappa T_{im}$, where T_{im} is the energy-momentum tensor for matter and κ is a gravitational constant proportional to Newton's. These are equations of broad but not yet general covariance.

$$G_{im} = -\varkappa \left(T_{im} - \frac{1}{2} g_{im} T \right),$$

$$\sum_{\rho\sigma} g^{\rho\sigma} T_{\rho\sigma} = \sum_{\sigma} T_{\sigma}^{\sigma} = T$$

Figure 1.2 In Einstein's communication to the Berlin Academy of November 25, 2015, the Einstein field equations make their first appearance. Three weeks earlier, Einstein had proposed the field equations $R_{im} = -\kappa T_{im}$, which retain their form under unimodular transformations (Fig. 1.1). The following week, he had shown that, as long as the trace T of the energy-momentum tensor for matter vanishes, these equations can be seen as generally covariant equations, $G_{im} = -\kappa T_{im}$, expressed in unimodular coordinates (in which $g = -1$ and $G_{im} = R_{im}$). To guarantee that $T = 0$, Einstein had assumed that all matter is ultimately made up out of electromagnetic fields. The addition of a term with the trace T in the November 25 communication obviated the need for this questionable assumption.

We now know that the two have nothing to do with each other.[3] In this sense, the formulation of generally covariant field equations in November 1915 was

---

[3] A spatial analogy will illustrate the difference. The shortest flight route between two cities is a segment of the great circle connecting them. Suppose we have a map of the region on which this route is represented by a straight line. By switching to a different map (most of them not very practical) we can turn any route between these two cities into a straight line. That is the analogue of general covariance. This does not mean, however, that all routes are equivalent. No matter what it looks like on any given map, the segment of the great circle remains the shortest route. General covariance, likewise, does not make all trajectories in space-time equivalent, which would have to be the case if all states of motion were equivalent.

For further discussion of Einstein's failed quest to generalize the principle of relativity see Janssen (2012, 2014) and references therein.

an empty victory. Despite its misleading name, however, general relativity was—and continues to be—a powerful theory of gravity, firmly based on the equivalence principle, which in its mature form states that both the spacetime geometry and gravity should be represented by the metric field (Einstein 1918b).[4]

In the third communication of November 1915, based on the field equations of the second but unaffected by their modification in the fourth, Einstein supplied the 43 seconds of arc per century missing in the Newtonian account of the perihelion motion of Mercury.[5]

This volume commemorates these milestones in Einstein's career.[6] Its centerpiece are the four November 1915 communications to the Berlin Academy (Einstein 1915a, 1915b, 1915c, 1915d). This flurry of papers was prompted by Einstein's realization in October 1915 that the 1913 *Entwurf* field equations were untenable.

Einstein's search for satisfactory field equations for the metric field, whose ten components served as the gravitational potentials in his new theory of gravity, began shortly after he moved from Prague to Zurich in July 1912 to take up a professorship at his *alma mater*, the *Eidgenössische Technische Hochschule* (ETH) in Zurich.[7] The mathematician Marcel Grossmann, his

---

[4] What Einstein made relative in general relativity is not motion but gravity. Two observers, one moving on a geodesic, the other moving on a non-geodesic, will disagree about how to split what is now often called the inertio-gravitational field, a guiding field affecting all matter the same way, into an inertial and a gravitational component (Janssen and Renn 2007, p. 839, note 1; Janssen 2012).

[5] This is the only one of the four that Einstein presented in person in a session of the Berlin Academy (Gutfreund and Renn 2015, p. 30). This volume by Hanoch Gutfreund and one of us also contains an account written for a general audience of "Einstein's intellectual odyssey to general relativity" (ibid., pp. 7–36).

[6] As we noted in the preface, the introductory essay and the annotation of most of the documents presented in this volume are based on Vols. 1–2 of *The Genesis of General Relativity* (Renn 2007a), especially our joint chapter in Vol. 2, "Untying the knot ..." (Janssen and Renn 2007). The analysis in *Genesis*, in turn, builds on a pair of classic papers by two of its co-authors, John Stachel (1989) and John Norton (1984). Renn (2006) makes this analysis accessible to a broader audience and looks at the development of general relativity from the wider perspective of the transformation of systems of knowledge in general. See also Renn (2020a, especially, pp. 134–141).

[7] In his book *Einstein in Bohemia*, Michael Gordin takes earlier commentators to task for giving short shrift to Einstein's work on gravity in Prague:

> That [Einstein] would eventually solve the problem of general relativity [read: find a satisfactory new theory of gravity] was not obvious in 1912, yet implicitly historians and physicists have tended to assume this by concentrating on Einstein's path once he arrived in Zurich—that is, when he was on a course that would eventually manifest as "the right track"—thus relegating the Prague theory to the shadows as an unsuccessful distraction. If what you are interested in is understanding the theory Einstein unveiled in Berlin in November 1915, this approach makes sense. If, on the other hand, we want to appreciate Einstein as a physicist like many others, wandering into mistaken cul-de-sacs and admitting his mistakes, and also to understand how the path toward one of the most successful physical theories

former classmate and new colleague at the ETH, assisted him in this search.[8] Notes of their collaboration, which resulted in their joint *Entwurf* paper, have been preserved in the now famous Zurich notebook.[9] This notebook shows that Einstein, before settling on the *Entwurf* field equations, had already considered and rejected the field equations he published in the first paper of November 1915 (as a comparison of Fig. 1.1 to Fig. 1.3 shows).[10]

Shortly after publication of the *Entwurf* theory and in collaboration with Michele Besso, his close friend and scientific confidant, Einstein calculated the sun's contribution to the motion of Mercury's perihelion predicted by the theory. The fruits of their labor have been preserved in the so-called Einstein-Besso manuscript (CPAE4, Doc. 14).[11] The theory as it stood could only account for 18 of the 43 missing seconds of arc of Mercury's perihelion motion, but Einstein put the experience he gained with this type of calculation to good use in his perihelion paper of November 1915.

> of all time was deeply rooted in the opportunities and constraints imposed by Prague, we must treat the static theory as Einstein did: very seriously (Gordin 2020, p. 52).

Pace Gordin, our goal is not "understanding the theory Einstein unveiled in Berlin in November 1915" but answering the question how Einstein arrived at that theory. The answer, we will argue, is that it was built like an arch on the scaffold provided by the theories that preceded it, both the *Entwurf* theory developed in Zurich (Einstein and Grossmann 1913) and the static theory developed in Prague (Einstein 1912a, 1912b). Here are two concrete examples for the latter. The static theory was responsible for an important prejudice on Einstein's part about the form of the metric for a static field that he only overcame when he calculated the perihelion motion of Mercury in November 1915 (cf. Ch. 3 and Ch. 5, note 12). The static theory also provided the blueprint for how Einstein handled various issues related to energy-momentum conservation in these later theories, such as the notion that the quantity representing gravitational energy-momentum should enter the field equations in the exact same way as the energy-momentum tensor for matter (Einstein 1912b, pp. 457–458; cf. Sec. 3.2). Gordin does not lose one word about these roots of general relativity in Einstein's work on gravity in Prague. If we really want to take this work seriously, we cannot ignore two of the most obvious ways in which it affected the genesis of general relativity.

[8] See Sauer (2015) for discussion of Grossmann's contributions to general relativity.

[9] An annotated transcription of the pages on gravity is presented as Doc. 10 in CPAE4. A facsimile and a transcription of all pages and a detailed *Commentary* on the pages on gravity constitute the core of Vols. 1–2 of *Genesis* (Renn 2007a). High-resolution scans of all pages are available at *Einstein Archives Online* (see note 2; search for 'Zurich Notebook' in the Archival Database). Unfortunately, the pages of the notebook are numbered differently in these three sources. We will use the page numbers (1L/1R through 43L with L and R for left and right) used in *Genesis* and add the page numbers (1 through 92) used in EAO in parentheses. For more information about the notebook, see Pt. II, Sec. 1.3.

[10] For another recent review of the saga of how Einstein abandoned and then returned to these field equations, see Weinstein (2018). We also want to draw attention to the chapter on the early history of general relativity in a recent book on the foundations of general relativity (Landsman 2021, Ch. 1).

[11] For information about physical appearance, contents, and access to facsimiles and annotated transcriptions of the Einstein-Besso manuscript, see Pt. II, Sec. 2.3.

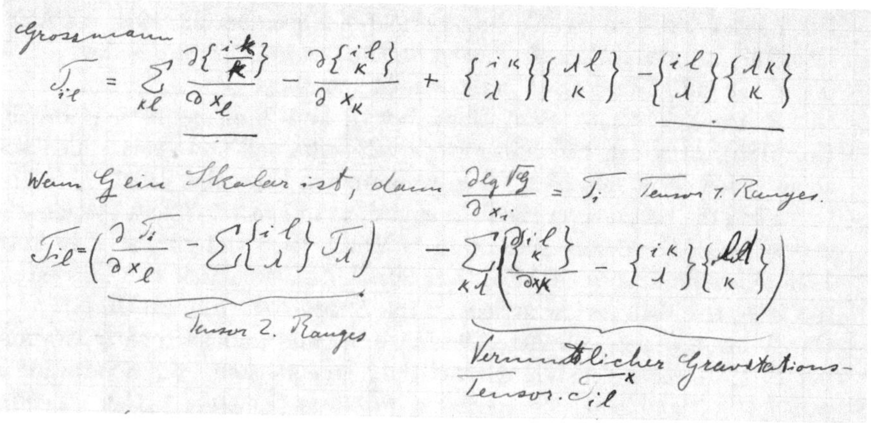

Grossmann

$$T_{il} = \sum_{\kappa l} \frac{\partial \left\{ {ik \atop \kappa} \right\}}{\partial x_l} - \frac{\partial \left\{ {il \atop \kappa} \right\}}{\partial x_\kappa} + \left\{ {i\kappa \atop \lambda} \right\}\left\{ {\lambda l \atop \kappa} \right\} - \left\{ {il \atop \lambda} \right\}\left\{ {\lambda\kappa \atop \kappa} \right\}$$

Wenn G ein Skalar ist, dann $\frac{\partial \lg \sqrt{G}}{\partial x_i} = T_i$ Tensor 1. Ranges.

$$T_{il} = \left(\frac{\partial T_i}{\partial x_l} - \sum \left\{ {il \atop \lambda} \right\} T_\lambda\right) - \sum_{\kappa\lambda}\left(\frac{\partial \left\{ {il \atop \kappa} \right\}}{\partial x_\kappa} - \left\{ {i\kappa \atop \lambda} \right\}\left\{ {\lambda l \atop \kappa} \right\}\right)$$

Tensor 2. Ranges

Vermutlicher Gravitations-Tensor. T_{il}^{x}

Figure 1.3 At the top of page 22R (p. 45) of the Zurich notebook (cf. note 9), Einstein wrote down the Ricci tensor, T_{il}, under the heading "Grossmann," who presumably introduced him to it. "If [the determinant of the metric] G is a scalar" (*Wenn G ein Skalar ist*), Einstein noted, the first two terms transform as a tensor. It follows that the second part also transforms as a tensor under what today are called unimodular transformations. Underneath this second part, Einstein wrote: "probable gravitation tensor" (*Vermutlicher Gravitationstensor*). Einstein had thus already considered the field equations he proposed in the first of his four November 1915 papers (Fig. 1.1) in 1912–13 (Einstein Archive, Hebrew University, Jerusalem).

In addition to the four papers of November 1915 and excerpts from the Zurich notebook and the Einstein-Besso manuscript, this volume presents (excerpts) from three additional papers of 1914 and 1916 and four of Einstein's letters of 1915–16 that are important for understanding the breakthrough of November 1915 and the consolidation of the new theory in 1916. This material is presented in Pt. II of this volume and divided into ten chapters:

- Ch. 1 has the most important pages from the Zurich notebook of 1912–13 (CPAE4, Doc. 10; cf. note 9).
- Ch. 2 has the most important pages from the Einstein-Besso manuscript of 1913 (CPAE4, Doc. 14).
- Ch. 3 has the section on the field equations from Einstein's review article on the *Entwurf* theory of November 1914 (Einstein 1914c, CPAE6, Doc. 9).
- Ch. 4 has a letter to Erwin Freundlich of September 30, 1915 (CPAE8, Doc. 123), in which Einstein reported that the *Entwurf* theory fails to make rotation relative.
- Ch. 5 has a letter to Hendrik A. Lorentz of October 12, 1915 (CPAE8, Doc. 129), in which Einstein reported that the uniqueness argument for the *Entwurf* field equations given in the 1914 review article fails.
- Ch. 6 has the four papers of November 1915 (Einstein 1915a; 1915b; 1915c; 1915d, CPAE6, Docs. 21, 22, 24, and 25).

- Ch. 7 has a letter to Arnold Sommerfeld of November 28, 1915 (CPAE8, Doc. 153), in which Einstein gave a detailed account of why he had abandoned the old and adopted the new field equations.
- Ch. 8 has a letter to Paul Ehrenfest of late January 1916 (CPAE8, Doc. 185), in which Einstein gave a self-contained and streamlined account of his derivation of the field equations in November 1915.
- Ch. 9 has the sections on the field equations in Einstein's first systematic exposition of the theory submitted in March 1916 and published in May 1916 (Einstein 1916b, CPAE6, Doc. 30). This 1916 article can be seen as replacing the 1914 review article on the *Entwurf* theory (see Ch. 3).[12]
- Ch. 10 has a paper, submitted late October and published early November 1916, in which Einstein derived the field equations from a variational principle without choosing special coordinates (Einstein 1916d, CPAE6, Doc. 41).

The manuscript material in Chs. 1–2 is presented in facsimile with an annotated transcription. For the (excerpts from) papers in Chs. 3, 6, 9 and 10, facsimiles of the German original are accompanied by annotated English translations. For the letters in Chs. 4–5 and 7–8, transcriptions of the German original are accompanied by annotated English translations.[13,14] In addition, the letter in Ch. 8 is presented in facsimile, as it is difficult to convey the disposition of the material on two of its three pages otherwise. Of these four letters, only the last one is part of the Einstein Archive in Jerusalem (see note 2). The archive, however, does contain copies of all four.[15]

The annotation of the papers on general relativity in CPAE6 is notoriously sparse. In our efforts to make sure that readers familiar with the basics of general relativity can follow Einstein's arguments without having to take out pen and paper, we may occasionally have erred in the other direction. Robert

[12] For commentary on the 1916 review article, see Janssen (2005), Sauer (2005b), and Gutfreund and Renn (2015). This last volume, written for a general audience, contains facsimile reproductions of both the article and the original hand-written manuscript as well as an English translation.

[13] The translations follow those in the Companion volumes to the Einstein edition, though we will occasionally and mostly silently deviate from them. In these translations, obvious typos in the German originals are silently corrected (cf. the remarks by Alfred Engel and Engelbert Schucking in the "Translator's Preface" to the Companion volume to CPAE6).

[14] Editorial notes are indicated by numbers in square brackets in the margins of the facsimiles of the papers in Pt. II, Chs. 3, 6, 9 and 10 and embedded in the text of the translations of these papers as well as in the transcriptions and translations of the letters in Pt. II, Chs. 4, 5, 7 and 8. Note numbers in square brackets have also been added to the transcription of the pages of the Zurich notebook in Pt. II, Ch. 1. No such numbers have been used in the case of the Einstein-Besso manuscript in Pt. II. Ch. 2. Instead commentary is provided per page.

[15] The format typically used to cite material from this archive is EA *x y*, where EA stands for Einstein Archive, *x* is the microfilm reel number and *y* a three-digit document number. The numbers for the items presented here are: 3 006 (Ch. 1); 79 896 (Ch. 2); 80 061 (Ch. 4); 16 442 (Ch. 5); 21 382, 21 382.1 (Ch. 7); 9 369 (Ch. 8).

Schulmann, former director of the Einstein Papers Project, once nicknamed a visiting editor *Ritter von Offenbar*, as many of his proposed editorial notes—invariably starting with "Apparently ..."—simply restated Einstein's perfectly lucid prose in this editor's own words. We realize there is no shortage of passages in Pt. II of our book that, especially for readers well-versed in general relativity, may bring to mind Captain Obvious.

The introductory essay that takes up the remainder of Pt. I posed a different challenge. We tried to write it in such a way that the reader can follow the conceptual developments without having to worry about the details of the mathematics. We realize, however, that for those comfortable with the relevant mathematics, it would actually be helpful to be more specific. The compromise we settled on was to put some of these specifics in footnotes (see, in particular, notes 7, 8 and 15 in Ch. 3 and notes 15, 16 and 21 in Ch. 5). The reader can skip these footnotes without loss of continuity when reading the introductory essay but we will refer back to them in various places in our commentary on the source material in Pt. II.

In this introductory essay, we trace Einstein's search for satisfactory field equations for his new theory of gravity from the Zurich notebook of 1912–13 to the four papers of November 1915, briefly cover the consolidation of the final result in 1916 and end with a short excursion into further developments in 1917–18. Our discussion of Einstein's search for field equations is organized around the question which of two strategies that can be discerned in his work during the period 1912–15 (and, as we will see, beyond) was more important for the breakthrough of November 1915, the so-called *mathematical strategy*, as John Norton (1984, 2000) has argued, or the so-called *physical strategy*, as we have argued (Janssen and Renn 2007, 2015a).

We use the terms 'mathematical strategy' and 'physical strategy' the way they are defined in *Genesis* (Renn 2007a, Vol. 1, p. 10; Vol. 2, pp. 500–501, p. 840). The field equations set an expression with second-order derivatives of the metric field on the left-hand side equal to an expression with the energy-momentum tensor on the right-hand side. The two strategies differ in the way in which Einstein generated and evaluated candidates for the left-hand side.

Following what we call the mathematical strategy, Einstein started from a suitable quantity from differential geometry. This gave him candidate field equations of broad covariance, which therefore, he thought, automatically met the demands of the equivalence principle and the relativity principle he was hoping to implement with his new theory. He then checked whether they also met two basic demands coming from physics, i.e., that they agreed with Newtonian theory in some appropriate limit and that they satisfied energy-momentum conservation. If necessary, he sacrificed some of the covariance of his candidate field equations to make sure these physical requirements were also met.

Following what we call the physical strategy, Einstein modeled his candidate field equations on the field equations of the electrodynamics of Maxwell and Lorentz, rewritten in the four-dimensional formalism of special relativity by Hermann Minkowski, Sommerfeld, and Max Laue (cf. Ch. 3, note 11). This construction guaranteed that these candidate field equations satisfied the physical requirements (Newtonian limit, energy-momentum conservation). He needed to check, however, whether their covariance was broad enough to meet the demands of the equivalence and relativity principles as well.

Einstein continued to alternate between these two strategies even after the search for field equations in the period 1912–15. As we will indicate in Ch. 7, the final chapter of this introductory essay, Einstein's (1916c, 1918a) work on gravitational waves was a natural extension of the analogy with electrodynamics that had been central to the physical strategy (Renn and Sauer 2007; Kennefick 2007, 2014), while Einstein's (1917) paper on cosmology and the related debate with the Leiden astronomer Willem de Sitter[16] should be seen in the context of the last of several attempts by Einstein to make all motion relative, an effort closely related to the mathematical strategy (Janssen 2014).[17]

The balance of Pt. I is organized as follows:

- Ch. 2 provides a short preliminary account of the transition from the *Entwurf* field equations to the Einstein field equations and raises the question whether the breakthrough of November 1915 was the result of the mathematical or the physical strategy.
- Ch. 3 covers Einstein's search for satisfactory field equations recorded in the Zurich notebook, vacillating between the mathematical and the physical strategy. We show how Einstein came to abandon the mathematical strategy even though it led him to the field equations he would resurrect in his first paper of 1915 and how the physical strategy eventually led him to the *Entwurf* field equations.
- Ch. 4 covers the consolidation phase of the *Entwurf* theory in the period 1913–14, focusing on the efforts to get a better handle on the intractable covariance properties of its field equations.
- Ch. 5 details how the variational formalism built around the *Entwurf* field equations in 1914 turned into the scaffold used to construct new field equations. We show how the physical strategy led Einstein back to field equations to which the mathematical strategy had already led him in the Zurich notebook.
- Ch. 6 covers Einstein's calculations for the perihelion motion of Mercury, both in November 1915 and, in collaboration with Besso, in 1913.
- Ch. 7 provides thumbnail sketches of Einstein's work on cosmology and gravitational waves in the early years of general relativity (1916–18).

---

[16] See the editorial note, "The Einstein-De Sitter-Weyl-Klein Debate," in CPAE, Vol. 8, pp. 351–357.

[17] We are grateful to Siska De Baerdemaeker for suggesting this way of extending the use of the distinction between these two strategies.

Chapter 2
From the *Entwurf* field equations to the Einstein field equations: a first pass

How did Einstein find the Einstein field equations? In his later years, he insisted that "the only way to find them was through a formal principle (general covariance)."[1] Such statements mainly served to justify his strategy in the search for a unified field theory during the second half of his career. As a description of how he found the field equations of general relativity, they are highly misleading (Van Dongen 2010, pp. 32–35).

Similarly misleading statements can be found in the first of Einstein's November 1915 papers. In the introduction of this paper, he made it sound as if he had gone from the old to the new field equations by leveling one building and erecting a new one on its ruins in a completely different style. The former was built according to principles of physics; the latter, or so Einstein would have his readers believe, according to principles of mathematics. "Hardly anybody who has truly understood the theory will be able to avoid coming under its spell," he rhapsodized (even though the field equations he proposed were not yet generally covariant), "it is a real triumph of the method of the general differential calculus developed by Gauss, Riemann, Christoffel, Ricci and Levi-Civita" (Einstein 1915a, p. 779). Careful examination of the November papers and his correspondence at the time, we argue, suggests a different architectural metaphor. Einstein used the elaborate framework he had built around the field equations of 1913 as a scaffold on which he carefully placed the arch stones of the Einstein field equations (Janssen and Renn 2015a).[2]

2.1 "With a heavy heart"

Einstein had already considered the November 4 equations three years earlier in the course of his collaboration with Grossmann. As Einstein recalled in the introduction of his first November 1915 paper, they had given up the

[1] Einstein to Louis de Broglie, February 8, 1954. Quoted and discussed in van Dongen (2010, pp. 2–3).

[2] For a more general discussion of the use of the 'arches and scaffolds' metaphor to capture instances of theory change, see Janssen (2019, see pp. 129–134 for the example of the field equations of general relativity). We elaborate on this specific use of the metaphor in Sec. 5.5. See note 23 in that section for a sketch of an alternative counterfactual account of this episode using the same metaphor.

M. Janssen, J. Renn, *How Einstein Found His Field Equations*, Classic Texts in the Sciences, https://doi.org/10.1007/978-3-030-97955-3_2

search for field equations of broad covariance based on the Riemann tensor "with a heavy heart" (Einstein 1915a, p. 778). What had defeated them were problems with the physical interpretation of such equations.[3] The field equations they eventually adopted and published in the *Entwurf* paper were constructed specifically to ensure their compatibility with Newtonian theory in the appropriate limit and with energy-momentum conservation.

Except for its field equations, the *Entwurf* theory already has all the basic elements of the mathematical formalism of general relativity. Einstein nonetheless cautiously referred to it as a *generalized* rather than a *general* theory of relativity. This was because it remained unclear whether the *Entwurf* equations are covariant under a sufficiently broad class of transformations for the generalization of the principle of relativity that Einstein hoped to achieve with his theory (cf. Ch. 1, note 4).

By August 1913, Einstein had found an ingenious (though ultimately fallacious) argument, known as the hole argument, that convinced him that generally covariant field equations were not to be had (see Sec. 4.1). By late 1914, he had come to believe (erroneously as it turned out) that the severely limited covariance of the *Entwurf* field equations was as broad as could be without running afoul of the hole argument and broad enough for the generalization of the relativity principle. He wrote a lengthy review article on what he now confidently referred to as the *general* theory of relativity. As a newly minted member of the Prussian Academy in Berlin—in March 1914, Einstein had left Zurich to take up a prestigious new position as a salaried member of the Berlin Academy—he dutifully submitted this paper for publication in its *Proceedings* (Einstein 1914c).[4]

2.2 "Every year he retracts what he wrote the year before"

In 1915 Einstein was invited to give the Wolfskehl lectures in Göttingen. From June 28 to July 5, he lectured on the *Entwurf* theory and captured the imagination of his host, the mathematician David Hilbert. The only surviving notes of these lectures (CPAE6, Appendix B) do not mention the field

[3] As Einstein himself put it shortly afterwards, in a letter to his friend Heinrich Zangger of December 9, 1915: "I am badly overworked because of the extraordinary exertions of the past few months. But the success is magnificent. What is interesting is that the first assumptions [*die ersten Ansätze*] I made with Grossmann, which realize the most radical theoretical demands, are now confirmed. Back then we were only missing a few relations of a formal nature without which the connection of the formulas to already known laws cannot be established. The matter is also starting to dawn in the heads of my colleagues [*Die Sache fängt an auch in den Köpfen der Kollegen aufzudämmern*]. In 10 or 20 years it will be something totally obvious [*eine Selbstverständlichkeit*]" (CPAE8, Doc. 161a, in CPAE10). On Einstein and Zangger, see Schulmann (2012).

[4] For expressions of his strong confidence in the theory at this point, see Ch. 4, note 16.

equations, but it is safe to assume that Einstein covered them along the lines of his 1914 review article.

Not long after these lectures, Einstein's confidence in these equations began to crumble. Worried that Hilbert might soon hit upon the same flaws he had spotted, Einstein rushed new field equations into print. He sent proofs of the paper to Hilbert[5] with a note in which he wrote that he had heard from Sommerfeld that Hilbert had also found "a hair in my soup[6] that completely spoiled it for you" (CPAE8, Doc. 136).

One week after his first communication of November 1915, Einstein proposed a modification of his new equations, only to retract and replace it with a different modification two weeks later. Fortunately, this modification did not affect the result for the perihelion of Mercury that Einstein had reported in between these two communications. Confident that the third set of field equations he had proposed in the span of only three weeks were the correct ones, he could afford to be self-deprecating about a few last stumbles on the home stretch. "Unfortunately, I have immortalized my final errors ... in the Academy papers," he wrote to Sommerfeld on November 28 (CPAE8, Doc. 153; Pt. II, Ch. 7). And a month later, on December 26, he told Paul Ehrenfest: "It is convenient with that fellow Einstein, every year he retracts what he wrote the year before" (CPAE8, Doc. 173).

A few months later, Einstein replaced the premature review article on the *Entwurf* theory by a new review article on the theory of November 1915 (Einstein 1916b). The papers of November 1915, especially the first and the fourth, relied heavily on the 1914 review article and proved difficult to understand even for the small group around Lorentz and Ehrenfest in Leiden who had closely followed Einstein's work (see Ch. 4, note 5). Especially Ehrenfest persisted in pressuring Einstein for clarification of various points. In late January 1916, Einstein finally broke down and wrote his friend a long letter with a self-contained version of the argument that had led him to the final form of the field equations. In closing, he asked Ehrenfest to return the letter when he was done with it as "as I do not have these things together in such a nice form anywhere else" (CPAE8, Doc. 185; Pt. II, Ch. 8).

We do not know whether Ehrenfest obliged[7] but the letter provides a nice bridge between the introduction of the field equations and the proof of their

---

[5] These page proofs were recently brought to light by Tilman Sauer. See note 10 for some preliminary comments on Hilbert's marginalia in this document.

[6] This is a standard German idiom roughly equivalent to the English idiom 'a fly in the ointment'.

[7] That the letter is now part of the Einstein Archive may suggest that Ehrenfest did but does not settle the matter. The Boerhaave Museum in Leiden and the Hebrew University in Jerusalem agreed to swap their holdings of the Einstein-Ehrenfest correspondence so that Einstein's letters are now with the Einstein papers in Jerusalem and Ehrenfest's letters are now with the Ehrenfest papers in Leiden. It can no longer be established whether this particular letter was ever part of the collection of letters from Einstein to Ehrenfest at the Boerhaave Museum. We are grateful to Ad Maas, curator at the Boerhaave Museum, for checking the museum's records for information about this letter.

compatibility with energy-momentum conservation in the November 1915 papers and the coverage of this material in the relevant sections of the review article (Einstein 1916b, secs. 12–18; see Pt. II, Ch. 9).[8] In these sections, Einstein still used the special (unimodular; see Figs. 1.1–1.3) coordinates he had used in November 1915. The manuscript of the article contains a draft of this section in general coordinates, which turned into an appendix and was then cut altogether.[9] In October 1916, he recycled this material for a paper in which he gave a variational derivation of the Einstein field equations in their generally covariant form (Einstein 1916d, see Pt. II, Ch. 10). By that time, as Einstein duly acknowledged, both Hilbert and Hendrik A. Lorentz had already published their variational derivations of the generally covariant equations. As we will see in Pt. II, Ch. 10, however, Einstein's approach and purpose were different from theirs and closely followed the general formalism laid out in Einstein's own 1914 review article (cf. Ch. 5, note 23).

2.3 "Nostrified"

In hindsight, Einstein could have taken his time in November 1915 and need not have worried about Hilbert beating him to the punch. The two men exchanged some letters during that hectic month but essentially talked past each other.[10] Neither had a clear idea of what the other was up to. Einstein

[8] In the past, we have called this letter a "blueprint" for these sections (Janssen and Renn 2007, p. 896; see also the introduction to CPAE8, p. l). That may be too strong. The letter is part of an extensive three-way correspondence about various aspects of Einstein's new theory between Einstein, Ehrenfest and Lorentz in late 1915 and early 1916 (see Ch. 4, note 5). In the course of this correspondence, Einstein gradually reworked his presentation of the field equations and their relation to energy-momentum conservation. This clearly paved the way but should perhaps not be called a blueprint for the presentation in the review article completed about a month and a half later (cf. Pt. II, Sec. 8.4).

[9] A transcription was published as CPAE6, Doc. 31. A facsimile can be found in Gutfreund and Renn (2015, pp. 130–138).

[10] In the page proofs of Einstein's first November 1915 paper mentioned in note 5, Hilbert corrected various typos and added some comments. Tellingly, he circled the Lagrangian Einstein (1915a) introduced on p. 784 (eq. (17)) with an arrow pointing to it and put a curly bracket with a question mark next to a passage on p. 785 in which Einstein examined the connection between energy-momentum conservation and the covariance of his field equations. Both are adapted from similar elements in the review article of the year before (Einstein 1914c). The Lagrangian, modeled on the one for the electromagnetic field, looks very different from the Riemann curvature scalar R Hilbert used as his Lagrangian. The vacuum field equations that follow from Einstein's Lagrangian set two of the four terms of the Ricci tensor (modern notation: $R_{\mu\nu}$) equal to zero, the ones that follow from Hilbert's Lagrangian set what is now called the Einstein tensor (modern notation: $G_{\mu\nu} = R_{\mu\nu} - \frac{1}{2}g_{\mu\nu}R$) equal to zero. In his fourth paper of November 1915, Einstein, as we saw (see Fig. 1.2), added a term with the trace T of the energy-momentum tensor to the right-hand side of his field equations, which by now had the full Ricci tensor on the left-hand side. Today we know that adding this term with

nonetheless complained to his two closest friends in Switzerland that Hilbert was poaching on his preserves. On November 26, he wrote to Zangger that a certain colleague had "nostrified" (i.e., co-opted, if not downright plagiarized) his theory (CPAE8, Doc. 152). Four days later, he wrote to Michele Besso that the behavior of unnamed colleagues had been appalling (CPAE8, Doc. 154).[11] It was probably at least in part to secure priority that, in between these two letters, Einstein sent Sommerfeld a blow-by-blow account of the developments of that turbulent month (CPAE8, Doc. 154; Pt. II, Ch. 7; see also Sec. 5.5 below). Whatever bad blood there was between Einstein and Hilbert that month, they quickly buried the hatchet. On December 20, Einstein sent Hilbert a conciliatory note (CPAE8, Doc. 167), even though this did not stop him from repeating his earlier complaint about the behavior of his colleagues the very next day in a postcard to Besso (CPAE8, Doc. 168).

Hilbert (1915) presented his field equations for the metric field in Göttingen on November 20, five days before Einstein presented his final result in Berlin. A footnote in the perihelion paper (Einstein 1915c, p. 831), submitted two days before Hilbert's presentation makes it clear, however, that Einstein had already made the final modification in his field equations at that point. More importantly, page proofs of Hilbert's paper bearing a date stamp of December 6 show that, despite the prominent presence of the Riemann tensor in these proofs,[12] the theory Hilbert originally proposed is conceptually closer to the *Entwurf* theory than to Einstein's new theory (as we will explain in more detail at the end of Sec. 4.3). By the time the paper was finally published in March 1916, Hilbert had switched to the new theory, giving Einstein full credit.

That Hilbert had almost scooped him, however, pursuing a purely mathematical line of reasoning, was probably at least partly responsible for a noticeable shift in Einstein's methodology (Norton 2000). In the letter to Sommerfeld mentioned above, Einstein still defended his own more physical approach. "It is easy, of course," he wrote, clearly referring to Hilbert, "to

---

T on the right-hand side is equivalent to adding the term with R to the Ricci tensor on the left-hand side. There is no indication that Einstein realized this in late 1915. And as we shall see (see Sec. 5.4 and Pt. II, Ch. 6), Einstein's addition of the term with T had nothing to do with the equivalent term with R in the Euler-Lagrange equations for Hilbert's Lagrangian but followed naturally from his own earlier work.

[11] In another letter to Zangger, dated before December 4, 1915, Einstein similarly wrote: "I am having some curious experiences with my dear colleagues these days. Except for one of them, they all try to find fault with my discovery or refute it if only in the most superficial manner. Only one recognizes it by "nostrifying" it with great fanfare, after I, with great effort, initiated him in the spirit of the theory" (CPAE8, Doc. 159a, in CPAE10).

[12] The presence of the Riemann tensor might have been even more prominent had it not been for part of a page being cut out (see Sauer 2005a). These page proofs garnered a lot of attention after Corry *et al.* (1997) reported their discovery in *Science* (Rowe 2006). For English translations of both the page proofs and the published version of Hilbert's paper, see Renn (2007a, Vol. 4, pp. 989–1015). For further discussion of Hilbert's work and its relation to Einstein's, see Sauer (1999, 2005a) and Renn and Stachel (2007).

write down these generally-covariant field equations but difficult to see that they are a generalization of the Poisson equation [of Newtonian theory] and not easy to see that they satisfy the conservation laws." In the search for a classical field theory unifying gravity and electromagnetism that took up much of the second half of his career, however, Einstein himself adopted a purely mathematical approach. To justify this approach, as carefully documented by Jeroen van Dongen (2010), he routinely claimed that this was how he had found the field equations of general relativity.

We agree with van Dongen (ibid., pp. 32–35) that Einstein succumbed here to a case of selective amnesia (Janssen and Renn 2007, p. 841).[13] As we saw above, however, there are passages even in the November 1915 communications indicating that the decisive step for Einstein had been to set aside the physical considerations that led him to the *Entwurf* equations and to return to considerations of mathematical elegance instead. Such passages, in our view, have not only misled some of the best modern commentators,[14] they also helped solidify Einstein's own distorted memory of the breakthrough of November 1915 as a triumph of mathematics. There is no question that this breakthrough required both mathematical and physical insights. Yet, it also seems clear that the crucial insight must have come from the physics. Einstein had known about the equations he published in his first November paper for three years. It was only in 1915, however, that he recognized that these equations made sense from a physics point of view. This is not to say that the mathematics was unimportant. In the end, it was probably the remarkable convergence of mathematical and physical lines of reasoning, which he had long hoped for but which had eluded him so far, that gave Einstein the necessary confidence that he got it right this time.

In Chs. 3–5, we cover Einstein's search for gravitational field equations during the period 1912–5 in more detail. The relative importance of mathematical and physical considerations in Einstein's breakthrough of November 1915 will be a central theme in our discussion.

---

[13] A more precise technical term for this phenomenon is *consistency bias*, defined as "incorrectly remembering one's past attitudes and behavior as resembling present attitudes and behavior" (see https://en.wikipedia.org/wiki/List_of_memory_biases). We thank Alex Wheeler for pointing this out to us.

[14] Norton (2000), e.g., takes them at face value.

Chapter 3
The Zurich Notebook: How Einstein found the *Entwurf* field equations

Shortly after his move from Prague to Zurich in August 1912, which reunited him with Grossmann, Einstein started working on a theory of gravity in which the metric tensor field, $g_{\mu\nu}(x^{\alpha})$,[1] represents both the geometry of what in general will be a curved space-time and the potentials of the gravitational field. Einstein's research notes on gravity from 1912–13, preserved in the Zurich notebook, begin after this decisive insight. This will also be our starting point.[2,3]

On the first few pages of his notes on gravity in the Zurich notebook, Einstein explored the connection between the metric theory he was now considering and the scalar theory for static gravitational fields he had been working on in Prague.[4] In this theory, a variable speed of light played the role of the gravitational potential (Einstein 1912a, 1912b). In the new theory, he expected weak static fields to be described by a metric tensor in which only the time-time component (g_{44}) is variable. All other components, he assumed,

---

[1] Einstein used Latin and Greek indices interchangeably, both of them running from 1 to 4, with 4 referring to the time component. He only adopted the standard downstairs/upstairs convention to distinguish (albeit not always consistently) between covariant and contravariant indices in his November 1914 review article on the *Entwurf* theory (Einstein 1914c), but he continued to write $\frac{\partial}{\partial x_\mu}$ instead of $\frac{\partial}{\partial x^\mu}$ and $\left\{ \begin{matrix} \mu\nu \\ \alpha \end{matrix} \right\}$ instead of $\left\{ \begin{matrix} \alpha \\ \mu\nu \end{matrix} \right\}$. Before that, he wrote all indices downstairs but used (albeit not always consistently) Latin letters for covariant components and Greek letters for contravariant ones (e.g., Einstein wrote $\gamma_{\mu\nu}$ instead of $g^{\mu\nu}$). Although he occasionally omitted summation signs before, he did not introduce the Einstein summation convention until the review article on the November 1915 version of his theory (Einstein 1916b, p. 781, see Pt. II, Sec. 8.4, note 4, and Sec. 9.3, note 1). Both in the introductory essay in Pt. I and in our commentary on the sources in Pt. II, we will use modern notation, including the Einstein summation convention, commas for ordinary derivatives and semi-colons for covariant derivatives.

[2] The earlier developments are covered by Stachel (2002b) and Renn (2007b) and, addressing a broader audience, Renn (2006, Ch. 5).

[3] See Räz and Sauer (2015) for an analysis of the collaboration between Einstein and Grossmann within a framework based on the conception of the application of mathematics of Bueno and Colyvan (2011).

[4] See p. 39L (p. 1) of the Zurich notebook and the commentary on this passage in Renn 2007, Vol. 2, pp. 504–505.

M. Janssen, J. Renn, *How Einstein Found His Field Equations*, Classic Texts in the Sciences, https://doi.org/10.1007/978-3-030-97955-3_3

would have the constant values they have in arbitrary Lorentz frames in the flat Minkowski space-time of special relativity. Given the relation between this one variable component, the velocity of light and the Newtonian gravitational potential, this is how he expected his new metric theory to reduce to his old scalar theory and, more importantly, to Newtonian theory in the case of weak static fields. For ease of reference, we call this metric the *weak static metric*. It is a diagonal metric of the form $g_{\mu\nu} = \mathrm{diag}(-1,-1,-1,g_{44}(x,y,z))$.

The goal of the research recorded in the Zurich notebook was to find the counterpart in the new theory of the Poisson equation for the gravitational potential in Newtonian theory.[5] Early on, Einstein decided that the energy-momentum tensor for matter, $T_{\mu\nu}$, was the natural candidate for the right-hand side of these field equations. The time-time component of this tensor, T_{44}, is the mass density on the right-hand side of the Poisson equation. The problem that takes up the bulk of the Zurich notebook was to find the left-hand side, a two-index expression involving second-order derivatives of the metric field that would take over the role of the Laplacian acting on the potential in the Poisson equation. In addition to reproducing the results of Newtonian theory in the appropriate limit, candidate field equations had to be compatible with energy-momentum conservation as well as with the relativity of arbitrary motion and the closely related equivalence of acceleration and gravity that Einstein hoped to implement with his new theory.

3.1 The mathematical strategy

In the notebook, we see Einstein vacillating between two strategies in his search for field equations, a mathematical strategy and a physical strategy (see Ch. 1). He tried the physical strategy first, constructing quantities with second-order derivatives of the metric modeled directly on the Laplacian acting on the Newtonian potential. As a complementary mathematical strategy, he tried to connect these quantities to two generally covariant scalars, the so-called Beltrami invariants, which Grossmann probably told him about. Although this first attack on the problem led to some insights that were put to good use later (see Sec. 3.2), it did not produce any viable candidates for the left-hand side of the field equations. The first real advance was made when the four-index Riemann tensor arrived on the scene, along with its two-index subordinate, the Ricci tensor. Grossmann was almost certainly responsible for these reinforcements. His name appears next to the first appearance of the Riemann tensor on p. 14L (p. 28) of the notebook. This page marks a

[5] The Newtonian potential φ satisfies the Poisson equation $\Delta\varphi = 4\pi G\rho$, where $\Delta \equiv \partial^2/\partial x^2 + \partial^2/\partial y^2 + \partial^2/\partial z^2$ is the Laplacian, ρ is the mass density, and G is Newton's gravitational constant. The potential φ of the field $\mathbf{E} = -\nabla\varphi$ generated by a static charge distribution ρ, of course, satisfies an equation of the same form: $\Delta\varphi = -\rho/\varepsilon_0$ (where ε_0 is the dielectric constant in vacuum). The minus sign on the right-hand side reflects that like charges repel whereas masses always attract each other.

clear turn to the mathematical strategy. Einstein immediately charged ahead and checked whether the Ricci tensor could serve as the left-hand side of the field equations. Within a few lines, he ran into the problem that would eventually make him give up on the mathematical strategy in the notebook and return to the physical strategy.

Einstein found that the Ricci tensor contains four terms with second-order derivatives of the metric only one of which reduces to the Laplacian acting on the metric for weak static fields. How to get rid of the other three? Today, this is trivial. To compare generally covariant field equations, i.e., equations that have the same form in all space-time coordinate systems, to the Poisson equation of Newtonian theory, which only has this particular form in inertial frames, we need to consider the generally covariant field equations in a similarly restricted class of coordinate systems. We do this by imposing four extra conditions on the metric field, so-called coordinate conditions. These conditions allow us to eliminate the unwanted terms with second-order derivatives from the field equations. The resulting truncated equations retain only one such term, which reduces to its Newtonian counterpart for weak static fields, plus a number of terms that are quadratic in first-order derivatives of the metric and negligible for weak fields.

In *'Subtle is the Lord ...'*, his well-known scientific biography of Einstein, Abraham Pais (1982, p. 222) suggested that Einstein rejected the Ricci tensor in 1913 because he was unaware at the time of this freedom to impose coordinate conditions. On p. 19L (p. 38) of the Zurich notebook, however, as was pointed out by Norton (1984, p. 117) in his ground-breaking paper on the notebook, Einstein actually eliminated the unwanted terms in the Ricci tensor by imposing what a modern relativist would immediately recognize as the so-called harmonic coordinate condition. This raises the question as to why Einstein nonetheless rejected the Ricci tensor. Following a suggestion by Stachel, Norton (ibid., pp. 115–119) conjectured that it was because the harmonic coordinate condition does not allow the weak static metric.

As long as p. 19L of the notebook is considered in isolation, this is an eminently plausible answer. The following pages of the notebook, however, cast serious doubt on it. These pages show unequivocally that Einstein used coordinate conditions in a way that is fundamentally different from the way they are used today (Renn 2007a, Vol. 2, p. 605, note 199). We now recognize that a coordinate condition has the status of a gauge condition. It selects one or more representatives from each equivalence class of metric fields. Which ones is a matter of convenience: different problems, different coordinate conditions. By contrast, Einstein used coordinate conditions—not just in the Zurich notebook but throughout the reign of the *Entwurf* theory—in a one-size-fits-all fashion: the same coordinate condition for all problems. In *Genesis* (2007a, Vol. 1, p. 11; Vol. 2, p. 497), we introduced the term *coordinate restriction* to characterize this idiosyncratic use of such conditions. Unlike coordinate conditions, coordinate restrictions are an integral part of the theory in which they are used. The fundamental field equations are no longer the generally covariant field equations but the truncated version of these equations one is left with after eliminating various terms with the help of such coordinate restrictions. The covariance of these truncated equations is determined by the covariance of the coordinate restrictions used to do the truncating. An additional difference with the modern approach is that Einstein expected both the compatibility with Newtonian theory and energy-momentum conservation to call for extra conditions on the metric field.

This last element clearly played a role in Einstein's rejection of the Ricci tensor—more accurately, his rejection of the Ricci tensor truncated with the harmonic coordinate restriction. Examining these candidate field equations in linear approximation, Einstein found a conflict between three elements: the harmonic coordinate restriction, energy-momentum conservation and the weak static metric. On p. 20L (p. 40), he resolved the conflict between the first two elements by adding a term with the trace of the energy-momentum tensor for matter to the field equations. Hence, the Zurich notebook already contains, albeit only in linear approximation, the field equations of Einstein's fourth and last communication of November 1915 (cf. Fig. 1.2). Einstein did not explore these equations any further in the notebook because the conflict with the third element remained. Field equations with a trace term do not allow the weak static metric. In all of this, it remains unclear whether Einstein realized that the first and the third element are also in direct conflict with each other, i.e., it remains unclear whether he realized that the harmonic coordinate restriction does not allow the weak static metric.

Einstein was not ready to turn his back on the Ricci tensor. He carefully explored two other candidates for the left-hand side of the field equations extracted from the Ricci tensor with the help of other coordinate restrictions. We focus on the first of these.[6] In both cases, he began by imposing the modest restriction to unimodular transformations. This allowed him to split the Ricci tensor into two parts, both of which transform as tensors under such transformations (p. 22R [p. 45]). "Probable gravitation tensor," he wrote underneath the second part (Fig. 1.3).[7] This part returned as the left-hand

---

[6] See Renn (2007a, Vol. 2, pp. 652–679) for the second, dubbed the ϑ-restriction.

[7] The quantities $\{\dots\}$ occurring in Figs. 1.1 and 1.3 are the *Christoffel symbols* (of the second kind), defined as (in modern notation; cf. note 1):

$$\begin{Bmatrix} \rho \\ \mu\nu \end{Bmatrix} \equiv \tfrac{1}{2} g^{\rho\sigma} \left(g_{\sigma\mu,\nu} + g_{\sigma\nu,\mu} - g_{\mu\nu,\sigma} \right)$$

Einstein used the relation

$$\begin{Bmatrix} \rho \\ \mu\rho \end{Bmatrix} = \frac{\partial\sqrt{-g}}{\partial x^{\mu}}$$

to split the Ricci tensor in two parts. Since g transforms as a scalar under unimodular transformations, $A_{\mu} \equiv (\sqrt{-g})_{,\mu}$ transforms as a vector under these transformations. Hence, $A_{\mu;\nu}$, the covariant derivative of this vector, transforms as a second-rank tensor under unimodular transformations. $A_{\mu;\nu}$ can be written as:

side of the field equations proposed in the first paper of November 1915 (Fig. 1.1). We therefore call it the *November tensor.*

The restriction to unimodular transformations eliminated some of the unwanted terms with second-order derivatives of the metric in the Ricci tensor. To eliminate the remaining ones, Einstein imposed the further restriction that the four-divergence of the metric vanishes. For ease of reference (and for no other reason than Einstein's discussion of it in a letter to Paul Hertz of August 22, 1915 [CPAE8, Doc. 111]), we call this the *Hertz condition/restriction.* This condition also returned in the November 1915 papers. An important point in favor of the Hertz condition/restriction was that it allows the weak static metric.

Einstein had already encountered the Hertz restriction earlier in the notebook. He had considered adding it as a further restriction to the field equations extracted from the Ricci tensor with the harmonic coordinate restriction to ensure energy-momentum conservation, at least in linear approximation. Even earlier, he had investigated its covariance properties in connection with one of his first crude attempts to find satisfactory field equations. From a modern point of view, there would seem to be nothing to investigate. The Hertz condition, the vanishing of the four-divergence of the metric, is covariant only under linear transformations. Its covariance, however, is broader if we allow what Einstein a few months later, in a letter to Lorentz of August 14, 1913 (CPAE5, Doc. 467), called "non-autonomous" (*unselbständige*) transformations and then, in publications in 1914, "justified" transformations between "adapted" coordinate systems (adapted to the metric field) (Einstein and Grossmann 1914; Einstein 1914c, see Sec. 4.2 and Pt. II, Sec. 3.3, note 8). In ordinary coordinate transformations—for which Ehrenfest, as Einstein told Lorentz, introduced the term *autonomous* transformations—the new coordinates are simply functions of the old ones (schematically: $x^{\mu} \rightarrow x'^{\mu}(x^{\rho})$). If equations for the metric field are invariant under an autonomous transformation, *any* solution of the equations in the old coordinates will be a solution in the new coordinates. In non-autonomous transformations, the new coordinates are functions of the old coordinates *and the metric field in those old coordinates* (schematically: $x^{\mu} \rightarrow x'^{\mu}(x^{\rho}, g_{\alpha\beta}(x^{\rho}))$). If equations for the metric field are invariant under a non-autonomous transformation, *only specific* solutions in the old coordinates will be solutions in the new coordinates.[8]

$$A_{\mu;\nu} = A_{\mu,\nu} - \begin{Bmatrix} \rho \\ \mu\nu \end{Bmatrix} A_{\rho} = \frac{\partial}{\partial x^{\nu}} \begin{Bmatrix} \rho \\ \mu\rho \end{Bmatrix} - \begin{Bmatrix} \rho \\ \mu\nu \end{Bmatrix} \begin{Bmatrix} \sigma \\ \rho\sigma \end{Bmatrix}.$$

This is just what Einstein called $S_{\mu\nu}$ in 1915. Since the sum of $S_{\mu\nu}$ and $R_{\mu\nu}$ is a generally covariant tensor, it follows that $R_{\mu\nu}$ also transforms as a tensor under unimodular transformations.

[8] The notions of *non-autonomous transformations* and *coordinate restrictions* are not nearly as sophisticated as this terminology may suggest. They are the sort of thing one would expect from a novice in tensor calculus. A simple example will illustrate this. Consider the ordinary (autonomous) transformation of the Hertz expression (set to zero in the Hertz condition/restriction) from coordinates x^{μ} to coordinates x'^{μ}:

When investigating covariance properties in the Zurich notebook, Einstein invariably started from the standard diagonal metric describing the geometry of Minkowski space-time in an arbitrary inertial Lorentz frame. Since its components are all constants, this metric trivially satisfies whatever equations for the metric field Einstein was considering. He then checked whether it remains a solution in a linearly accelerated or rotating frame. This is how Einstein tried to implement both the relativity of non-uniform motion and the equivalence principle as he then understood it. To relativize rotation, for instance, it sufficed, as far as Einstein was concerned, to show that the Minkowski metric in rotating coordinates is a vacuum solution of the field equations. This, in turn, could be done by showing that this *rotation metric*, as we will call it, is allowed by whatever coordinate restriction was used to extract those field equations from generally covariant ones. The equivalence principle would also be satisfied in that case. As long as the rotation metric is a vacuum solution, Einstein thought, the inertial forces of rotation can be interpreted as gravitational forces.

This then, in all likelihood, is why Einstein abandoned the November tensor in the Zurich notebook. The rotation metric does not satisfy the Hertz restriction. It follows that it is not a vacuum solution of the field equations extracted from the November tensor with the help of Hertz restriction. These field equations thus would not allow Einstein to interpret the inertial forces of rotation as gravitational forces.

The notions of coordinate restrictions and non-autonomous transformations fell by the wayside in Einstein's final theory and play no role in modern mainstream general relativity.[9] Both in the Zurich notebook and in his pa-

---

$$\frac{\partial g'^{\mu\nu}}{\partial x'^{\nu}} = \frac{\partial x^{\alpha}}{\partial x'^{\nu}} \frac{\partial}{\partial x^{\alpha}} \left(\frac{\partial x'^{\mu}}{\partial x^{\rho}} \frac{\partial x'^{\nu}}{\partial x^{\sigma}} g^{\rho\sigma} \right).$$

We can write the right-hand side as a sum of two terms:

$$\frac{\partial g'^{\mu\nu}}{\partial x'^{\nu}} = \frac{\partial x^{\alpha}}{\partial x'^{\nu}} \frac{\partial x'^{\mu}}{\partial x^{\rho}} \frac{\partial x'^{\nu}}{\partial x^{\sigma}} \frac{\partial g^{\rho\sigma}}{\partial x^{\alpha}} + \frac{\partial x^{\alpha}}{\partial x'^{\nu}} \frac{\partial}{\partial x^{\alpha}} \left(\frac{\partial x'^{\mu}}{\partial x^{\rho}} \frac{\partial x'^{\nu}}{\partial x^{\sigma}} \right) g^{\rho\sigma}.$$

Rewriting the first term—using that $\frac{\partial x^{\alpha}}{\partial x'^{\nu}} \frac{\partial x'^{\nu}}{\partial x^{\sigma}} = \delta^{\alpha}_{\sigma}$ and $\delta^{\alpha}_{\sigma} \frac{\partial g^{\rho\sigma}}{\partial x^{\alpha}} = \frac{\partial g^{\rho\alpha}}{\partial x^{\alpha}}$ and relabeling the summation index α in this last expression ν—we find:

$$\frac{\partial g'^{\mu\nu}}{\partial x'^{\nu}} = \frac{\partial x'^{\mu}}{\partial x^{\rho}} \frac{\partial g^{\rho\nu}}{\partial x^{\nu}} + \frac{\partial x^{\alpha}}{\partial x'^{\nu}} \frac{\partial}{\partial x^{\alpha}} \left(\frac{\partial x'^{\mu}}{\partial x^{\rho}} \frac{\partial x'^{\nu}}{\partial x^{\sigma}} \right) g^{\rho\sigma}.$$

Hence the Hertz expression transforms as a four-vector if the second term on the right-hand side vanishes. In more hifalutin terms: It transforms as a four-vector under the *non-autonomous transformations* defined by imposing the *coordinate restriction* that sets that term equal to zero. The transformation matrices $\partial x'^{\mu}/\partial x^{\nu}$ satisfying this restriction will depend on the metric $g_{\mu\nu}$.

[9] For a discussion of non-autonomous transformations in modern general relativity, see Bergmann and Komar (1972). These authors, however, treat these transformations as active point transformations and not (as we did, following Einstein) as passive coordinate transformations.

pers on the *Entwurf* theory, however, one of Einstein's central concerns was to make sure that his coordinate restrictions were covariant under a broad enough class of non-autonomous transformations to implement a relativity principle for arbitrary motion, a principle which itself turned out to be illusory.

3.2 The physical strategy

Einstein was not satisfied with any of the coordinate restrictions with which he tried to extract candidate field equations from the Ricci tensor in the Zurich notebook. Although these restrictions took care of the unwanted second-order derivatives of the metric in the Ricci tensor, it remained at best unclear whether they allowed the Minkowski metric in rotating and linearly accelerated frames and whether they were compatible with the exact conservation of energy-momentum.

These problems eventually drove Einstein back to the physical strategy, using the four-dimensional formulation of the Maxwell-Lorentz theory of the electromagnetic field as a template for his theory of the gravitational field. His pursuit of the mathematical strategy had given him a clearer picture of the structure of the left-hand side of suitable field equations. It should be the sum of one term with second-order derivatives of the metric and a number of terms quadratic in first-order derivatives, which all vanish for weak fields. As before, the right-hand side should be (minus κ times) the energy-momentum tensor for matter. Suppressing all indices, we can represent the sought-after format of the field equations schematically as:[10]

$$\frac{\partial^2 g}{\partial x^2} + \sum \frac{\partial g}{\partial x} \cdot \frac{\partial g}{\partial x} = -\kappa \begin{pmatrix} \text{em} \\ \text{matter} \end{pmatrix} \tag{3.1}$$

(where the 'product', indicated by a dot, is actually the contraction of an index of the metric in the first 'factor' and an index of the metric in the second 'factor').

Einstein had learned from his 1912 scalar theory that gravitational energy-momentum should enter the field equations in the exact same way as all other energy-momentum (Einstein 1912b, pp. 457–458). By analogy with the four-dimensional formalism for electrodynamics,[11] he assumed that the

[10] For a more systematic use of this kind of schematic notation, see Renn and Sauer (2007).

[11] Einstein was a recent convert to this formalism when he developed his 1912 scalar theory (see CPAE4, Doc. 1). Once he had overcome his earlier reservations, he also recognized that this formalism, stripped of its electromagnetic particulars, provides a general framework for relativistic continuum mechanics, as developed by Laue (1911) in the first textbook on (special) relativity. See Janssen and Mecklenburg (2007), Janssen (2019, pp. 124–129) and Walter (1999, 2007) for discussion. The formalism was generalized from flat to curved space-time in a paper by Friedrich Kottler (1912), which

gravitational energy-momentum density is quadratic in first-order derivatives of the gravitational potential (ibid.). This assumption also carried over from the scalar to the tensor theory. The sum of terms with first-order derivatives of the metric in Eq. (3.1) should thus split into two parts

$$\sum \frac{\partial g}{\partial x} \cdot \frac{\partial g}{\partial x} = \sum{}' \frac{\partial g}{\partial x} \cdot \frac{\partial g}{\partial x} + \sum{}'' \frac{\partial g}{\partial x} \cdot \frac{\partial g}{\partial x}, \tag{3.2}$$

with one part representing (κ times) the energy-momentum density of the gravitational field:

$$\kappa \begin{pmatrix} \text{em} \\ \text{grav field} \end{pmatrix} \equiv \sum{}' \frac{\partial g}{\partial x} \cdot \frac{\partial g}{\partial x}. \tag{3.3}$$

The other part—the sum labeled by a double prime in Eq. (3.2)—might or might not vanish. Substituting Eqs. (3.2) and (3.3) into Eq. (3.1), we see that Einstein was looking for field equations of the form

$$\frac{\partial^2 g}{\partial x^2} + \sum{}'' \frac{\partial g}{\partial x} \cdot \frac{\partial g}{\partial x} = -\kappa \begin{pmatrix} \text{em} \\ \text{matter} \end{pmatrix} + \begin{pmatrix} \text{em} \\ \text{grav field} \end{pmatrix}. \tag{3.4}$$

On pp. 26L-R (pp. 52–53) of the notebook, he set out to construct field equations of this form under the constraints that they satisfy energy-momentum conservation and reduce to the Poisson equation for weak static fields.

Einstein's construction is based on the energy-momentum balance equation he had found on p. 5R (p. 11) of the notebook (cf. note 15 below). Drawing on the analogy with the four-dimensional formalism for electrodynamics,[12] he interpreted this balance equation as setting the four-divergence of the energy-momentum tensor for matter equal to the density of the four-force exerted by the gravitational field on matter. Schematically:

$$4\text{div} \begin{pmatrix} \text{em} \\ \text{matter} \end{pmatrix} = 4\text{force density}. \tag{3.5}$$

This four-force density gives the energy-momentum transfer from gravitational field to matter, which is minus the energy-momentum transfer from matter to field. The transfer in the latter direction can be written as the

---

is cited both in the *Entwurf* paper (Einstein and Grossmann 1913, p. 19, p. 23) and in Einstein (1913, p. 1257), the published text of his Vienna lecture (to which we will return in Ch. 4). When Einstein delivered this lecture, he "asked whether Kottler was in the audience. A young man rose. Einstein asked him to remain standing—so that all could see the man whose help had been so useful" (Clark 1971, p. 156).

[12] Einstein first covered the electromagnetic case in the 1912 manuscript mentioned in the preceding note (CPAE4, Doc. 1, [pp. 62–63]). Various sets of lecture notes that have survived show that Einstein also covered this material in courses on electrodynamics and special relativity in Zurich (1913–14) and Berlin (1914–15, 1918–19): see CPAE4, Doc. 19, [p. 1]; CPAE6, Doc. 7, [p. 19]; and CPAE7, Doc. 12, [p. 9], respectively. He also covered it in his Vienna lecture (Einstein 1913, p. 1253; see Ch. 4 below) and in his review article on the *Entwurf* theory the following year (Einstein 1914c, pp. 1054–1056).

divergence of a quantity representing the energy-momentum density of the gravitational field. The energy-momentum balance in Eq. (3.5) can thus also be expressed as

$$\text{4div}\begin{pmatrix}\text{em}\\ \text{grav field}\end{pmatrix} = -\text{4force density}, \tag{3.6}$$

which has the same form as the analogous equation for the electromagnetic field. In the electromagnetic case, the four-force on the right-hand side is the four-force exerted by the electromagnetic field on its source, the charge-current density.[13]

Combining Eqs. (3.5) and (3.6), we can write the energy-momentum balance equation as the vanishing of the four-divergence of the sum of the energy-momentum tensor for matter and the corresponding quantity for gravitational energy-momentum.[14]

$$\text{4div}\begin{pmatrix}\text{em} & + & \text{em}\\ \text{matter} & & \text{grav field}\end{pmatrix} = 0. \tag{3.7}$$

Another constraint on the construction of the field equations is that the same quantity is playing the role of the energy-momentum density of the gravitational field in Eqs. (3.4) and (3.7) (Janssen and Renn 2007, pp. 860–861).

In electrodynamics, the density of the four-force exerted by the electromagnetic field on its source is a 'product' of field and source (the contraction of the index of the charge-current density four-vector with one of the indices of the electromagnetic field tensor). In electrostatics, it is simply the product of charge and electric field. Drawing once again on the analogy with electrodynamics, Einstein interpreted the expression for the density of the four-force exerted by the gravitational field on its source, the energy-momentum tensor for matter, as *minus* a 'product' of field and source (the minus sign reflecting that masses attract while like charges repel):

$$\text{4force density} = -\text{grav field}\cdot\begin{pmatrix}\text{em}\\ \text{matter}\end{pmatrix}. \tag{3.8}$$

[13] This is the definition of the four-force used in the passages cited in the preceding note, except for the 1913 Vienna lecture and the 1914 review article, where it is defined as the force exerted *on* the field, which is minus the force exerted *by* the field.

[14] This quantity cannot be a tensor. The energy-momentum balance equation is given by the vanishing of the covariant divergence of the energy-momentum tensor for matter (see note 15), which transforms as a generally covariant vector. The ordinary divergence of the sum of two tensors, however, only transforms as a vector under linear (autonomous) transformations (cf. note 8). In late 1913, Einstein argued that energy-momentum conservation therefore restricted the covariance of his theory to linear transformations. He explicitly mentioned that this presupposed the energy-momentum density of the gravitational field transforming in the same way as the energy-momentum density of matter (Einstein 1913, pp. 1257–1258; Einstein 1914a). Early the following year, he realized his mistake and withdrew the argument (Einstein and Grossmann 1914, p. 218; see Norton 1984, pp. 126–127, for discussion).

Inserting Eq. (3.8) into Eq. (3.5), we arrive at:

$$4\text{div}\begin{pmatrix}\text{em}\\ \text{matter}\end{pmatrix} = -\text{grav field}\cdot\begin{pmatrix}\text{em}\\ \text{matter}\end{pmatrix}. \tag{3.9}$$

This provides a way of checking whether candidate field equations are compatible with energy-momentum conservation: substitute the left-hand side of the field equations for the energy-momentum tensor of matter on both the left- and the right-hand side of Eq. (3.9) and check whether the resulting equation holds as an identity. If we were to use the field equations and the definition of the gravitational field of the final theory, we would find the famous contracted Bianchi identities in this way. Einstein, however, was still looking for a suitable candidate for the left-hand side of the field equations. All he knew was that there should be one term with second-order derivatives of the metric that reduces to the Laplacian acting on the metric for weak static fields. He hit upon a clever strategy to use that knowledge to find the other terms.

Unfortunately, he overlooked an important ambiguity. There are two ways in which the gravitational four-force density can be written as a 'product' of the gravitational field and the energy-momentum tensor for matter, resulting in different definitions of the gravitational field. It can be defined as one term that is essentially minus the gradient of the metric or as the sum of three such terms that together give minus the Christoffel symbols.[15] The Christoffel symbols occur both in the Riemann tensor and in the geodesic equation (determining the curves of maximal length in space-time, which

---

[15] The vanishing of the covariant divergence of the energy-momentum tensor for matter, $T^{\alpha}_{\mu;\alpha}=0$, can be written as

$$T^{\alpha}_{\mu,\alpha}+\begin{Bmatrix}\beta\\ \alpha\beta\end{Bmatrix}T^{\alpha}_{\mu}-\begin{Bmatrix}\beta\\ \mu\alpha\end{Bmatrix}T^{\alpha}_{\beta}=0$$

(see note 7 for a definition of the Christoffel symbols). Multiplying by $\sqrt{-g}$ and using that $\sqrt{-g}\begin{Bmatrix}\beta\\ \alpha\beta\end{Bmatrix}=(\sqrt{-g})_{,\alpha}$ (cf. note 7), we can rewrite this as

$$\left(\sqrt{-g}\,T^{\alpha}_{\mu}\right)_{,\alpha}-\sqrt{-g}\begin{Bmatrix}\beta\\ \mu\alpha\end{Bmatrix}T^{\alpha}_{\beta}=0.$$

Introducing the mixed tensor density $\mathfrak{T}^{\alpha}_{\mu}\equiv\sqrt{-g}\,T^{\alpha}_{\mu}$, we can also write this as $\mathfrak{T}^{\alpha}_{\mu,\alpha}=\begin{Bmatrix}\beta\\ \mu\alpha\end{Bmatrix}\mathfrak{T}^{\alpha}_{\beta}$. This relation has the form of Eq. (3.9) if the gravitational field is defined as

$$\text{grav field}\equiv-\begin{Bmatrix}\beta\\ \mu\alpha\end{Bmatrix}.$$

However, this is not the only option. Note that

$$\begin{Bmatrix}\beta\\ \mu\alpha\end{Bmatrix}\mathfrak{T}^{\alpha}_{\beta}=\tfrac{1}{2}g^{\beta\rho}(g_{\rho\mu,\alpha}+g_{\rho\alpha,\mu}-g_{\mu\alpha,\rho})\mathfrak{T}^{\alpha}_{\beta}=\tfrac{1}{2}(g_{\rho\mu,\alpha}-g_{\mu\alpha,\rho})\mathfrak{T}^{\alpha\rho}+\tfrac{1}{2}g^{\beta\rho}g_{\rho\alpha,\mu}\mathfrak{T}^{\alpha}_{\beta}.$$

are the trajectories of free test particles in the theory). Mathematics thus suggests the latter definition. Physics, however, suggests the former. Given that the metric field plays the role of the gravitational potential, it is only natural that minus its gradient should play the role of the gravitational field. Einstein chose the definition suggested by the physics, a decision he would come to regret. Einstein appears to have been led in particular by the analogy with electrostatics in which the electric field is minus the gradient of the electrostatic potential, $\mathbf{E} = -\nabla\varphi$. In general, of course, the electromagnetic field tensor, $F_{\mu\nu} = A_{\mu,\nu} - A_{\nu,\mu}$ is a combination of two terms that both have the form of a (four-)gradient of the (four-)vector potential, A_μ.

That the field equations should reduce to the Poisson equation for weak static fields gave Einstein the first term of the left-hand side of Eq. (3.1). To find the remaining terms, he formed the 'product' of this one term with the gravitational field, using the definition suggested by electrostatics, and rewrote the resulting expression as a sum of two kinds of terms, ones that are the four-divergence of expressions quadratic in first-order derivatives and ones that are 'products' of the gravitational field and expressions quadratic in first-order derivatives. He thus arrived at an identity of the form:[16]

$$\text{grav field} \cdot \left(\frac{\partial^2 g}{\partial x^2}\right) = -\,4\text{div}\left(\sum{}' \frac{\partial g}{\partial x} \cdot \frac{\partial g}{\partial x}\right) - \text{grav field} \cdot \left(\sum{}' \frac{\partial g}{\partial x} \cdot \frac{\partial g}{\partial x} + \sum{}'' \frac{\partial g}{\partial x} \cdot \frac{\partial g}{\partial x}\right). \quad (3.10)$$

Regrouping terms, we can rewrite this as:

$$\text{grav field} \cdot \left(\frac{\partial^2 g}{\partial x^2} + \sum{}' \frac{\partial g}{\partial x} \cdot \frac{\partial g}{\partial x} + \sum{}'' \frac{\partial g}{\partial x} \cdot \frac{\partial g}{\partial x}\right) = -\,4\text{div}\left(\sum{}' \frac{\partial g}{\partial x} \cdot \frac{\partial g}{\partial x}\right). \quad (3.11)$$

Comparison of this identity to

$$-\,\text{grav field} \cdot \kappa \begin{pmatrix}\text{em}\\ \text{matter}\end{pmatrix} = -\,4\text{div}\,\kappa \begin{pmatrix}\text{em}\\ \text{grav field}\end{pmatrix}, \quad (3.12)$$

The first term on the right-hand side vanishes (it is a contraction of a part that is anti-symmetric in ρ and α and a part that is symmetric in those same indices). So $T^\alpha_{\mu;\alpha} = 0$ can also be written as $\mathfrak{T}^\alpha_{\mu,\alpha} = \frac{1}{2} g^{\beta\rho} g_{\rho\alpha,\mu} \mathfrak{T}^\alpha_\beta$, which suggests the alternative definition,

$$\text{grav field} \equiv -\tfrac{1}{2} g^{\beta\rho} g_{\rho\alpha,\mu}.$$

[16] The derivation of this identity can be found on pp. 26L–R (pp. 52–53) of the Zurich notebook, the last two pages of the notebook related to the search for gravitational field equations, as well as in Einstein and Grossmann (1913, pp. 14–16, pp. 37–28). For detailed reconstructions see the *Commentary* in *Genesis* (Renn 2007a, sec. 6, pp. 706–711) and Norton (1984, sec. 4.3, pp. 122–126).

which follows from Eqs. (3.9) and (3.7), allowed Einstein to read off of Eq. (3.11) both the expression for (κ times) the gravitational energy-momentum density,

$$\kappa \begin{pmatrix} \text{em} \\ \text{grav field} \end{pmatrix} = \sum{}' \frac{\partial g}{\partial x} \cdot \frac{\partial g}{\partial x}, \tag{3.13}$$

and the full field equations

$$\frac{\partial^2 g}{\partial x^2} + \sum{}' \frac{\partial g}{\partial x} \cdot \frac{\partial g}{\partial x} + \sum{}'' \frac{\partial g}{\partial x} \cdot \frac{\partial g}{\partial x} = -\kappa \begin{pmatrix} \text{em} \\ \text{matter} \end{pmatrix}. \tag{3.14}$$

Note that Eqs. (3.13) and (3.14) have the form of Eqs. (3.3) and (3.4), respectively, which Einstein was looking for. Their construction guarantees that the gravitational four-force density can be rewritten as the divergence of gravitational energy-momentum density and that the four-divergence of the total energy-momentum density vanishes (cf. Eqs. (3.7)–(3.9)). Moreover, the expression for gravitational energy-momentum density that occurs in the law of energy-momentum conservation law enters the field equations in the exact same way as the energy-momentum density for matter.

As our schematic reconstruction of the derivation of these field equations shows, it relies heavily on the analogy with the four-dimensional formalism for electrodynamics (cf. note 11). This analogy was central to Einstein's physical strategy.

These then are the field equations that Einstein and Grossmann published in their *Entwurf* paper of June 1913 (Einstein and Grossmann 1913). The paper shows a strict division of labor: Einstein is listed as the author of the first, physical, part; Grossmann as the author of the second, mathematical, part.

Chapter 4
Consolidating the *Entwurf* Theory

The *Entwurf* field equations satisfy the two main requirements imposed by physical considerations (Newtonian limit, energy-momentum conservation). Einstein even seems to have been under the impression that these two requirements determine the equations uniquely. Their covariance properties, however, were unclear and appeared to be intractable. The covariance of field equations extracted from generally covariant ones with some coordinate restriction is determined by the covariance of that coordinate restriction, which will involve expressions much simpler than the left-hand side of the field equations. The *Entwurf* equations, however, were not found this way. Einstein therefore had to check directly, for instance, whether the rotation metric is a vacuum solution of these equations. In a letter of August 1913, Ehrenfest told Lorentz that Einstein had meanwhile done the calculation "five or six times," finding "a different result almost every time" (Janssen 2007, p. 833). Over the next two years, Einstein changed his mind repeatedly about whether it is a solution or not.[1] Much of his time and effort during this period went into clarifying the covariance properties of the *Entwurf* field equations. An elaborate analogy with electromagnetism sustained and guided him in these efforts.

In September 1913, in a masterful plenary lecture, "The current state of the problem of gravity," at the *Naturforscherversammlung*, the annual conference of the *Gesellschaft Deutscher Naturforscher und Ärzte* (Society of German Natural Scientists and Physicians), held in Vienna that year, Einstein (1913) put great emphasis on the analogy between gravity and electromagnetism (Janssen 1991). The analogy even made it into a report on the lecture in a Viennese newspaper. *Die Neue Freie Presse* of September 24, 1913, carried a summary of Einstein's lecture the day before. This summary is a slightly reworded version of the introduction in the printed version of the lecture.[2] After Hertz's detection of electromagnetic waves had shown that

---

[1] On the problem of rotation, see Pt. II, Chs. 2 and 4 and Janssen (1999, 2007, 2012, 2014).

[2] It is not clear whether Einstein supplied this text himself. What speaks for it is that in the lecture it says that action-at-a-distance is replaced by "partial differential equations," while in the summary it says that action-at-a-distance is replaced by "an

M. Janssen, J. Renn, *How Einstein Found His Field Equations*, Classic Texts in the Sciences, https://doi.org/10.1007/978-3-030-97955-3_4

action-at-a-distance theories of electromagnetism were untenable, Einstein told his audience,

> it was only natural that confidence in the correctness of Newton's action-at-a-distance theory of gravity was also shaken and gave way to the conviction that Newton's law of gravity does not encompass all gravitational phenomena, just as Coulomb's law of electrostatics and magnetostatics do not encompass all electromagnetic phenomena. That Newton's law sufficed earlier for calculating the motions of celestial bodies is because the velocities and accelerations of those motions are small. In fact, it is easy to demonstrate that celestial bodies moving under the influence of electrical forces coming from electrical charges on those bodies would not reveal Maxwell's laws of electrodynamics as long as their velocities and accelerations are of the same order of magnitude as those in the motions of our ordinary celestial bodies. Those motions could have been described with great accuracy on the basis of Coulomb's law.

In the body of the lecture, Einstein (1913, p. 1258) recast the *Entwurf* equations in a form that matches Maxwell's equations for the electromagnetic field even more closely than in the Zurich notebook and the *Entwurf* paper. Like Maxwell's equations, the *Entwurf* field equations now set the four-divergence of the field equal to its source. For the gravitational field, the source is the sum of the energy-momentum density of matter and of the gravitational field itself. Schematically, the *Entwurf* field equations can thus be rewritten as:[3]

$$4\text{div}\begin{pmatrix}\text{grav}\\ \text{field}\end{pmatrix} = -\kappa\begin{pmatrix}\text{em} & + & \text{em}\\ \text{matter} & & \text{grav field}\end{pmatrix}. \tag{4.1}$$

The printed text of the Vienna lecture mentions two arguments that reconciled Einstein with the lack of general covariance of his field equations. The first was a short-lived argument from energy-momentum conservation that seemed to show that the field equations can only be covariant under linear transformations (see Ch. 3, note 14). In a footnote, Einstein mentioned that he had recently found a more general argument showing that generally covariant field equations are inadmissible.

4.1 The hole argument

The argument against general covariance that Einstein alluded to in his Vienna lecture is the famous hole argument.[4] The hole argument was first published in the comments Einstein added to the reprint of the *Entwurf* paper

---

action from point to point." What speaks against it is that the summary at one point refers to Maxwell's laws of "thermodynamics" rather than electrodynamics.

[3] Einstein achieved this simplification compared to Eqs. (3.4) and (3.14) by multiplying the gravitational field and the energy-momentum densities of matter and field by $\sqrt{-g}$, i.e., by switching from tensors to tensor densities (cf. Ch. 3, note 15).

[4] Einstein probably added the footnote referring to the argument to page proofs of the printed text of the lecture in early September, even though he only delivered the lecture September 23 (Janssen 2007, pp. 792–793). The earliest extant version of what

in a mathematics journal in January 1914 (Einstein 1914a). Before the end of 1914, Einstein would rehearse the argument in print three more times (Einstein 1914b; Einstein and Grossmann 1914; Einstein 1914c).

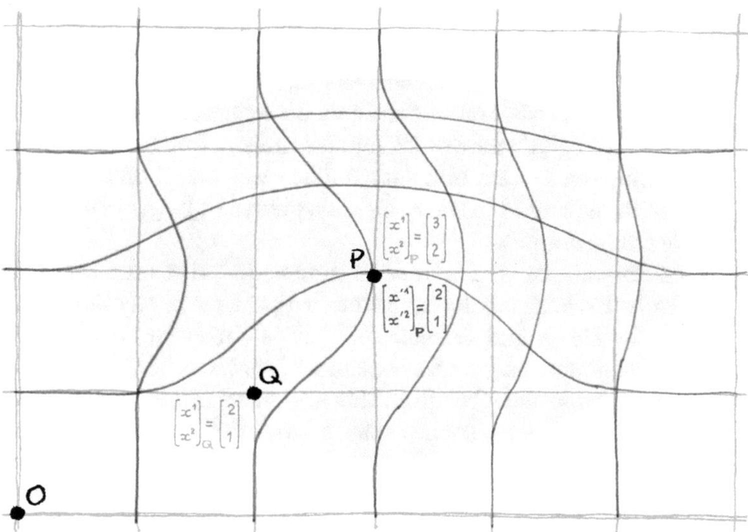

Figure 4.1 The hole argument purporting to show that generally covariant field equations cannot uniquely determine the metric field for a given matter distribution (drawing by Laurent Taudin in Janssen, 2014, p. 189).

Fig. 4.1 shows two space-time coordinate systems—one with grey straight lines and coordinates (x^1,x^2), one with darker squiggly lines and coordinates (x'^1,x'^2) (two dimensions are suppressed). They coincide everywhere except in the grey oval in the figure, representing the matter-free region from which the hole argument derives its name. Suppose that the metric field $g(x^1,x^2)$ is a solution of the field equations in the x-coordinates (the indices of the metric tensor are suppressed). If the field equations are generally-covariant, that same metric field expressed in the x'-coordinates, $g'(x'^1,x'^2)$, is also a solution. Consider the point P in the hole with coordinates $(x^1,x^2) = (3,2)$ and $(x'^1,x'^2) = (2,1)$. The solutions $g(x^1,x^2)$ and $g'(x'^1,x'^2)$ assign the same curvature—say, C—to P.

The key move in the hole argument is to note that $g'(x')$ remains a solution *if we read the values for x' as coordinates with respect to the straight lines of the x-coordinate system rather than with respect to the squiggly lines of the x'-coordinate system.* The equations, after all, remain the same regardless of

is essentially the hole argument can be found in notes by Besso dated August 28, 1913 (ibid., p. 789 [facsimile]; pp. 819–830 [transcription and analysis]).

which coordinate system we take x' to be referring to. Moreover, this metric field, which, suppressing all indices, we can write as $g'(x)$, is a solution *for the same matter distribution*. After all, there is no matter inside the hole, and outside the hole the two coordinate systems coincide. The metric field $g'(x)$, however, appears to be different from the one we had before: whereas $g'([x'^1, x'^2] = [2,1])$ assigns the curvature C to point P, $g'([x^1, x^2] = [2,1])$ assigns that same curvature to point Q (see Fig. 4.1).

Pais (1982, p. 222, p. 244) trivialized Einstein's argument, suggesting he mistook the fact that the *same* field can be expressed in *different* coordinates (e.g., $g(x)$ and $g'(x')$ above) for indeterminism. Stachel (1989), however, showed that Einstein's claim of indeterminism was based on his construction of *different* fields expressed in the *same* coordinates (e.g., $g(x)$ and $g'(x)$ above; cf. Pt. II, Sec. 3.3, note 5).

Although Einstein made the hole argument in print four times, he published generally covariant field equations without even mentioning it. Only when queried by Besso and Ehrenfest did Einstein come up with an escape from the hole argument, a counter-argument known as the *point-coincidence argument.*[5] On a superficial reading, this argument turns on the observation that all point coincidences, such as the intersection of two world lines, are uniquely determined, even though the metric field is not. Since all we ever observe are intersections of world lines, the argument then continues, the indeterminism lurking in the hole argument is harmless.[6] On a more profound

[5] See Einstein to Ehrenfest, December 26, 1915 and January 5, 1916 (CPAE8, Docs. 173 and 180), Ehrenfest to Einstein, January 1, 1916 (CPAE8, Doc. 177a, in CPAE13) and Einstein to Besso, January 3, 1916 (CPAE8, Doc. 178). In a paper with Ole Engler, one of us has argued—against Howard and Norton (1993), who argue that Einstein got it from Kretschmann (1915)—that the point-coincidence argument resulted from a conversation Einstein had with Moritz Schlick sometime in December 1915 (Engler and Renn 2013, pp. 142–152 and the foldout sheet following p. 156). The issue also plays a prominent role in fourteen letters Lorentz and Ehrenfest exchanged between late December 1915 and late January 1916 about the latest version of Einstein's theory. In these letters, they not only discussed Einstein's papers of November 1915 but also Einstein's letters to both of them in response to their queries (Kox 2018, Docs. 174–187, cf. Pt. II, Sec. 8.4, note 1). This three-way corrrespondence is discussed in Kox (1988). In the first of his letters to Lorentz, written December 23, 1915, Ehrenfest is already wondering what happened to the hole argument: "In 1914 Einstein developed an almost philosophical proof for the necessity of the "adaptedness" of the coordinate system [a reference to Einstein (1914c, §,12, see Pt. II, Ch. 3) follows] *Is this proof correct*?" (Kox 2018, Doc. 175). In a letter to Ehrenfest of January 9, 1916 (Doc. 179), Lorentz developed his own version of the hole argument. In a letter to Lorentz that same day, Ehrenfest wrote that "a long discussion" with his wife, Tatyana Alexeyevna Afanasyeva, had convinced him that the point-coincidence argument in Einstein's letter of January 5, which he had received the day before, was "*completely* right" (Doc. 180). He enclosed Einstein's letter with his own and in a long letter written January 10–11, Lorentz reported that he "had read only part of [Einstein's letter] when it dawned on [him] that [Einstein] is entirely right." In a postscript, he wrote: "I have now congratulated Einstein on his brilliant results" (*Ik heb Einstein nu met zijne schitterende uitkomsten gelukgewenscht*) (Doc. 181).

[6] This is how Einstein's (1916b, pp. 776–777) presented the point-coincidence argument in his review article on his new theory (for discussion, see Howard 1999). It is also how

reading, the argument turns on the observation that points (bare-manifold points in modern parlance) only become space-time points once values for the metric field have been assigned to them. Space-time points, the argument then continues, cannot be individuated independently of these values. In that case, the mathematical distinction between the points P and Q above does not correspond to a physical distinction, and generally-covariant field equations do not give rise to indeterminism to begin with.[7]

As Norton (2005, 2007) has pointed out, the hole argument is incompatible with the modern understanding of coordinate conditions as gauge conditions. So if Einstein had this modern understanding, the argument should have looked fishy to him even before he could put his finger on where exactly it goes wrong. Coordinate conditions pick out one or more representatives of every possible equivalence class of metric fields. Since $g(x)$, $g'(x')$, $g''(x'')$, etc. represent the same metric field, a good coordinate condition *should not allow all of them.* Field equations, not coordinate conditions, pick out which equivalence classes of metric fields are allowed. Since, according to the hole argument, $g(x)$, $g'(x)$, $g''(x)$, etc. represent different metric fields, a good coordinate condition *should allow all of them.* Since members of pairs $g'(x)$ & $g'(x')$, $g''(x)$ & $g''(x'')$, etc., have the same form, *no coordinate condition can allow one but not the other* (Janssen 2007, p. 829).

4.2 A variational formalism for the *Entwurf* field equations

Following the suggestion of one of their Zurich colleagues, the mathematician Paul Bernays,[8] Einstein and Grossmann finally got a better handle on the covariance properties of the *Entwurf* field equations by deriving them from a variational principle. It is easier to study the covariance of the Lagrangian—a single function of the metric and its first derivatives that determines the field equations in this variational approach—than the covariance of the many-component field equations themselves. Einstein and Grossmann argued that the field equations simply inherit their covariance properties from the Lagrangian, although this was disputed by Tullio Levi-Civita, one of the ac-

Lorentz understood the argument, which made a deep impression on him. As he put it in the letter to Ehrenfest of January 10–11, 2016 mentioned in note 5: "We cannot claim that [the field equations] determine the $g_{\mu\nu}$. Instead we must say that when for the $g_{\mu\nu}$ we take arbitrary functions that satisfy the equations, and then calculate all phenomena with these $g_{\mu\nu}$, we will reproduce all coincidences correctly" (Kox 2018, Doc. 181, p. 465). Lorentz (1916–17) made this the basis for a geometric formulation of general relativity in a series of four papers, the first of which he submitted on February 26, 1916 (for discussion, see Janssen 1992).

[7] For further discussion of the hole argument and its resolution, see the entries on the hole argument in two online resources, the *Stanford Encyclopedia of Philosophy* and *Living Reviews in Relativity*, written by John Norton and John Stachel, respectively. Our discussion is based on Janssen (2014, pp. 188–192).

[8] On Bernays, see Rowe (2001, pp. 391–392).

knowledged masters of Riemannian geometry.[9] They presented their results in their second and last joint paper (Einstein and Grossmann 1914). By the time it was published in May 1914, Einstein had already moved from Zurich to Berlin.

The Lagrangian for the *Entwurf* field equations has the same structure as the Lagrangian for the source-free Maxwell equations. In both cases, it is essentially the square of the fields. Einstein and Grossmann found four conditions, written as $B_\mu = 0$, on the metric field and its derivatives, which not only guarantee the vanishing of the four-divergence of the total energy-momentum density (of matter and gravitational field) but also determine under which non-autonomous transformations the *Entwurf* Lagrangian retains its form. They introduced a new name for these transformations, calling them "justified" (*berechtigte*) transformations between "adapted" (*angepaßte*) coordinate systems (adapted to a specific metric field) (Einstein and Grossmann 1914, p. 221). On the basis of some general argument, Einstein convinced himself that the class of non-autonomous transformations allowed by the $B_\mu = 0$ conditions was as broad as was allowed by the hole argument and that it included such important test cases as rotation in Minkowski space-time. Since B_μ actually does *not* vanish for the rotation metric (Janssen 1999, pp. 150–151, note 47), this general argument has to be fallacious. Einstein sketched the argument in a letter to Lorentz of January 23, 1915 (CPAE8, Doc. 47) but it remains unclear exactly how it was supposed to work (Janssen 2007, p. 816).

Einstein used this same variational approach in the review article on the *Entwurf* theory published in November 1914 (Einstein 1914c). Instead of specifying the Lagrangian right away, he now left open how the Lagrangian depends on the metric and its first-order derivatives. He analyzed both the covariance of this more general Lagrangian and its compatibility with energy-momentum conservation. To ensure energy-momentum conservation, he derived two sets of conditions that the Lagrangian and its derivatives with respect to the metric and derivatives of the metric had to satisfy. As he showed in a letter to Lorentz the following year (CPAE8, Doc. 129; Pt. II, Ch. 4), the first set, written as $S_\nu^\mu = 0$, can be interpreted as setting two quantities playing the role of gravitational energy-momentum density in different places in the general formalism equal to one another: one entering into the field equations as a source term, the other entering into the law of energy-momentum conservation written as the vanishing of the four-divergence of the total energy-momentum density (cf. Eqs. (4.1) and (3.7)).

The vanishing of this four-divergence is guaranteed by the conjunction of the field equations and the second set of conditions, written as $B_\mu = 0$. This condition is obtained by taking the four-divergence of the left-hand side of the field equations (cf. Eq. (4.1)). Schematically,[10]

---

[9] See Cattani and De Maria (1989) for analysis of the relevant correspondence between Einstein and Levi-Civita.

[10] Somewhat more explicitly, we would write: $\frac{\partial}{\partial x^\mu}\frac{\partial}{\partial x^\nu}\begin{pmatrix}\text{grav}\\ \text{field}\end{pmatrix}^{\mu\nu}$.

$$4\text{div}\,4\text{div}\begin{pmatrix}\text{grav}\\ \text{field}\end{pmatrix} = 0. \tag{4.2}$$

As in the special case of the *Entwurf* Lagrangian considered in Einstein's second paper with Grossmann, the conditions $S^{\mu}_{\nu} = 0$ and $B_{\mu} = 0$ also determine the covariance of the Lagrangian. This connection between covariance and energy-momentum conservation can be seen as an instantiation *avant la lettre* of one of Noether's theorems connecting symmetries and conservation laws.[11]

Emmy Noether (1918) actually found the general theorems as part of a concerted effort in Göttingen to clarify the status of energy-momentum conservation in general relativity (Kossmann-Schwarzbach 2011; Rowe 1999, 2001, 2021). The way Einstein looked upon his result in late 1914 was that energy-momentum conservation is ultimately what restricts the covariance of the field equations. This is a more sophisticated version of the argument given, for instance, in the Vienna lecture, that energy-momentum conservation restricts the covariance of the field equations to linear transformations (see Ch. 3, note 14). "What can be more beautiful," he had written to Ehrenfest in early November 1913, referring to this earlier argument, "than that the necessary specialization follows from the conservation laws?" (CPAE5, Doc. 481). That same sentiment applied to this new argument as well.[12]

---

[11] For insightful discussions of Noether's theorems, see Brading (2002), Brown and Brading (2002), and Kossmann-Schwarzbach (2011).

[12] Our assessment of Einstein's result contrasts sharply with that of Brian Pitts in an article in which he discusses many of the same papers discussed here—including Einstein and Grossmann (1914) and Einstein (1914c)—from an unabashedly modern perspective. In his estimation, the result qualifies as a "surprising blunder reflecting ignorance not only of the literature of his day [Pitts cites Born (1914), Herglotz (1911), and Mie (1912, 1913)] but also of results known already to Lagrange et al. [i.e., the connection between translation invariance in time and space and the conservation of energy and momentum]" (Pitts 2016, p. 54). He concludes that Einstein

> used the wrong symmetry, rigid strictly linear coordinate transformations, with fanfare on his way to inferring conservation laws. Far from anticipating a special case of Noether's first theorem (one that was already in the literature of his day and that had simpler analogs that were very old), Einstein had a surprisingly poor grasp of the connection between symmetries and conservations (p. 60).

Illustrating the dangers inherent in his approach, Pitts never mentions the distinction between autonomous and non-autonomous coordinate transformations that was key to Einstein's belief at the time that the covariance of the *Entwurf* field equations goes beyond "rigid strictly linear coordinate transformations" and includes (non-autonomous) transformations to accelerating frames of reference (see also Ch. 5, note 2). Although Pitts (sec. 7, pp. 60–63) is right that the connection between translation invariance and energy-momentum conservation had been recognized well before the 1910s, there does not seem to have been a general awareness of the connection between symmetries and conservation laws at the time. In her monograph on the Noether theorems, Kossmann-Schwarzbach (2011, p. 85) quotes a telling passage from an interview Kuhn conducted with Heisenberg for the *Archive for the History of Quantum Physics* project in 1963.

The $S^{\mu}_{\nu} = 0$ conditions, Einstein thought, were satisfied only by the *Entwurf* Lagrangian. "We have now," he declared, "in a completely formal manner, i.e., without direct use of our physical knowledge about gravity, arrived at very definite field equations" (Einstein 1914c, p. 1076). Even though he exaggerated the mathematical purity of his new derivation of the *Entwurf* equations, one readily understands Einstein's enthusiasm. Finally, he seemed to have found the convergence of mathematical and physical lines of reasoning that had eluded him in the Zurich notebook.

4.3 The *Entwurf* theory, the Nordström theory, and Hilbert's theory

With the derivation of the *Entwurf* field equations in the 1914 review article, Einstein had *almost* done for the *Entwurf* theory what he had done earlier for its main competitor, the special-relativistic theory of gravity proposed in 1912 by the young Finnish physicist Gunnar Nordström (1912, 1913a, 1913b).[13] In this theory, the gravitational potential is represented by a scalar field in flat Minkowski space-time. Einstein had taken a strong interest in this theory. In his Vienna lecture, he had presented it alongside his own *Entwurf* theory, allotting roughly equal space to both (at least in the printed version). In February 1914, he had co-authored a paper on it with Adriaan D. Fokker, one of Lorentz's students. Einstein and Fokker (1914) reformulated Nordström's special-relativistic theory along the lines of Einstein's generalized theory of relativity, making the metric field represent both the geometry of some curved space-time and the gravitational potential.

This new derivation of the field equation of the Nordström theory is a picture-perfect application of the mathematical strategy developed in the Zurich notebook. Einstein and Laue had already convinced Nordström that

---

Talking about the notorious violation of energy conservation in the so-called BKS theory (Bohr, Kramers, and Slater 1924), Heisenberg said:

> Much later, of course, the physicists recognized that the conservation laws and the group theoretical properties were the same. And therefore, if you touch the energy conservation, then it means that you touch the translation in time. And that, of course, nobody would have dared to touch. But at that time, this connection was not so clear. Well, it was apparently clear to Noether, but not for the average physicist. Also in Göttingen it was not clear. The Noether paper has been written in Göttingen, I understand [in a follow-up remark, also quoted by Kossmann-Schwarzbach, Heisenberg notes that "it was actually formulated in connection with general relativity"]. But it was not popular among the physicists, so I certainly wouldn't learn that from Born in Göttingen. By the way, do you recall when the Noether paper had been written? I think it must have been also around '23 or so.

Kuhn's response (recall: this is 1963) is also telling: "I've heard of that paper, but never looked at it."

[13] See Renn (2007a, Vol. 3) for translations of these three papers and a reprint of a paper by Norton (1992) with historical analysis of the Nordström theory (see also Norton, 1993, and Giulini and Straumann, 2006).

the right-hand side of the field equation of his theory should be the trace of the energy-momentum tensor for matter. This is a generally covariant scalar. It was only natural to extract the left-hand side from the curvature scalar, the generally covariant scalar formed out of the Riemann tensor. The curvature scalar itself was unacceptable given the hole argument's injunction against generally covariant field equations. What was needed was a coordinate restriction to extract a suitable candidate from the curvature scalar. Einstein and Fokker imposed the restriction that only those coordinates are allowed in which the metric is the product of a function of the space-time coordinates and the standard diagonal metric of Minkowski space-time.[14] This function is just the gravitational potential of the original formulation of the Nordström theory and satisfies the original version of its field equation. In the conclusion of their paper, Einstein and Fokker expressed the hope that the Riemann tensor, given its role in their derivation of the Nordström field equations, "may also open the way for a derivation of the Einstein-Grossmann gravitation equations that is independent of physical assumptions" (p. 328).

Although this is not quite what Einstein accomplished in the 1914 review article on the *Entwurf* theory, he came close. He had both the field equations and the coordinate restriction, the $B_\mu = 0$ conditions, with which these field equations could presumably be extracted from some generally covariant object. He did not know that object but was confident it must exist.[15] Despite Levi-Civita's criticism of his mathematics (see note 9), Einstein appears to have been happy with this state of affairs well into 1915.[16] There is no

[14] In the usual parlance, space-time is conformally flat in the Nordström theory and the gravitational potential is the conformal factor.

[15] This can be gathered from Einstein's remarks in a paper written in response to criticism of the *Entwurf* theory in the discussion following the Vienna lecture by Gustav Mie, whose own theory of gravity Einstein had not covered in his lecture (Einstein et al. 1913). Einstein wrote: "When one has equations relating certain quantities that only hold in certain coordinate systems, one has to distinguish between two cases: 1. There are generally-covariant equations corresponding to the equations ... ; 2. There are no generally-covariant equations that can be found on the basis of the equations given for a particular choice of reference frame. In case 2, the equations do not tell us anything about the things represented by these quantities; they only restrict the choice of reference frame. If the equations tell us anything at all about the things represented by these quantities, we are always dealing with case 1" (Einstein 1914b, pp. 177–178).

[16] Strong statements of Einstein's confidence in the *Entwurf* theory (and in the hole argument) in late 1914/early 1915 can be found in letters to Zangger. In letter dated after December 27, 1914, he wrote: "Recently I have been working nonstop again on graviation and have now arrived at a splendid clarity about how things are connected. Formally, the matter is now so compelling that it must be hard for someone who truly and deeply appreciates it to escape from the magic of this edifice [*dass es für einen, der wirklich tief hineinsieht, schwer sein muss, sich dem Zauber dieses Gebäude zu entziehen*]. If all physical events are completely determined by other observable physical events, the path I have chosen turns into a necessity" (CPAE8, Doc. 41a, in CPAE10). As we saw, Einstein made a very similar comment about the successor to the *Enwurf* theory in his first paper of November 1915: "Hardly anybody who has truly understood the theory will be able to avoid coming under its spell" [*Dem Zauber dieser Theorie*

indication he showed any reservations about the *Entwurf* theory when he lectured on it in Göttingen in late June/early July. In fact, the theory Hilbert presented in November 1915 has the same structure as the *Entwurf* theory, or, for that matter, the Nordström theory as reformulated by Einstein and Fokker.

Hilbert used the curvature scalar as the Lagrangian for his theory. While this is the obvious choice from the point of view of mathematics, a *prima facie* objection from the point of view of physics is that the curvature scalar depends not just on first-order derivatives of the metric but on second-order ones as well. At first blush, one would therefore expect the corresponding field equations to contain terms with third-order derivatives. On closer inspection—and this is basic textbook knowledge nowadays—one sees that the field equations remain of second order. Einstein used this same Lagrangian in his variational derivation of the Einstein field equations in November 1916 (Einstein 1916d, see Pt. II, Ch. 10). The $B_\mu = 0$ conditions for this particular Lagrangian do not restrict the allowed coordinates in any way but hold as identities. They are just the contracted Bianchi identities. Although he did not make the connection with the Bianchi identities,[17] this was exactly what Einstein expected at that point: the general covariance of the field equations guarantees energy-momentum conservation.

The big surprise of the discovery of the page proofs of Hilbert's paper (see Ch. 2, note 12) was that this was *not* how Hilbert saw things, at least not in November 1915. Hilbert, at that point in time, accepted the hole argument against generally covariant field equations. To his "axiom II," which says that the Lagrangian is invariant under arbitrary transformations of what he calls "world parameters," he therefore added "axiom III," which restricts the covariance of the theory to transformations between "space-time coordinates," defined as those world parameters that satisfy four conditions on the metric suggestively called the "energy theorem."[18] The similarity of Hilbert's original theory to the *Entwurf* theory is striking. In both theories, energy-momentum conservation imposes a restriction on the covariance of the field equations. Axiom III, the energy theorem, was dropped in the published version of Hilbert's paper.

---

wird sich kaum jemand entziehen können, der sie wirklich erfaßt hat] (Einstein 1915a, p. 779). At the beginning of 1915, however, on January 11, Einstein had written to Zangger about the *Entwurf* theory: "I assume you have received the article on general relativity [i.e., Einstein (1914c)]; it is the happy conclusion of my efforts in this area" (CPAE8, Doc. 45a, in CPAE10).

[17] Three correspondents alerted Einstein to these identities in letters of 1917–18 but the recipient showed no interest in them (see Pt. II, Sec. 10.3, note 9, and CPAE8, Vol. 8, p. li).

[18] This same constellation can be discerned in a letter from Hilbert to Einstein of November 13, 1915 (CPAE8, Doc. 140).

Chapter 5
The *Entwurf* field equations as the scaffold for the Einstein field equations

That the theory in the page proofs of Hilbert's paper has the same structure as the *Entwurf* theory is a strong indication that Einstein's confidence in the *Entwurf* theory remained unshaken when he lectured on it in Göttingen in the summer of 1915. By late summer, however, nagging doubts must have set in. Why else would he have checked yet again whether or not the rotation metric is a vacuum solution of the *Entwurf* field equations? He had changed his mind a number of times on this score, but had eventually decided that it is. This is a necessary (though by no means sufficient) condition for the relativity of rotation, which he had hailed as one of the signature achievements of the *Entwurf* theory in the introduction of his review article (Einstein 1914c, pp. 1031–1032). Now he discovered to his dismay that he had been wrong. The rotation metric is not a vacuum solution (cf. Ch. 4, note 1). "This is a blatant contradiction," he wrote on September 30 to his protégé Erwin Freundlich (CPAE8, Doc. 123; Pt. II, Ch. 4), the young astronomer who had been exploring various avenues to test the *Entwurf* theory astronomically.[1]

Shortly thereafter, the uniqueness argument for the *Entwurf* Lagrangian fell apart. On October 12, Einstein explained its fatal flaw in a letter to Lorentz (CPAE8, Doc. 129; Pt. II, Ch. 5). The conditions $S_\nu^\mu = 0$ central to the argument had come out of his analysis of energy-momentum conservation. He now realized that, like the $B_\mu = 0$ conditions, one encounters these same conditions in the analysis of covariance properties. In that context, they express the condition that the Lagrangian be invariant under linear transformations. The *Entwurf* Lagrangian is hardly unique in meeting that minimal requirement. In letters written shortly after the breakthrough of November 1915, Einstein listed the illusory character of this uniqueness argument as the third of three reasons for abandoning the *Entwurf* equations.[2]

1. Freundlich had led an ill-fated eclipse expedition to the Crimea in 1914 to measure the bending of light in gravitational fields and had also become interested in possible evidence for a gravitational redshift in light from both the Sun and Jupiter (Hentschel 1992, 1994; Crelinsten 2006; Kennefick 2019).

2. Einstein to Sommerfeld, November 28, 1915 (CPAE8, Doc. 153; Pt. II, Ch. 7) and Einstein to Lorentz, January 1, 1916 (CPAE8, Doc. 177). In the paper mentioned above (see Ch. 4, note 12), Pitts (2016, p. 59–60) clearly identifies the error in the uniqueness argument for the *Entwurf* Lagrangian in the 1914 review article (Einstein 1914c) but does not mention the role Einstein's belated discovery of the error played in the development of his theory.

M. Janssen, J. Renn, *How Einstein Found His Field Equations*, Classic Texts in the Sciences, https://doi.org/10.1007/978-3-030-97955-3_5

5.1 The "fateful prejudice" and the "key to the solution"

The unraveling of the uniqueness argument for the *Entwurf* field equations was undoubtedly a setback for Einstein but it also opened the door to new possibilities. Einstein could now consider different Lagrangians while keeping the general formalism of the 1914 review article intact.[3] A natural thing to try would be to exploit the freedom in reading off the expression for the gravitational field from the energy-momentum balance equation (cf. Sec. 3.2 and note 15 therein). Retaining the electrodynamically inspired structure of the *Entwurf* Lagrangian but redefining the gravitational field occurring in it as minus the Christoffel symbols, one arrives at vacuum field equations in which one readily recognizes what we called the November tensor, one of the two tensors covariant under unimodular transformations into which Einstein had split the Ricci tensor in the Zurich notebook (see Fig. 1.3 and Ch. 3, note 7).

If one restricts the use of the variational formalism to unimodular transformations, in which case various factors of $\sqrt{-g}$ can be omitted, the vacuum field equations simply set the November tensor equal to zero. Setting it equal to (minus κ times) the energy-momentum tensor for matter, one obtains the field equations of Einstein's first communication of November 1915. The restriction to unimodular transformations complicates the use of the variational formalism but that must have seemed to be a small price to pay for this striking convergence of physical and mathematical lines of reasoning.[4] We registered Einstein's excitement about this kind of convergence in the 1914 review article. One would expect Einstein to have been even more excited about this new even more remarkable convergence. Of course, this is all on the assumption that Einstein actually found the field equations of his first November 1915 paper in the way we just described—by changing the definition of the gravitational field in the *Entwurf* Lagrangian. Did he?

Two comments he made in November 1915 strongly suggest an affirmative answer. In his first November communication, he wrote that the energy-momentum balance equation

> has led me in the past to look upon [one term with a gradient of the metric] as the natural expressions of the components of the gravitational field, even though the formulas of the absolute differential calculus suggest the Christoffel symbols [...] instead. *This was a fateful prejudice* (Einstein 1915a, p. 782; our emphasis).

[3] Two years later, Einstein (1916d) published a paper, which can be seen as a return to the 1914 review article but now with the generally covariant curvature scalar as the Lagrangian (see Pt. II, Ch. 10). There is no evidence to suggest that Einstein considered this option in 1914 and, even if he did, it would have been ruled out by the hole argument.

[4] In hindsight, Einstein would have saved himself a lot of trouble had he not restricted the covariance of his variational formalism and chosen the curvature scalar as his Lagrangian, as he would two years later (cf. notes 3 and 23).

And in the letter to Sommerfeld of November 28, he wrote about the final field equations:

> *The key to this solution* was my realization that not [one term with a gradient of the metric] but the related Christoffel symbols [...] are to be regarded as the natural expression for the "components" of the gravitational field (CPAE8, Doc. 153; Pt. II, Ch. 7; our emphasis).

These two passages are the smoking guns for our reconstruction of Einstein's breakthrough of October–November 1915. By changing one element in the variational formalism developed for the *Entwurf* field equations, which he had found pursuing the physical strategy, Einstein was led back to the November tensor, which he had found pursuing the mathematical strategy in the Zurich notebook. This convergence, we conjecture, convinced him that he had finally hit upon the correct field equations, even though a number of problems remained to be solved.

5.2 Untying the knot

Einstein had put himself in a bind in the Zurich notebook by making his coordinate restrictions do triple duty. A coordinate restriction had to determine the covariance properties of the field equations, guarantee their compatibility with energy-momentum conservation and reduce them to the Poisson equation of Newtonian theory in the case of weak static fields. A coordinate condition in the modern sense only has to perform the third of these tasks.

With the variational formalism he developed in 1914, Einstein only tightened the knot. The first two tasks were now handled by the $B_\mu = 0$ conditions. As long as the Lagrangian was the one leading to the *Entwurf* equations, the third task was automatically taken care of: the Newtonian limit had been built into the *Entwurf* field equations. As soon as the *Entwurf* Lagrangian was changed, however, the third task called for renewed attention. By redefining the gravitational field in the *Entwurf* Lagrangian, Einstein may well have pulled the thread untying the knot into which he had tied the strands corresponding to these three tasks.

Most obviously affected is the third strand which becomes separated from the other two. The $B_\mu = 0$ conditions no longer take care of the Newtonian limit. This may have suggested to Einstein that, while energy-momentum conservation still called for a coordinate restriction, the Newtonian limit could be handled with a coordinate condition in the modern sense. In the first November 1915 paper Einstein actually does seem to use a coordinate condition rather than a coordinate restriction for this purpose. In the final section of the paper, he imposed the Hertz condition to show that his new field equations have the correct Newtonian limit (p. 786). To conclude that section, he then noted that (autonomous) transformations to rotating coordinates are allowed in the new theory since they are unimodular. The rotation metric, however, does not satisfy the Hertz condition. As we argued in Sec. 3.1, this is probably why Einstein abandoned the field equations of the first November paper when he first considered them in the Zurich notebook. That it did not bother him anymore that the Hertz condition does not allow the rotation metric suggests—on the face of it—that Einstein was no longer using the Hertz condition as a coordinate restriction but as a coordinate condition. Einstein, however, only demanded that the Hertz restriction be satisfied in a first-order approximation, which the rotation metric does (cf. Pt, II, Sec. 1.3, note 20, and Sec. 6.3, note 21). It thus remains unclear whether Einstein had arrived at our modern understanding of coordinate conditions at this point.

What further complicates matters is that, in the other three November 1915 papers, Einstein recovered the results of Newtonian theory without imposing the usual four coordinate conditions. His perihelion calculations showed him that it was enough to use unimodular coordinates along with four special assumptions about the metric field of the Sun.[5] The introduction of unimodular coordinates had nothing to do with the Newtonian limit and only requires one condition on the metric ($g = -1$) instead of the usual four. In the 1916 review article, Einstein also used unimodular coordinates and did not introduce a further coordinate condition to recover the results of Newtonian theory.

The Hertz condition only returned in Einstein's papers on gravitational waves, the first one published in June 1916, the second in February 1918 (Einstein 1916c, 1918a).[6] In the latter, Einstein, for the first time, explained the modern notion of coordinate conditions. After adding the Hertz condition to the field equations, he wrote:

> At first sight, it may seem strange that [the ten field equations for the ten components of the metric tensor] should allow for four additional and arbitrary conditions without becoming overdetermined. That this procedure is justified can be seen as follows ... If I introduce a new coordinate system, then the $g_{\mu\nu}$ in the new system depend on four arbitrary functions, which define the transformation of the coordinates. These four functions can be chosen in such a way that the $g_{\mu\nu}$ in the new system satisfy four arbitrarily chosen relations (Einstein 1918a, pp. 155–156).

Why would Einstein belabor this point in 1918 if it had already been clear to him in 1915?

That this nonetheless had been clear to Einstein in 1915 is suggested by the hole argument. As the consideration at the end of Sec. 4.1 was meant to show, it is hard to imagine that anyone would recognize the freedom to apply coordinate conditions in the modern sense and still insist on the validity of the hole argument. Einstein did not lose a word about the hole argument when he published generally covariant field equations in November 1915 and only put his finger on exactly where it went wrong when pressed on the issue by Besso and Ehrenfest the following month. Yet somehow he must have convinced

[5] These assumptions are: (*i*) the metric does not depend on time; (*ii*) it is spherically symmetric; (*iii*) its mixed time-space components all vanish; (*iv*) it is Minkowskian at spatial infinity (Einstein 1915c, p. 833).

[6] For the history of theoretical work on gravitational waves, see Kennefick (2007). We return to Einstein's early work on gravitational waves in Sec. 7.2.

himself that the argument is invalid. Had he recognized the freedom to apply coordinate conditions in November 1915 and was *that* what told him the hole argument had to be wrong? Or did the resolution of the hole argument give him a better understanding of coordinate conditions? There does not seem to be enough evidence to answer these questions but what does seem to be clear is that Einstein's transition from coordinate restrictions to coordinate conditions did not happen overnight but took place gradually. We will return to this point in Sec. 5.3.[7]

So much for the third strand in the knot that we suggested Einstein untied in October/November 1915. The effect of the redefinition of the gravitational field on the other two strands is more subtle. They too are untwined but only to be tied back together in a slightly different way.

Ever since his second paper with Grossmann, Einstein's interpretation of the $B_\mu = 0$ conditions had been that energy-momentum conservation restricts the covariance of the field equations. With his new Lagrangian, however, there appeared to be a mismatch between the covariance of the field equations and the restrictions imposed by energy-momentum conservation. The new field equations retain their form under arbitrary and autonomous unimodular transformations, whereas the $B_\mu = 0$ conditions seem to restrict their covariance to a much narrower class of non-autonomous transformations.

This appearance is deceptive. It turns out that the $B_\mu = 0$ conditions can be replaced in this case by a modest strengthening of the restriction to unimodular transformations. Unimodular transformations require that the determinant g of the metric transform as a scalar. Einstein showed that the field equations guarantee that all *four* $B_\mu = 0$ conditions are satisfied if *one* expression involving derivatives of g is set equal to the trace T of the energy-momentum tensor for matter. In his first November 1915 paper, Einstein substituted this one condition on g for the four $B_\mu = 0$ conditions. The purpose of Einstein's subsequent amendments to the field equations, both the one proposed in the second November paper and the one proposed in the fourth, was to change this condition on g so that he could set $g = -1$, thereby choosing unimodular coordinates. This not only allowed him to look upon the November tensor as the generally covariant Ricci tensor expressed in unimodular coordinates (the other part of the Ricci tensor vanishes in these coordinates), it also restored the connection Einstein had come to expect between energy-momentum conservation and covariance. Einstein, however, now started to read this connection in the opposite direction: not as energy-momentum conservation *restricting* the covariance of the field equations but as the covariance of the field equations *guaranteeing* energy-momentum conservation.

[7] Norton (2005, 2007) has argued that Einstein had the modern understanding of coordinate conditions all along but only overcame the obstacles to using them in late 1915. We find it much more plausible that Einstein did not have this modern understanding before 1915 and developed all or most of it in late 1915, but our overall account does not hinge on whether one follows Norton on this point or us. For discussion, see Räz (2015).

In the November 1915 papers and in the 1916 review article, Einstein only showed that covariance of the field equations under *unimodular transformations* guarantees energy-momentum conservation in *unimodular coordinates.* It was not until November 1916, in the paper in which he gave a variational derivation of the Einstein field equations in general coordinates (Einstein 1916d), that he showed how the *general* covariance of the field equations guaranteed energy-momentum conservation in *arbitrary* coordinates. In the conclusion of that paper, he emphasized that the generally covariant law of energy-momentum conservation followed from the general covariance of the gravitational field equations. He told several correspondents that the main point of the paper had, in fact, been "to bring out the connection between relativity and the energy principle."[8]

5.3 Physics, mathematics or both?

In his first paper of November 1915, Einstein presented the field equations as the product of the mathematical strategy. In the introduction he explained the failure of his uniqueness argument for the *Entwurf* equations and concluded:

> For these reasons I completely lost confidence in the field equations I had constructed and looked for a way that would constrain the possibilities in a natural manner. I thus returned (*gelangte zurück*) to the demand of a more general covariance of the field equations (Einstein 1915a, p. 778).

Writing to Sommerfeld shortly after the fourth November 1915 paper, he distanced himself from the *Entwurf* equations and the *Entwurf* theory in even stronger terms:

> After all confidence in the result and the method of the earlier theory had thus given way, I saw clearly that a satisfactory solution could only be found if I could establish a connection to the general theory of covariants, i.e., to [the Riemann tensor] (CPAE8, Doc. 153; Pt. II, Ch. 7).

We already quoted the last paragraph of the introduction of the first November 1915 paper, in which Einstein called the new theory a triumph of mathematics (see Sec. 2). The next few pages of the paper strongly confirm this impression.

In sec. 1, Einstein showed how the restriction to unimodular transformations simplifies elements of the differential geometry presented in the mathematical part of the 1914 review article. He then split the Ricci tensor, $G_{\mu\nu}$, into two parts, $R_{\mu\nu}$ and $S_{\mu\nu}$, both transforming as tensors under unimodular transformations (Fig. 1.1). In sec. 2, he argued that the Christoffel symbols, which occur both in the Ricci tensor and in the geodesic equation, are the

---

[8] Einstein to Ehrenfest, October 24, 1926 (CPAE8, Doc. 269; cf. Pt. II, Sec. 10.3, note 13).

natural candidates to represent the components of the gravitational field.[9] "After what has been said so far," he proclaimed at the beginning of sec. 3, "it is only natural to posit field equations of the form $R_{\mu\nu} = -\kappa T_{\mu\nu}$" (p. 783).

The remainder of sec. 3 is taken up with a variational derivation of these equations and the proof that a modest strengthening of the restriction to unimodular transformations suffices to guarantee their compatibility with energy-momentum conservation. This proof relies heavily on the variational formalism of the 1914 review article on the *Entwurf* theory, thus putting the lie to Einstein's claim in the letter to Sommerfeld that he had lost confidence both "in the result *and the method* of the earlier theory."

Einstein's 1914 formalism was a product of the physical strategy. It had been designed to investigate the covariance properties of the *Entwurf* field equations. In the original derivation of the *Entwurf* equations, Einstein had made extensive use of the analogy between gravity and electromagnetism (see Sec. 3.2). In particular, he had defined the gravitational field in such a way that the gravitational force is equal to a product of field and source, as in electromagnetism. In sec. 2 of his first November 1915 paper, Einstein conceded that it would have been a problem had it turned out that this relation no longer holds with the new mathematically more natural definition of the gravitational field. In developing the variational formalism for the *Entwurf* theory, Einstein had once again been guided by the analogy between gravity and electromagnetism. In particular, he had modeled the Lagrangian for the gravitational field on the Lagrangian for the electromagnetic field. This part of the analogy also survives the redefinition of the gravitational field.

In the first November 1915 paper, the Lagrangian for the new theory appears right after the first introduction of the field equations at the beginning of sec. 3. Einstein did not mention here or anywhere else that one obtains this new Lagrangian simply by substituting the new mathematically motivated definition for the gravitational field into the old electrodynamically motivated *Entwurf* Lagrangian. We argued that this is how Einstein, following the physical strategy, found this new Lagrangian and the field equations that follow from it. Our smoking guns were Einstein's comments about "a fateful prejudice" and "the key to the solution." However, John Norton (1984, p. 145), according to whom Einstein found his field equations following the mathematical strategy, has dismissed these comments as having "all the flavor of an-after-the fact rationalization." Conversely, we dismissed as highly misleading what Norton took at face value, viz. Einstein's statements sug-

[9] The Christoffel symbols also represent the affine connection in general relativity. This is perhaps the strongest argument for identifying them as the components of the gravitational or, more accurately, the inertio-gravitational field. They represent the guiding field that makes test particles subject to no other interaction but gravity move on affine geodesics, the straightest possible paths in what in general will be a curved space-time (which in general relativity coincide with metric geodesics, paths of extremal length). This, in turn, is the natural way to implement the equivalence principle. The concept of an affine connection, however, was not introduced until 1916–17 (Stachel 2007).

gesting an abrupt turn from the physical to the mathematical strategy. Such different assessments are only to be expected. Since Einstein made conflicting statements about the relative importance of mathematical and physical considerations in finding his field equations, both at the time and in his later years, *any* reconstruction will involve putting extra weight on some and explaining away other statements.

A switch from the physical to the mathematical strategy is suggested by the following pair of quotes from Einstein's letters to Besso (Castagnetti et al. 1993, pp. 55–56). On December 10, 1915, he wrote about finding the Einstein field equations:

> This time the obvious [*das Nächstliegende*] was correct; however, Grossmann and I believed that the conservation laws would not be satisfied and that Newton's law would not come out in first approximation (CPAE8, Doc. 162)

Around March 1914, he had written about finding a result in the *Entwurf* theory:

> The general theory of invariants only proved to be an obstacle. The direct route proved to be the only feasible one. The only thing that is incomprehensible is that I had to feel my way around for so long before I found the obvious [*das Nächstliegende*] (CPAE5, Doc. 514).

In 1914, "the obvious" thus referred to the physics, while in 1915, "the obvious" referred to the mathematics.[10] Less than three years later, however, in another letter to Besso, Einstein distanced himself from what he told Besso in December 1915. On August 28, 1918, he wrote:

> Rereading your last letter I find something that almost makes me angry: that speculation has proved itself to be superior to empiricism. You are thinking here about the development of relativity theory. I find that this development teaches something else, which is almost the opposite, namely that a theory, to deserve our trust, must be built upon generalizable facts ... In the case of general relativity: equality of inertial and gravitational mass ...Never has a truly useful and profound theory really been found purely speculatively (CPAE8, Doc. 607).

So we must be careful not to cherry-pick Einstein's writings in support of our favorite reconstruction of how he arrived at general relativity.

With that in mind, let us compare our reconstruction to Norton's. Suppose that sometime in October 1915, as Norton (2000, p. 148, pp. 151–152) suggests, Einstein turned his back on the entire edifice he had meanwhile built around the *Entwurf* theory and decided he was going to build a new theory around general covariance and the Riemann tensor. One obvious question facing this scenario is: why did he initially forsake general covariance and settle for covariance under unimodular transformations? In other words, why did he use the Ricci tensor "as a half and not a whole" (to quote Jerney Kaagman in the song "Weekend" by the Dutch band Earth & Fire)? We already gave our answer: redefinition of the gravitational field in the *Entwurf*

[10] See Ch. 4, note 16 for a similar pair of quotes.

Lagrangian pointed to the November tensor rather than to the Ricci tensor. Norton's answer is that field equations based on the Ricci tensor were still not an option for Einstein because the harmonic coordinate condition needed to reduce it to the Poisson equation of Newtonian theory for weak static fields ruled out the simple form that Einstein expected the metric field to take in that case. The November tensor in conjunction with the Hertz condition, Norton's answer continues, was the next best thing.

Both our answer and Norton's are speculative, though Norton's considerably more so than ours. Whereas Einstein emphasized the importance of the redefinition of the gravitational field, the harmonic coordinate condition is not mentioned anywhere. There is no evidence that Einstein was even aware of the conflict between the harmonic coordinate condition and the weak static metric.[11] There is strong evidence, however, for the other key element of Norton's answer, viz. Einstein's prejudice about the weak static metric.[12] Einstein had to overcome that prejudice before he could use unimodular coordinates (rather than just unimodular *transformations*) in his second November 1915 paper (the determinant of the metric cannot be a constant if only one of its components is variable) or add a term to his field equations with the trace of the energy-momentum tensor for matter in his fourth November 1915 paper (the prejudice about the weak static metric had made him reject such a term in the Zurich notebook). In both Norton's account and ours, Einstein overcame this prejudice when he calculated the perihelion motion of Mercury and found a metric with several variable components for the weak static field of the Sun. He realized that this poses no problem for the recovery of the results of Newtonian theory since only g_{44} enters into Mercury's equation of motion in first-order approximation.[13]

Suppose Einstein did find his way back to the November tensor along the lines of Norton's reconstruction. His next step would have been to find a way around the physical objections that had made him abandon the November tensor in the Zurich notebook. Suppose that he had already come to the

[11] There is no indication that he was aware of the problem when he first considered the harmonic coordinate condition/restriction in the Zurich notebook (see note 16 of the commentary on p. 20L (p. 40) in Pt. II, Sec. 1.3).

[12] It is not entirely fair to simply call this a prejudice. First of all, it had been a natural assumption to make to connect the new metric theory developed in the Zurich notebook with the old scalar theory developed in Prague (see Ch. 3). Furthermore, on p. 21R (p. 42) of the Zurich notebook, Einstein went through an ingenious though ultimately fallacious argument that convinced him that all components of the metric for a weak static field must be constant except for g_{44} (see Renn 2007, Vol. 2, pp. 640–642, for a reconstruction and a critique of this argument). In a letter to Freundlich of March 19, 1915 (CPAE8, Doc. 63), drawing on results from Laue's (1911) relativistic continuum mechanics (cf. Ch. 3, note 11), Einstein argued that g_{44} is the only variable component of a metric field generated by a static source, thus further strengthening his conviction about the form of the metric for weak static fields.

[13] Strong evidence that the perihelion calculation did indeed play this role can be found in three postcards Einstein sent to Besso, on December 10 and 21, 1915, and January 3, 1916 (CPAE8, Docs. 162, 168 and 178), to which we will return briefly in Ch. 6.

realization that the Newtonian limit does not call for a coordinate restriction but can be handled with a coordinate condition. Recall that Norton believes he understood the distinction all along (see note 7). That leaves the problem of energy-momentum conservation. Here Norton's reconstruction and ours merge once again. The mathematical turn of Norton's scenario is followed by a physical turn. Einstein knew of only one way to demonstrate the compatibility of his field equations with energy-momentum conservation and that was with the help of the variational formalism developed for the *Entwurf* equations. He had to adjust the formalism carefully because of the restriction to unimodular transformations and he had to find the Lagrangian for the new field equations based on the November tensor. This last task turned out to be easier than he had any right to expect: he could hold on to the electrodynamically-inspired *Entwurf* Lagrangian and replace the electrodynamically-inspired definition of the gravitational field by the only other physically sensible definition, which, as luck would have it, was also more natural from a mathematical point of view.

This nicely illustrates the main lesson we believe the comparison of our account to Norton's teaches us. There is no conclusive evidence to decide whether Einstein found his way back to the November tensor following Norton's mathematical route or our physical route, though the preponderance of evidence points to our reconstruction.[14] But this may not matter much in the end. The more important point is that Einstein needed *both* routes to these new field equations to make sure they met all requirements. The convergence of these two routes is probably what made these new field equations compelling for him in the first place. With two routes to the same equations, Einstein had the luxury of a choice in how to present them in his November 1915 papers. Understandably, he chose the simpler mathematical route and only used stretches of the messier physical route in his discussion of the relation between field equations, energy-momentum conservation and covariance in sec. 3 of the paper.

We briefly return to that section. After writing down the Lagrangian and verifying that it gives the right field equations, Einstein derived the expression for the energy-momentum density of the gravitational field and rewrote

---

[14] One other probably irrelevant bit of evidence is a remark by Einstein in a letter to Zangger of October 15, 1915: "I wrote a supplementary article [*eine ergänzende Arbeit*] to my investigation about general relativity last year" (CPAE8, Doc. 130). Last year's article is undoubtedly the review article on the *Entwurf* theory (Einstein 1914c). But what is the *ergänzende Arbeit*? The most likely candidate, as suggested by Tilman Sauer, is (an early version of) a paper on electromagnetism in the presence of gravity, submitted to the Berlin Academy on February 3, 1916 (Einstein 1916a). A footnote on its first page clearly identifies this article as an addendum to the review article on the *Entwurf* theory. And Einstein had already mentioned some of its results in a letter to Lorentz of September 23, 1915 (CPAE8, Doc. 122). Yet it is tempting to speculate that the *ergänzende Arbeit* was about a new derivation of the *Entwurf* field equations after Einstein's discovery (reported to Lorentz three days before this letter to Zangger) that the uniqueness argument for these equations in Einstein (1914c) failed. This paper then turned, after major revision, into the first November 1915 paper (Janssen 1999, p. 152).

the field equations so that the right-hand side becomes the sum of the energy-momentum density of field and matter. The vanishing of the divergence of the left-hand side then guarantees energy-momentum conservation. This corresponds to the $B_\mu = 0$ conditions in the general formalism, even though the relevant quantity is not called B_μ here. B_μ is the four-divergence of a one-component expression.[15] Instead of making its four-divergence vanish, Einstein imposed the stronger condition that this expression itself vanish. He then compared this one condition to the fully contracted version of the field equations and noted that the former can be replaced by one condition on the determinant g of the metric over and above the requirement that g transform as a scalar (as it must under unimodular transformations).[16]

These maneuvers are readily understandable in light of the relation between energy-momentum conservation and covariance that Einstein had come to expect on the basis of his general variational formalism (see Sec. 5.2) and hard to make sense of any other way. The concluding paragraph of this section, in fact, still shows clear traces of the way Einstein thought about this relation in the context of the *Entwurf* theory in terms of justified transformations between adapted coordinate systems. Commenting on the extra condition on g, he explained that "it would not be valid in a new coordinate system resulting from the original one by an unallowed transformation. The [condition] therefore shows how the coordinate system has to be adapted to the manifold" (Einstein 1915a, p. 785). Compare this statement to the passage of his response to Mie quoted in Ch. 4, note 15 and the passage from his second paper on gravitational waves quoted in Sec. 5.2. The statement, we submit, is much closer to the former than to the latter.

5.4 Finishing touches

The one condition on the determinant g of the metric that Einstein had to add to ensure energy-momentum conservation in his first paper of November

---

[15] In the new theory, the conditions $B_\mu = 0$ become $\frac{\partial}{\partial x^\mu}\left(\frac{\partial^2 g^{\alpha\beta}}{\partial x^\alpha \partial x^\beta} - \kappa t\right) = 0$, where t is the trace of the gravitational energy-momentum density (Janssen and Renn 2007, pp. 883–884).

[16] The expression in parentheses in note 15 also occurs in the fully contracted version of the field equations

$$\frac{\partial^2 g^{\alpha\beta}}{\partial x^\alpha \partial x^\beta} - \kappa t + \frac{\partial}{\partial x^\alpha}\left(g^{\alpha\beta}\frac{\partial}{\partial x^\beta}\ln\sqrt{-g}\right) = -\kappa T$$

(p. 890, Eqs. (86)–(87)). Hence, by imposing the condition

$$\frac{\partial}{\partial x^\alpha}\left(g^{\alpha\beta}\frac{\partial}{\partial x^\beta}\ln\sqrt{-g}\right) = -\kappa T$$

(ibid., p. 885, Eq. (81)), Einstein achieved that it follows from the field equations that $B_\mu = 0$. For further details, see Pt. II, Sec. 6.1.3, notes 17 and 18.

1915 was a peculiar one. It set an expression involving derivatives of g equal to the trace T of the energy-momentum tensor for matter (see note 16). Since the energy-momentum tensor for a swarm of particles has a non-vanishing trace, this rules out unimodular coordinates (for which $g = -1$[17]). Here was a theory covariant under unimodular transformations in which one could not use unimodular coordinates, the natural coordinates to use in such a theory.[18] Einstein understandably looked for ways around this strange prohibition.

The amendments of the second and fourth paper of November 1915 amount to two different proposals to avoid the condition that g cannot be a constant. In the second paper, he set $T = 0$; in the fourth, he added a term with T to the field equations. The second proposal is mathematically much more elegant than the first. Recall that the condition on g resulted from a combination of two relations, the fully contracted field equations and a condition guaranteeing the vanishing of the four-divergence of the total energy-momentum density. The addition of a trace term to the field equations changes these two relations in such a way that the resulting condition on g no longer involves T at all. The second proposal, as we shall see shortly, also has a much more convincing physical justification than the first.

As mentioned above, the perihelion calculation published in the third paper of November 1915 freed Einstein from his prejudice about the weak static metric and thereby removed the final obstacle both to the use of unimodular coordinates and to the addition of a trace term. Given that Einstein already introduced unimodular coordinates in the second November 1915 paper, it is safe to assume that he had already done much of the perihelion calculation at that point.

To justify setting $T = 0$, Einstein briefly flirted with the electromagnetic worldview, championed by Mie (1912, 1913)[19] and by Hilbert, who thought of his theory as a synthesis of Einstein's theory for the gravitational field and Mie's theory for the electromagnetic field. According to the electromagnetic worldview, all matter is ultimately made of electromagnetic fields governed by some non-linear generalization of Maxwell's equations. In that case, Einstein argued, T will probably vanish, as it does for ordinary electromagnetic fields. The non-vanishing trace of the phenomenological energy-momentum tensor for a swarm of particles, he speculated, might then be due to the non-vanishing of the trace of the energy-momentum density of the gravitational field associated with these particles. Gravity would thus play an essential role in the constitution of matter.

---

[17] The rotation metric, of course, also has determinant -1, but this is no problem since the Minkowski metric is a vacuum solution.

[18] John Stachel and Kaća Bradonjić have suggested that Einstein perhaps should have left well enough alone and forego the step from unimodular to general covariance, in which case general relativity would have started out as the unimodular theory of gravity they prefer (see, e.g., Bradonjić 2014).

[19] For discussion of Mie's theory, see CPAE8, Doc. 346, note 2, and Smeenk and Martin (2007).

Einstein soon had second thoughts about this hypothesis. In the announcement of his next paper, he tried to pass off the perihelion result as evidence for this hypothesis but in the paper itself he included a footnote saying that one can do without it. The following year he clearly distanced himself from the electromagnetic worldview. In a letter to Hermann Weyl of November 23, 1916, for instance, he called Hilbert's assumptions about matter "infantile, in the sense of a child innocent of the tricks of the real world" (CPAE8, Doc. 278; cf. Pt. II, Sec. 10.3, note 2). In 1919, however, he resurrected the hypothesis of his second November 1915 communication in a paper entitled, "Do gravitational fields play an essential role in the structure of the elementary particles of matter?" (Einstein 1919).[20] His dismissive attitude in 1916 may reflect some lingering embarrassment over his overhasty endorsement of the electromagnetic worldview in November 1915.

The big payoff of this endorsement in his second November 1915 paper was that he could now set $g=-1$. In that case, the Ricci tensor $G_{\mu\nu}$ reduces to the November tensor $R_{\mu\nu}$ ($S_{\mu\nu}$ vanishes if $g=-1$). The field equations of the first November paper could thus be seen as generally covariant equations, $G_{\mu\nu}=-\kappa T_{\mu\nu}$, expressed in unimodular coordinates.

Einstein quickly realized that he had overpaid for general covariance. As he already indicated in a footnote at the beginning of the perihelion paper, there was no need to set $T=0$ (Einstein 1915c, p. 831). In the fourth November 1915 paper, he showed that the addition of a trace term to the field equations allowed him to set $g=-1$, regardless of the value of T.[21,22] Moreover, he had a strong physical argument for this modification of his field equations. The trace term had to be there to make sure that the energy-momentum density of matter enters the field equations in the exact same way as the energy-

[20] For discussion, see Earman (2003).

[21] With this extra trace term, the conditions $B_\mu=0$ in note 15 change to

$$\frac{\partial}{\partial x^\mu}\left(\frac{\partial^2 g^{\alpha\beta}}{\partial x^\alpha \partial x^\beta}-\kappa(t+T)\right)=0$$

and the fully contracted version of the field equations (see note 16) changes to

$$\frac{\partial^2 g^{\alpha\beta}}{\partial x^\alpha \partial x^\beta}-\kappa(t+T)+\frac{\partial}{\partial x^\alpha}\left(g^{\alpha\beta}\frac{\partial}{\partial x^\beta}\ln\sqrt{-g}\right)=0$$

(Janssen and Renn 2007, p. 890, Eqs. (88)–(89)). The condition on g in note 16 thus changes to:

$$\frac{\partial}{\partial x^\alpha}\left(g^{\alpha\beta}\frac{\partial}{\partial x^\beta}\ln\sqrt{-g}\right)=0$$

(ibid., p. 889, Eq. (84)). This condition allows setting $g=-1$. For further details, see Pt. II, Sec. 6.4.3, note 1.

[22] When Einstein considered adding a trace term to the field equations in linear approximation in the Zurich notebook, this was also, at least in part, to get around the problem that a condition on the metric implied that the energy-momentum tensor had to be traceless (see p. 20L [p. 40] of the notebook and its annotation in Pt. II, Ch. 1).

momentum density of the gravitational field. This requirement had already played an important role in his 1912 theory for static gravitational fields as well as in the *Entwurf* theory.

Note that both modifications Einstein made to the field equations of his first November 1915 paper—setting $T = 0$ and adding a term with T—were driven by his dissatisfaction with the condition setting derivatives of g equal to T. His correspondence with Hilbert played no role in these developments (with the possible exception of his short dalliance with the electromagnetic worldview). The page proofs of Hilbert's paper show that Hilbert did not use any insights from his correspondence with Einstein either. Hilbert only took from Einstein what he got out of his Wolfskehl lectures in the summer of 1915. The correspondence between Einstein and Hilbert of November 1915 thus conjures up Longfellow's image of ships passing in the night.

5.5 Scaffold and arch

On December 29, 1915, Einstein wrote to his Polish colleague Władysław Natanson: "I once again toppled my house of cards and built a new one" (CPAE8, Doc. 175). This statement confirms the impression created by the first November 1915 paper, in which Einstein made it sound as if he had gone from the *Entwurf* equations to field equations based on the Ricci tensor by leveling one building and erecting a new one on its ruins in a completely different style. This is pretty much how Einstein would come to remember what had happened during that turbulent month in 1915. Back then he knew better. In the letter to Natanson, for instance, he immediately qualified his house-of-cards metaphor: "at least, the middle section is new."

As we already noted in Ch. 2, our analysis suggests a different architectural metaphor. The elaborate framework built around the *Entwurf* field equations served as the scaffold on top of which the arch stones of the Einstein field equations were carefully placed (Janssen and Renn 2015a, Janssen 2019, pp. 129–134). The Zurich notebook shows that Einstein had already drawn up plans for this arch in 1912–13 (Sec. 3.1). As the term *Entwurf* in the title of his 1913 paper with Grossmann suggests, he was hoping that this paper would lay the foundation on which to build this arch (Sec. 3.2). When he published the review article on the theory in November 1914, he was clearly under the impression that he had now done so (Sec. 4.2). That arch, however, soon began to show some serious cracks (see the introduction of Ch. 5). Finally, in November 1915, using the *Entwurf* theory as his scaffold, Einstein turned one of the mathematical designs in the Zurich notebook into a structure sturdy enough to meet the exacting demands of physics and stand the test of time (Secs. 5.1–5.4). As he wrote to Hilbert on November 18, 1915:

> The difficulty did not lie in finding generally-covariant equations ... this is easily done with the help of the Riemann tensor. Rather it was difficult to recognize that

> these equations formed a generalization of Newton's laws and indeed a simple and natural generalization (CPAE8, Doc. 148).

Oddly, Einstein did not mention the problem with energy-momentum conservation in this letter, which he did mention in a letter to Sommerfeld ten days later (CPAE8, Doc. 153; Pt. II, Ch. 7) as well as in a postcard to Besso of December 10, 1915 (CPAE8, Doc. 162). As in the case of this letter to Sommerfeld, we have to keep in mind that Einstein had an obvious ulterior motive in making this point. Presumably, he first and foremost wanted to make it clear that he had needed no help from Göttingen to put his house in order. His point is nonetheless well taken. Mathematicians, Hilbert himself liked to point out, tend to build their houses *before* securing their foundations (Rowe 2006, p. 507; Janssen 2019, pp. 103–104, p. 131).

The extant papers, notebooks and letters allowed us to reconstruct in considerable detail how the scaffold was used to build the arch in this case. The formalism of the *Entwurf* theory contains expressions for and relations between the field equations, the four-force, the energy-momentum density, and the Lagrangian for the gravitational field, all of which mimic corresponding expressions and relations for the electromagnetic field in the four-dimensional formalism for electrodynamics in special relativity. It also features a strong connection between energy-momentum conservation and covariance (Secs. 3.2 and 4.2). These expressions and relations provided the structure holding the scaffold together. This whole constellation survives intact when the basic building block is exchanged for a new one, i.e., when the gravitational field is redefined from (essentially) minus the gradient of the metric to minus the Christoffel symbols (Sec. 5.1). The new building blocks, the stones of the new arch, are kept in place by the same structure that held the scaffold together. Parts of the scaffold could be discarded, notably the ungainly concept of non-autonomous transformations. The hole argument was dropped somewhat prematurely, resurrected under pressure from two of Einstein's critics, and eventually replaced by the point-coincidence argument (Sec. 4.1). Parts were moved and reconfigured. Ropes were untied and retied in new ways. The link between covariance and energy-momentum conservation was inverted from *conservation restricting covariance* to *covariance guaranteeing conservation.* Coordinate restrictions turned into coordinate conditions. The Hertz condition was put in to secure the Newtonian limit, then taken out again when it proved to be redundant (Sec. 5.2).

Despite Einstein's efforts to hide the scaffold, the arch unveiled in the first November 1915 paper still shows clear traces of it (Secs. 5.3–5.4). Einstein was at pains to display its mathematical splendor but could not hide that his structure was still not quite up to code. To prevent a violation of energy-momentum conservation from bringing it down, the edifice had to be propped up with some lumber from the *Entwurf* scaffold, a support beam cut to the size of unimodular transformations. Unfortunately, this support beam did not leave enough room for another critical piece of scaffolding, the restriction to unimodular coordinates. It is only when that restriction is put in place that one can fully appreciate the magnificence of the arch as a structure that looks the same from all angles, even though, during construction, it could only be viewed from some. To make room for unimodular coordinates, Einstein brought in the heavy machinery of the electromagnetic worldview. He used this equipment to build an unsightly buttress (gravity is part of the cement that holds elementary particles together), which came down as fast as it went up. He quickly realized he did not need the electromagnetic worldview. He just had to place the keystone in the arch, a term with the trace of the energy-momentum tensor for matter (Einstein 1915a; 1915b; 1915d, see Pt. II, Ch. 6)

This is how Einstein left the construction site in November 1915. A few months later, he returned to it, cleared away the debris of the first two November 1915 papers, and made a half-hearted attempt at removing the restriction to unimodular coordinates. He reported the results in his review article published in May 1916 (Einstein 1916b, see Pt. II, Ch. 9). It was only in October 1916 that he finally lifted the reliance on special coordinates (Einstein 1916d, see Pt. II, Ch. 10). The arch was self-supporting at last, a marvelous sight to behold for generations of physicists to come.[23]

[23] In the unkind glare of hindsight, we can see that the restriction to unimodular transformations and unimodular coordinates in the papers of November 1915 (Einstein 1915a, 1915b, 1915d) and the review article of early 1916 (Einstein 1916b) served both as a bridge *and as a hurdle* in getting from the *Entwurf* field equations to the generally covariant Einstein field equations. In this introductory essay, we focused on its role as a bridge. Changing the definition of the gravitational field while retaining the *Entwurf* theory's expression for the Lagrangian in terms of the gravitational field only leads to the field equations of November 1915 if the general covariance of the formalism of the November 1914 review article on the *Entwurf* theory (Einstein 1914c) is restricted to unimodular transformations. The article of November 1916 (Einstein 1916d), however, shows that there is at least one other path that leads from the *Entwurf* field equations to the Einstein field equations and that does not require any restrictions on the formalism of the November 1914 review article. All Einstein had to do, in hindsight, was to change the Lagrangian for the *Entwurf* field equations to the Riemann curvature scalar (of course, this also would have required Einstein to set aside the hole argument against generally covariant field equations). Regardless of which of these two pathways we consider, the *Entwurf* theory can be seen as a scaffold for the arch of the generally covariant theory. In both cases, one element of the scaffold (the expression for the gravitational field or the expression for the Lagrangian) is replaced while the rest of the structure is kept intact. The actual route, of course, was the first and not the second one. But given how closely Einstein's paper November 1916 follows the one of November 1914, it is not too farfetched to imagine the alternative scenario, in which he leapfrogged the messy developments of November 1915 and early 1916 altogether and went directly from "Formale Grundlage ..." (Einstein 1914c) to "Hamiltonsches Prinzip ..." (Einstein 1916d) (cf. Pt. II, Sec. 10.3). Note in this context that a discarded appendix to the 1916 review article already contains many of the elements of this November 1916 paper (cf. note 9).

Chapter 6
Mercury's perihelion: From 18″ in the *Entwurf* theory to 43″ in general relativity

On November 18, 1915, Einstein delivered a lecture on the perihelion motion of Mercury to the Prussian Academy of Sciences. The paper based on this lecture was submitted that same day and published one week later (Einstein 1915c). Einstein was able to show that his new theory added 43″ to the more than 500″ per century that could be attributed to the perturbation of other planets, thereby closing the gap of 45″ ± 5″ between theory and observation.[1] As he told Sommerfeld on December 9, 1915 (CPAE8, Doc. 161), this gave him a whole new appreciation of "the pedantic precision of astronomy that I secretly used to make fun of in the past."[2] Years later, he told Fokker that the result had given him heart palpitations (Pais 1982, p. 253). His excitement also comes through in a letter to Zangger of November 15, 1915: "*I have now derived the up to this point unexplained anomalies in the motion of the planets from the theory.* Imagine my good fortune! [*Stellen Sie sich mein Glück vor!*]" (CPAE8, Doc. 144a, in CPAE10). The day he submitted the paper, he shared the good news with his Göttingen competitor. "Congratulations on conquering the perihelion motion," an astonished Hilbert wrote back the very next day,

> If I could calculate as fast as you can, the electron would be forced to surrender to my equations and the hydrogen atom would have to bring a note from home to be excused for not radiating (CPAE8, Doc. 149).

Einstein did not bother to tell his colleague that he had done very similar calculations before. He probably enjoyed giving Hilbert a taste of his own medicine. A few months later, he characterized Hilbert's style as "creating the impression of being superhuman by obfuscating one's methods."[3]

---

[1] For the history of the problem of the anomalous motion of Mercury's perihelion, see Roseveare (1982), Earman and Janssen (1993), and Smith (2014, pp. 307–317).

[2] The German astronomy community did not endear itself to Einstein by making life difficult for his protégé Freundlich (see Pt. II, Sec. 4.3, note 1, and Sec. 7.3, note 17). As he wrote to Zangger in a letter dated before December 4, 1915 (from which we already quoted in note 11 in Ch. 2): "The astronomers behave like an anthill disturbed in its mindless activity by a hiker who has inadvertently stepped in it: they start biting at the hiker without making a dent in his shoes" (CPAE8, Doc. 159a, in CPAE10).

[3] Einstein to Ehrenfest, May 24, 1916 (CPAE8, Doc. 220). For the passage leading up to this characterization, see Pt. II, Sec. 10.3, note 2. In a letter a week later, on May

M. Janssen, J. Renn, *How Einstein Found His Field Equations*, Classic Texts in the Sciences, https://doi.org/10.1007/978-3-030-97955-3_6

In the late 1980s, the Einstein Papers Project obtained a copy of a manuscript that Besso's son Vero had given to the editor of the correspondence between Einstein and Besso (1972). This manuscript consists of about 50 pages of scratchpad calculations, showing that Einstein and Besso used the *Entwurf* field equations to calculate the perihelion motion of Mercury. Most of these notes can be dated to June 1913.[4] Einstein scholars were surprised to learn that Einstein had already calculated the perihelion motion of Mercury in 1913. That he had done so in collaboration with Besso surprised them even more. Besso had always been thought of as an important sounding board for Einstein, never as a serious scientific collaborator. Einstein eventually handed over the project to Besso. In a letter that can be dated to early 1914, he told his friend: "Here you finally have your manuscript package. It is really a shame if you do not bring the matter to completion" (CPAE5, Doc. 499). It is safe to assume that this refers to the Einstein-Besso manuscript and the perihelion problem. It is only because the manuscript ended up in Besso's hands rather than Einstein's, who almost certainly would have discarded it, that we can belatedly call Einstein's bluff trying to put one over on Hilbert.

Einstein's exhortation did not fall on deaf ears. Material made available to Einstein scholars by the Besso family in 1998 shows that Besso initially planned an ambitious paper on the perihelion problem.[5] It is not clear when exactly he abandoned the project but one surmises he had well before he and Einstein were scooped in December 1914. In a paper submitted that month, Johannes Droste (1915, p. 1010), another one of Lorentz's students, reported that De Sitter, using equations of motion derived by Lorentz, had found that the *Entwurf* field equations predict an additional advance of Mercury's perihelion of only 18″ per century (Röhle 2007, pp. 197–200).[6]

The calculations of 1913 undoubtedly helped Einstein in his calculation of the perihelion motion in November 1915. A comparison between the relevant pages of the Einstein-Besso manuscript and the perihelion paper shows that there are many similarities between the two sets of calculations (see Pt. II, Chs. 2 and 6). Yet, there are also important differences. The most important one perhaps is that, in 1915, Einstein determined the field of the sun in terms of the Christoffel symbols, whereas, in 1913, he determined it in terms of the metric tensor (Earman and Janssen 1993, secs. 5–7). This difference between the calculations of 1913 and 1915, we like to think, reflects his epiphany that the Christoffel symbols rather than the gradient of the metric field represents

---

30, 1916, Einstein also complained about this to Hilbert himself, albeit in more polite terms: "Why do you make it so hard for poor mortals by withholding the technique behind your thinking?" (CPAE8, Doc. 223, quoted in Landsman 2021, p. 18).

[4] For more information on the Einstein-Besso manuscript, see Pt. II, Sec. 2.3

[5] For analysis of the pages Besso added to the Einstein-Besso manuscript in 1914, see Janssen (2007, pp. 790–806).

[6] On an undated page of Einstein's so-called Scratch Notebook (CPAE3, Appendix A, [p. 61]), a value of 17″ is given (CPAE4, p. 345). See our commentary on [p. 28] of the Einstein-Besso manuscript in Pt. II, Sec. 2.3, for discussion.

the gravitational field. These differences help to underscore that it remains an impressive feat that Einstein was able to produce the 1915 perihelion paper as fast as he did. Still, given the undeniable importance of his earlier calculations, one can legitimately ask why Einstein did not invite Besso as a co-author. As it happened, Besso did not even get an acknowledgment. We suspect that this is largely because of the race Einstein perceived himself to be in with Hilbert. He was in a hurry[7] and it probably never even occurred to him to ask Besso to write a joint paper on the topic.

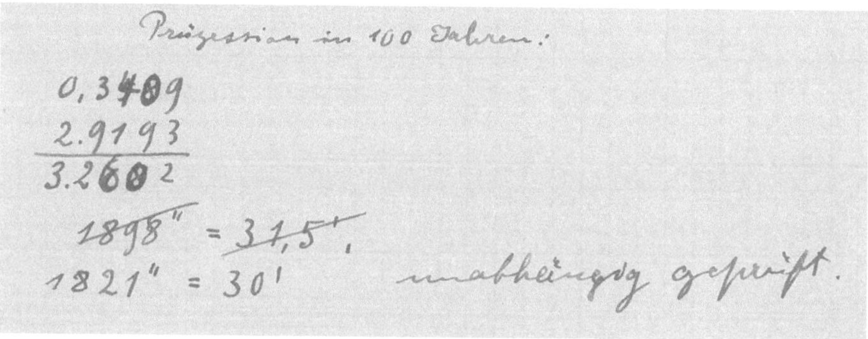

Figure 6.1 On [p. 28] of the Einstein-Besso manuscript, Einstein recorded the value for the "precession in 100 years" (*Präzession in 100 Jahren*) of Mercury's perihelion produced by the field of the Sun that he and Besso found in June 1913 using the *Entwurf* theory. This theory predicts 5/12 the size of the effect predicted by general relativity, which comes to about 18″ per century. So the value that Einstein here claims was "independently checked" (*unabhängig geprüft*) is too large by a factor of 100. If this value were correct, the contribution of the field of the Sun to the perihelion motion of Mercury would be more than three times the effect of about 530″ per century coming from the perturbations of all other planets combined! Einstein almost certainly realized that his result was off by a factor of 100 but it looks as if Besso first found the source of the error (Figs. 6.2 and 6.3).

In fact, none of the three postcards of late 1915 in which Einstein enthusiastically reported his success with Mercury to Besso explicitly refer to their earlier collaboration.[8] The second and the third at least implicitly do. "You will be surprised," Einstein wrote in the second, "about the occurrence of g_{11} ... g_{33}." In the third, he similarly wrote: "it is only because in first-order approximation g_{11}–g_{33} do not occur in the equations of motion for a point particle that the Newtonian theory is so simple." Einstein reiterated this point

[7] An unusually high number of typos in the paper suggests that it was written in haste (see Pt. II, Sec. 6.3.3).

[8] Einstein to Besso, November 17, December 10, and December 21, 1915 (CPAE8, Docs. 147, 162 and 168)

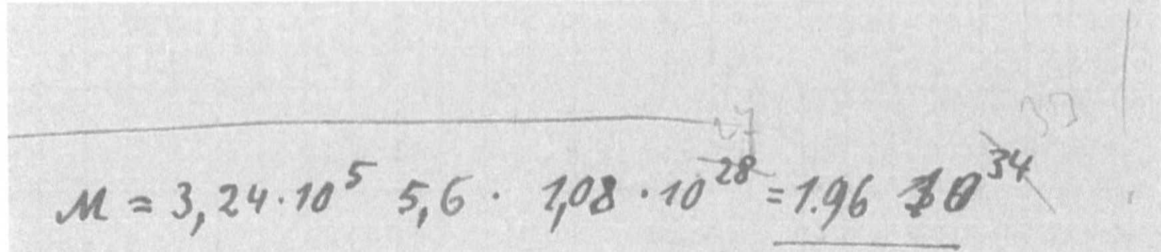

Figure 6.2 On [p. 35] of the Einstein-Besso manuscript, Besso located the source of the error in the value Einstein reported on [p. 28] for the perihelion motion of Mercury (Fig. 6.1). The material in pen, the numerical calculation of M, the mass of the Sun, is in Einstein's hand but Besso added the corrections and exclamation point in pencil. The value for M is found as the product of three factors, the ratio of the Sun's mass to the Earth's mass, the Earth's density and the Earth's volume. Besso noticed that the value for the Earth's volume was off by a factor of 10. Since M occurs squared in the formula for the perihelion shift, this throws off the end result by a factor of 100.

Figure 6.3 On [p. 30] of the Einstein-Besso manuscript, we find the same calculation of M that we found on [p. 35] (Fig. 6.2). It also has the same correction, this time in Einstein's hand. The exclamation point on [p. 35] suggests that Besso found the mistake first.

in a letter of January 3, 1916, in which he finally did explicitly refer to their earlier calculations:[9]

> That the effect is so much larger than in *our calculation* is because in the new theory the g_{11}–g_{33} appear in first order and thus contribute to the perihelion motion (CPAE8, Doc. 178; our emphasis).

When Einstein and Besso collaborated in 1913, the metric they found as a first-order solution of the *Entwurf* field equations had only one variable component, viz. g_{44}, in accordance with Einstein's prejudice about the metric for weak static fields at the time. The new perihelion calculations freed him from this prejudice (see Sec. 5.3 and note 12 therein).

Besso made substantial contributions to the calculations in 1913. Einstein solved the *Entwurf* field equations for a point mass, using the same iterative approximation procedure he used in 1915. He was surprised when, only a month later, Karl Schwarzschild sent him the exact solution, which was

[9] In a letter to Besso dated after December 6, 1916, Einstein once again mentions Mercury's perihelion without referring to their joint efforts. In this letter Einstein dismisses two attempts to account for Mercury's perihelion motion without general relativity, a new one by Emil Wiechert (1916) and an old one by Paul Gerber (1917), republished on the initiative of Ernst Gehrcke (CPAE8, Doc. 287a, in CPAE10). For Gehrcke's role in the anti-relativity movement, see Wazeck (2009).

published early the following year (Schwarzschild 1916).[10] Once Einstein had found the field, Besso derived the differential equation for the angle traversed by the radius of a planet moving in this field as a function of its distance to the Sun. Einstein integrated this equation to find the angle between the minimum and maximum value of this distance, i.e., between perihelion and aphelion. The deviation of this angle from π gives the perihelion motion in radians per half a revolution. Converting this result to seconds of arc per century, he found a value of 1821 seconds or 30 *minutes* of arc! Next to this bizarre result, which is more than three times the total secular motion of Mercury's perihelion, he wrote: "independently checked" (Fig. 6.1). The correct value of 18″ is nowhere to be found in the manuscript. Both Besso and Einstein, however, located the error. The value they used for the mass of the Sun, which occurs squared in the final formula, was too large by a factor of 10 (Figs. 6.2 and 6.3).

In the letter to Sommerfeld of November 28, 1915 (CPAE8, Doc. 153; Pt. II, Ch. 7), Einstein listed the value of 18″ as one of his reasons for abandoning the *Entwurf* theory. It was generally assumed that he got that number from Droste's paper. The Einstein-Besso manuscript shows that he and Besso had already found it in 1913. Evidently, Einstein had not been too worried at the time about the discrepancy with the missing $45'' \pm 5''$.[11]

The Einstein-Besso manuscript thus allows us to answer a question that, before its discovery, would have been purely hypothetical. What would Einstein have done had his theory predicted the wrong result for the perihelion motion of Mercury? If he had followed the prescriptions of Karl R. Popper (1959), he should have accepted that his theory had been falsified and gone back to the drawing board. Einstein did nothing of the sort. He kept quiet about the 18″ and continued to work on the *Entwurf* theory. This was perfectly rational. The theory would have been decisively refuted had it truly predicted 1800″; that it could only account for less than half the missing seconds was undoubtedly disappointing but not particularly troublesome. It

---

[10] See Schwarzschild to Einstein, December 22, 1915 (CPAE8, Doc. 169) and Einstein to Schwarzschild, December 29, 1915 and January 9, 1916 (CPAE8, Docs. 176 and 181). "I would not have thought that the exact solution of the point problem would be so simple," he told Schwarzschild in the first of these two letters. Einstein used this exact solution (following a derivation given by Hermann Weyl 1918, pp. 202–205) when he covered the calculation of the perihelion advance of Mercury in lectures on general relativity in Berlin and Zurich in 1919 (see CPAE7, Docs. 19 and 20 and Janssen and Schulmann 1998). In his book about the 1919 eclipse expedition, Matthew Stanley offers a different perspective: "Einstein's calculation of the orbit of Mercury was only an approximation. This is a standard practice for physics problems that require large amounts of calculation. When it becomes clear that an exact answer will be labor-intensive, there is a strong incentive to find a shortcut" (Stanley 2019, p. 158).

[11] Einstein and Besso explored whether the *Entwurf* theory predicted additional effects on the perihelion motion that could account for another 25″ or so but came up empty handed. In particular, they considered the effect of the rotation of the Sun. Even though they overestimated it by at least two orders of magnitude, this effect was much too small. Worse yet, it was in the wrong direction (CPAE4, pp. 354–356).

does seem to diminish the importance of getting it exactly right, however, if being off by more than 50% is no big deal. The older Einstein might well have agreed with that assessment. In a letter to Max Born of May 12, 1952, the sage from Princeton had this to say on the three classic tests of general relativity:

> Even if there had been no deflection of light, no perihelion motion and no redshift, the gravitational equations would still be convincing because they avoid the inertial system (the phantom that affects everything but is not itself affected). It is actually rather curious that humans are mostly deaf to the strongest arguments, while they always tend to overestimate the accuracy of measurements (Einstein and Born 1969, Doc. 99, p. 192).

Chapter 7
Beyond the search for field equations

That general relativity could explain the well-known anomaly in the perihelion motion of Mercury made for good advertisement for Einstein's new theory of gravity. The real importance of the new theory, however, was not that it described the orbits of the planets in the solar system even more accurately than Newton's theory, as emphasized by George Smith (2014, pp. 317–320), but that it predicted qualitatively new phenomena, such as gravitational lensing[1] and gravitational waves, and opened up entirely new fields of research, such as relativistic cosmology.

In this section we sketch Einstein's first foray into cosmology and his first exploration of gravitational waves. We will connect these two efforts to the two strategies used in Einstein's search for field equations, the work on cosmology to the mathematical strategy, the work on gravitational waves to the physical strategy.

7.1 Cosmology as a continuation of the mathematical strategy

With the replacement of the *Entwurf* field equations by the Einstein field equations in November 1915, Einstein's theory was finally and entirely generally covariant.[2] Equating general covariance with general relativity (cf. Ch. 1, note 3), Einstein was laboring under the illusion that the theory thereby automatically extended the principle of relativity for uniform motion of special relativity to arbitrary motion—in other words, that it was now immediately obvious that his theory lived up to its billing as a general theory of relativity. In a letter to Besso of July 31, 1916, for instance, he expressed his relief that one no longer had to do any explicit calculations making sure the rotation

[1] As can be inferred from an entry in the so-called Scratch Notebook (CPAE3, Appendix A, [pp. 43–48]) from his days in Prague, Einstein had already pondered the possibility of gravitational lensing in 1912 (Renn, Sauer, and Stachel 1997; Renn and Sauer 2003).

[2] This subsection is based on the editorial note, "The Einstein-De Sitter-Weyl-Klein debate," in CPAE8, pp. 351–357, and on Janssen (2014, sec. 5, pp. 198–208). Since then, Jan Guichelaar has brought to light an entry in one of De Sitter's notebooks documenting the beginning of the Einstein-De Sitter debate in September 1916 in Leiden (see Fig. 7.1 for a facsimile and a transcription of the first part of this entry and below for a full English translation). See Smeenk (2014) for an account of these developments placed in the broader context of early-twentieth-century astronomy.

M. Janssen, J. Renn, *How Einstein Found His Field Equations*, Classic Texts in the Sciences, https://doi.org/10.1007/978-3-030-97955-3_7

metric is a solution to the field equations to know that the theory makes rotation relative. Such calculations, he noted, are[3]

> of interest only if one does not know whether rotation-transformations are among the 'allowed' ones, i.e., if one is not clear about the transformation properties of the equations, a stage which, thank God, has definitively been overcome (CPAE 8, Doc. 245).

Only a few months later, during a visit by Einstein to his Leiden colleagues in September 1916, De Sitter disabused Einstein of the illusion that general covariance means general relativity.[4] As De Sitter pointed out, Einstein, in applying the theory (to the problem of Mercury's perihelion, for instance), used Minkowskian boundary values for the metric at spatial infinity (see Ch. 5, note 5). The theory thus retained a remnant of absolute space. Absolute motion had not been completely eradicated.

De Sitter recorded Einstein's initial response to this objection in one of his notebooks (see Fig. 7.1). The heading of the relevant entry, "Conversation with Einstein, Ehrenfest and Nordström, September 28 (1916) about the relativity of rotation," makes it clear that Ehrenfest and Nordström also participated in these discussions. Nordström was working with Lorentz and Ehrenfest in Leiden in 1916–8. De Sitter wrote:

> *Einstein* wants the *hypothesis of the boundedness of the world.*
>
> What he means by that is that he makes the hypothesis (aware that this is an unprovable hypothesis) that at infinity (i.e., at very large, mathematically finite distance but farther away than any observable material body, hence outside the Milky Way etc.) there are masses—not observable though not in principle unobservable—such that the g_{ij} at infinity *take on particular values* (these do not have to be zero: that's impossible to say a priori), *the same* in *all* coordinate systems. This will certainly not be possible in *all* coordinate systems. E.g., if one transforms $x'_1 = x_4$ $x'_4 = x_1$ [i.e., interchanging the time axis with one of the spatial axes], it is not. He is therefore prepared to give up the complete freedom of transformations and, e.g., to restrict transformations to those in which there is always one time coordinate and three spatial coordinates, and in which a space axis never turns into a time axis etc. ∞^4 transformations should always remain possible—the restriction can only be imposed through inequalities.
>
> Whether it is possible to find a set of degenerate values of g_{ij} which are invariant under a group of transformations that is not too restrictive is a question that can be settled mathematically.
>
> If the answer is *no* (as Ehrenfest and I suspect), then Einstein's hypothesis of boundedness is false. If the answer is yes, then the hypothesis is not in conflict with relativity theory. Still, I maintain that *even in that case* it is in conflict with the *spirit* of the relativity principle. And Einstein concedes that I have that option. Rejection of the hypothesis is *completely* allowed in relativity theory.

---

[3] For further discussion of this letter, see Janssen (1999, pp. 153–155; 2012, pp. 169–172; 2014, p. 197).

[4] Mie (1914) had tried in vain to make Einstein see this (cf. Ch. 4, note 15) and would try again in 1917–18, drawing Einstein's attention to Kretschmann's (1917) famous paper on the topic. See Mie (1917) and Mie to Einstein, May 30, 1917 and February 17–19, 1918 (CPAE8, Docs. 346 and 465); see also note 13.

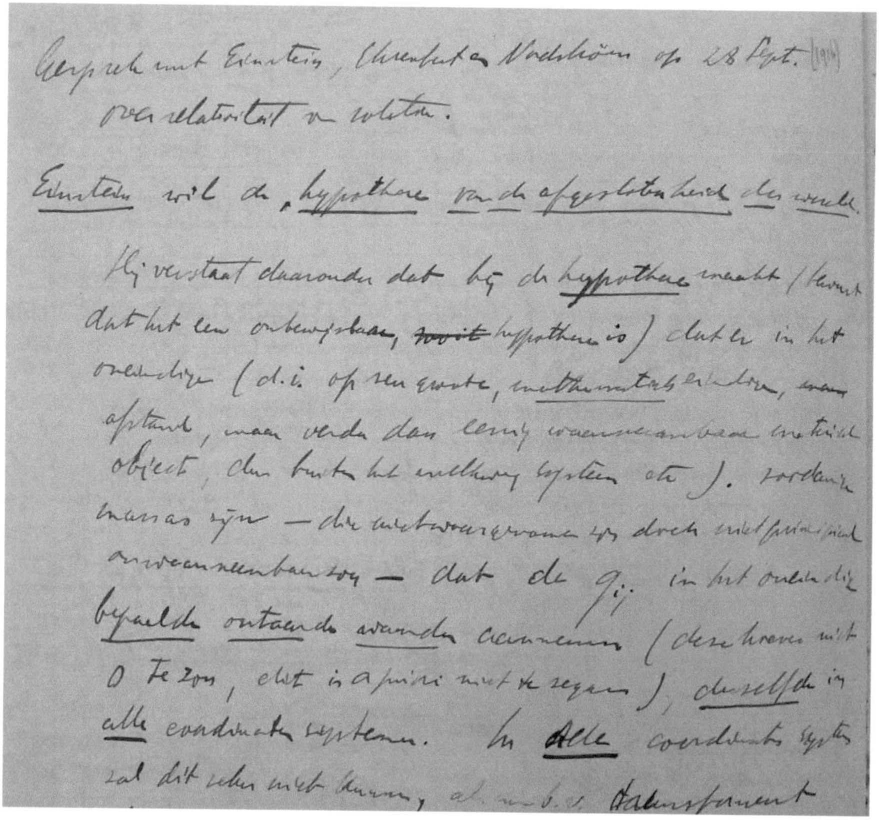

Figure 7.1 De Sitter's notes about a conversation with Einstein, Ehrenfest, and Nordström on September 28, 1916 (De Sitter, studieschrift S12, p. 238; picture by Jan Guichelaar). The Dutch text reads (see Sec. 7.1 for an English translation): "Gesprek met Einstein, Ehrenfest en Nordström op 28 Sept (1916) over relativiteit van rotatie. *Einstein* wil de *hypothese van de afgeslotenheid der wereld.* Hij verstaat daaronder dat hij de *hypothese* maakt (bewust dat het een onbewijsbaar [sic] ⟨nooit⟩ hypothese is) dat er in het oneindige (d.i. op zeer groote, *mathematisch* eindige ⟨maar⟩ afstand, maar verder dan eenig waarneembaar materieel object, dus buiten het melkweg systeem etc). zoodanige massas zijn—die niet waarneembaar zijn doch niet principieel onwaarneembaar zijn—dat de g_{ij} in het oneindige *bepaalde ontaarde waarden* aannemen (deze hoeven niet 0 te zijn, dit is a priori niet te zeggen), *dezelfde* in *alle* coordinaten systemen. In *alle* coordinaten systemen zal dit zeker niet kunnen."

If the answer were yes, it would raise the practical question how to determine those masses in such a way as to satisfy all conditions. That question would be difficult to answer. One would get a very complicated system of masses. A spherical shell, e.g., would certainly not do and the masses would have to be oriented, probably in the direction of the ecliptic etc. and this condemns the system, just as the epicycle theory is condemned, even though it was kinematically correct.[5]

29 Sept. Einstein says it can be done as follows:

---

[5] To the left of this paragraph, De Sitter put a left curly bracket and the comment: "Remark of mine of September 29." To the right of the second half of this paragraph

The degenerate values are $\begin{cases} 0 & 0 & 0 & \infty \\ 0 & 0 & 0 & \infty \\ 0 & 0 & 0 & \infty \\ \infty & \infty & \infty & \infty^2 \end{cases}$

i.e., at infinity the *time-cone* has been flattened and the velocity of light is ∞.

One then has to limit oneself to transformation in which the time axis always remains within the time-cone.

Let this be the case, even then I do not accept the hypothesis because it would make the world *finite*.[6]

Einstein, in short, wanted to ensure the relativity of arbitrary motion by having degenerate values for the metric at spatial infinity and masses outside the visible universe to turn those degenerate values into the Minkowskian boundary values at the edge of the visible universe. De Sitter sharply criticized this proposal, both in a letter to Einstein of November 1, 1916 (CPAE8, Doc. 272) and in two publications, a paper in the Proceedings of the Amsterdam Academy (De Sitter 1916b, pp. 531–532, footnote 2) and the second part of the trilogy in *Monthly Notices* that introduced a British audience to general relativity (De Sitter 1916c, sec. 30). As far as De Sitter was concerned, what Einstein proposed was a cure worse than the disease. What if better telescopes became available and the visible universe got bigger? Was Einstein then going to move his masses further out?

Einstein eventually came to accept this criticism. As he wrote to De Sitter on February 2, 1917:

> I have completely abandoned my views, rightfully contested by you, on the degeneration of the $g_{\mu\nu}$. I am curious to hear what you will have to say about the somewhat crazy idea I am considering now (CPAE8, Doc. 293).

Less than a week later, on February 8, Einstein submitted a paper entitled "Cosmological considerations in the general theory of relativity" to the Berlin Academy (Einstein 1917). On February 14, he sent a copy of this paper to Ehrenfest with the request to show it to Lorentz and De Sitter as well (CPAE8, Doc. 298). Einstein now explicitly rejected the idea of degenerate boundary values for the metric field (pp. 145–146). Instead he developed the "crazy idea" mentioned in the letter to De Sitter quoted above.

This idea was actually quite ingenious. To get rid of boundary values at spatial infinity Einstein simply got rid of spatial infinity. To this end, he considered a universe that is spatially closed. He focused on a particularly simple model with a spherical spatial geometry of constant radius. Suppressing two of the three spatial dimensions, we can picture this model as a circle of fixed radius persisting through time, from $t = -\infty$ to $t = +\infty$. This Einstein universe, the first relativistic cosmological model, is therefore also known as

---

(starting with the clause "e.g., would certainly not do"), he put a right curly bracket and the comment: "That is nonsense (1917)."

6 De Sitter underlined the word "finite" twice. He later added in pencil: "namely *bounded* [underlined twice]. This is what I am against and is also what I meant back then: to finiteness without boundedness I have no objection (May 1917)."

the *cylinder universe*. For this model to serve his purposes, Einstein obviously had to make sure it is allowed by his theory. Unfortunately, the metric field describing the space-time geometry of the cylinder universe is not a solution of the Einstein field equations for a static mass distribution. It is a static solution, however, of slightly modified equations as long as an additional condition is satisfied. The modification was to add a term $-\lambda g_{\mu\nu}$, now known as the cosmological term, to the left-hand side of the Einstein field equations (p. 151, Eq. (13a); cf. Fig. 1.2). The additional condition was that the constant λ, now known as the cosmological constant, the radius R and the mass density ρ of the cylinder universe satisfy (p. 152, Eq. (14)):

$$\lambda = 1/R^2 = \kappa\rho/2.$$

The cosmological term provides the anti-gravity needed to prevent the cylinder world from collapsing under the mutual attraction of all the mass it contains. It seems to have gone unnoticed for well over a decade that the equilibrium thus achieved in Einstein's model is unstable (Eddington 1930).

Einstein, in all likelihood, went through this sequence of steps in essentially the same order in which we just presented them. He probably hit upon the simple cylindrical model of a universe without spatial boundaries first and only subsequently modified his field equations to allow this simple model as a static solution. This is why we claim that Einstein's first foray into cosmology can be seen as a continuation of what we called the mathematical strategy in his search for satisfactory field equations. Generalizing our definition in Ch. 1 somewhat, we can say that the mathematical strategy was to start from a mathematical object such that the relativity principle, as Einstein understood it, is automatically satisfied and worry about the additional physical requirements the object has to satisfy later. In the context of the Zurich notebook, the mathematical object was a set of equations of broad covariance with second-order derivatives of the metric and the physical requirements were compatibility with energy-momentum conservation and, in the appropriate limit, Newton's gravitational theory. In the context of Einstein's first foray into cosmology, the mathematical object was the cylinder universe (which was meant to implement the general relativity principle by eliminating boundary conditions at spatial infinity) and the basic physical requirement was that it be allowed by the field equations for a static matter distribution. To satisfy this requirement, Einstein had to add the cosmological term to the field equations. That this was the logical, if not also the temporal order, in which Einstein hit upon the cylinder universe and the cosmological term is suggested by what he told De Sitter in letter dated by the recipient as March 12, 1917:

> From the standpoint of astronomy, I have, of course, just built a spacious castle in the air. It was a burning question for me, however, whether the relativity thought can be followed through to its conclusion, or whether one runs into contradictions. I am satisfied that I can think it through to the end without running into contradictions. Now the problem no longer bothers me, while before it gave me no rest.

> Whether the model I constructed corresponds to reality is another matter, about which we will probably never gain information (CPAE8, Doc. 311).[7]

In the cosmology paper, the cosmological term is presented before the cylinder universe, i.e., the order of presentation is just the opposite of the logical order suggested by this letter to De Sitter. Einstein starts with an analogy with Newtonian cosmology to justify the addition of the cosmological term to the Einstein field equations. Only then does he go on to show that the cylinder universe is a solution of those amended field equations. As one of us has argued (Janssen 2014, p. 201), Einstein probably wanted to argue for the cosmological term independently of the cylinder universe to preempt De Sitter's predictable criticism that he had only introduced the cosmological term—like the hypothetical masses outside the visible universe earlier—because of his philosophical predilections. If we take him at his word in the letter to De Sitter quoted above, Einstein's main goal in the paper was to rescue the relativity of arbitrary motion after De Sitter had made it clear to him that general covariance does not automatically mean general relativity. This goal is not stated explicitly in the paper. A reader can certainly be forgiven for thinking, as many have done, that the goal of the paper was to open up a new and natural area of research for Einstein's new theory of gravity, the investigation of the large scale space-time structure of the universe. The new physics presented in the paper, however, was subordinate to the mathematics with which Einstein tried to implement the general relativity of motion and retroactively justify the name of his new theory of gravity.

The remainder of the debate with De Sitter, which was joined in 1918 by Hermann Weyl and Felix Klein, confirms our reading of the cosmology paper. We will only sketch these developments very briefly here.[8]

De Sitter's response to the letter in which Einstein revealed the hidden agenda of his cosmology paper nicely illustrates the phrase 'hot as Dutch love'. On March 15, 1917, De Sitter wrote to Einstein:

> Well, as long as you do not want to force your conception on reality, we are in agreement. As a consistent train of thought, I have nothing against it and I admire it. I cannot give you my final approval before I have had a chance to calculate with it (CPAE8, Doc. 312).

Five days later, De Sitter had done his calculations and found an alternative to Einstein's cylinder world. Suppressing two of the three spatial dimensions, we can picture De Sitter model as the two-dimensional surface of a hyperboloid embedded in 2+1 dimensional Minkowski space-time. This De Sitter universe is therefore also called the *hyperboloid world.* De Sitter credited Ehrenfest with suggesting the model to him. De Sitter's calculations showed

---

[7] In the remainder of the letter, however, Einstein speculates about a way to determine the radius R of the cylinder universe

[8] For more detailed discussion, see the editorial note, "The Einstein-De Sitter-Weyl-Klein debate," the annotation of the relevant correspondence in CPAE8, Janssen (2014, sec. 5, pp. 198–208), and the dissertation of Stefan Röhle (2007).

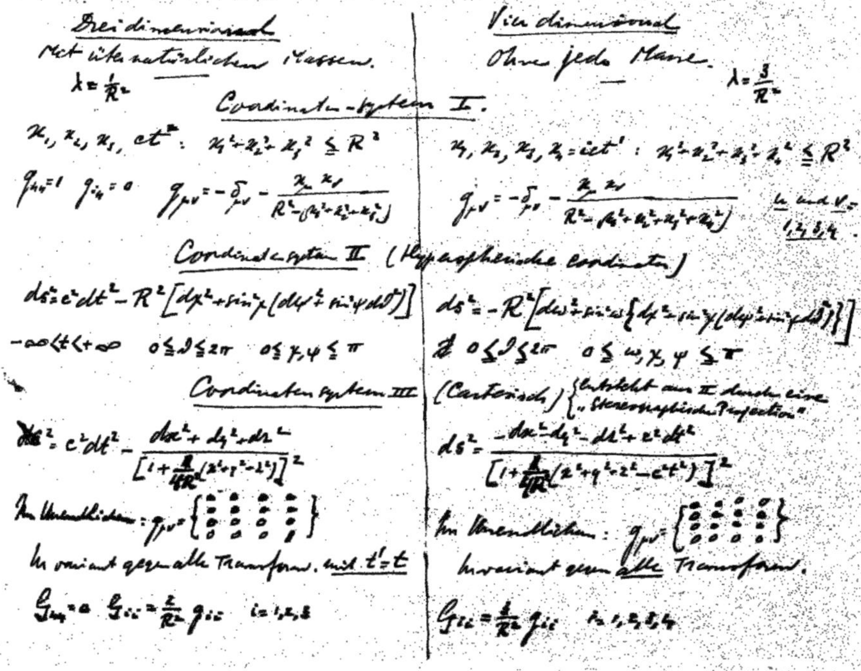

Figure 7.2 A comparison of Einstein's cylinder world (on the left: "three-dimensional," "with supernatural masses") and De Sitter's hyperboloid world (on the right: "four-dimensional," "without any mass") in three different coordinate system. De Sitter to Einstein, March 20, 1917 (CPAE8, Doc. 313).

that the metric field describing the geometry of the hyperboloid world is a solution of Einstein's amended field equations as long as the constant λ, the constant radius R of the hyperboloid, and the mass density ρ satisfy two conditions, $\lambda = 3/R^2$ and $\rho = 0$. De Sitter had thus found a vacuum solution of the field equations with cosmological term. The hyperboloid world is completely empty. There is no mass giving a gravitational attraction counteracting the repulsion given by the cosmological term.

In a letter of March 20, 1917 (CPAE8, Doc. 313), De Sitter sent Einstein a side-by-side comparison of the latter's cylinder world and his own hyperboloid world (see Fig. 7.2). The same comparison can be found in the paper on the new solution, which De Sitter (1917a) submitted to the Amsterdam Academy on March 31.

The absence of matter made the hyperboloid world unpalatable to Einstein. As he wrote in response to De Sitter's letter on March 24, 1917:

> It would be unsatisfactory, in my opinion, if a world without matter were possible. Rather, the $g_{\mu\nu}$-field should be *determined by matter and not be able to exist without it.* This is the core of what I mean by the requirement of the relativity of inertia (CPAE8, Doc. 317; it is not entirely clear whether the emphasis is Einstein's or De Sitter's).

De Sitter got Einstein's permission to quote this passage in a postscript to his paper (De Sitter 1917a). Einstein's dissatisfaction is understandable. Although he had not said so explicitly when he added the cosmological term to his field equations, his hope had been that the amended field equations would not allow space-times without matter. If that requirement, for which he would introduce the name Mach's principle the following year (Einstein 1918b), were satisfied, any talk about motion with respect to space-time could be seen as a *façon de parler* about motion with respect to the matter fully determining the metric field of that space-time (Maudlin 1990, p. 561).

Einstein tried to discredit the De Sitter solution by looking for singularities in some coordinate representation of the hyperboloid world. He fastened onto the so-called static form of the solution, which De Sitter communicated to him in a letter of June 20, 1917 (CPAE8, Doc. 355) and published shortly thereafter (De Sitter 1917b, 1917c). In early 1918, Einstein convinced himself that he had found what we would now call an intrinsic singularity in the static form of the De Sitter solution. He thereupon published a critical note on the De Sitter solution in which he argued that the basic difference between the cylinder world and the hyperboloid world was simply that matter was distributed evenly throughout the universe in the former and concentrated on a singular plane in the latter (Einstein 1918c). With this counter-example to the prohibition against vacuum solutions defeated (or so he thought), Einstein (1918b) published another short paper in which he officially introduced the name *Mach's principle* for what he had referred to as "the core of what I mean by the requirement of the relativity of inertia" in his first reaction to De Sitter's vacuum solution.

What Einstein and De Sitter—and even the great mathematician Weyl—overlooked in all of this is that any singularity in a coordinate representation of a completely regular hypersurface such as that of a hyperhyperboloid embedded in a 4+1 dimensional Minkowski space-time has to be a coordinate singularity and cannot be an intrinsic singularity. It was another leading mathematician, Klein, who made Einstein see this basic point. Once Klein, in a letter of June 16, 1918 (CPAE8, Doc. 566), had explained it to him in terms he understood, Einstein immediately conceded that his criticism of the De Sitter solution was unfounded (see Einstein to Klein, June 20, 1918 [CPAE8, Doc. 567]). He still rejected the De Sitter solution, but now on the much weaker argument that one cannot represent it in a static form without introducing a singularity. Initially, Einstein's assumption that the universe is static—to which De Sitter, in a letter to Einstein of April 1, 1917 (CPAE8, Doc. 321), had taken strong exception—had been an innocuous simplifying assumption. By June 1918, it had become crucial to Einstein's defense of Mach's principle.

Einstein's enthusiasm for Mach's principle was already waning the following year. By 1919, he seems to have realized that it was predicated on an antiquated billiard-ball ontology (Janssen 2014, p. 171; CPAE7, p. xxxiii). As he put it the year before he died:[9]

> Mach's principle ... dates back to the time in which one thought that the "ponderable bodies" are the only physically real entities and that all elements of the theory which are not completely determined by them should be avoided. (I am well aware of the fact that I myself was long influenced by this *idée fixe*) (Einstein to Felix Pirani, February 2, 1954).

In the age of field theories, Mach's principle boiled down to the requirement that one field be reduced to another. Given Einstein's disenchantment with Mach's principle, it need not surprise us that in the early 1920s he switched from reducing one field to another to unifying the electromagnetic and metric fields. Despite this change of objective, his preferred method for achieving the objective stayed the same. As carefully documented by van Dongen (2010), Einstein's approach in his work on unified field theory was clearly an extension of his mathematical strategy in the 1910s.

7.2 Gravitational waves as a continuation of the physical strategy

In the discussion following his lecture in Vienna in September 1913, Einstein made the offhand comment that, in linear approximation, the *Entwurf* field equations allow gravitational waves traveling at the speed of light (Einstein et al. 1913, p. 1266). Einstein only started looking in earnest for gravitational-wave solutions after he had replaced the *Entwurf* field equations by the Einstein field equations in November 1915. Within a year, he had already flip-flopped once on whether or not there are such solutions.[10]

On a postcard to Schwarzschild of February 19, 1916, in response to a letter or postcard no longer extant, he wrote

> I confirmed your calculation. My comment [in the first November paper (Einstein 1915a)] no longer applies given the new condition $\sqrt{-g} = 1$, as I already realized. The choice of coordinate system according to the condition $\partial g^{\mu\nu}/\partial x^{\nu} = 0$ [the Hertz condition used to recover the Newtonian limit in the first November paper (see Sec. 5.2)] is not compatible with $\sqrt{-g} = 1$ [used in the other three November papers (Einstein 1915b, 1915c, 1915d)]. Since then, I have handled Newton's case differently according to the final theory [see the perihelion paper (Einstein 1915c, p. 833) and the review article (Einstein 1916b, pp. 818–819)]—*Thus there are no gravitational waves analogous to light waves.* This is probably related to the one-sidedness of the sign of the scalar T [i.e., mass, unlike electric charge, is always positive]. (Nonexistence of the "dipole.") (CPAE8, Doc. 194; our emphasis).

From a letter from Schwarzschild to Sommerfeld of February 17, 1916, it can be inferred that Schwarzschild had been looking for wave solutions of the

[9] See Hoefer (1994, p. 330) and Renn (2007c, p. 61) for further discussion.

[10] He would do so again in the process of submitting, revising, and resubmitting Einstein and Rosen (1937). See Kennefick (2005, 2007, pp. 81–97) for discussion of this episode.

field equations of the first November paper, using the coordinate condition $\partial g^{\mu\nu}/\partial x^{\nu} = 0$ (Gutfreund and Renn 2017, p. 96). Schwarzschild told Sommerfeld: "I am continuing to rummage through Einstein's field equations. Today I am totally bewildered ... No wave motion but infinite speed of propagation" (Sommerfeld 2000, Doc. 238).

The two coordinate conditions, $\sqrt{-g} = 1$ and $\partial g^{\mu\nu}/\partial x^{\nu} = 0$, go with two different approximation schemes (Kennefick 2007, Chs. 2–3). To compare general relativity to Newton's theory, it is easier to use an approximation scheme with unimodular coordinates ($\sqrt{-g} = 1$) and make the field equations look like the Poisson equation. To compare general relativity to special relativity, it is easier to use an approximation scheme with isotropic coordinates ($\partial g^{\mu\nu}/\partial x^{\nu} = 0$) and make the field equations look like Maxwell's equations. It is this connection to Maxwell's equations that makes it easier to find gravitational-wave solutions in isotropic coordinates. It need not surprise us that Einstein could not find gravitational waves when he was looking for them in unimodular coordinates.

A letter to De Sitter of June 22, 1916 documents Einstein's switch from unimodular to isotropic coordinates in his search for gravitational waves. He wrote:

> Your letter [no longer extant[11]] pleased me very much and greatly inspired me. I found that one can solve the gravitation equations in first approximation exactly by means of retarded potentials, if one abandons the condition for [unimodular coordinates]. Your solution for the mass point [in isotropic coordinates, published in De Sitter (1916a)] is then the result upon specialization to this case. Obviously your solution differs from my old one in the choice of coordinate system, but not intrinsically (CPAE8, Doc. 227).

After switching to isotropic coordinates, Einstein (1916c), in short order, did find gravitational-wave solutions. These solutions were peculiar in that they described three types of waves, two of which do not carry any energy. By switching back to unimodular coordinates, Einstein was able to transform these two types away. From this he concluded that unimodular coordinates are physically privileged after all. His letter to De Sitter quoted above continues:

> One might think that the coordinate choice $\sqrt{-g} = 1$ is not very natural at all. However, I have found a highly interesting physical justification for them. I denote the $\sqrt{-g} = 1$ system as K, the generalized de Sitter system [satisfying the condition $\partial g^{\mu\nu}/\partial x^{\nu} = 0$ for isotropic coordinates] as K'. We now ask about plane gravitational waves. In system K' I find 3 types of waves, only one of which, however, is connected to energy transportation. In system K, by contrast, only this one energy-carrying type is present. What does this mean? It means that the first two types of waves obtained with K' do not exist in reality but are simulated by the coordinate system's wavelike motions in the Galilean space [read: Minkowski space-time]. The ($\sqrt{-g} = 1$)-system thus excludes reference systems exhibiting a

[11] De Sitter's reply to this letter of June 22 did turn up after CPAE8 was published. See CPAE12, Doc. 227a.

wavelike motion simulating energyless gravitational waves. System K' is nonetheless useful for the integration of the field equations in first-order approximation (CPAE8, Doc. 227).

In June 1916, Einstein had thus back-slided to a position similar to the one he held in the first November 1915 paper, when he thought that the covariance of the field equations had to be restricted to unimodular *transformations*. His argument for the privileged nature of unimodular *coordinates* in this letter to De Sitter was included in the appendix of the paper on gravitational waves submitted that same day. Einstein (1916c, p. 696) concluded that this argument gives unimodular coordinates "a deep-seated physical justification."[12]

At the same time, Einstein's use of isotropic coordinates in *finding* gravitational-wave solutions underscores the importance of the analogy with electromagnetism for his paper. The same analogy had been central to the physical strategy he had relied on in his search for gravitational field equations (see Sec. 3.2 and Ch. 5). Einstein's paper on gravitational waves turned out to be the opening salvo in a debate that lasted for several decades over analogies and disanalogies between electromagnetic and gravitational waves (Kennefick 2007, 2014).

Two long letters from Nordström—the first written September 22–28, 1917 (CPAE8, Doc. 382), the second on October 23, 1917 (CPAE8, Doc. 393)—finally convinced Einstein once and for all that there are no physically privileged coordinates (Kennefick 2007, pp. 61–65). As can be inferred from the first of the two letters, Nordström had calculated the energy of the sun's gravitational field, using the formalism of Einstein (1916d). In this paper, the connection between covariance and energy-momentum conservation was finally given in full generality and the last vestiges of the restriction to unimodular coordinates were finally removed (see Sec. 5.5 and Pt. II, Ch. 10). Nordström found that the sun's gravitational field carries no energy. Puzzled, he double-checked his result, this time using the formalism of Einstein's (1916c) paper on gravitational waves. He now found a non-zero energy for the sun's gravitational field. He could not find a mistake in either calculation and eventually realized that the two results were perfectly compatible with one another. They were just done in different coordinate systems, the first in unimodular coordinates, the second in isotropic coordinates. Redoing the first calculation in isotropic coordinates, however, Nordström arrived at an expression for the energy of the gravitational field that still differed from the expression given in Einstein's (1916c) gravitational waves paper. Nordström concluded that Einstein's expression had to be wrong and wrote to Einstein to alert him to the error.

Unfortunately, we do not have Einstein's reply but his reaction to this letter can be reconstructed from Nordström's second letter, sent a month after the first, and from Einstein's (1918a) second paper on gravitational waves published in February the following year. Einstein had a hard time

[12] See Kennefick (2007, pp. 51–58) for further discussion.

accepting that the energy of the sun's gravitational field could be zero in the unimodular coordinates he believed at the time were physically privileged. He only granted this when Nordström had gone over his calculation in detail in his second letter. He came to recognize that Nordström's calculation in isotropic coordinates was also correct. He eventually found the error in his own calculation and submitted a corrected version of his paper on gravitational waves to the Berlin Academy, only partly acknowledging Nordström's critical intervention (Einstein 1918a, p. 159).

In this new paper, Einstein corrected "the regrettable calculational error" to which he had been alerted by Nordström and derived the famous quadrupole formula for gravitational radiation—except for a mistake of a factor of 2 corrected four years later by Arthur S. Eddington (Kennefick 2007, p. 76). In his first paper on gravitational waves, Einstein (1916c) had found monopole radiation, which in and of itself should have made him suspicious of his derivation; in the second he found neither monopole nor dipole radiation. However, he still found three types of waves, two of which carry no energy. He now showed that those two solutions are nothing but the metric field of Minkowski space-time in "wavy" coordinates (Einstein 1918a, p. 161). Hence he could dismiss them as unphysical without privileging unimodular coordinates. With a nod to the point-coincidence argument (see Sec. 4.1), he emphatically stated his newly (re)found position that no coordinates are physically privileged in a letter to Mie of February 8, 1918:[13]

> All physical descriptions that yield the same observable relations (coincidences) are fundamentally equivalent, provided both descriptions are based on the same physical laws. The choice of coordinates can be of great practical importance from the point of view of clarity of description; but in essence it is totally meaningless (CPAE8, Doc. 460; quoted and discussed in Kennefick 2007, pp. 58–60).

It is probably no coincidence that, as we noted in Sec. 5.2, Einstein's second paper on gravitational waves also contains his first explicit statement of the modern understanding of coordinate conditions (Einstein 1918a, pp. 155–156).

[13] The point appears to have been lost on Mie. In a letter to Einstein of May 6, 1918 (CPAE8, Doc. 532), Mie wrote that he had just read Einstein's (1916c) first paper on gravitational waves and drew attention to the similarity between Einstein's spurious waves and an example he had used in his Wolfskehl lectures in Göttingen (Mie 1917, Pt. III, p. 599) as part of his argument against equating general covariance and general relativity, namely that given an appropriate choice of coordinates a straight rod could be made to look like a slithering snake (cf. note 4). For further discussion, see CPAE8, p. lii, and Kennefick (2007, pp. 58–60).

Part II
Sources

Chapter 1
The Zurich Notebook

M. Janssen, J. Renn, *How Einstein Found His Field Equations*, Classic Texts in the Sciences, https://doi.org/10.1007/978-3-030-97955-3_8

1.1 Facsimile

[p. 14L]

$$\left[{\mu\nu \atop l}\right] = \frac{1}{2}\left(\frac{\partial g_{\mu l}}{\partial x_\nu} + \frac{\partial g_{\nu l}}{\partial x_\mu} - \frac{\partial g_{\mu\nu}}{\partial x_l}\right) \qquad \frac{\partial}{\partial x_\kappa}\left[{il \atop m}\right] - \frac{\partial}{\partial x_i}\left[{\kappa l \atop m}\right]$$

$$(i\kappa, lm) = \frac{1}{2}\left(\frac{\partial^2 g_{im}}{\partial x_\kappa \partial x_l} + \frac{\partial^2 g_{\kappa l}}{\partial x_i \partial x_m} - \frac{\partial^2 g_{il}}{\partial x_\kappa \partial x_m} - \frac{\partial^2 g_{\kappa m}}{\partial x_i \partial x_l}\right) + \sum_{\rho\sigma} \gamma_{\rho\sigma}\left(\left[{im \atop \sigma}\right]\left[{\kappa l \atop \rho}\right] - \left[{il \atop \sigma}\right]\left[{\kappa m \atop \rho}\right]\right)$$

Grossmann Tensor vierter Mannigfaltigkeit

$$\sum \gamma_{\kappa l}(i\kappa, lm)$$

$$\sum \gamma_{\kappa l}\left[{\kappa l \atop \rho}\right] = \sum \gamma_{\kappa l}\left[\frac{\partial g_{\kappa\rho}}{\partial x_l} + \frac{\partial g_{l\rho}}{\partial x_\kappa} - \frac{\partial g_{\kappa l}}{\partial x_\rho}\right]$$

$$= -\frac{\partial \lg G}{\partial x_\rho} + 2\sum_{\kappa l} \gamma_{\kappa l}\frac{\partial g_{\kappa\rho}}{\partial x_l}$$

$$\frac{1}{4}\sum \gamma_{\rho\sigma}\left(\frac{\partial g_{i\sigma}}{\partial x_m} + \frac{\partial g_{m\sigma}}{\partial x_i} - \frac{\partial g_{im}}{\partial x_\sigma}\right)\left[-\frac{\partial \lg G}{\partial x_\rho} + 2\sum_{\kappa l}\gamma_{\kappa l}\frac{\partial g_{\kappa\rho}}{\partial x_l}\right]$$

$$\sum \gamma_{\kappa l}\gamma_{\rho\sigma}\left(\left[{im \atop \rho}\right]\left[{\kappa l \atop \sigma}\right] - \left[{il \atop \sigma}\right]\left[{\kappa m \atop \rho}\right]\right)$$

$$= \sum_\rho -\left\{{im \atop \rho}\right\}\cdot\frac{\partial \lg G}{\partial x_\rho} + 2\sum_{\kappa l\rho}\left\{{im \atop \rho}\right\}\cdot\gamma_{\kappa l}\frac{\partial g_{\kappa\rho}}{\partial x_l} - \sum_{\rho\kappa l}\left\{{il \atop \rho}\right\}\left(\frac{\partial g_{\kappa\rho}}{\partial x_m}\right)\gamma_{\kappa l} + \sum_{\rho l}\left\{{il \atop \rho}\right\}\cdot\left\{{\rho m \atop l}\right\}$$

$$\sum_\kappa\left(\frac{\partial^2 g_{\kappa\kappa}}{\partial x_i \partial x_m} - \frac{\partial^2 g_{i\kappa}}{\partial x_\kappa \partial x_m} - \frac{\partial^2 g_{m\kappa}}{\partial x_\kappa \partial x_i}\right) = 0$$

Sollte verschwinden.

[p. 19L]

Nochmalige Berechnung des Elemententensors

$$\frac{1}{2}\left(\frac{\partial^2 g_{im}}{\partial x_\kappa \partial x_l} + \frac{\partial^2 g_{\kappa l}}{\partial x_i \partial x_m} - \frac{\partial^2 g_{il}}{\partial x_\kappa \partial x_m} - \frac{\partial^2 g_{\kappa m}}{\partial x_i \partial x_l}\right) \Bigg| \gamma_{\kappa l}$$

$$-\frac{1}{4}\gamma_{\varrho\sigma}\left(\frac{\partial g_{i\varrho}}{\partial x_l}+\frac{\partial g_{l\varrho}}{\partial x_i}-\frac{\partial g_{il}}{\partial x_\varrho}\right)\left(\frac{\partial g_{\kappa\sigma}}{\partial x_m}+\frac{\partial g_{m\sigma}}{\partial x_\kappa}-\frac{\partial g_{m\kappa}}{\partial x_\sigma}\right)$$

$\frac{1}{2}\gamma_{\kappa l}\frac{\partial^2 g_{im}}{\partial x_\kappa\partial x_l}$ bleibt stehen.

$$\gamma_{\kappa l}\begin{bmatrix}\kappa l\\ i\end{bmatrix} = \gamma_{\kappa l}\left(2\frac{\partial g_{il}}{\partial x_\kappa}-\frac{\partial g_{\kappa l}}{\partial x_i}\right)=0 \quad \Bigg| \frac{\partial}{\partial x_m}$$

$$\gamma_{\kappa l}\begin{bmatrix}\kappa l\\ m\end{bmatrix} \quad \gamma_{\kappa l}\left(2\frac{\partial g_{m\kappa}}{\partial x_l}-\frac{\partial g_{\kappa l}}{\partial x_m}\right)=0 \quad \Bigg| \frac{\partial}{\partial x_i}$$

$$2\gamma_{\kappa l}\left(\frac{\partial^2 g_{il}}{\partial x_\kappa\partial x_m}+\frac{\partial^2 g_{m\kappa}}{\partial x_i\partial x_l}-\frac{\partial^2 g_{\kappa l}}{\partial x_i\partial x_m}\right)+\frac{\partial\gamma_{\kappa l}}{\partial x_m}\left(2\frac{\partial g_{il}}{\partial x_\kappa}-\frac{\partial g_{\kappa l}}{\partial x_i}\right)+\frac{\partial\gamma_{\kappa l}}{\partial x_i}\left(2\frac{\partial g_{m\kappa}}{\partial x_l}-\frac{\partial g_{\kappa l}}{\partial x_m}\right)=0$$

$$-\frac{1}{2}\gamma_{\kappa l}(\quad) = \frac{1}{4}\left|\frac{\partial\gamma_{\kappa l}}{\partial x_m}\left(2\frac{\partial g_{il}}{\partial x_\kappa}-\frac{\partial g_{\kappa l}}{\partial x_i}\right)+\frac{\partial\gamma_{\kappa l}}{\partial x_i}\left(2\frac{\partial g_{m\kappa}}{\partial x_l}-\frac{\partial g_{\kappa l}}{\partial x_m}\right)\right.$$

zweites Glied:

$$-\frac{1}{4}\gamma_{\varrho\sigma}\frac{\partial g_{l\varrho}}{\partial x_i}\frac{\partial g_{\kappa\sigma}}{\partial x_m}\gamma_{\kappa l} \qquad +\frac{1}{4}\frac{\partial\gamma_{\varrho\sigma}}{\partial x_i}\frac{\partial g_{\kappa\sigma}}{\partial x_m}g_{l\varrho}\gamma_{l\kappa} \qquad \frac{1}{4}\frac{\partial\gamma_{\varrho\sigma}}{\partial x_i}\frac{\partial g_{\varrho\sigma}}{\partial x_m}$$

$$-\frac{1}{4}\gamma_{\varrho\sigma}\left(\frac{\partial g_{i\varrho}}{\partial x_l}-\frac{\partial g_{il}}{\partial x_\varrho}\right)\left(\frac{\partial g_{m\sigma}}{\partial x_\kappa}-\frac{\partial g_{m\kappa}}{\partial x_\sigma}\right)\gamma_{\kappa l}$$

$$= -\frac{1}{2}\gamma_{\varrho\sigma}\gamma_{\kappa l}\frac{\partial g_{i\varrho}}{\partial x_l}\frac{\partial g_{m\sigma}}{\partial x_\kappa}+\frac{1}{2}\gamma_{\varrho\sigma}\gamma_{\kappa l}\frac{\partial g_{il}}{\partial x_\varrho}\frac{\partial g_{m\sigma}}{\partial x_\kappa}$$

Der mit 2 multiplizierte Elemententensor erhält also die Form

$$\gamma_{\kappa l}\frac{\partial^2 g_{im}}{\partial x_\kappa\partial x_l}-\frac{1}{2}\frac{\partial\gamma_{\kappa l}}{\partial x_m}\frac{\partial g_{\kappa l}}{\partial x_i}+\frac{\partial\gamma_{\kappa l}}{\partial x_m}\frac{\partial g_{il}}{\partial x_\kappa}+\frac{\partial\gamma_{\kappa l}}{\partial x_i}\frac{\partial g_{m\kappa}}{\partial x_l}$$

$$-\gamma_{\varrho\sigma}\gamma_{\kappa l}\frac{\partial g_{i\varrho}}{\partial x_l}\frac{\partial g_{m\sigma}}{\partial x_\kappa}+\gamma_{\varrho\sigma}\gamma_{\kappa l}\frac{\partial g_{il}}{\partial x_\varrho}\frac{\partial g_{m\sigma}}{\partial x_\kappa}$$

Resultat sicher. Gilt für Koordinaten, die der Gl. $\Delta\varphi = 0$ genügen.

-37-

[p. 19R]

Für die erste Annäherung lautet unsere Nebenbedingung.

$$\sum_\kappa \gamma_{\kappa\kappa}\left(2\frac{\partial g_{i\kappa}}{\partial x_\kappa} - \frac{\partial g_{\kappa\kappa}}{\partial x_i}\right) = 0$$

Zerfällt vielleicht in

$$\sum \gamma_{\kappa\kappa}\frac{\partial g_{i\kappa}}{\partial x_\kappa} = 0 \quad \text{u.} \quad \sum \gamma_{\kappa\kappa} g_{\kappa\kappa} = \text{Konst.}$$

Gleichungen

$$\sum \gamma_{\kappa\kappa}\frac{\partial^2 g_{im}}{\partial x_\kappa^2} = K\varrho_0 \frac{dx_i}{d\tau}\frac{dx_m}{d\tau} g_{ii}\, g_{mm}$$

$$\sum_{\kappa i m}\gamma_{\kappa\kappa}\frac{\partial^2 g_{im}}{\partial x_\kappa^2}\frac{\partial g_{im}}{\partial x_6} = \sum_{\kappa i m}\gamma_{\kappa\kappa}\left[\frac{\partial}{\partial x_\kappa}\left(\frac{\partial g_{im}}{\partial x_\kappa}\frac{\partial g_{im}}{\partial x_6}\right) - \frac{1}{2}\frac{\partial}{\partial x_6}\left(\frac{\partial g_{im}}{\partial x_\kappa}\right)^2\right]$$

Energie- u. Impulssatz gilt mit der in Betr. kommenden Annäherung. Eindeutigkeit u. Nebenbedingungen

$$\square g_{im} = K\varrho_0 \underbrace{\frac{dx_i}{d\tau}}_{\mathfrak{w}_i}\underbrace{\frac{dx_m}{d\tau}}_{\mathfrak{w}_m} \qquad ic\,dt = du$$

Kontinuitätsbedingung $\frac{\varrho_0}{\sqrt{1-\frac{q^2}{c^2}}}$ Dichte materieller Punkte

$$-\frac{\partial}{ic\,\partial t}\left(\frac{\varrho_0 ic}{\sqrt{1-\frac{q^2}{c^2}}}\right) = \frac{\partial}{\partial x}\left(\frac{\varrho_0 q_x}{\sqrt{\ }}\right) + \cdot + \cdot$$

$$\frac{\partial}{\partial x}(\varrho_0 \mathfrak{w}_x) + \frac{\partial}{\partial y}(\varrho_0 \mathfrak{w}_y) + \cdot + \frac{\partial}{\partial u}(\varrho_0 \mathfrak{w}_u) = 0$$

Beide obige Bedingungen sind aufrecht zu erhalten.

$$\frac{\partial}{\partial x}(\varrho_0 \mathfrak{w}_x \mathfrak{w}_x) + \frac{\partial}{\partial y}(\varrho_0 \mathfrak{w}_x \mathfrak{w}_y) + \cdot + \cdot$$

$$- \varrho_0 \mathfrak{w}_x \frac{\partial \mathfrak{w}_x}{\partial x} - \varrho_0 \mathfrak{w}_y \frac{\partial \mathfrak{w}_x}{\partial y} - \cdot - \cdot = 0$$

$$-\varrho\frac{D\mathfrak{w}_x}{D\tau}$$

$$dt\sqrt{1-\frac{q^2}{c^2}} = d\tau$$

$$\underbrace{\frac{\dot{x}}{\sqrt{\ }}}_{\mathfrak{w}_x} \quad \frac{\dot{y}}{\sqrt{\ }} \cdot \underbrace{\frac{ic}{\sqrt{\ }}}_{\mathfrak{w}_u}$$

$$2\sum_m \varrho_0 \mathfrak{w}_m \frac{\partial \mathfrak{w}_i}{\partial x_m} - \sum$$

[p. 20L]

$$\sum \frac{\partial g_{ik}}{\partial x_k} = 0 \qquad \sum \frac{\partial g^{x}_{kk}}{} = 0$$

$$\sum_{\sigma} \frac{\partial^2 g_{ik}}{\partial x_\sigma^2} = \varrho_0 \frac{dx_i}{d\tau}\frac{dx_k}{d\tau} - \left(\frac{1}{4}\varrho_0 \sum \frac{dx_\kappa}{d\tau}\frac{dx_\kappa}{d\tau}\right)$$

für gleiche i, k.

$$\sum\left(\frac{\partial g_{ik}}{\partial x_k} - \frac{1}{2}\frac{\partial g_{kk}}{\partial x_i}\right) = 0 \qquad \Delta g_{im} =$$

$$\sum g_{kk} = U$$

Gravitationsgleichungen

$$\Delta(g_{11} - \tfrac{1}{2}U) = T_{11} \qquad \Delta g_{12} = T_{12} \quad . \quad \Delta g_{14} = T_{14}$$

— — — — — — — — —

— — — — — — — — — —

— — — — — — — — — —

$$2\Delta U = \sum T_{kk}$$

Hieraus Gleichungen

$$\Delta g_{11} = T_{11} + \tfrac{1}{2}\sum T_{kk} \qquad \Delta g_{12} = T_{12} \quad . \quad \Delta g_{14} = T_{14}$$

— — — — — — — — — —

— — — — — — — — — —

— — — — — — — — — — ...

$$-\frac{1}{2}\sum \Delta U \frac{\partial g_{kk}}{\partial x_\sigma} = -\frac{1}{2}\sum \frac{\partial^2 g_{\alpha\alpha}}{\partial x_\beta \partial x_\beta}\frac{\partial g_{kk}}{\partial x_\sigma} = -\frac{1}{2}\sum \Delta U \frac{\partial U}{\partial x_\sigma}$$

$$= -\frac{1}{2}\sum\left(\frac{\partial^2 U}{\partial x^2} + \cdot + \cdot + \cdot\right)\frac{\partial U}{\partial x_\sigma}$$

Darstellbar in der verl. Form.

- 39 -

[p. 22R]

Grossmann

$$T_{il} = \sum_{\kappa l} \frac{\partial \left\{ {i\kappa \atop \kappa} \right\}}{\partial x_l} - \frac{\partial \left\{ {il \atop \kappa} \right\}}{\partial x_\kappa} + \left\{ {i\kappa \atop \lambda} \right\}\left\{ {\lambda l \atop \kappa} \right\} - \left\{ {il \atop \lambda} \right\}\left\{ {\lambda\kappa \atop \kappa} \right\}$$

Wenn G ein Skalar ist, dann $\frac{\partial \lg \sqrt{G}}{\partial x_i} = T_i$ Tensor 1. Ranges.

$$T_{il} = \left(\frac{\partial T_i}{\partial x_l} - \sum \left\{ {il \atop \lambda} \right\} T_\lambda \right) - \sum_{\kappa\lambda} \left(\frac{\partial \left\{ {il \atop \kappa} \right\}}{\partial x_\kappa} - \left\{ {i\kappa \atop \lambda} \right\}\left\{ {\lambda l \atop \kappa} \right\} \right)$$

Tensor 2. Ranges

Vermutlicher Gravitations-Tensor. T_{il}^x

Weitere Umformung des Gravitationstensors

$$\frac{\partial \left\{ {il \atop \kappa} \right\}}{\partial x_\kappa} = \frac{1}{2}\frac{\partial}{\partial x_\kappa}\left(\gamma_{\kappa\alpha} \left(\frac{\partial g_{i\alpha}}{\partial x_l} + \frac{\partial g_{l\alpha}}{\partial x_i} - \frac{\partial g_{il}}{\partial x_\alpha} \right) \right)$$

Wir setzen voraus $\sum_\kappa \frac{\partial \gamma_{\kappa\alpha}}{\partial x_\kappa} = 0$, dann ist dies gleich

$$-\frac{1}{2}\sum \gamma_{\kappa\alpha} \frac{\partial^2 g_{il}}{\partial x_\alpha \partial x_\kappa} + \frac{1}{2}\sum \left(\frac{\partial \gamma_{\kappa\alpha}}{\partial x_l} \frac{\partial g_{i\alpha}}{\partial x_\kappa} + \frac{\partial \gamma_{\kappa\alpha}}{\partial x_i} \frac{\partial g_{l\alpha}}{\partial x_\kappa} \right)$$

Ferner

$$\left\{ {i\kappa \atop \lambda} \right\}\left\{ {\lambda l \atop \kappa} \right\} = \frac{1}{4}\gamma_{\lambda\alpha}\gamma_{\kappa\beta}\left(\frac{\partial g_{i\alpha}}{\partial x_\kappa} - \frac{\partial g_{i\kappa}}{\partial x_\alpha} + \frac{\partial g_{\alpha\kappa}}{\partial x_i} \right)\left(\frac{\partial g_{l\beta}}{\partial x_\lambda} - \frac{\partial g_{l\lambda}}{\partial x_\beta} + \frac{\partial g_{\lambda\beta}}{\partial x_l} \right)$$

$$= -\frac{1}{4}\gamma_{\lambda\alpha}\gamma_{\kappa\beta}\left(\frac{\partial g_{i\alpha}}{\partial x_\kappa} - \frac{\partial g_{i\kappa}}{\partial x_\alpha} \right)\left(\frac{\partial g_{l\lambda}}{\partial x_\beta} - \frac{\partial g_{l\beta}}{\partial x_\lambda} \right) + \frac{1}{4}\gamma_{\lambda\alpha}\gamma_{\kappa\beta}\frac{\partial g_{\alpha\kappa}}{\partial x_i}\frac{\partial g_{\lambda\beta}}{\partial x_l}$$

$\alpha\ \kappa\ \lambda\ \beta$

$\alpha\ \beta\ \kappa\ \lambda$

$$-\frac{\partial \gamma_{\lambda\alpha}}{\partial x_i}\frac{\partial g_{\lambda\alpha}}{\partial x_l}$$

oder $-\frac{\partial \gamma_{\lambda\alpha}}{\partial x_l}\frac{\partial g_{\lambda\alpha}}{\partial x_i}$

Hieraus

$$-2T_{il}^x = \sum\left(\gamma_{\alpha\beta}\frac{\partial^2 g_{il}}{\partial x_\alpha \partial x_\beta} - \frac{1}{2}\gamma_{\alpha\kappa}\gamma_{\beta\lambda}\left(\frac{\partial g_{i\alpha}}{\partial x_\beta} - \frac{\partial g_{i\beta}}{\partial x_\alpha} \right)\left(\frac{\partial g_{l\kappa}}{\partial x_\lambda} - \frac{\partial g_{l\lambda}}{\partial x_\kappa} \right) \right)$$

$$+ \frac{1}{2}\sum\left(\frac{\partial \gamma_{\alpha\beta}}{\partial x_i}\left[{\alpha\beta \atop l} \right] + \frac{\partial \gamma_{\alpha\beta}}{\partial x_l}\left[{\alpha\beta \atop i} \right] \right) + \sum \frac{1}{4}\frac{\partial \gamma_{\alpha\beta}}{\partial x_i}\frac{\partial g_{\alpha\beta}}{\partial x_l}$$

– 44 –

[p. 22L]

$$\sum \frac{\partial \gamma'_{\mu\nu}}{\partial x'_\nu} = 0 \qquad |p_{\mu\nu}| = 1$$

$$\sum \pi_{\nu i} \frac{\partial}{\partial x_i}\{p_{\mu\alpha} p_{\nu\beta} \gamma_{\alpha\beta}\} = 0 \qquad \sum$$

$$= \sum p_{\mu\alpha} \frac{\partial \gamma_{\alpha i}}{\partial x_i} + \underbrace{\sum \gamma_{\alpha\beta}\pi_{\nu i}\frac{\partial p_{\mu\alpha}p_{\nu\beta}}{\partial x_i}}$$

$$\sum \gamma_{\alpha\beta}\pi_{\nu i}\left\{p_{\mu\alpha}\frac{\partial p_{\nu\beta}}{\partial x_i} + p_{\nu\beta}\frac{\partial p_{\mu\alpha}}{\partial x_i}\right\}$$

$$= \sum \gamma_{\alpha i}\frac{\partial p_{\mu\alpha}}{\partial x_i} + \sum \gamma_{\alpha\beta}\pi_{\nu i}p_{\mu\alpha}\frac{\partial p_{\nu\beta}}{\partial x_i}$$ verschwindet, wenn Funkt. Det. = 1.

$$= \sum \frac{\partial}{\partial x_i}(\gamma_{\alpha i} p_{\mu\alpha}) - p_{\mu\alpha}\frac{\partial\gamma_{\alpha i}}{\partial x_i} + \frac{\partial}{\partial x_i}(\gamma_{\alpha i}p_{\mu\alpha}) - \gamma_{\alpha\beta}p_{\mu\alpha}p_{\nu\beta}\frac{\partial \pi_{\nu i}}{\partial x_i}$$

$$\text{(}p_{\mu\alpha}\frac{\partial\gamma_{\alpha i}}{\partial x_i} = 0\text{)} \qquad -\frac{\partial}{\partial x_i}(\gamma_{\alpha i}p_{\mu\alpha})$$

$$\sum \gamma_{\alpha i}\frac{\partial p_{\mu\alpha}}{\partial x_i} + \gamma_{\alpha\beta}p_{\mu\alpha}\pi_{\nu i}\frac{\partial p_{\nu\beta}}{\partial x_i}$$

$$\sum \gamma_{\kappa l}\left\{\frac{\partial^2 g_{\kappa i}}{\partial x_l \partial x_m} + \frac{\partial^2 g_{\kappa m}}{\partial x_l\partial x_i}\right\}$$

$$= - \cdot + \frac{\partial}{\partial x_m}\sum \gamma_{\kappa l}\frac{\partial g_{\kappa i}}{\partial x_l}$$

Genügt, wenn $\sum \frac{\partial \gamma_{\kappa l}}{\partial x_l}$ verschwindet.

$$\sum_{lm} \gamma_{lm} T_{iklm}$$

g_{ik}

γ_{ik}

$$\frac{\partial g}{\partial x_i \partial x_k}$$

1.2 Transcription

[p. 14 L]

$$\begin{bmatrix}\mu\nu\\ l\end{bmatrix} = \frac{1}{2}\left(\frac{\partial g_{\mu l}}{\partial x_\nu} + \frac{\partial g_{\nu l}}{\partial x_\mu} - \frac{\partial g_{\mu\nu}}{\partial x_l}\right) \qquad \frac{\partial}{\partial x_\kappa}\begin{bmatrix}i\ l\\ m\end{bmatrix} - \frac{\partial}{\partial x_i}\begin{bmatrix}\kappa\ l\\ m\end{bmatrix}$$

$$(i\kappa, lm) = \frac{1}{2}\left(\frac{\partial^2 g_{im}}{\partial x_\kappa \partial \kappa_l} + \frac{\partial^2 g_{\kappa l}}{\partial x_i \partial x_m} - \frac{\partial^2 g_{il}}{\partial x_\kappa \partial x_m} - \frac{\partial^2 g_{\kappa m}}{\partial x_i \partial x_l}\right)$$
$$+\sum_{\rho\sigma}\gamma_{\rho\sigma}\left(\begin{bmatrix}i\ m\\ \sigma\end{bmatrix}\begin{bmatrix}\kappa\ l\\ \rho\end{bmatrix} - \begin{bmatrix}i\ l\\ \sigma\end{bmatrix}\begin{bmatrix}\kappa\ m\\ \rho\end{bmatrix}\right)$$

Grossmann Tensor vierter Mannigfaltigkeit [1]

$\sum \gamma_{\kappa l}(i\kappa, lm)$? [2]

$$\sum \gamma_{\kappa l}\begin{bmatrix}\kappa\ l\\ \rho\end{bmatrix} = \sum \gamma_{\kappa l}\left[\frac{\partial g_{\kappa\rho}}{\partial x_l} + \frac{\partial g_{l\rho}}{\partial x_\kappa} - \frac{\partial g_{\kappa l}}{\partial x_\rho}\right]$$ [3]

$$=\langle 2\rangle - \frac{\partial \lg G}{\partial x_\rho} + 2\sum_{\kappa l}\gamma_{\kappa l}\frac{\partial g_{\kappa\rho}}{\partial x_l}$$

$$\frac{1}{4}\sum \gamma_{\rho\sigma}\left(\frac{\partial g_{i\sigma}}{\partial x_m} + \frac{\partial g_{m\sigma}}{\partial x_i} - \frac{\partial g_{im}}{\partial x_\sigma}\right)\left[-\frac{\partial \lg G}{\partial x_\rho} + 2\sum_{\kappa l}\gamma_{\kappa l}\frac{\partial g_{\kappa\rho}}{\partial x_l}\right]$$

$$\sum \gamma_{\kappa l}\gamma_{\rho\sigma}\left(\begin{bmatrix}i\ m\\ \sigma\end{bmatrix}\begin{bmatrix}\kappa\ l\\ \rho\end{bmatrix} - \begin{bmatrix}i\ l\\ \sigma\end{bmatrix}\begin{bmatrix}\kappa\ m\\ \rho\end{bmatrix}\right)$$

$$= \sum_\rho -\begin{Bmatrix}i\ m\\ \rho\end{Bmatrix}\cdot\frac{\partial \lg G}{\partial x_\rho} + 2\sum_{\kappa l\rho}\begin{Bmatrix}i\ m\\ \rho\end{Bmatrix}\cdot\gamma_{\kappa l}\frac{\partial g_{\kappa\rho}}{\partial x_l} - \sum_{\rho l\kappa}\begin{Bmatrix}i\ l\\ \rho\end{Bmatrix}\left(\frac{\partial g_{\kappa\rho}}{\partial x_m}\right)\gamma_{\kappa l}$$
$$+\sum_{\rho l}\begin{Bmatrix}i\ l\\ \rho\end{Bmatrix}\cdot\begin{Bmatrix}\rho\ m\\ l\end{Bmatrix}$$

$$\sum_\kappa\left(\frac{\partial^2 g_{\kappa\kappa}}{\partial x_i \partial x_m} - \frac{\partial^2 g_{i\kappa}}{\partial x_\kappa \partial x_m} - \frac{\partial^2 g_{m\kappa}}{\partial x_\kappa \partial x_i}\right) = 0$$

Sollte verschwinden.

[p. 19 L]

Nochmalige Berechnung des Ebenentensors

$$\frac{1}{2}\left(\frac{\partial^2 g_{im}}{\partial x_\kappa \partial x_l} + \frac{\partial^2 g_{\kappa l}}{\partial x_i \partial x_m} - \frac{\partial^2 g_{il}}{\partial x_\kappa \partial x_m} - \frac{\partial^2 g_{\kappa m}}{\partial x_i \partial x_l}\right)$$

$$-\frac{1}{4}\gamma_{\rho\sigma}\left(\frac{\partial g_{i\rho}}{\partial x_l} + \frac{\partial g_{l\rho}}{\partial x_i} - \frac{\partial g_{il}}{\partial x_\rho}\right)\left(\frac{\partial g_{\kappa\sigma}}{\partial x_m} + \frac{\partial g_{m\sigma}}{\partial x_\kappa} - \frac{\partial g_{m\kappa}}{\partial x_\sigma}\right) \quad \Bigg| \quad \gamma_{\kappa l}$$ [4]

$$\frac{1}{2}\gamma_{\kappa l}\frac{\partial^2 g_{im}}{\partial x_\kappa \partial x_l}$$ bleibt stehen. [5]

$$\gamma_{\kappa l}\begin{bmatrix}\kappa & l\\ & i\end{bmatrix} = \gamma_{\kappa l}\left(2\frac{\partial g_{il}}{\partial x_\kappa} - \frac{\partial g_{\kappa l}}{\partial x_i}\right) = 0 \quad \Bigg| \quad \frac{\partial}{\partial x_m}$$ [6]

$$\gamma_{\kappa l}\begin{bmatrix}\kappa & l\\ & m\end{bmatrix} \quad \gamma_{\kappa l}\left(2\frac{\partial g_{m\kappa}}{\partial x_l} - \frac{\partial g_{\kappa l}}{\partial x_m}\right) = 0 \quad \Bigg| \quad \frac{\partial}{\partial x_i}$$

$$2\gamma_{\kappa l}\left(\frac{\partial^2 g_{il}}{\partial x_\kappa \partial x_m} + \frac{\partial^2 g_{m\kappa}}{\partial x_i \partial x_l} - \frac{\partial^2 g_{\kappa l}}{\partial x_i \partial x_m}\right) + \frac{\partial \gamma_{\kappa l}}{\partial x_m}\left(2\frac{\partial g_{il}}{\partial x_\kappa} - \frac{\partial g_{\kappa l}}{\partial x_i}\right) + \frac{\partial \gamma_{\kappa l}}{\partial x_i}\left(2\frac{\partial g_{m\kappa}}{\partial x_l} - \frac{\partial g_{\kappa l}}{\partial x_i}\right) = 0$$

$$-\frac{1}{2}\gamma_{\kappa l}(\quad) = \frac{1}{4}\Bigg| \frac{\partial \gamma_{\kappa l}}{\partial x_m}\left(2\frac{\partial g_{il}}{\partial x_\kappa} - \frac{\partial g_{\kappa l}}{\partial x_i}\right) + \frac{\partial \gamma_{\kappa l}}{\partial x_i}\left(2\frac{\partial g_{m\kappa}}{\partial x_l} - \frac{\partial g_{\kappa l}}{\partial x_i}\right)$$

zweites Glied:

$$-\frac{1}{4}\gamma_{\rho\sigma}\frac{\partial g_{l\rho}}{\partial x_i}\frac{\partial g_{\kappa\sigma}}{\partial x_m}\gamma_{\kappa l} \qquad +\frac{1}{4}\frac{\partial \gamma_{\rho\sigma}}{\partial x_i}\frac{\partial g_{\kappa\sigma}}{\partial x_m}g_{l\rho}\gamma_{l\kappa}$$

$$\frac{1}{4}\frac{\partial \gamma_{\rho\sigma}}{\partial x_i}\frac{\partial g_{\kappa\sigma}}{\partial x_m}$$

$$-\frac{1}{4}\gamma_{\rho\sigma}\left(\frac{\partial g_{i\rho}}{\partial x_l} - \frac{\partial g_{il}}{\partial x_\rho}\right)\left(\frac{\partial g_{m\sigma}}{\partial x_\kappa} - \frac{\partial g_{m\kappa}}{\partial x_\sigma}\right)\gamma_{\kappa l}$$

$$= -\frac{1}{2}\gamma_{\rho\sigma}\gamma_{\kappa l}\frac{\partial g_{i\rho}}{\partial x_l}\frac{\partial g_{m\sigma}}{\partial x_\kappa} + \frac{1}{2}\gamma_{\rho\sigma}\gamma_{\kappa l}\frac{\partial g_{il}}{\partial x_\rho}\frac{\partial g_{m\sigma}}{\partial x_\kappa}$$

Der mit 2 multiplizierte Ebenentensor erhält also die Form [7]

$$\langle\tfrac{1}{2}\rangle\gamma_{\kappa l}\frac{\partial^2 g_{im}}{\partial x_\kappa \partial x_l} - \frac{1}{2}\frac{\partial \gamma_{\kappa l}}{\partial x_m}\frac{\partial g_{\kappa l}}{\partial x_i} + \frac{\partial \gamma_{\kappa l}}{\partial x_m}\frac{\partial g_{il}}{\partial x_\kappa} + \frac{\partial \gamma_{\kappa l}}{\partial x_i}\frac{\partial g_{m\kappa}{}^{l}}{\partial x_{l}{}_{\kappa}}$$

$$-\gamma_{\rho\sigma}\gamma_{\kappa l}\frac{\partial g_{i\rho}}{\partial x_l}\frac{\partial g_{m\sigma}}{\partial x_\kappa} + \gamma_{\rho\sigma}\gamma_{\kappa l}\frac{\partial g_{il}}{\partial x_\rho}\frac{\partial g_{m\sigma}}{\partial x_\kappa}$$

Resultat sicher. Gilt für Koordinaten, die der Gl. $\Delta\varphi = 0$ genügen. [8]

[p. 19 R]

Für die erste Annäherung lautet unsere Nebenbedingung. [9]

$$\sum_\kappa \gamma_{\kappa\kappa}\left(2\frac{\partial g_{i\kappa}}{\partial x_\kappa} - \frac{\partial g_{\kappa\kappa}}{\partial x_i}\right) = 0$$

Zerfällt vielleicht in

$$\sum \gamma_{\kappa\kappa}\frac{\partial g_{i\kappa}}{\partial x_\kappa} = 0 \qquad \text{u} \qquad \sum \overset{\gamma_{\kappa\kappa}}{g_{\kappa\kappa}} = \text{konst.}$$

Gleichungen [10]

$$\sum \gamma_{\kappa\kappa}\frac{\partial^2 g_{im}}{\partial x_\kappa^2} = K\rho_0\frac{dx_i}{ds}\frac{dx_m}{ds}g_{ii}g_{mm}$$

$$\sum_{\kappa i m} \gamma_{\kappa\kappa}\frac{\partial^2 g_{im}}{\partial x_\kappa^2}\frac{\partial g_{im}}{\partial x_\sigma} = \sum_{\kappa i m}\gamma_{\kappa\kappa}\left[\frac{\partial}{\partial x_\kappa}\left(\frac{\partial g_{im}}{\partial x_\kappa}\frac{\partial g_{im}}{\partial x_\sigma}\right) - \frac{1}{2}\frac{\partial}{\partial x_\sigma}\left(\frac{\partial^2 g_{im}^2}{\partial x_\kappa}\right)^{2}\right]$$

Energie- u Impulssatz gilt mit der in Betr. kommenden Annäherung.
Eindeutigkeit u Nebenbedingungen [11]

$$\Box g_{im} = K\rho_0\overset{\mathfrak{w}_i}{\frac{dx_i}{dx_\tau}}\overset{\mathfrak{w}_m}{\frac{dx_m}{dx_\tau}} \qquad\qquad icdt = du$$

Kontinuitätsbedingung $\frac{\rho_0}{\sqrt{1-\frac{q^2}{c^2}}}$ Dichte materieller Punkte

$$-\frac{\partial}{ic\,\partial t}\left(\frac{\rho_0 ic}{\sqrt{1-\frac{q^2}{c^2}}}\right) = \frac{\partial}{\partial x}\left(\frac{\rho_0 q_x}{\sqrt{}}\right) + \cdot + \cdot$$

$$\frac{\partial}{\partial x}(\rho_0\mathfrak{w}_x) + \frac{\partial}{\partial y}(\rho_0\mathfrak{w}_y) + \cdot + \frac{\partial}{\partial u}(\rho_0\mathfrak{w}_u) = 0$$

Beide obige Bedingungen sind aufrecht zu erhalten.

$$\frac{\partial}{\partial x}(\rho_0\mathfrak{w}_x\mathfrak{w}_x) + \frac{\partial}{\partial y}(\rho_0\mathfrak{w}_x\mathfrak{w}_y) + \cdot + \cdot$$

$$-\rho_0\mathfrak{w}_x\frac{\partial\mathfrak{w}_x}{\partial x} - \rho_0\mathfrak{w}_y\frac{\partial\mathfrak{w}_x}{\partial y} - \cdot - \cdot = 0$$

$$-\rho\frac{D\mathfrak{w}_x}{D\tau}$$

$$dt\sqrt{1-\frac{q^2}{c^2}} = d\tau$$

$$\underset{\mathfrak{w}_x}{\frac{\dot{x}}{\sqrt{}}}\quad\frac{\dot{y}}{\sqrt{}}\quad\cdot\quad\underset{\mathfrak{w}_u}{\frac{ic}{\sqrt{}}}$$

$$2<\frac{\partial}{\partial x_m}>\sum\frac{\partial}{\partial x_m}(\rho_0\mathfrak{w}_i\mathfrak{w}_m)$$

$$-\frac{\partial}{\partial x_i}\sum\left(\frac{\partial\rho_0\mathfrak{w}_m\mathfrak{w}_m}{\partial x_i}\right)$$

$$2\sum_m\rho_0\mathfrak{w}_m\frac{\partial\mathfrak{w}_i}{\partial x_m} - \sum$$

[p. 20 L]

$$\sum \frac{\partial g_{i\kappa}}{\partial x_\kappa} = 0 \qquad \sum g^x_{\kappa\kappa} = 0 \qquad [12]$$

$$\sum_\sigma \frac{\partial^2 g_{i\kappa}}{\partial x_\sigma^2} = \rho_0 \frac{dx_i}{d\tau}\frac{dx_\kappa}{d\tau} - \left(\frac{1}{4}\rho_0 \sum \frac{dx_\kappa}{d\tau}\frac{dx_\kappa}{d\tau}\right) \qquad [13]$$

für gleiche i u κ.

$$\sum \left(\frac{\partial g_{i\kappa}}{\partial x_\kappa} - \frac{1}{2}\frac{\partial g_{\kappa\kappa}}{\partial x_i}\right) = 0 \qquad [14] \qquad \sum^{\partial} \Delta g_{im} =$$

$$\sum g_{\kappa\kappa} = U$$

Gravitationsgleichungen

$$\Delta\left(g_{11} - \frac{1}{2}U\right) = T_{11} \quad \Delta g_{12} = T_{12} \quad \cdot \quad \Delta g_{14} = T_{14} \qquad [15]$$

— — — — — — — — — — —

— — — — — — — — — — —

— — — — — — — — — — —

$$2\Delta U = \sum T_{\kappa\kappa}$$

Hieraus Gleichungen

$$\Delta g_{11} = T_{11} + \frac{1}{2}\sum T_{\kappa\kappa} \quad \Delta g_{12} = T_{12} \quad \cdot \quad \Delta g_{14} = T_{14}$$

— — — — — — — — — — —

— — — — — — — — — — —

— — — — — — — — — — —

$$-\frac{1}{2}\sum \Delta U \frac{\partial g_{\kappa\kappa}}{\partial x_\sigma} = -\frac{1}{2}\sum \frac{\partial^2 g_{\alpha\alpha}}{\partial x_\beta \partial x_\beta}\frac{\partial g_{\kappa\kappa}}{\partial x_\sigma} = -\frac{1}{2}\sum \Delta U \frac{\partial U}{\partial x_\sigma} \qquad [16]$$

$$= -\frac{1}{2}\sum\left[\frac{\partial^2 U}{\partial x^2} + \cdot + \cdot + \cdot\right]\frac{\partial U}{\partial x_\sigma}$$

Darstellbar in der verl. Form.

[p. 22 R]

Grossmann

$$T_{il} = \sum_{\kappa l} \frac{\partial \begin{Bmatrix} i\,\kappa \\ \kappa \end{Bmatrix}}{\underline{\partial x_l}} - \frac{\partial \begin{Bmatrix} i\,l \\ \kappa \end{Bmatrix}}{\partial x_\kappa} + \begin{Bmatrix} i\,\kappa \\ \lambda \end{Bmatrix}\begin{Bmatrix} \lambda\,l \\ \kappa \end{Bmatrix} - \underline{\begin{Bmatrix} i\,l \\ \lambda \end{Bmatrix}\begin{Bmatrix} \lambda\,\kappa \\ \kappa \end{Bmatrix}}$$ [17]

Wenn $\underline{G}$ ein Skalar ist, dann $\frac{\partial \lg \sqrt{G}}{\partial x_i} = T_i$ Tensor 1. Ranges. [18]

$$T_{il} = \underbrace{\left(\frac{\partial T_i}{\partial x_l} - \sum \begin{Bmatrix} i\,l \\ \lambda \end{Bmatrix} T_\lambda\right)}_{\text{Tensor 2. Ranges}} \quad \underbrace{- \sum_{\kappa l}\left(\frac{\partial \begin{Bmatrix} i\,l \\ \kappa \end{Bmatrix}}{\partial x_\kappa} - \begin{Bmatrix} i\,\kappa \\ \lambda \end{Bmatrix}\begin{Bmatrix} l\,\lambda \\ \kappa \end{Bmatrix}\right)}_{\text{Vermutlicher Gravitations-Tensor. } T_{il}^x}$$

Weitere Umformung des Gravitationstensors [19]

$$\frac{\partial \begin{Bmatrix} i\,l \\ \kappa \end{Bmatrix}}{\partial x_\kappa} = \frac{1}{2}\frac{\partial}{\partial x_\kappa}\left(\gamma_{\kappa\alpha}\left(\frac{\partial g_{i\alpha}}{\partial x_l} + \frac{\partial g_{l\alpha}}{\partial x_i} - \frac{\partial g_{il}}{\partial x_\alpha}\right)\right)$$

Wir setzen voraus $\sum_\kappa \frac{\partial \gamma_{\kappa\alpha}}{\partial x_\kappa} = 0$, dann ist dies gleich

$$-\overset{\frac{1}{2}}{\sum} \gamma_{\kappa\alpha}\frac{\partial^2 g_{il}}{\partial x_\alpha \partial x_\kappa} - \overset{\frac{1}{2}}{\sum}\left(\frac{\partial \gamma_{\kappa\alpha}}{\partial x_l}\frac{\partial g_{i\alpha}}{\partial x_\kappa} + \frac{\partial \gamma_{\kappa\alpha}}{\partial x_i}\frac{\partial g_{l\alpha}}{\partial x_\kappa}\right)$$

Ferner

$$\begin{Bmatrix} i\,\kappa \\ \lambda \end{Bmatrix}\begin{Bmatrix} \lambda\,l \\ \kappa \end{Bmatrix} = \overset{\frac{1}{4}}{\gamma_{\lambda\alpha}}\gamma_{\kappa\beta}\left(\frac{\partial g_{i\alpha}}{\partial x_\kappa} - \frac{\partial g_{i\kappa}}{\partial x_\alpha} + \frac{\partial g_{\alpha\kappa}}{\partial x_i}\right)\left(\frac{\partial g_{l\beta}}{\partial x_\lambda} - \frac{\partial g_{l\lambda}}{\partial x_\beta} + \frac{\partial g_{\lambda\beta}}{\partial x_l}\right)$$

$$= -\overset{\frac{1}{4}}{\gamma_{\lambda\alpha}}\gamma_{\kappa\beta}\left(\frac{\partial g_{i\alpha}}{\partial x_\kappa} - \frac{\partial g_{i\kappa}}{\partial x_\alpha}\right)\left(\frac{\partial g_{l\lambda}}{\partial x_\beta} - \frac{\partial g_{l\beta}}{\partial x_\lambda}\right) + \overset{\frac{1}{4}}{\gamma_{\lambda\alpha}}\gamma_{\kappa\beta}\frac{\partial g_{\alpha\kappa}}{\partial x_i}\frac{\partial g_{\lambda\beta}}{\partial x_l}$$

$\alpha\ \kappa\ \lambda\ \beta$

$\alpha\ \beta\ \kappa\ \lambda$

$$-\frac{\partial \gamma_{\lambda\alpha}}{\partial x_i}\frac{\partial g_{\lambda\alpha}}{\partial x_l}$$

oder $-\frac{\partial \gamma_{\lambda\alpha}}{\partial x_l}\frac{\partial g_{\lambda\alpha}}{\partial x_i}$

Hieraus

$$-{}^2T_{il}^x = \sum\left(\gamma_{\alpha\beta}\frac{\partial^2 g_{il}}{\partial x_\alpha \partial x_\beta} - \overset{\frac{1}{2}}{\gamma_{\alpha\kappa}}\gamma_{\beta\lambda}\left(\frac{\partial g_{i\alpha}}{\partial x_\beta} - \frac{\partial g_{i\beta}}{\partial x_\alpha}\right)\left(\frac{\partial g_{l\kappa}}{\partial x_\lambda} - \frac{\partial g_{l\lambda}}{\partial x_k}\right)\right)$$

$$+\overset{\frac{1}{2}}{\sum}\left(\frac{\partial \gamma_{\alpha\beta}}{\partial x_i}\begin{bmatrix} \alpha\,\beta \\ l \end{bmatrix} + \frac{\partial \gamma_{\alpha\beta}}{\partial x_l}\begin{bmatrix} \alpha\,\beta \\ i \end{bmatrix}\right) + \sum \frac{1}{4}\frac{\partial \gamma_{\alpha\beta}}{\partial x_i}\frac{\partial g_{\alpha\beta}}{\partial x_l}$$

[p. 22 L]

$$\sum \frac{\partial \gamma_{\mu\nu}'}{\partial x_\nu'} = 0 \qquad |p_{\mu\nu}| = 1 \qquad [20]$$

$$\sum \pi_{\nu i} \frac{\partial}{\partial x_i} \{p_{\mu\alpha} p_{\nu\beta} \gamma_{\alpha\beta}\} = 0 \qquad \sum$$

$$= \sum p_{\mu\alpha} \frac{\partial \gamma_{\alpha i}}{\partial x_i} + \underbrace{\sum \gamma_{\alpha\beta} \pi_{\nu i} \frac{\partial p_{\mu\alpha} p_{\nu\beta}}{\partial x_i}}$$

$$\sum \gamma_{\alpha\beta} \pi_{\underline{\nu} i} \left\{ p_{\mu\alpha} \frac{\partial p_{\nu\beta}}{\partial x_i} + p_{\underline{\nu}\beta} \frac{\partial p_{\mu\alpha}}{\partial x_i} \right\}$$

$$= \underline{\sum \gamma_{\alpha i} \frac{\partial p_{\mu\alpha}}{\partial x_i}} + \sum \gamma_{\alpha\beta} \pi_{\nu i} p_{\mu\alpha} \frac{\partial p_{\nu\beta}}{\partial x_i}$$

verschwindet, wenn Funkt. Det. = 1.

$$= \sum \frac{\partial}{\partial x_i} (\gamma_{\alpha i} p_{\mu\alpha}) - p_{\mu\alpha} \frac{\partial \gamma_{\alpha i}}{\partial x_i} + \frac{\partial}{\partial x_i} (\gamma_{\alpha i} p_{\mu\alpha}) - \gamma_{\alpha\beta} p_{\mu\alpha} p_{\nu\beta} \frac{\partial \pi_{\nu i}}{\partial x_i}$$

$$p_{\mu\alpha} \frac{\partial \gamma_{\alpha i}}{\partial x_i} = 0 \qquad -\frac{\partial}{\partial x_i} (\gamma_{\alpha i} p_{\mu\alpha})$$

$$\sum \gamma_{\alpha i} \frac{\partial p_{\mu\alpha}}{\partial x_i} + \gamma_{\alpha\beta} p_{\mu\alpha} \pi_{\nu i} \frac{\partial p_{\nu\beta}}{\partial x_i}$$

---

[21]

$$\sum \gamma_{\kappa l} \left\{ \frac{\partial^2 g_{\kappa i}}{\partial x_l \partial x_m} + \frac{\partial^2 g_{\kappa m}}{\partial x_l \partial x_i} \right\}$$

$$= - \cdot + \frac{\partial}{\partial x_m} \sum \gamma_{\kappa l} \frac{\partial g_{\kappa i}}{\partial x_l}$$

Genugt, wenn $\sum \frac{\partial \gamma_{\kappa l}}{\partial x_l}$ verschwindet.

$$\sum_{lm} \gamma_{lm} \underline{T_{i\kappa lm}}$$

$g_{i\kappa}$

$\gamma_{i\kappa}$

$$\underline{\frac{\partial G}{\partial x_i \partial x_\kappa}}$$

1.3 Commentary

The Zurich notebook contains 84 pages of notes by Einstein on various topics. Most pages (57) are on gravity. For brief characterizations of their full scope, see CPAE4, p. 202, and *Genesis* (Renn 2007a, Vol. 1, pp. 314–315). The notebook can be dated to 1912–13, as all entries are related to Einstein's research interests and teaching duties during the first part of his tenure at the ETH in Zurich. This is also why it has become known as the Zurich notebook.

The notebook is reproduced in full—in facsimile and in transcription—in *Genesis*, Vol. 1, pp. 316–487. High-resolution scans are available to the public at *Einstein Archives Online* (EAO). Its designation in the Einstein Archive is EA 3 006 (cf. Pt. I, Ch. 1, notes 2 and 15). The part dealing with gravity is presented in transcription in CPAE4, Doc. 10. Unfortunately, the pages of the notebook are numbered differently in these three sources (see Pt. I, Ch. 1, note 9). We use the page numbers used in *Genesis* with the EAO page numbers in parentheses. At the beginning of our annotation of each of the six pages of the notebook selected for inclusion in this volume we list all three page numbers.

For further analysis of the notes on gravity in the notebook, see (a) CPAE4, the editorial note, "Einstein's research notes on a generalized theory of relativity" (pp. 192–199) and the annotation of Doc. 10 (pp. 201–269), and (b) *Genesis*, Vol. 2, pp. 489–714, "A commentary on the notes on gravity in the Zurich notebook," which reiterates, extends, and in a few places corrects the analysis in CPAE4. The annotation in this section is based on the latter, which will be referred to in the remainder of this chapter simply as the *Commentary*. A road map to the research on gravity recorded in the notebook is provided by the overall introduction of the *Commentary* (sec. 1, pp. 492–501) and the introductions for the two major groups of pages into which the *Commentary* divides the notes on gravity (sec. 4.1, pp. 523–526, and secs. 5.1–5.2, pp. 603–610). The six pages presented here are from the second group, documenting two of Einstein's attempts to extract gravitational field equations from the Riemann curvature tensor following the "mathematical strategy" (see Pt. I, Sec. 3.1). After both these and at least one other attempt failed, he returned to the "physical strategy," which led him to the field equations Einstein and Grossmann (1913) published in their joint *Entwurf* paper (see Pt. I, Sec. 3.2).

[P. 14L] = p. 27 in CPAE = p. 28 in EAO.

See Commentary, *sec. 5.3.1, pp. 610–614.*

On p. 14L, Einstein first considered the Ricci tensor as a candidate for the left-hand side of the gravitational field equations. He noted that it contains four terms with second-order derivatives of the metric whereas the Poisson equation of Newtonian theory only contains one such term.

1. The quantity $(i\kappa, lm)$ (modern notation: $R_{\rho\mu\sigma\nu}$) is the fully covariant form of the Riemann curvature tensor (both Latin and Greek indices run from 1 to 4, where 1–3 are the spatial indices and 4 is the temporal index). It is called a "tensor of the fourth multiplicity [*Mannigfaltigkeit*]." In the notebook, the modern term "rank" occurs only on p. 22R (see note 17). This is the first appearance of the Riemann tensor in the notebook. Presumably, Grossmann, whose name appears right next to it, introduced Einstein to it. The same expression—down to the labeling of almost all indices—can be found in Grossmann's part of the *Entwurf* paper (p. 35, Eq. 43). In his first paper of November 1915, Einstein (1915a, p. 781, Eq. 10, see Ch. 6) first introduced the Riemann tensor in this form but then referred the reader to the introduction of the

Riemann tensor in his 1914 review article on the *Entwurf* theory (Einstein 1914c, p. 1053). (In his 1916 review article, Einstein (1916b, 799–800) followed the 1914 review article on this point.)

The fully covariant Riemann tensor is written as the sum of four terms with second-order derivatives of the metric and two terms with products of first-order derivatives. The quantities $\gamma_{\rho\sigma}$ (modern notation: $g^{\mu\nu}$) and $\begin{bmatrix} im \\ \sigma \end{bmatrix}$ (modern notation: $[\mu\nu,\alpha]$) in these two terms are the contravariant components of the metric and the Christoffel symbols of the first kind, respectively. The latter are defined at the top of the page. In modern notation: $[\mu\nu,\alpha] \equiv \frac{1}{2}(g_{\alpha\mu,\nu} + g_{\alpha\nu,\mu} - g_{\mu\nu,\alpha})$, where commas indicate coordinate derivatives.

In the notebook, Einstein distinguished between covariant and contravariant tensor components by using Latin letters for one and Greek letters for the other. He only started using "downstairs" indices for covariant and "upstairs" indices for contravariant components in the 1914 review article (Einstein 1914c). In the notebook, Latin tend to be used for contravariant components and Greek for covariant ones. In the *Entwurf* paper and other early publications on the *Entwurf* theory it is just the other way around (*Commentary*, p. 521, note 67; p. 536, note 95). Both in the notebook and in these early publications, however, $g_{\mu\nu}$ and $g^{\mu\nu}$ are written as $g_{\mu\nu}$ and $\gamma_{\mu\nu}$, respectively.

It was not until the 1916 review article that Einstein (1916b, p. 781, see Sec. 9.3, note 1) officially introduced the summation convention (i.e., repeated indices—one covariant, one contravariant—should always be summed over). In the Zurich notebook, Einstein occasionally omitted summation signs, thus implicitly adopting the summation convention. In our annotation, we will omit all summation signs.

2. The quantity $\gamma_{\kappa l}(i\kappa, lm)$ is the Ricci tensor, the contraction of the Riemann tensor over two of its indices. In modern notation: $R_{\mu\nu} \equiv R^{\sigma}{}_{\mu\sigma\nu} = g^{\sigma\rho} R_{\rho\mu\sigma\nu}$.

   Einstein considered field equations of the form (in modern notation) $R_{\mu\nu} \propto -T_{\mu\nu}$, where $T_{\mu\nu}$ is the (covariant) energy-momentum tensor. For weak static fields in a Lorentz frame, $g_{\mu\nu}$ takes the form $\eta_{\mu\nu} + \delta g_{\mu\nu}$, where $\eta_{\mu\nu} \equiv \mathrm{diag}(-1,-1,\ -1,1)$ is the standard diagonal Minkowski metric and $\delta g_{\mu\nu}$ depends only on the spatial coordinates and has absolute values much smaller than 1. In that case, $\gamma_{\kappa l}\, g_{im,\kappa l}$, the first term in Einstein's expression for the Ricci tensor, reduces to $-\Delta g_{\mu\nu}$, where $\Delta \equiv \partial^2/\partial x^2 + \partial^2/\partial y^2 + \partial^2/\partial z^2$ is the Laplacian. Einstein assumed at this point that in weak static fields only the 44-component of the metric differs from its constant Minkowskian value (see Pt. I, Ch. 3). For matter at rest, the only non-vanishing component of the energy-momentum tensor is the 44-component: $T_{44} \propto \rho$ (where ρ is the mass density). If the Ricci tensor were to reduce to $\gamma_{\kappa l}\, g_{im,\kappa l}$ for weak static fields, the only non-trivial component of the field equations $R_{\mu\nu} \propto -T_{\mu\nu}$ would thus be $\Delta g_{44} \propto \rho$. These field equations would then be compatible with both Newton's theory and Einstein's (1912a, 1912b) own theory for static gravitational fields, in which a variable velocity of light—which turns into the 44-component of the metric in the new theory—plays the role of the gravitational potential.

   Alas, the Ricci tensor gives four terms with second-order derivatives of the metric, not just one (the terms with products of first-order derivatives of the metric can all be neglected for weak fields). The remaining three terms with second-order derivatives are listed at the bottom of the page. "Should vanish," Einstein wrote underneath them.

3. On the remainder of this page, Einstein rewrote the terms in the Ricci tensor $\gamma_{\kappa l}(i\kappa, lm)$ with products of first-order derivatives. In the course of this, he introduced Christoffel symbols of the second kind, which are obtained through contraction of the contravariant metric with Christoffel symbols of the first kind. In Einstein's notation: $\begin{Bmatrix} im \\ \rho \end{Bmatrix} \equiv \gamma_{i\sigma} \begin{bmatrix} \sigma m \\ \rho \end{bmatrix}$. In modern notation: $\begin{Bmatrix} \alpha \\ \mu\nu \end{Bmatrix} \equiv g^{\alpha\beta}[\beta\mu,\nu]$. He used G to denote

the determinant of the metric (modern notation: g or $-g$). Earlier in the notebook (on p. 6L), he had derived the standard relation: $\gamma_{\kappa l}\frac{\partial g_{\kappa l}}{\partial x_\rho} = \frac{\partial \lg G}{\partial x_\rho}$ (see *Commentary*, sec. 4.2, p. 527). In modern notation: $g^{\mu\nu}\frac{\partial g_{\mu\nu}}{\partial x^\rho} = \frac{\partial \ln(-g)}{\partial x^\rho}$. Another relation he used is (in modern notation) $\left\{\begin{matrix}\mu \\ \mu\,\nu\end{matrix}\right\} = \frac{1}{2}\frac{\partial \ln(-g)}{\partial x_\nu}$. In the 1916 review article, Einstein (1916b, p. 796) included an elementary derivation of this relation.

[P. 19L] = p. 37 in CPAE = p. 38 in EAO.

See Commentary, *sec. 5.4.1, pp. 623–626.*

On p. 19L, Einstein showed that three of the four terms in the Ricci tensor that contain second-order derivatives of the metric can be eliminated with the help of the harmonic coordinate condition/restriction.

4. As indicated by the header, "Renewed calculation of the plane tensor (*Ebenentensor*)," Einstein once again considered the fully covariant Riemann tensor $(i\kappa,lm)$ and the Ricci tensor $\gamma_{\kappa l}(i\kappa,lm)$ (cf. p. 14L and notes 1 and 2).

   On p. 17L of the notebook (*Commentary*, p. 618), Einstein used the term "plane tensor of the fourth multiplicity" for $(i\kappa,lm)$ and called the contravariant form of the Ricci tensor, $\gamma_{ip}\gamma_{mq}\gamma_{\kappa l}(i\kappa,lm)$ (in modern notation: $R^{\alpha\beta} = g^{\alpha\mu}g^{\beta\nu}g^{\sigma\rho}R_{\rho\mu\sigma\nu}$), a "point tensor" (*Punkttensor*). On p. 13L (*Commentary*, sec. 4.6.1), he introduced the designations "plane" and "point" for covariant and contravariant indices, respectively. He never used these terms in print.

   The first line of the expression for $(i\kappa,lm)$ given here is identical to the one at the top of p. 14L. The second line can be rewritten as (cf. the definition of the Christoffel symbols of the first kind in note 1):

$$-\gamma_{\rho\sigma}\begin{bmatrix} il \\ \rho \end{bmatrix}\begin{bmatrix} \kappa m \\ \sigma \end{bmatrix} = -\gamma_{\rho\sigma}\begin{bmatrix} il \\ \sigma \end{bmatrix}\begin{bmatrix} \kappa m \\ \rho \end{bmatrix}, \tag{1.1}$$

   where we switched the summation indices ρ and σ and used that $\gamma_{\sigma\rho} = \gamma_{\rho\sigma}$. Comparing this expression to the one on the second line of the expression for $(i\kappa,lm)$ on p. 14L, one notes that there is a term missing in the expression for $(i\kappa,lm)$ on p. 19L:

$$\gamma_{\rho\sigma}\begin{bmatrix} im \\ \sigma \end{bmatrix}\begin{bmatrix} \kappa l \\ \rho \end{bmatrix}. \tag{1.2}$$

   However, as becomes clear two lines further down, Einstein imposed what today is known as the harmonic condition

$$\gamma_{\kappa l}\begin{bmatrix} \kappa l \\ i \end{bmatrix} = 0, \tag{1.3}$$

   which ensures that this missing term does not contribute to the Ricci tensor:

$$\gamma_{\kappa l}\gamma_{\rho\sigma}\begin{bmatrix} im \\ \sigma \end{bmatrix}\begin{bmatrix} \kappa l \\ \rho \end{bmatrix} = 0. \tag{1.4}$$

   As he indicated by drawing a vertical line to the right of the expression for the Riemann tensor $(i\kappa,lm)$ and writing $\gamma_{\kappa l}$ next to it, Einstein was interested in the Ricci tensor $\gamma_{\kappa l}(i\kappa,lm)$ at this point rather than in the Riemann tensor.

5. The harmonic condition (see note 4) allowed Einstein to eliminate three of the four terms in the expression for the Ricci tensor with second-order derivatives of the metric. "Remains," Einstein wrote next to the one term that is left: $\frac{1}{2}\gamma_{\kappa l}\frac{\partial^2 g_{im}}{\partial x_\kappa \partial x_l}$. This term reduces to $-\Delta g_{im}$ for weak static fields, which means that, in conjunction with the harmonic condition and Einstein's assumption about the form of the metric for weak static fields, the field equations $R_{\mu\nu} \propto -T_{\mu\nu}$ would be compatible with both Newton's theory and Einstein's own 1912 theory for static fields (see note 2).
6. To show that the harmonic condition can indeed be used to eliminate all but one of the terms with second-order derivatives in the Ricci tensor, Einstein rewrote the equation

$$\frac{\partial}{\partial x_m}\left(\gamma_{kl}\begin{bmatrix}kl\\ i\end{bmatrix}\right)+\frac{\partial}{\partial x_i}\left(\gamma_{kl}\begin{bmatrix}kl\\ m\end{bmatrix}\right)=0, \tag{1.5}$$

which holds on account of the harmonic condition, as:

$$-\frac{1}{2}\gamma_{\kappa l}\left(-\frac{\partial^2 g_{\kappa l}}{\partial x_i \partial x_m}+\frac{\partial^2 g_{il}}{\partial x_\kappa \partial x_m}+\frac{\partial^2 g_{\kappa m}}{\partial x_i \partial x_l}\right)$$

$$=\frac{1}{4}\left(\frac{\partial \gamma_{\kappa l}}{\partial x_m}\left(2\frac{\partial g_{il}}{\partial x_\kappa}-\frac{\partial g_{\kappa l}}{\partial x_i}\right)+\frac{\partial \gamma_{\kappa l}}{\partial x_i}\left(2\frac{\partial g_{m\kappa}}{\partial x_l}-\frac{\partial g_{\kappa l}}{\partial x_m}\right)\right). \tag{1.6}$$

In the notebook there is an error in the very last term: instead of x_m it has x_i. We also reordered the three terms in parentheses on the left-hand side to make it easier to compare them to the corresponding terms in the expression for $(i\kappa, lm)$ at the top of p. 19L. The equation above shows that these three terms with second-order derivatives of the metric can be replaced by a sum of terms consisting of products of two first-order derivatives of the metric. On the next few lines, Einstein wrote out the individual terms in this sum.
7. Einstein concluded that, when the harmonic condition is applied, "[t]he plane tensor [i.e., the Ricci tensor] multiplied by 2 thus takes the form" of one term with second-order derivatives of the metric, which reduces to the Laplacian acting on the metric for weak static fields, and a sum of terms consisting of products of two first-order derivatives of the metric, which can all be neglected in weak fields.
8. Einstein noted that the "result [i.e., the expression for the Ricci tensor he found by applying the harmonic condition] [is] certain. Holds for coordinates that satisfy the condition $\Delta\varphi = 0$." Such coordinates were known as "isothermal coordinates" in the contemporary literature. Today they are called "harmonic coordinates." The equation $\Delta\varphi = 0$ used to define these coordinates suggests that Einstein—or Grossmann—found these coordinates in a textbook on differential geometry such as Bianchi (1910, secs. 36–37) or Wright (1908, sec. 57). On p. 6L (p. 12) of the notebook, immediately following expressions for the so-called first and second Beltrami invariants, which can also be found in the books by Bianchi (secs. 22–24) and Wright (sec. 53), Einstein already wrote down the harmonic condition in a different form (in modern notation): $(\sqrt{-g}\,g^{\mu\nu})_{,\nu} = 0$ (*Commentary*, pp. 526–527).

   When p. 19L is considered in isolation, it looks as if Einstein simply used the harmonic condition as a modern coordinate condition, which is what Norton (1984, p. 115) concluded in his pioneering study of the Zurich notebook. In Renn (2007a, Vols. 1–2), however, it is argued that when p. 19L is considered in combination with several other pages (such as 22L-R and 23L-R) it is clear that Einstein's use of the harmonic condition on p. 19L is an example of him using what in *Genesis* and in this volume is called a *coordinate restriction* (see Pt. I, Ch. 3, and *Commentary*, p. 605, note 199, and sec. 5.5.4, specially the concluding remarks on pp. 658–659 about p. 23L).

[P. 19R] = p. 38 in CPAE = p. 39 in EAO.

See Commentary, *sec. 5.4.2, pp. 626–632.*

On p. 19R, Einstein ran into a problem with the field equations based on the Ricci tensor introduced on p. 19L. He imposed a linearized version of what we call the Hertz condition/restriction to ensure that these field equations are compatible with energy-momentum conservation, at least in the special case of pressureless dust in a weak field. The combination of the linearized Hertz condition/restriction and a linearized version of harmonic coordinate condition/restriction (imposed on p. 19L to ensure that these field equations reduce to the Poisson equation in the case of weak static fields) leads to the problematic condition that the trace of the metric for weak fields must be a constant.

9. Under the header, "For the first order our additional condition is," Einstein wrote down a linearized version of the harmonic condition/restriction introduced on p. 19L (see note 4). His notation is awkward. He used the same index for different summations and did not distinguish between the metric components $g_{\mu\nu}$ and their small deviations $\delta g_{\mu\nu}$ from the Minkowskian values $\eta_{\mu\nu}$ (cf. note 2). Further down on the page, Einstein introduced an imaginary time coordinate $u = ict$, in which case $\eta_{\mu\nu}$ is just the Kronecker delta, $\delta_{\mu\nu}$ (with $\delta_{\mu\nu} = 1$ if $\mu = \nu$ and $\delta_{\mu\nu} = 0$ if $\mu \neq \nu$). Using this imaginary time coordinate and modifying Einstein's notation to clarify matters, we can write the linearized harmonic condition as

$$\delta^{\kappa\kappa'}\begin{bmatrix}\kappa\,\kappa'\\ i\end{bmatrix} = \delta^{\kappa\kappa'}\left(\delta g_{i\kappa,\kappa'} + \delta g_{i\kappa',\kappa} - \delta g_{\kappa\kappa',i}\right) = \delta^{\kappa\kappa'}\left(2\delta g_{i\kappa,\kappa'} - \delta g_{\kappa\kappa',i}\right). \tag{1.7}$$

Einstein conjectured that this condition "splits *perhaps* into" (in our notation)

$$\delta^{\kappa\kappa'}\delta g_{i\kappa,\kappa'} = 0 \quad \text{and} \quad \delta^{\kappa\kappa'} g_{\kappa\kappa'} = \text{constant}. \tag{1.8}$$

The linearized harmonic condition/restriction follows from the conjunction of these two conditions. The latter says that the trace of the metric is constant. The former is the linearized form of the condition $\gamma_{\mu\nu,\nu} = 0$ (modern notation: $g^{\mu\nu}{}_{,\nu} = 0$), which we called the *Hertz condition/restriction* (Pt. I, Sec. 3.1, see especially note 8; *Commentary*, p. 524). It takes only a few lines to show this. Since $\gamma_{\kappa\kappa'} g_{i\kappa} = \delta_{\kappa' i}$ is a constant, it follows that

$$0 = (\gamma_{\kappa\kappa'} g_{i\kappa})_{,\kappa'} = \delta_{i\kappa}\,\delta\gamma_{\kappa\kappa',\kappa'} + \delta^{\kappa\kappa'}\delta g_{i\kappa,\kappa'}. \tag{1.9}$$

Hence, $\delta^{\kappa\kappa'}\delta g_{i\kappa,\kappa'} = 0$, the first of the two conditions above, is equivalent to the linearized Hertz condition/restriction, $\delta\gamma_{i\kappa,\kappa} = 0$. Einstein had introduced and studied the Hertz condition/restriction before, on pp. 10L–11L of the notebook (see *Commentary*, sec. 4.5). On p. 19R, as we will see below, he used it to ensure that the linearized version of his candidate field equations is compatible with energy-momentum conservation.

10. In our notation (see note 9), the weak-field equations are $\delta^{\kappa\kappa'}\dfrac{\partial^2\delta g_{im}}{\partial x^\kappa \partial x^{\kappa'}} = K\,T_{im}$, where $T_{im} = \delta_{ii'}\delta_{mm'}T^{i'm'}$ is the covariant form of $T^{im} = \rho_0 \dfrac{dx^i}{ds}\dfrac{dx^m}{ds}$, the contravariant energy-momentum tensor for pressureless dust. In first-order approximation (i.e., weak fields and slow-moving dust particles), the gravitational four-force density on these dust particles is given by $T_{im}\delta\gamma_{im,\sigma}$ (cf. Pt. I, Sec. 3.2, especially Eq. (3.8) and note 15). Inserting the left-hand side of the linearized field equations for T_{im}, we can write (K times) this four-force density as

$$\delta^{\kappa\kappa'}\frac{\partial^2\delta g_{im}}{\partial x^\kappa \partial x^{\kappa'}}\frac{\partial\,\delta\gamma_{im}}{\partial x^\sigma}. \tag{1.10}$$

This is the left-hand side of the equation on the next line on p. 19R (in our notation), except that the notebook erroneously has $g_{im,\sigma}$ instead of $\gamma_{im,\sigma}$. Einstein verified that this expression for the four-force density can be written as the divergence of a quantity representing the energy-momentum density of the gravitational field. This is how Einstein checked the compatibility of candidate field equations with energy-momentum conservation (cf. Pt. I, Sec. 3.2, Eqs. (3.7)–(3.9) and (3.12)). He concluded that "energy-momentum conservation holds in the relevant approximation."

11. Though Einstein wrote "uniqueness and additional conditions," the material on the remainder of p. 19R only bears on the acceptability of the three additional conditions on the metric field listed at the top of the page: the harmonic condition, the Hertz condition, and the trace condition. The third is a consequence of the combination of the first two. Einstein clearly saw the trace condition as problematic. On p. 20L he modified the weak-field equations to get around it (only to conclude that these modified equations are unacceptable for other reasons). The point of his calculations on the bottom half of p. 19R was to check whether the Hertz condition/restriction can be dropped so that the trace condition no longer follows. The Hertz condition/restriction could not be dropped. "*Both* conditions are to be maintained," Einstein concluded, the harmonic condition/restriction to ensure compatibility with Newtonian theory, the Hertz condition/restriction to ensure compatibility with energy-momentum conservation, both in the weak-field limit (Renn and Sauer 1999, sec. 3.2.2, pp. 109–111).

Einstein began the calculation that led him to this conclusion by once again writing down the weak-field equations for the special case of pressureless dust. In our notation (see notes 9 and 10):

$$\Box\,\delta\gamma_{im} = K\rho_0\,\mathfrak{w}^i\,\mathfrak{w}^m, \tag{1.11}$$

where $\Box \equiv \Delta + \partial^2/\partial u^2$ and $\mathfrak{w}^i \equiv dx^i/d\tau$. Note that the notebook erroneously has g_{im} instead γ_{im} and dx_τ instead of $d\tau$ (which refers to an infinitesimal amount of proper time just as ds refers to an infinitesimal amount of proper length). The linearized Hertz condition/restriction, $\delta\gamma_{im,m} = 0$, ensures that these weak-field equations imply that the ordinary four-divergence of the energy-momentum tensor vanishes, which expresses energy-momentum conservation in the first-order approximation considered here:

$$KT^{im}{}_{,m} = K\left(\rho_0\,\mathfrak{w}^i\,\mathfrak{w}^m\right)_{,m} = \left(\Box\,\delta\gamma_{im}\right)_{,m} = \Box\,\delta\gamma_{im,m} = 0. \tag{1.12}$$

The question, as explained above, is whether one can do without the Hertz restriction and hence without a conservation law of the form $T^{\mu\nu}{}_{,\nu} = 0$ in this particular case (pressureless dust in a weak field). This is not an option. The conservation law follows directly from the conjunction of the continuity equation and the geodesic equation. In first-order approximation, energy-momentum conservation for pressureless dust can be written as

$$\left(\rho_0\,\mathfrak{w}^i\,\mathfrak{w}^m\right)_{,m} = \rho_0\,\mathfrak{w}^i{}_{,m}\,\mathfrak{w}^m + \left(\rho_0\,\mathfrak{w}^m\right)_{,m}\mathfrak{w}^i. \tag{1.13}$$

The second term on the right-hand side vanishes on account of the continuity equation, $(\rho_0\,\mathfrak{w}^m)_{,m} = 0$. The expression $\mathfrak{w}^i{}_{,m}\,\mathfrak{w}^m$ in the first term can be rewritten as $\dfrac{\partial\mathfrak{w}^i}{\partial x^m}\dfrac{dx^m}{d\tau} = \dfrac{Dx^m}{D\tau}$, which vanishes on account of the geodesic equation.

[P. 20L] = p. 39 in CPAE = p. 40 in EAO.

See Commentary, *sec. 5.4.3, pp. 632–636.*

On p. 20L, Einstein addressed the problem he ran into on p. 19R. It follows from the conjunction of the linearized Hertz condition/restriction and the linearized harmonic condition/restriction that the trace of the metric for weak fields must be a constant.

This was unacceptable to Einstein for (at least) two reasons. First, through the weak-field equations, it would lead to the requirement that the energy-momentum tensor for pressureless dust be traceless (at least in first-order approximation), which clearly it is not. Second, it would not allow a metric of the form Einstein assumed for weak static fields. To avoid these problems, Einstein considered modifying the weak-field equations by adding a term with the trace of either the energy-momentum tensor or the metric. However, these modified equations do not allow a metric of the form Einstein assumed for weak static fields, which defeats the purpose of the modification.

12. Einstein once again wrote down the two conditions into which he tentatively split the linearized harmonic condition/restriction at the top of p. 19R (see note 9).
13. Through the linearized field equations, $\Box g_{i\kappa} = KT_{i\kappa}$, the condition that the trace of the metric for weak fields is constant forces the trace of the energy-momentum tensor, $T = g^{i\kappa}T_{i\kappa}$, to be zero in first-order approximation: $KT \approx \delta^{i\kappa}T_{i\kappa} = \delta^{i\kappa}\Box\delta g_{i\kappa} = \Box(\delta^{i\kappa}\delta g_{i\kappa})$. This last expression vanishes on account of the condition that the trace of the metric for weak fields is constant. The implication can be avoided if a term with the trace of the energy-momentum tensor is added to the weak-field equations:

$$\Box g_{i\kappa} = K(T_{i\kappa} - \tfrac{1}{4}\delta_{i\kappa}T). \tag{1.14}$$

In the notebook, this modification is given for the special case of pressureless dust, in which case $T^{i\kappa} = \rho_0 \dfrac{dx^i}{d\tau}\dfrac{dx^\kappa}{d\tau}$. Einstein erroneously used the contravariant rather than the covariant form of this energy-momentum tensor on the right-hand side; and instead of using the Kronecker delta, Einstein wrote underneath the second term on the right-hand side: "for the same i and κ." The contraction of $\delta^{i\kappa}$ with the right-hand side of these modified weak-field equations is identically zero (note that $\delta^{i\kappa}\delta_{i\kappa} = 4$). The condition on the trace of the metric for weak fields thus no longer results in a condition on the trace of the energy-momentum tensor. Einstein nonetheless crossed out these modified weak-field equations, probably because the modification does not solve the other problem with the condition that $\delta^{i\kappa}\delta g_{i\kappa}$ is a constant: it still rules out a metric of the form Einstein assumed for weak static fields.
14. Einstein once again wrote down the linearized harmonic condition/restriction given on p. 19R (see note 9). If a term with the trace of the metric, $U \equiv \delta^{\kappa\kappa'}g_{\kappa\kappa'}$, is added to the left-hand side of the weak-field equations, the linearized Hertz condition/restriction is no longer needed to guarantee compatibility with energy-momentum conservation in first-order approximation. The linearized harmonic condition/restriction will now serve this purpose (as we will verify in note 15). In that case, there also is no need anymore for the problematic condition that the trace of the metric for weak fields is a constant (Renn and Sauer 1999, sec. 3.2.3, pp. 111–114).
15. The components of weak-field equations with an extra term involving the trace of the metric are written down under the header "Gravitational equations" for the special case of static fields (so that the d'Alembertian $\Box$ reduces to the Laplacian Δ). A factor K is missing on the right-hand side. For arbitrary weak fields these modified weak-field equations can be written compactly as:

$$\Box(g_{im} - \tfrac{1}{2}\delta_{im}U) = KT_{im}. \tag{1.15}$$

Contracting both sides with δ^{im}, we find that $-\Box U = KT$ (in the notebook, $2\Delta U = T$ should be $-\Delta U = KT$). These new weak-field equations can thus also be written as

$$\Box g_{im} = K(T_{im} - \tfrac{1}{2}\delta_{im}T) \tag{1.16}$$

(in the notebook, the factor K is missing and the plus sign in the 11-component should be a minus sign).

In first-order approximation, energy-momentum conservation is expressed by $T^{im}{}_{,m} = 0$ (cf. note 11). If $\delta^{mm'} T_{im,m'} = 0$, then $T^{im}{}_{,m} = 0$ as well. That $\delta^{mm'} T_{im,m'} = 0$ follows directly from the modified weak-field equations and the linearized harmonic condition/restriction:

$$
\begin{aligned}
2K\,\delta^{mm'} T_{im,m'} &= 2\,\delta^{mm'} \left(\Box\left(g_{im} - \tfrac{1}{2}\,\delta_{im}\,U\right)\right)_{,m'} \\
&= \Box\left(2\,\delta^{mm'}\,\delta g_{im,m'} - \delta^{mm'}\,\delta_{im}\,\delta^{\kappa\kappa'}\,\delta g_{\kappa\kappa',m'}\right).
\end{aligned}
$$

Relabeling the summation indices in the first term on the second line, we can write it as: $2\,\delta^{\kappa\kappa'}\,\delta g_{i\kappa,\kappa'}$. The second term can be rewritten as:

$$
-\,\delta^{mm'}\,\delta_{im}\,\delta^{\kappa\kappa'}\,\delta g_{\kappa\kappa',m'} = -\delta_i^{m'}\,\delta^{\kappa\kappa'}\,\delta g_{\kappa\kappa',m'} = -\delta^{\kappa\kappa'}\,\delta g_{\kappa\kappa',i}. \tag{1.17}
$$

We thus arrive at:

$$
2K\,\delta^{mm'} T_{im,m'} = \Box\left(\delta^{\kappa\kappa'}\left(2\delta g_{i\kappa,\kappa'} - \delta g_{\kappa\kappa',i}\right)\right), \tag{1.18}
$$

which vanishes on account of the linearized harmonic condition/restriction (see note 9). Hence, $\delta^{mm'} T_{im,m'} = 0$ and $T^{im}{}_{,m} = 0$.

16. At the bottom of p. 20L, Einstein checked whether the weak-field equations with a term involving U, the trace of the metric, still allowed him to write the gravitational four-force density, $T_{im}\delta\,\gamma_{im\sigma}$, as the four-divergence of the energy-momentum density of the gravitational field (cf. note 10). He concluded that it "can be represented in the required form." He confirmed this conclusion in a more detailed calculation on p. 21L (*Commentary*, pp. 636–637).

Despite this encouraging result, Einstein did not pursue field equations with a trace term any further in the notebook. It was not until the last of his four November 1915 papers, that Einstein (1915d) once again included a trace term in the field equations (see Pt. I, Fig. 1.2 and Sec. 5.4, note 22).

The calculation on p. 21L suggests that one problem with the trace term was that the result that these modified weak-field equations still allowed him to express the gravitational four-force density as the four-divergence of gravitational energy-momentum density could not be made exact. Another problem was that these modified weak-field equations do not allow a metric of the form Einstein assumed for weak static fields. This is suggested by an argument on p. 21R with which Einstein reassured himself that the metric for weak static fields really cannot have any variable components besides the g_{44} component (see *Commentary*, pp. 640–642, and Pt. I, Sec. 5.3, note 12).

Yet another problem is that a metric of this form is not allowed by the harmonic coordinate condition/restriction. However, Einstein was probably unaware of this problem. He clearly was unaware of it when he introduced the harmonic coordinate condition/restriction on p. 19L, otherwise it is hard to see why he would have continued to investigate field equations extracted from the Ricci tensor with the help of this coordinate condition/restriction on pp. 19R–21R. There is no indication on those later pages either that he had come to recognize this problem. Problems with the metric for weak static fields were definitely an important part of the reason that Einstein abandoned field equations extracted from the Ricci tensor with the help of the harmonic condition/restriction, but the problem was *not*, as Norton (1984, p. 115, 119) conjectured, that the harmonic condition/restriction rules out a metric of the form Einstein assumed for weak static fields (see Pt. I, Sec. 5.3, note 11).

[P. 22R] = p. 44 in CPAE = p. 45 in EAO.

See Commentary, *sec. 5.5.1–5.5.2, pp. 647–650.*

On p. 22R, Einstein extracted new candidate field equations from the Ricci tensor. The Ricci tensor is split into two parts both of which transform as tensors under unimodular transformations. Restricting the allowed transformations to unimodular transformations, Einstein took one of these parts as his new candidate for the left-hand side of the field equations (see Pt. I, Ch. 3, note 7). We call it the November tensor *because Einstein (1915a) resurrected it in his first paper of November 1915 (see Pt. I, Figs. 1.1 and 1.3). Both in 1912–13 and in 1915, Einstein used the Hertz condition/restriction, the vanishing of the four-divergence of the contravariant metric, to eliminate unwanted terms with second-order derivatives of the metric from the November tensor. Einstein's investigation of the covariance properties of the Hertz condition/restriction on p. 22L of the notebook makes it clear that, at this point, he was using it as a coordinate* restriction *(see Pt. I, Sec. 3.1). It is unclear what its status was in 1915 (see Pt. I, Sec. 5.2). The Hertz condition/restriction has two advantages over the harmonic condition/restriction, which Einstein used on p. 19L to eliminate unwanted terms with second-order derivatives of the metric from the Ricci tensor. First, it allows a metric of the form Einstein assumed for weak static fields. Second, it sufficed all by itself to ensure compatibility of the field equations with energy-momentum conservation, at least in the weak-field approximation. Einstein thus avoided the problems he had run into on p. 19R.*

17. The appearance of his name at the top of p. 22R suggests that Grossmann handed Einstein the expression for the Ricci tensor given here, just as he had handed him the expression at the top of pp. 14L and 19L. On p. 19L, the Ricci was given in terms of the metric and its first- and second-order derivatives. Here it is given in terms Christoffel symbols of the second kind (see note 3) and their first-order derivatives (the summation over k and l should be over k and λ).

18. Einstein noted that "if G is a scalar, then $\dfrac{\partial \lg\sqrt{G}}{\partial x_i} = \mathscr{T}_i$ is a first-rank tensor." This is the only page in the notebook where Einstein used the term "rank" (*Rang*) of a tensor (*Commentary*, p. 646, note 279). On page 14L, he used the term "multiplicity" (*Mannigfaltigkeit*; see note 1). The determinant G of the metric transforms as a scalar under unimodular transformations, for which the determinant $|\partial x^\mu/\partial x^\nu| = 1$. The covariant derivative of $\mathscr{T}_i$, in modern notation, $\mathscr{T}_{i;l} = \mathscr{T}_{i,l} - \left\{ {\lambda \atop il} \right\} \mathscr{T}_\lambda$, thus transforms as a "second-rank tensor" under unimodular transformations. Since $\mathscr{T}_i \equiv \dfrac{\partial \ln\sqrt{-g}}{\partial x_i} = \left\{ {\kappa \atop \kappa i} \right\}$ (cf. note 3), $\mathscr{T}_{i;l}$ is given by the two underlined terms in the expression for the Ricci tensor $\mathscr{T}_{il}$ at the top of the page.

    Since the Ricci tensor is a generally covariant tensor and two of its four terms form a tensor under unimodular transformation, the remaining two terms must also form a tensor under unimodular transformation. This is the *November tensor*, Einstein's new candidate for the left-hand side of the field equations. As he wrote underneath it: "probable gravitation tensor $\mathscr{T}_{il}^{x}$."

19. Under the header "Further rewriting of the gravitation tensor," Einstein first used the Hertz restriction to eliminate unwanted terms with second-order derivatives of the metric from the November tensor and then, after drawing a horizontal line, rewrote the terms with products of first-order derivatives of the metric, which can be neglected in first-order approximation. We focus on terms with second-order derivatives (see *Commentary*, pp. 648–650, for discussion of Einstein's rewriting of first-order-derivative terms). The derivative of the Christoffel symbol $\left\{ {\kappa \atop il} \right\}$ in the

November tensor gives rise to three terms with second-order derivatives of the metric. Only one of these, $-\frac{1}{2}g^{\kappa\alpha}g_{il,\kappa\alpha}$, should appear in the field equations if these are to reduce to the Poisson equation of Newtonian theory for weak static fields (cf. p. 19L). The other two, $\frac{1}{2}g^{\kappa\alpha}g_{i\alpha,l\kappa}+\frac{1}{2}g^{\kappa\alpha}g_{l\alpha,i\kappa}$, should be eliminated.

On p. 22L, right below the horizontal line, Einstein explicitly wrote down these two terms except for the factor $\frac{1}{2}$ and with different labeling of the indices ($l\to m$, $\kappa\to l$, and $\alpha\to\kappa$, which gives $\frac{1}{2}g^{l\kappa}g_{i\kappa,ml}+\frac{1}{2}g^{l\kappa}g_{m\kappa,il}$). He convinced himself that with the help of the Hertz restriction, $g^{\mu\nu}{}_{,\nu}=0$, these terms could be turned into products of first-order derivatives of the metric and concluded: "suffices if $\sum\frac{\partial\gamma_{\kappa l}}{\partial x_l}$ vanishes."

On p. 22R, Einstein used the relation (in modern notation) $g^{\mu\alpha}g_{\alpha\nu,\beta}=-g^{\mu\alpha}{}_{,\beta}\,g_{\alpha\nu}$ (which follows directly from $0=\delta^{\mu}_{\nu,\beta}=(g^{\mu\alpha}g_{\alpha\nu})_{,\beta}$) to rewrite the two terms in the November tensor that contain unwanted second-order derivatives of the metric as products of first-order derivatives:

$$\tfrac{1}{2}(g^{\kappa\alpha}g_{i\alpha,l}+g^{\kappa\alpha}g_{l\alpha,i})_{,\kappa}=-\tfrac{1}{2}(g^{\kappa\alpha}{}_{,l}\,g_{i\alpha}+g^{\kappa\alpha}{}_{,i}\,g_{l\alpha})_{,\kappa}=-\tfrac{1}{2}(g^{\kappa\alpha}{}_{,l}\,g_{i\alpha,\kappa}+g^{\kappa\alpha}{}_{,i}\,g_{l\alpha,\kappa}). \quad (1.19)$$

The Hertz condition was used to set $g^{\kappa\alpha}{}_{,l\kappa}=g^{\kappa\alpha}{}_{,i\kappa}=0$. It looks as if Einstein initially forgot the factors $\frac{1}{2}$ and only inserted them later.

[P. 22L] = p. 43 in CPAE = p. 44 in EAO.

See Commentary, *sec. 5.5.3, pp. 650–652.*

On p. 22L, Einstein derived the condition for so-called non-autonomous unimodular transformations preserving the Hertz restriction. The field equations extracted from the November tensor with the help of the Hertz restriction will be invariant under this same group of transformations. The non-autonomous unimodular transformation to a rotating frame of reference in Minkowski space-time does not preserve the Hertz restriction, which is probably why Einstein did not pursue the field equations based on the November tensor and the Hertz restriction any further in the notebook.

20. At the top of p. 22L, Einstein wrote down the Hertz restriction in primed coordinates. Next to it, he wrote the condition that the determinant of $p_{\mu\nu}$—Einstein's notation for the transformation matrix $\partial x'^{\mu}/\partial x^{\nu}$ of the coordinate transformation $x^{\mu}\to x'^{\mu}$—is equal to 1. As this condition indicates, the goal of the calculation on the top half of p. 22L was to establish under which unimodular transformations the Hertz restriction is invariant. This makes it clear that Einstein was thinking in terms of a coordinate *restriction* rather than in terms of a coordinate *condition* at this point (cf. note 8).

As long as we limit ourselves to ordinary coordinate transformations, the Hertz restriction is invariant only under general linear transformations (unimodular and non-unimodular). In such coordinate transformations, the new coordinates are simply functions of the old ones [schematically: $x^{\mu}\to x'^{\mu}(x^{\rho})$]. However, if we also allow transformations in which the new coordinates are functions of the old coordinates *and the metric field in the old coordinates* [schematically: $x^{\mu}\to x'^{\mu}(x^{\rho},g_{\alpha\beta}(x^{\sigma}))$], the Hertz restriction will be invariant under a broader class of transformations. In a letter to Lorentz of August 14, 1913 (CPAE5, Doc. 467), Einstein introduced the terms 'autonomous' and 'non-autonomous' (*selbständig* and *unselbständig*) transformations to distinguish between these two kinds of transformations (*Commentary*, sec. 4.3, introduction, especially p. 535, note 94). In his second paper with Grossmann, he adopted the terms "justified transformations" (*berechtigte Transformationen*) between "adapted coordinate systems" (*angepaßte Koordinatensysteme*) for non-autonomous transformations, where 'adapted' means adapted to the metric field (Einstein and Grossmann 1914, p. 221).

On p. 10L (p. 20) of the notebook, Einstein had already derived the condition for infinitesimal non-autonomous transformations preserving the Hertz restriction (*Commentary*, sec. 4.5.1, see especially p. 565, note 146). On p. 11L (p. 22), he had checked whether infinitesimal transformations to rotating and uniformly accelerating frames in Minkowski space-time belong to this class of transformations. He had come to the conclusion that rotation does but uniform acceleration does not (*Commentary*, sec. 4.5.2).

Using modern notation but the same mix of Latin and Greek indices used on p. 22L, we can write the transformation law for the expression set to zero in the Hertz restriction as (cf. Pt. I, Sec. 3.1, note 8):

$$\frac{\partial g'^{\mu\nu}}{\partial x'^{\nu}} = \frac{\partial x^i}{\partial x'^{\nu}}\frac{\partial}{\partial x^i}\left(\frac{\partial x'^{\mu}}{\partial x^{\alpha}}\frac{\partial x'^{\nu}}{\partial x^{\beta}}g^{\alpha\beta}\right) = \frac{\partial x'^{\mu}}{\partial x^{\alpha}}\frac{\partial g^{\alpha\beta}}{\partial x^{\beta}} + \frac{\partial x^i}{\partial x'^{\nu}}\frac{\partial}{\partial x^i}\left(\frac{\partial x'^{\mu}}{\partial x^{\alpha}}\frac{\partial x'^{\nu}}{\partial x^{\beta}}\right)g^{\alpha\beta},$$

where we used that $\frac{\partial x^i}{\partial x'^{\nu}}\frac{\partial x'^{\mu}}{\partial x^{\alpha}}\frac{\partial x'^{\nu}}{\partial x^{\beta}} = \delta^i_{\beta}\frac{\partial x'^{\mu}}{\partial x^{\alpha}}$. If we rewrite these equations in Einstein's notation (with $p_{\mu\nu}$ for the transformation matrix $\frac{\partial x'^{\mu}}{\partial x^{\nu}}$, $\pi_{\mu\nu}$ for its inverse $\frac{\partial x^{\nu}}{\partial x'^{\mu}}$, $\gamma_{\mu\nu}$ for $g^{\mu\nu}$, and x_{μ} for x^{μ}) and relabel one of the summation indices, we arrive at the corresponding equations in the notebook:

$$0 = \frac{\partial\gamma'_{\mu\nu}}{\partial x'_{\nu}} = \pi_{\nu i}\frac{\partial}{\partial x_i}\left\{p_{\mu\alpha}\,p_{\nu\beta}\,\gamma_{\alpha\beta}\right\} = p_{\mu\alpha}\frac{\partial\gamma_{\alpha i}}{\partial x_i} + \pi_{\nu i}\frac{\partial}{\partial x_i}\left\{p_{\mu\alpha}\,p_{\nu\beta}\right\}\gamma_{\alpha\beta}.$$

A non-autonomous transformation thus preserves the Hertz restriction if the transformation matrix $p_{\mu\nu}$ and its inverse $\pi_{\mu\nu}$ satisfy

$$\gamma_{\alpha\beta}\,\pi_{\nu i}\left\{p_{\nu\beta}\frac{\partial p_{\mu\alpha}}{\partial x_i} + p_{\nu\beta}\frac{\partial p_{\mu\alpha}}{\partial x_i}\right\} = 0. \tag{1.20}$$

On the next few lines, Einstein tried to simplify this expression. On p. 10L, he had found that the term $\gamma_{\alpha\beta}\,\pi_{\nu i}\,p_{\nu\beta}\frac{\partial p_{\mu\alpha}}{\partial x_i}$ vanishes for *infinitesimal unimodular* transformations (*Commentary*, p. 567, note 148). This is probably why he wrote "vanishes if func[tional] det[terminant] = 1" next to this expression on p. 22L. However, the expression does not vanish for *finite unimodular* transformations.

Since its components are all constants, the metric $\eta_{\mu\nu} \equiv \text{diag}(-1,-1,-1,c^2)$ of Minkowski space-time in an arbitrary Lorentz frame with space-time coordinates $x^{\mu} \equiv (x,y,z,t)$ obviously satisfies both the Hertz restriction ($g^{\mu\nu}{}_{,\nu} = 0$) and the vacuum field equations extracted from the November tensor with the help of this coordinate restriction. However, as we will show below, the metric of Minkowski space-time in a rotating frame of reference—the *rotation metric* for short—only satisfies the Hertz restriction to first order in the angular velocity ω of the rotating frame with respect to a Lorentz frame. It follows that it is also only a solution of the vacuum field equations to first order in ω. This is true even though the November tensor vanishes exactly for the rotation metric, as a (non-autonomous) transformation to rotating coordinates in which the metric in the original coordinates is the standard diagonal Minkowski metric is unimodular. Hence, the field equations Einstein considered on pp. 22L–R are invariant under infinitesimal but not under finite non-autonomous transformations of this kind. This, in all likelihood, is why Einstein did not pursue these candidate field equations any further in the notebook. They were resurrected only in the first paper of November 1915. Einstein (1915c, p. 786, Eq. (22)) once again imposed the condition $g^{\mu\nu}{}_{,\nu} = 0$. The status of this condition in this 1915 paper is unclear (see Sec. 6.1.3, note 21).

In two places in the notebook—on pp. 12L-R (pp. 24–25) and on p. 42R (p. 70)—Einstein gave a simple derivation of the form of the rotation metric (*Commentary*, sec. 4.5.6, pp. 584–587, and sec. 5.5.7, pp. 668–669). We present this derivation in slightly modernized notation (Janssen 2012, p. 171). Consider a frame with pseudo-Cartesian coordinates $x^\mu = (x, y, z, t)$ rotating counterclockwise with angular velocity ω around the z-axis with respect to a Lorentz frame with pseudo-Cartesian coordinates $x'^\mu = (x', y', z', t')$. In the (primed) coordinates of the non-rotating frame, we have

$$ds^2 = \eta_{\mu\nu} dx'^\mu dx'^\nu = (c^2 - v'^2) dt'^2, \tag{1.21}$$

where $\vec{v}' \equiv \dot{\vec{x}}' \equiv \left(\frac{dx'}{dt}, \frac{dy'}{dt}, \frac{dz'}{dt} \right)$ is the velocity of a test particle with respect to the non-rotating frame. This velocity is the sum of $\vec{v} = \dot{\vec{x}}$, the particle's velocity with respect to the rotating frame, and $\vec{\omega} \times \vec{x}$ (where $\vec{\omega} \equiv (0, 0, \omega)$), the velocity of the position of the particle in the rotating frame with respect to the non-rotating frame:

$$\vec{v}' = \vec{v} + \vec{\omega} \times \vec{x} = (\dot{x} - \omega y, \dot{y} + \omega x, \dot{z}). \tag{1.22}$$

Substituting this expression for $\vec{v}'$ and dt for dt' into the expression for the line element above, we can write ds^2 in terms of the (unprimed) coordinates of the rotating frame:

$$ds^2 = (c^2 - \dot{x}^2 + 2\omega y \dot{x} - \omega^2 y^2 - \dot{y}^2 - 2\omega x \dot{y} - \omega^2 x^2 - \dot{z}^2) dt^2. \tag{1.23}$$

Regrouping terms, we find:

$$ds^2 = -dx^2 - dy^2 - dz^2 + 2\,\omega\, y\, dx\, c dt - 2\,\omega\, x\, dy\, c dt + \left(c^2 - \omega^2 \left(x^2 + y^2\right)\right) dt^2. \tag{1.24}$$

Using $ds^2 = g_{\mu\nu}\, dx^\mu dx^\nu$, we can read off the components $g_{\mu\nu}$ of the rotation metric.

$$g^{\text{rot}}_{\mu\nu} = \begin{pmatrix} -1 & 0 & 0 & \omega\, y \\ 0 & -1 & 0 & -\omega\, x \\ 0 & 0 & -1 & 0 \\ \omega\, y & -\omega\, x & 0 & c^2 - \omega^2 \left(x^2 + y^2\right) \end{pmatrix}. \tag{1.25}$$

Einstein repeatedly made the mistake of setting $g_{14} = g_{41} = 2\,\omega\, y$ and $g_{24} = g_{42} = -2\,\omega\, x$ (Janssen 1999, pp. 145–146). We find this mistake not only on p. 12R and p. 42R of the Zurich notebook, but also on [p. 41] of the Einstein-Besso manuscript (see Pt. II, Ch. 2).

Inverting the matrix above, we find the contravariant form of the rotation metric:

$$g^{\mu\nu}_{\text{rot}} = \begin{pmatrix} -1 + \frac{\omega^2}{c^2} y^2 & -\frac{\omega^2}{c^2} xy & 0 & \frac{\omega}{c^2} y \\ -\frac{\omega^2}{c^2} xy & -1 + \frac{\omega^2}{c^2} x^2 & 0 & -\frac{\omega}{c^2} x \\ 0 & 0 & -1 & 0 \\ \frac{\omega}{c^2} y & -\frac{\omega}{c^2} x & 0 & \frac{1}{c^2} \end{pmatrix}. \tag{1.26}$$

One readily verifies that $g^{\mu\rho}_{\rm rot}\, g^{\rm rot}_{\rho\nu} = \delta^{\mu}_{\nu}$.

As long as terms of order ω^2/c^2 are neglected, the rotation metric satisfies the Hertz restriction, $g^{\mu\nu}{}_{,\nu} = 0$. The field equations extracted from the November tensor are thus invariant under *infinitesimal* non-autonomous transformations to rotating coordinates in which the metric in the original coordinates is the standard diagonal Minkowski metric. This is not true for *finite* rotations of this kind. The rotation metric does not satisfy the Hertz restriction exactly:

$$g^{\mu\nu}_{{\rm rot},\nu} = \left(-\frac{\omega^2}{c^2}\,x, -\frac{\omega^2}{c^2}\,y, 0, 0\right). \tag{1.27}$$

21. The material on the bottom half of p. 22L, below the horizontal line, appears to be an early attempt to use the Hertz restriction to eliminate unwanted terms with second-order derivatives of the metric from the November tensor (see note 19 and *Commentary*, p. 648).

Chapter 2
The Einstein-Besso Manuscript on the Perihelion Motion of Mercury

M. Janssen, J. Renn, *How Einstein Found His Field Equations*, Classic Texts in the Sciences, https://doi.org/10.1007/978-3-030-97955-3_9

2.1 Facsimile

[p. 1] Einstein

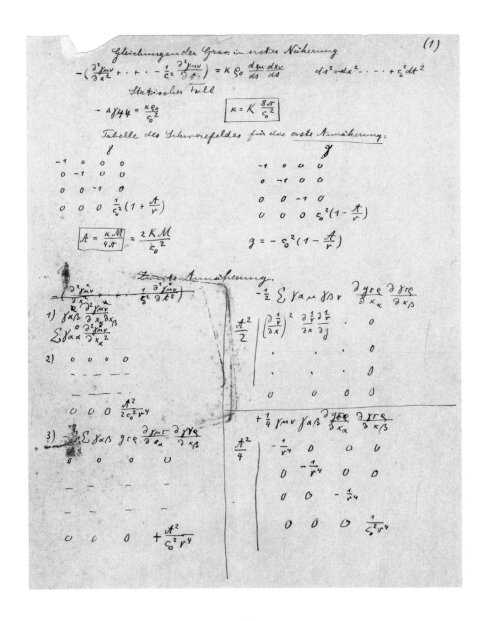

(1)

Gleichungen der Grav. in erster Näherung

$$-\left(\frac{\partial^2\gamma_{\mu\nu}}{\partial x^2}+\cdot+\cdot-\frac{1}{c^2}\frac{\partial^2\gamma_{\mu\nu}}{\partial t^2}\right)=\kappa\rho_0\frac{dx_\mu}{ds}\frac{dx_\nu}{ds}\qquad ds^2=-dx^2-\cdots+c_0^2dt^2$$

Statischer Fall

$$-\Delta\gamma_{44}=\frac{\kappa\rho_0}{c_0^2}\qquad \boxed{\kappa=K\frac{8\pi}{c_0^2}}$$

Tabelle des Schwerefeldes für die erste Annäherung:

γ

$$\begin{matrix}-1&0&0&0\\0&-1&0&0\\0&0&-1&0\\0&0&0&\frac{1}{c_0^2}\left(1+\frac{A}{r}\right)\end{matrix}$$

g

$$\begin{matrix}-1&0&0&0\\0&-1&0&0\\0&0&-1&0\\0&0&0&c_0^2\left(1-\frac{A}{r}\right)\end{matrix}$$

$$\boxed{A=\frac{\kappa M}{4\pi}}=\frac{2KM}{c_0^2}\qquad g=-c_0^2\left(1-\frac{A}{r}\right)$$

~~Zweite~~ Annäherung.

$$-\left(\frac{\partial^2\gamma_{\mu\nu}}{\partial x^2}+\cdot+\cdot\frac{1}{c_0^2}\frac{\partial^2\gamma_{\mu\nu}}{\partial t^2}\right)$$

1) $\gamma_{\alpha\beta}\frac{\partial^2\gamma_{\mu\nu}}{\partial x_\alpha\partial x_\beta}$

$\sum\gamma_{\alpha\alpha}\frac{\partial^2\gamma_{\mu\nu}}{\partial x_\alpha^2}$

2)
$$\begin{matrix}0&0&0&0\\-&-&-&-\\-&-&-&-\\0&0&0&\frac{A^2}{2c_0^2r^4}\end{matrix}$$

3) $-\frac{1}{2}\sum\gamma_{\alpha\beta}\,g_{\tau\rho}\frac{\partial\gamma_{\mu\tau}}{\partial x_\alpha}\frac{\partial\gamma_{\nu\rho}}{\partial x_\beta}$

$$\begin{matrix}0&0&0&0\\-&-&-&-\\-&-&-&-\\0&0&0&+\frac{A^2}{c_0^2r^4}\end{matrix}$$

$$-\frac{1}{2}\sum\gamma_{\alpha\mu}\gamma_{\beta\nu}\frac{\partial g_{\tau\rho}}{\partial x_\alpha}\frac{\partial\gamma_{\tau\rho}}{\partial x_\beta}$$

$$\frac{A^2}{2}\left|\begin{matrix}\left(\frac{\partial\frac{1}{r}}{\partial x}\right)^2&\frac{\partial\frac{1}{r}}{\partial x}\frac{\partial\frac{1}{r}}{\partial y}&\cdot&0\\\cdot&\cdot&\cdot&0\\\cdot&\cdot&\cdot&0\\0&0&0&0\end{matrix}\right.$$

$$+\frac{1}{4}\gamma_{\mu\nu}\gamma_{\alpha\beta}\frac{\partial g_{\tau\rho}}{\partial x_\alpha}\frac{\partial\gamma_{\tau\rho}}{\partial x_\beta}$$

$$\frac{A^2}{4}\left|\begin{matrix}-\frac{1}{r^4}&0&0&0\\0&-\frac{1}{r^4}&0&0\\0&0&-\frac{1}{r^4}&\\0&0&0&\frac{1}{c_0^2r^4}\end{matrix}\right.$$

[p. 7] Einstein

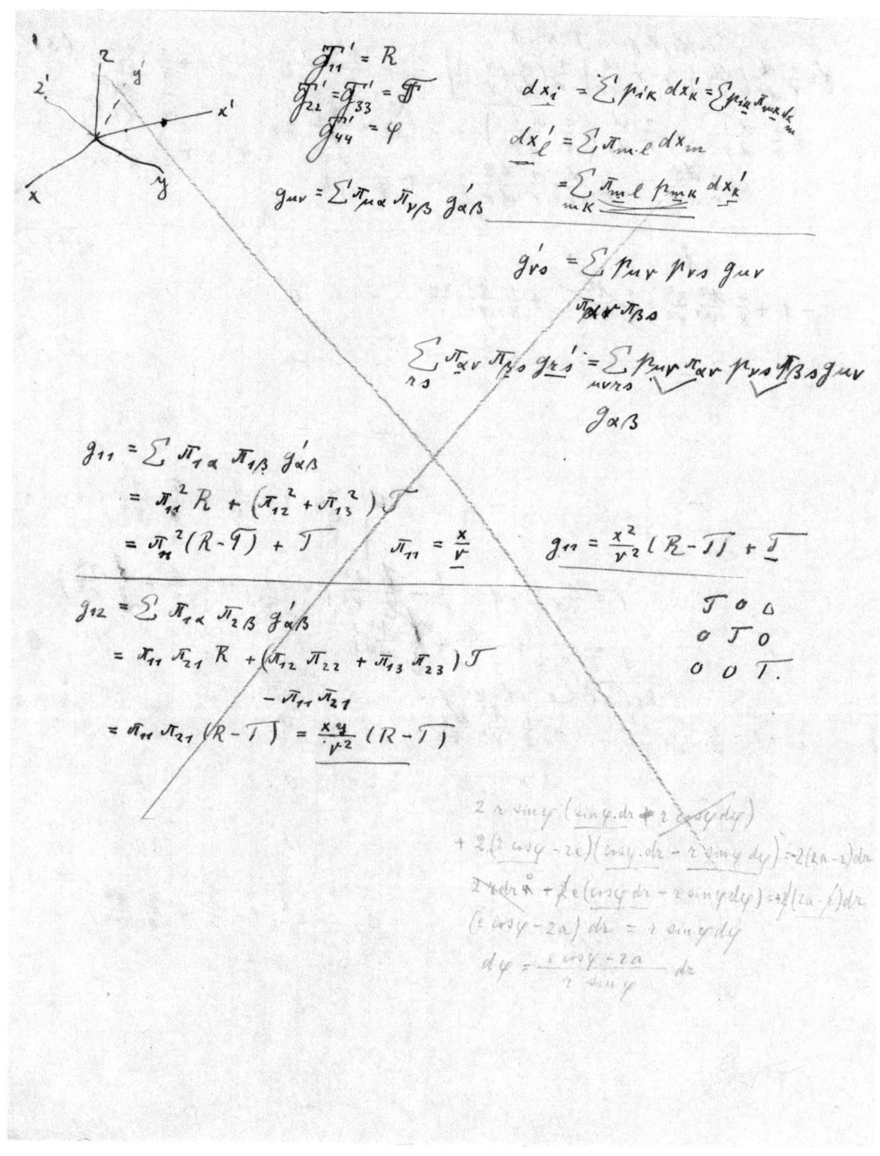

$$g'_{11} = R$$
$$g'_{22} = g'_{33} = \mathfrak{T}$$
$$g'_{44} = \varphi$$
$$g_{\mu\nu} = \sum' \pi_{\mu\alpha} \pi_{\nu\beta} g'_{\alpha\beta}$$

$$dx_i = \sum' p_{ik} dx'_k = \sum p_{ik} \pi_{mk} dx_m$$
$$dx'_l = \sum \pi_{ml} dx_m$$
$$= \sum_{mk} \pi_{ml} p_{mk} dx'_k$$

$$g'_{rs} = \sum p_{\mu r} p_{\nu s} g_{\mu\nu}$$
$$\pi_{\alpha r} \pi_{\beta s}$$
$$\sum_{rs} \pi_{\alpha r} \pi_{\beta s} g'_{rs} = \sum_{\mu\nu rs} p_{\mu r} \pi_{\alpha r} p_{\nu s} \pi_{\beta s} g_{\mu\nu}$$
$$g_{\alpha\beta}$$

$$g_{11} = \sum \pi_{1\alpha} \pi_{1\beta} g'_{\alpha\beta}$$
$$= \pi_{11}^2 R + (\pi_{12}^2 + \pi_{13}^2) \mathfrak{T}$$
$$= \pi_{11}^2 (R - \mathfrak{T}) + \mathfrak{T} \qquad \pi_{11} = \frac{x}{r}$$

$$g_{11} = \frac{x^2}{r^2} (R - \mathfrak{T}) + \mathfrak{T}$$

$$\begin{matrix} \mathfrak{T} & 0 & 0 \\ 0 & \mathfrak{T} & 0 \\ 0 & 0 & \mathfrak{T}. \end{matrix}$$

$$g_{12} = \sum \pi_{1\alpha} \pi_{2\beta} g'_{\alpha\beta}$$
$$= \pi_{11} \pi_{21} R + (\pi_{12} \pi_{22} + \pi_{13} \pi_{23}) \mathfrak{T}$$
$$- \pi_{11} \pi_{21}$$
$$= \pi_{11} \pi_{21} (R - \mathfrak{T}) = \frac{xy}{r^2} (R - \mathfrak{T})$$

[p. 3] Einstein

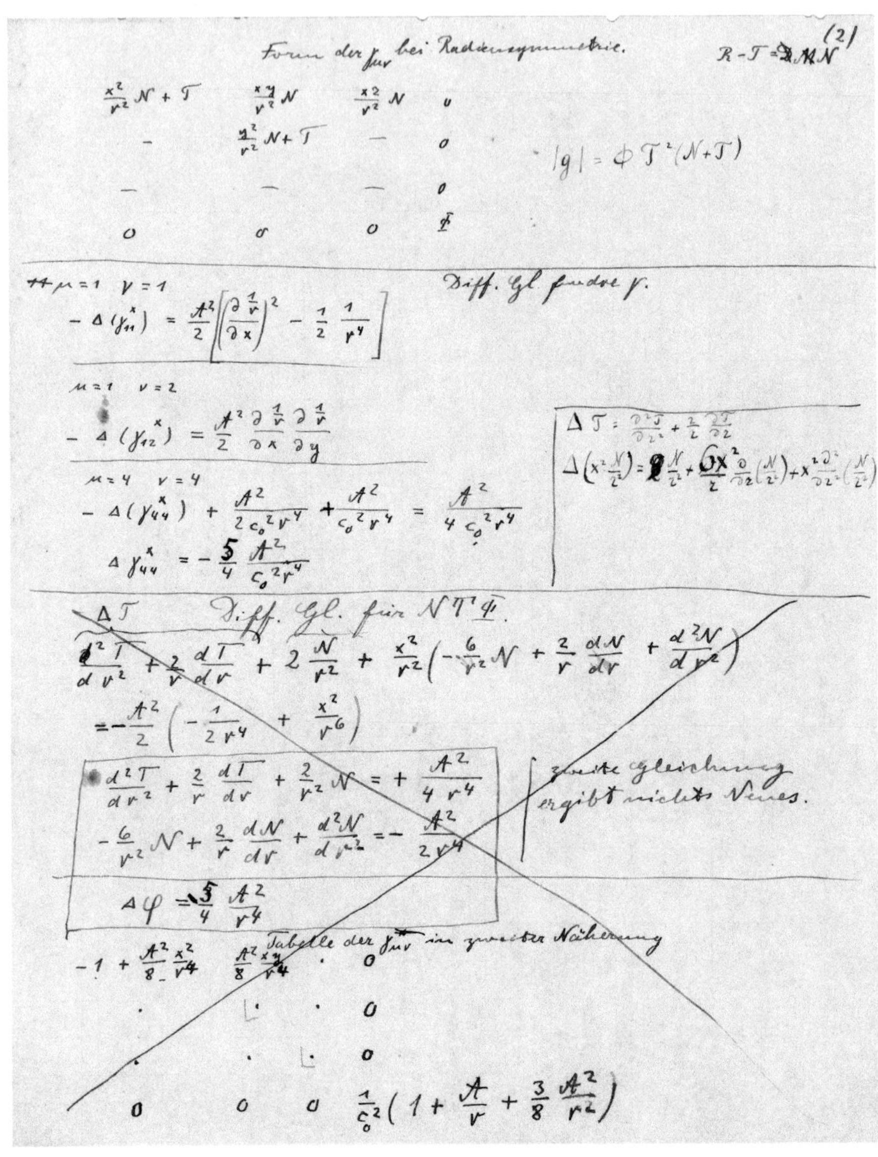

[p. 4] Einstein

2a)

$-\frac{1}{2}$

$\varphi = \frac{\alpha}{r^2}$ $\quad -6 + 8$

$+\frac{3}{4} + 7$

$\Delta\left(T + x^2 \frac{N}{r^2}\right)$

$\frac{\partial T}{\partial x} = x \frac{1}{r}\frac{dT}{dr}$

$\frac{d^2T}{dx^2} = \frac{1}{r}\frac{dT}{dr} + x^2 \cdot \frac{1}{r}\frac{d}{dr}\left(\frac{1}{r}\frac{dT}{dr}\right)$

$\Delta T = \frac{3}{r}\frac{dT}{dr} + r \frac{d}{dr}\left(\frac{1}{r}\frac{dT}{dr}\right)$

$\frac{\partial}{\partial x}\left(x^2\frac{N}{r^2}\right) = 2x\frac{N}{r^2} + x^3\cdot\frac{1}{r}\frac{d}{dr}\left(\frac{N}{r^2}\right) \qquad 2\frac{x^2}{r}\frac{d}{dr}\left(\frac{N}{r^2}\right)$

$\frac{\partial^2}{\partial x^2} = 2\frac{N}{r^2} + 5x^2\cdot\frac{1}{r}\frac{d}{dr}\left(\frac{N}{r^2}\right) + x^4\cdot\frac{1}{r}\frac{d}{dr}\left(\frac{1}{r}\frac{d}{dr}\left(\frac{N}{r^2}\right)\right)$

$\frac{\partial}{\partial y} = x^2 y\cdot\frac{1}{r}\frac{d}{dr}\left(\frac{N}{r^2}\right)$

$\frac{\partial^2}{\partial y^2} = x^2\cdot\frac{1}{r}\frac{d}{dr}\left(\frac{N}{r^2}\right) + x^2y^2\cdot\frac{1}{r}\frac{d}{dr}\left(\frac{1}{r}\frac{d}{dr}\left(\frac{N}{r^2}\right)\right)$

$\Delta\left(x^2\frac{N}{r^2}\right) = 2\frac{N}{r^2} + 7x^2\frac{1}{r}\frac{d}{dr}\left(\frac{N}{r^2}\right) + x^2 r\frac{d}{dr}\left(\frac{1}{r}\frac{d}{dr}\left(\frac{N}{r^2}\right)\right)$

$\frac{3}{r}\frac{dT}{dr} + \frac{1}{r}\frac{d}{r}\left(\frac{1}{r}\frac{dT}{dr}\right) + 2\frac{N}{r^2} + x^2\left(7\frac{1}{r}\frac{d}{dr}\left(\frac{N}{r^2}\right) + r\frac{d}{dr}\left(\frac{1}{r}\left(\frac{d}{dr}\frac{N}{r^2}\right)\right)\right)$

$+\frac{A^2}{4}\frac{1}{r^4} \qquad -\frac{A^2}{2}\frac{1}{r^6}$

$+\frac{A^2}{2}\left(\frac{1}{2}\frac{1}{r^4} - \frac{x^2}{r^6}\right)$

$N = \frac{1}{8}\frac{A^2}{r^2}$

$\frac{N}{r^2} = \frac{1}{8}\frac{A^2}{r^4}$

$\frac{d}{dr}(\quad) = -\frac{1}{2}\frac{A^2}{r^5}$

$\frac{1}{r}(\quad) = -\frac{1}{2}\frac{A^2}{r^6} \quad | \, 7$

$\frac{d}{dr}(\quad) = 3\frac{A^2}{r^7}$

$r(\quad) = 3\frac{A^2}{r^6} \quad | \, 1$

[p. 6] Einstein

Diff gl. für T N u Φ (3)

$$7 \cdot \frac{1}{r}\frac{d}{dr}\left(\frac{N}{r^2}\right) + r\frac{d}{dr}\left[\frac{1}{r}\left(\frac{d}{dr}\left(\frac{N}{r^2}\right)\right)\right] = -\frac{A^2}{2}\cdot\frac{1}{r^6}$$

$$3\,\frac{1}{r}\frac{dT}{dr} + \frac{2}{r}\frac{d}{dr}\left(\frac{1}{r}\frac{dT}{dr}\right) + 2\frac{N}{r^2} = \frac{A^2}{4}\cdot\frac{1}{r^4}$$

$$3\cdot\frac{1}{r}\frac{d\Phi}{dr} + \frac{2}{r}\frac{d}{dr}\left(\frac{1}{r}\frac{d\Phi}{dr}\right) = +\frac{5}{4}\frac{A^2}{c_0^2 r^4}$$

$$N = +\frac{1}{8}\frac{A^2}{r^2}$$

$$T = \frac{1}{4}\frac{A^2}{r^2}\ 0$$

$$\Phi = +\frac{1}{c_0^2}\frac{5}{8}\frac{A^2}{r^2}$$

$$-\frac{1}{8}+\frac{3}{4}+\frac{5}{8}$$

γ

$$-1 + \frac{1}{8}\frac{A^2}{r^2}\frac{x^2}{r^2} + \frac{1}{4}\frac{A^2}{r^2} \qquad +\frac{1}{8}\frac{A^2}{r^2}\frac{xy}{r^2} \qquad \cdot \qquad 0$$

$$- \qquad - \qquad - \qquad -$$

$$- \qquad - \qquad - \qquad -$$

$$- \qquad - \qquad 0 \quad \Big| \quad \frac{1}{c_0^2}\left(1 + \frac{A}{r} + \frac{5}{8}\frac{A^2}{r^2}\right)$$

$$\gamma = \frac{1}{c_0^2}\left\{-1 - \frac{A}{r} - \frac{1}{2}\frac{A^2}{r^2}\right\} = -\frac{1}{c_0^2}\left(1 + \frac{A}{r} + \frac{1}{2}\frac{A^2}{r^2}\right)$$

$$g = -c_0^2\left\{1 - \frac{A}{r} + \frac{1}{2}\frac{A^2}{r^2}\right\}$$

Tabelle der g.

$$-1 - \frac{1}{8}\frac{A^2}{r^2}\frac{x^2}{r^2} \qquad -\frac{1}{8}\frac{A^2}{r^2}\frac{xy}{r^2} \qquad \cdot \qquad 0$$

$$- \qquad - \qquad - \qquad 0$$

$$- \qquad - \qquad - \qquad 0$$

$$0 \qquad 0 \qquad 0 \qquad c_0^2\left(1 - \frac{A}{r} + \frac{3}{8}\frac{A^2}{r^2}\right)$$

[p. 8] Besso

4)

Die Bewegungsgleichungen des materiellen Punktes lauten:

(Bess) $\frac{dJ_x}{dt} = \mathfrak{R}_x$ also (bei $m=1$): $-\frac{d}{dt}\frac{g_{11}\dot{x}_1 + \ldots + g_{14}}{\frac{ds}{dt}} = -\frac{1}{2}\sum_{\mu\nu}\frac{\partial g_{\mu\nu}}{\partial x_1}\cdot\frac{dx_\mu}{dt}\cdot\frac{dx_\nu}{dt}\cdot\frac{dt}{ds}$ (Seite 7, Gl. 7 & 8)

$\mathfrak{E} = +m\left(g_{41}\frac{dx_1}{ds} + \ldots + g_{44}\frac{dx_4}{ds}\right)$ also: $\mathfrak{E} = -g_{44}\frac{dt}{ds}$. (Seite 7, Gl. 9)$_{III}$

Dabei ist (aus Gl. 5 S. 6 f, auch S. 7 Gl. 5), wenn man ~~relativ kleine Grössen~~, die Grössenordnung von der Ordnung $\frac{A}{r}$ und $\frac{q^2}{c_0^2} = \frac{\dot{x}^2+\dot{y}^2+\dot{z}^2}{c_0^2}$ als unendlich klein erster Ordnung betrachtet (worin $\frac{A}{r} = \frac{M}{4\pi}\cdot\frac{8\pi}{c_0^2}\cdot\frac{K}{r} = \frac{2M}{c_0^2 r}K$, worin M die Sonnenmasse, c_0 die Lichtgeschwindigkeit für r - Abstand von der Sonne - unendlich), K die Gravitationskonstante ist - Kraft $= K\frac{Mm}{r^2}$ bedeutet - und (ferner) berücksichtigt dass sich am Felde mit der Zeit nichts ändert, also $g_{\nu 4}$ für $\nu = 1, 2, 3$ $= 0$ zu setzen sind, wenn man auch wieder $x_1 = x$, $x_2 = y$, $x_3 = z$, $x_4 = t$ setzt

$$\frac{ds}{dt} = \sqrt{g_{11}\dot{x}^2 + \ldots + 2g_{12}\dot{x}_1\dot{x}_2 + \ldots 2g_{14}\dot{x} + \ldots + g_{44}} = \sqrt{g_{11}\dot{x}^2 + g_{22}\dot{y}^2 + g_{33}\dot{z}^2 + 2g_{12}\dot{x}\dot{y} + \ldots + g_{44}}$$

Die $g_{\nu\nu\neq 4}$ sind aber, bis auf unendl. kl. zweiter Ordnung $= -1$, die $g_{\nu\neq\mu} = 0$, $g_{44} = c_0^2\left(1 - \frac{A}{r} + \frac{3}{8}\frac{A^2}{r^2}\right)$. Daher reduziert sich ~~die Gleichungen~~ der Ausdruck für $\frac{ds}{dt}$ auf

(Bess) $\frac{ds}{dt} = \sqrt{c_0^2 - q^2 - c_0^2\frac{A}{r} + \frac{3}{8}c_0^2\frac{A^2}{r^2}}$; die Gleichungen werden

(Bess) $\frac{d}{dt}\frac{\dot{x}}{\frac{ds}{dt}} = -\frac{1}{2}\cdot\frac{\partial g_{44}}{\partial x}\cdot\frac{dt}{dt}\cdot\frac{dt}{dt}\cdot\frac{1}{\frac{ds}{dt}} = -\frac{1}{2}\frac{dg_{44}}{dr}\frac{\partial r}{\partial x}\Big/\frac{ds}{dt} = -\frac{1}{2}\frac{dg_{44}}{dr}\frac{x}{r}\Big/\frac{ds}{dt}$

4) $\mathfrak{E} = +g_{44}\Big/\frac{ds}{dt}$.

Aus y · Gl. 1) − x · Gl. 2) ergibt sich der Flächensatz

$$y\frac{d}{dt}\left(\dot{x}\Big/\frac{ds}{dt}\right) - x\frac{d}{dt}\left(\dot{y}\Big/\frac{ds}{dt}\right) = \frac{d}{dt}\frac{y\dot{x} - x\dot{y}}{\frac{ds}{dt}} = 0$$

Nimmt man die Bahnebene als xy Ebene, so ergibt sich daraus

$2\dot{f} = y\dot{x} - x\dot{y} = \mathfrak{B}\frac{ds}{dt}$ (Flächensatz Konstante $= \mathfrak{B}c$) und
f = Flächengeschwindigkeit

$g_{44} = \mathfrak{E}\frac{ds}{dt}$ ($\mathfrak{E} \sim c_0$)

Da nun $\frac{ds}{dt}$ eine Wurzel ist, so ist es bequemer mit den quadrierten Gleichungen (in Polarcoordinaten) zu operieren:

[p. 9] Besso

5)

$$2\dot{f} = \dot{\varphi} r^2 = \mathcal{B}W$$

$$\mathcal{E} = \frac{g_{44}}{W}$$

$$\mathcal{E}\dot{\varphi} r^2 = \mathcal{B} g_{44} = \mathcal{B} c_0^2 \left(1 - \frac{A}{r} + \frac{3}{8}\frac{A^2}{r^2}\right)$$

$$\dot{\varphi} r^2 = \frac{F^2}{\frac{\mathcal{B}c_0^2}{\mathcal{E}}}\left(1 - \frac{A}{r}\right)$$

$$c_0^4\left(1 - 2\frac{A}{r} + \frac{7}{4}\frac{A^2}{r^2}\right) = \mathcal{E}^2 c_0^2\left(1 - \frac{q^2}{c_0^2} - \frac{A}{r} + \frac{3}{8}\frac{A^2}{r^2}\right)$$

$$1 - \frac{A}{r} + \frac{q^2}{c_0^2} + \frac{11}{8}\frac{A^2}{r^2} + \left(\frac{q^2}{c_0^2} + \frac{A}{r}\right)^2 = \frac{\mathcal{E}^2}{c_0^2} - 1$$

$$1 - \frac{\mathcal{E}^2}{c_0^2} = \varepsilon \qquad \frac{\mathcal{E}^2}{c_0^2} = 1 - \varepsilon \qquad \frac{c_0^2}{\mathcal{E}^2} = 1 + \varepsilon \qquad \frac{c_0^2}{\mathcal{E}^2} - 1 = \varepsilon$$

$$\frac{1}{1-\alpha} = 1 + \alpha + \alpha^2$$

$$\frac{1}{\mathcal{E}^2} c_0^2\left(1 - 2\frac{A}{r} + \frac{7}{4}\frac{A^2}{r^2}\right) = \left(1 - \frac{q^2}{c_0^2} - \frac{A}{r} + \frac{3}{8}\frac{A^2}{r^2}\right)$$

$$\frac{\mathcal{E}^2}{c^2}\left(1 - \frac{q^2}{c_0^2} - \frac{A}{r} + \frac{3}{8}\frac{A^2}{r^2}\right) = 1 - 2\frac{A}{r} + \frac{7}{4}\frac{A^2}{r^2}$$

$$\frac{q^2}{c_0^2} = 1 - \frac{c_0^2}{\mathcal{E}^2} + \frac{A}{r}\left(2\frac{c_0^2}{\mathcal{E}^2} - 1\right) + \frac{A^2}{r^2}\left(\frac{3}{8} - \frac{7}{4}\frac{c_0^2}{\mathcal{E}^2}\right) \qquad 2\left(\frac{c_0^2}{\mathcal{E}^2} - 1\right) + 1$$

$$= -\varepsilon + \frac{A}{r}(2\varepsilon + 1) + \frac{A^2}{r^2}\left(-\frac{11}{8}\right)$$

$$\frac{dr^2 + r^2 d\varphi^2}{dt^2} = c_0^2\left[-\varepsilon + \frac{A}{r}(1 + 2\varepsilon) - \frac{A^2}{r^2}\cdot\frac{11}{8}\right]$$

$$r^4\frac{d\varphi^2}{dt^2} = F^2\left(1 - 2\frac{A}{r}\right)$$

$$\frac{1}{c_0^2}\left(dr^2 + r^2 d\varphi^2\right) F^2\left(1 - 2\frac{A}{r}\right) = -r^4 d\varphi^2 \cdot \left(\varepsilon - (1 + 2\varepsilon)\frac{A}{r} + \frac{11}{8}\frac{A^2}{r^2}\right)$$

$$\frac{1}{c_0^2} dr^2 F^2\left(1 - 2\frac{A}{r}\right) = d\varphi^2\left\{-\varepsilon r^4 + (1 + 2\varepsilon)A r^3 - \left[\frac{11}{8}A^2 + \frac{F^2}{c_0^2}\right]r^2 + 2\frac{F^2}{c_0^2}A r\right\}$$

$$d\varphi = \frac{\frac{F}{c_0}\left(1 - \frac{A}{r}\right) dr}{\sqrt{-\varepsilon r^4 + (1 + 2\varepsilon)A r^3 - \left(\frac{11}{8}A^2 + \frac{F^2}{c_0^2}\right) r^2 + 2\frac{F^2}{c_0^2}A r}} =$$

$$\int d\varphi = \frac{F}{c_0\sqrt{\varepsilon}}\int\frac{1 - \frac{A}{r}}{\sqrt{-r^4 + \left(\frac{1}{\varepsilon} + 2\right)A r^3 - \frac{1}{\varepsilon}\left(\frac{F^2}{c_0^2} + \frac{11}{8}A^2\right) r^2 + \frac{2}{\varepsilon}\frac{F^2}{c_0^2}A r}}\, dr$$

[p. 10] Einstein

$$\int \frac{dr}{\sqrt{(r-r')(r-r'')(r-r_1)(r_2-r_1)}} \qquad \cancel{= \frac{2\pi}{\sqrt{r_1 r_2}}\left[1 + \frac{1}{2}\left(\frac{1}{r_1}+\frac{1}{r_2}\right)\right]}$$

$$= \int \frac{dr}{r\sqrt{(r-r_1)(r_2-r_1)}\sqrt{1-\frac{r'}{r}}\sqrt{1-\frac{r''}{r}}} \qquad = \frac{2\pi}{\sqrt{r_1 r_2}}\left[1 + \frac{1}{4}\left(\frac{1}{r_1}+\frac{1}{r_2}\right)\underbrace{(r'+r'')}_{-\frac{d}{c}}\right]$$

$$= \int \frac{\left(1+\frac{r'+r''}{2}\frac{1}{r}\right)dr}{r\sqrt{(r-r_1)(r-r_2)}}$$

$$\int \frac{dr}{r\sqrt{(r_2-r_1)(r-r_2)}} = \frac{2\pi}{\sqrt{r_1 r_2}}$$

$$= -\int_{+} \frac{1}{-i\sqrt{r_1 r_2}} \cdot \frac{dr}{r} = \frac{1}{i\sqrt{r_1 r_2}} \underbrace{\int i\, d\varphi}_{2\pi i}$$

J_1 $\qquad \frac{1}{-i\sqrt{r_1 r_2}}$ $\qquad -2\pi$, $-\pi$, $-\frac{3}{2}\pi$

$$\int \frac{dr}{r^2\sqrt{(r-r_1)(r_2-r)}} \qquad \left| \frac{1}{-i\sqrt{r_1 r_2}\sqrt{\left(1-\frac{r}{r_1}\right)\left(1-\frac{r}{r_2}\right)}} = \frac{1}{-i\sqrt{r_1 r_2}}\left(1+\frac{1}{2}\left(\frac{1}{r_1}+\frac{1}{r_2}\right)r\right)\right.$$

$$-\int_{+\,0\text{-Kreis}} \frac{1}{-i\sqrt{r_1 r_2}} \frac{1}{2}\left(\frac{1}{r_1}+\frac{1}{r_2}\right) \underbrace{\int \frac{dr}{r}}_{2\pi i}$$

$$\int \frac{dr}{\sqrt{-\alpha r^4 + b r^3 + c r^2 + d r + e}}$$

Unendlich kleine Wurzeln aus der Gleichung

$$c r^2 + d r + e = 0 = c(r-r')(r-r'')$$

$$d = -c(r'+r'')$$

$$r'+r'' = -\frac{d}{c}$$

$$\alpha r^4 - b r^3 - c r^2 - d r - e = \alpha(r-r_1)(r-r_2)(r-r')(r-r'')$$

$$= \alpha \left| \begin{array}{l} r^4 \\ -(r_1+r_2+r'+r'')r^3 \\ +(r_1 r_2 + r_1 r' + r_2 r' + \cdot + \cdot + r' r'')r^2 \end{array}\right.$$

Für endliche Grössen

$$r_1 + r_2 = \frac{b}{\alpha}$$

$$r_1 r_2 = -\frac{c}{\alpha}$$

[p. 11] Einstein

6)

$$\int \frac{dr}{\sqrt{(r-r')(\ )(\ )(\ )}} = \frac{2\pi}{\sqrt{-\frac{c}{a}}}\left[1 + \frac{1}{4}\,\frac{\frac{b}{a}}{-\frac{c}{a}} \cdot -\frac{d}{c}\right]$$

$$= \frac{2\pi}{\sqrt{-c}}\left[1+\frac{1}{4}\,\frac{bd}{c^2}\right]$$

$$\int \frac{dr\,(1+\alpha\frac{1}{r}+\beta\frac{1}{r^2})}{\sqrt{(r-r_1)(r_2-r)(r-r')(r-r'')}} = I$$

$$= \int\frac{dr}{r\sqrt{(r-r_1)(r_2-r)}}\left\{(1-\frac{r'}{r})^{-\frac{1}{2}}(1-\frac{r''}{r})^{-\frac{1}{2}}(1+\alpha\frac{1}{r}+\beta\frac{1}{r^2})\right)$$

Faktor auf ∞ kl. erster Ordnung berechnet

$$\left[1+\frac{1}{2}(r'+r'')\cdot\frac{1}{r}\right](1+\alpha\frac{1}{r})$$

$$1+\underbrace{(\alpha+\frac{r'+r''}{2})}_{\alpha'}\frac{1}{r}$$

$$I = \int\frac{dr\,(1+\alpha'\frac{1}{r})}{r\sqrt{(r-r_1)(r_2-r)}} = \int\frac{dr\,(1+\alpha'\frac{1}{r})}{r\sqrt{(r-r_1)(r_2-r)}}$$

$$= \frac{2\pi}{\sqrt{r_1r_2}}\left[1+\frac{1}{2}\left(\frac{1}{r_1}+\frac{1}{r_2}\right)\left(\alpha+\frac{r'+r''}{2}\right)\right]$$

$$I = \int\frac{1+\alpha\frac{1}{z}}{\sqrt{-z^4+az^3+bz^2+cz+d}}\,dz$$

$a = z_1+z_2+z'+z''$

$b = z_1z_2+(z_1+z_2)(z'+z'')$

$c = z_1z_2(z'+z'')$

$d = 0$

$$c_0 B\int\frac{dr\,(1-A\frac{1}{r})}{\sqrt{(\mathfrak{C}^2-c_0^2)r^4+A(2c_0^2-\mathfrak{C}^2)r^3+\left[A^2(\frac{7}{4}c_0^2-\frac{3}{8}\mathfrak{C}^2)-B^2c_0^2\right]r^2+2B^2c_0^2Ar-\frac{3}{4}B^2c_0^2A^2}}$$

$$= \frac{c_0B}{\sqrt{c_0^2-\mathfrak{C}^2}}\cdot\frac{2\pi}{\sqrt{r_1r_2}}\left[1+\frac{1}{2}\left(\frac{1}{r_1}+\frac{1}{r_2}\right)\left(-A+-\frac{1}{2}\underbrace{\frac{d}{c}}_{-2A}\right)\right]$$

$$\sqrt{r_1r_2} = \sqrt{-c} = \sqrt{\frac{c_0^2B^2}{c_0^2-\mathfrak{C}^2}}$$

Faktor $\frac{1}{2}$ vor dem ganzen Integral vergessen.

[p. 14] Besso

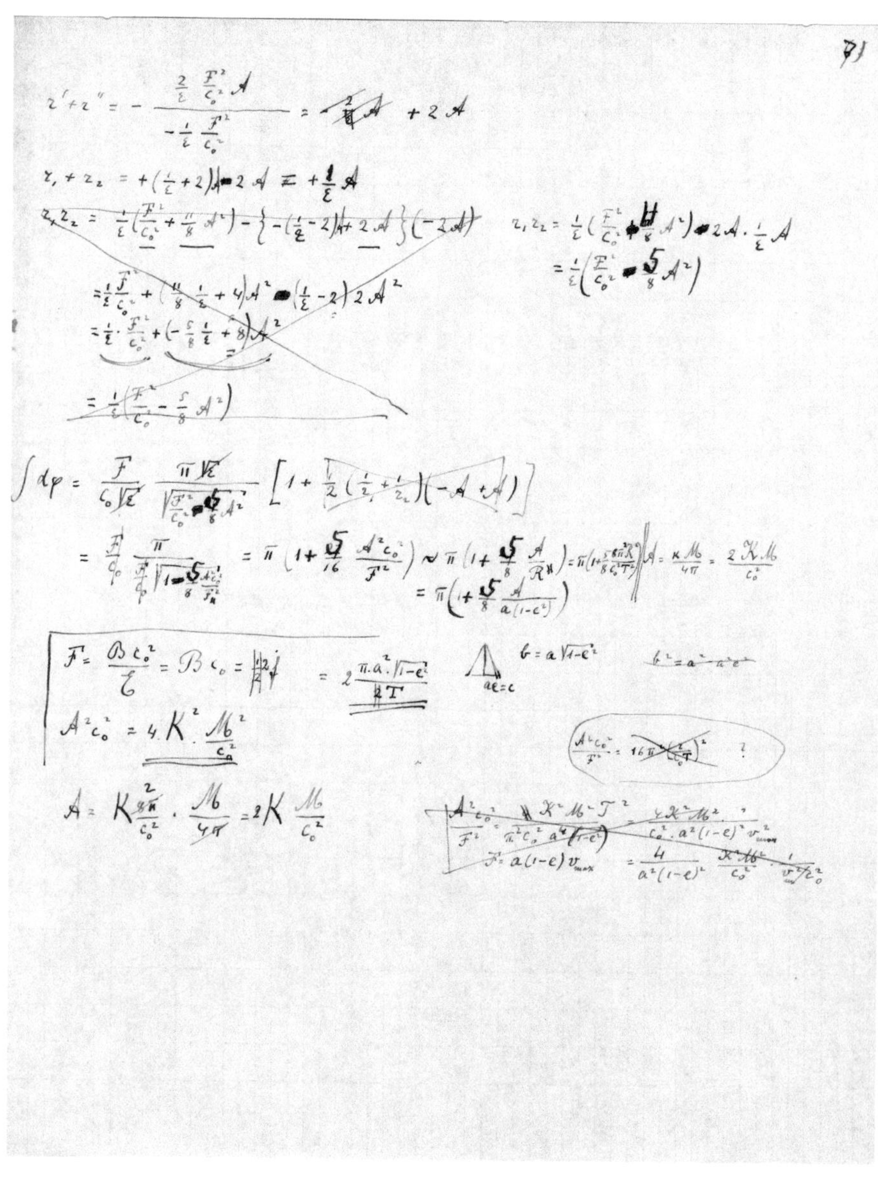

[p. 26] Einstein

I 5a)

$r' + r'' = +2A$

$r_1 + r_2 = \left(\frac{1}{\varepsilon} + 2\right)A - 2A = \frac{1}{\varepsilon}A$

[crossed out]

$r_1 r_2 = \frac{1}{\varepsilon}\left(\frac{F^2}{c_0^2} + \frac{11}{8}A^2\right) - \frac{1}{\varepsilon}A \cdot 2A = \frac{1}{\varepsilon}\left(\frac{F^2}{c_0^2} - \frac{5}{8}A^2\right)$

$\int d\varphi = \ldots \sqrt{1 - \frac{5}{8}\frac{A^2 c_0^2}{F^2}} = \pi\left(1 + \frac{5}{16}\frac{A^2 c_0^2}{F^2}\right)$

$F = \frac{\pi a^2 \sqrt{1-e^2} \cdot 2}{T}$

$A = \frac{2KM}{c_0^2}$

$K = \frac{1}{3862^2}$

$lg K = -2 \cdot 3.58681 = -7.17362$

$lg M = 5.51108$

28.78265

28.29373

| 5.6 = Dichte d. E

$lg V = 28.03456$

$lg 5.6 = 24819$

$lg M_e = 28.78265$

$\frac{A c_0}{F} = K \cdot \frac{2M}{c_0^2} \cdot \frac{c_0 \, T}{\pi a^2 \sqrt{1-e^2} \cdot 2}$

$lg T = lg\, 82.97 \cdot 24 \cdot 60 \cdot 60$

= 1.94433

1.38021

3.55630

6.88084

$lg c_0 = 10.47712$

$lg \pi = 0.49715$

$lg a = .58782 - 1$

13.17164

12.15946

$lg a^2 = 25.51892$

$lg(1-e^2) = lg = 1.2056$; 0.7944-1

$lg \sqrt{1-e^2} = 9906-1$

Mittlerer Abst. = grosse Halbachse gesetzt.

lg Präzession pro halben Umlauf.

$2 \cdot \frac{T_{Erde}}{T_{Merkur}} \cdot 100$

Winkelgraden

[p. 28] Einstein

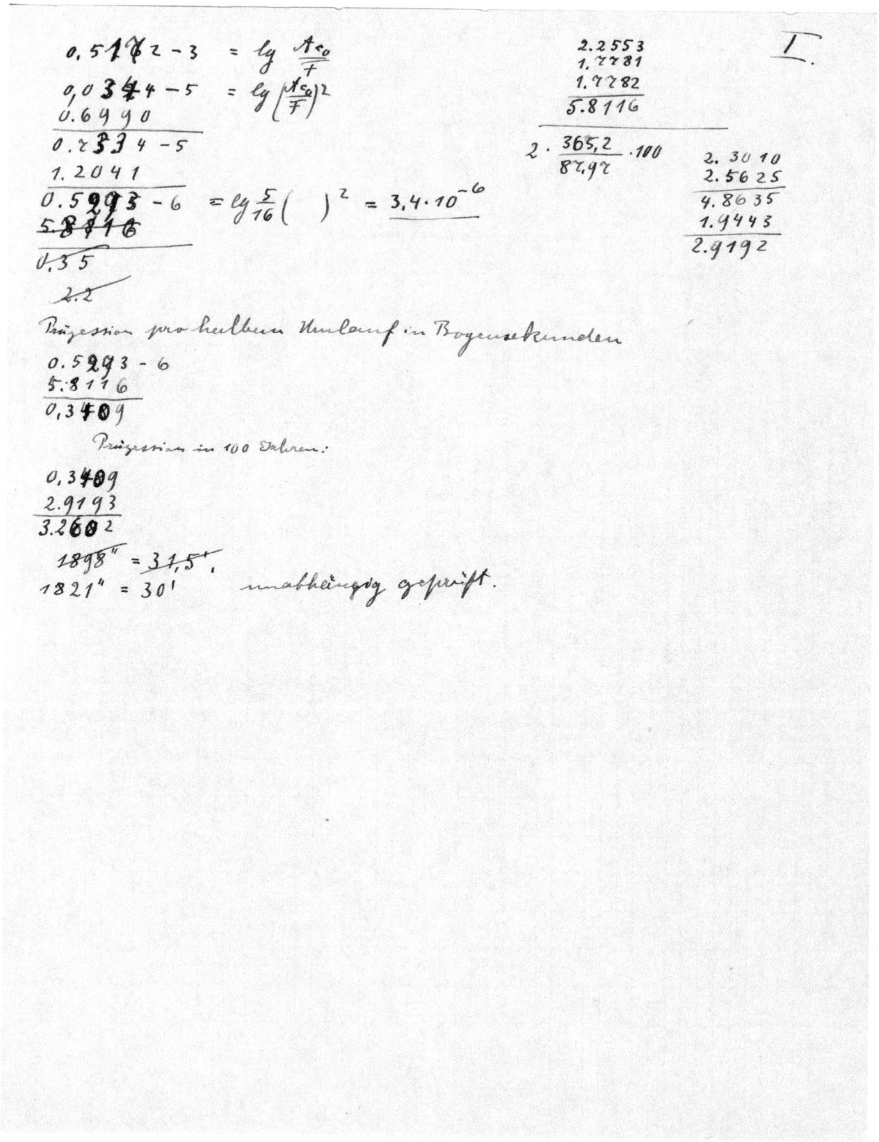

I.

$0.5172 - 3 = \lg \frac{Ac_0}{F}$

$0.0344 - 5 = \lg \left(\frac{Ac_0}{F}\right)^2$

0.6990

0.7334 − 5

1.2041

$0.5293 - 6 = \lg \frac{5}{16} (\ \ )^2 = \underline{3.4 \cdot 10^{-6}}$

~~5.8116~~

0.35

~~2.2~~

2.2553
1.7781
1.7782
5.8116

$2 \cdot \frac{365,2}{87,97} \cdot 100$

2.3010
2.5625
4.8635
1.9443
2.9192

Präzession pro halben Umlauf in Bogensekunden

0.5293 − 6
5.8116
0,3409

Präzession in 100 Jahren:

0,3409
2.9193
3.2602

~~1898" = 31,5'~~

1821" = 30' unabhängig geprüft.

[p. 41] Einstein (top half) and Besso (bottom half)

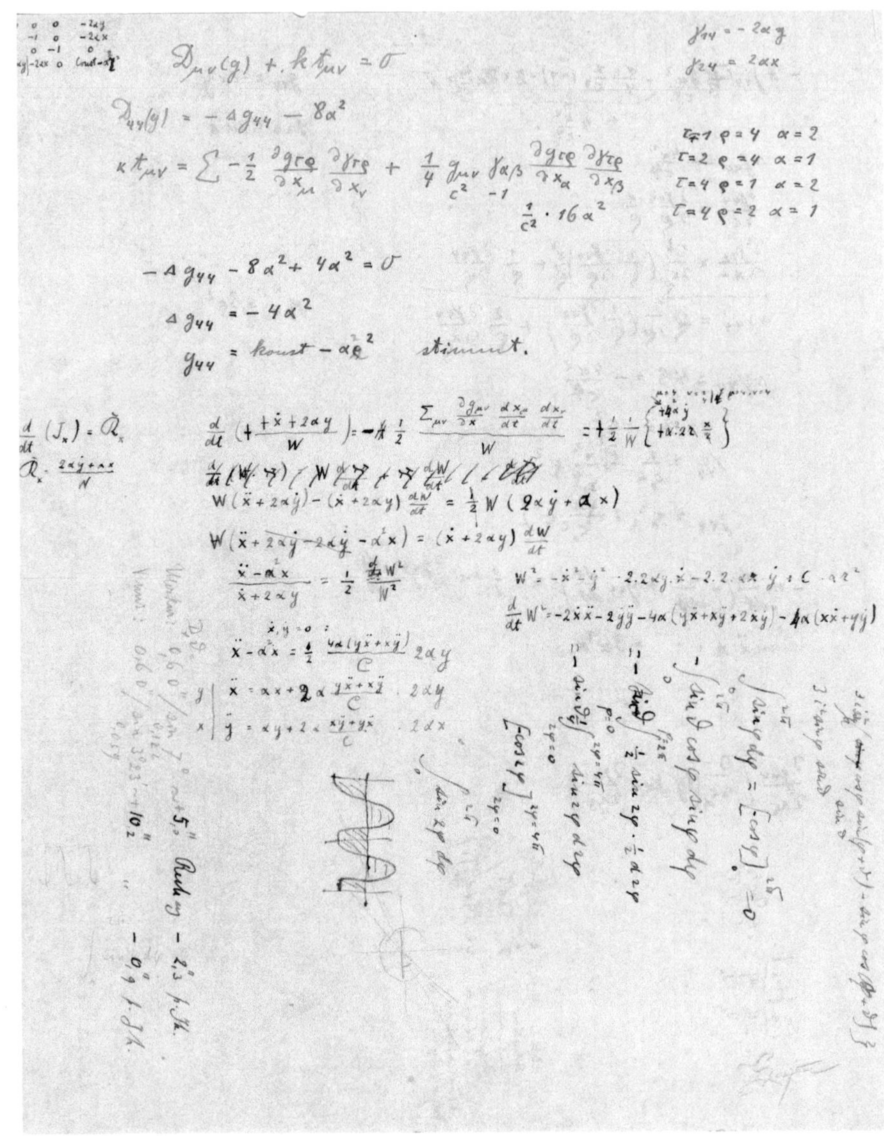

[p. 42] Einstein (left side) and Besso (right side and bottom)

2.2 Transcription

[p. 1] Einstein

Gleichungen der Grav. in erster Näherung (1)

[eq. 1] $$-\left(\frac{\partial^2\gamma_{\mu\nu}}{\partial x^2}+\cdot+\cdot-\frac{1}{c^2}\frac{\partial^2\gamma_{\mu\nu}}{\partial t}\right)=\kappa\rho_0\frac{dx_\mu}{\partial s}\frac{dx_\nu}{\partial s} \qquad ds^2=-dx^2-.-.+c_0^2dt^2$$

Statischer Fall

[eq. 2] $$-\Delta\gamma_{44}=\frac{\kappa\rho_0}{c_0^2} \qquad \boxed{\kappa=K\frac{8\pi}{c_0^2}}$$ [eq. 3]

Tabelle des Schwerefeldes für die <u>erste Annäherung:</u>

[eq. 4] $$\gamma \quad \begin{matrix}-1&0&0&0\\0&-1&0&0\\0&0&-1&0\\0&0&0&\frac{1}{c_0^2}(1+\frac{A}{r})\end{matrix} \qquad g \quad \begin{matrix}-1&0&0&0\\0&-1&0&0\\0&0&-1&0\\0&0&0&c_0^2(1-\frac{A}{r})\end{matrix}$$ [eq. 5]

[eq. 6] $$\boxed{A=\frac{\kappa M}{4\pi}}=\frac{2KM}{c_0^2}$$ [eq. 7] $$g=-c_0^2(1-\frac{A}{r})$$

Zweite Annäherung.

$$-\left(\frac{\partial^2\gamma^x_{\mu\nu}}{\partial x^2}+\cdot+\cdot-\frac{1}{c_0^2}\frac{\partial^2\gamma^x_{\mu\nu}}{\partial t^2}\right)$$ (struck out)

1) $$\gamma^{<0>}_{\alpha\beta}\frac{\partial^2\gamma^{<x>}_{\mu\nu}}{\partial x_\alpha\partial x_\beta}$$

$$\sum\gamma^0_{\alpha\alpha}\frac{\partial^2\gamma^x_{\mu\nu}}{\partial x_a^2}$$

2) $$\begin{matrix}0&0&0&0\\-&-&-&-\\-&-&-&-\\0&0&0&\frac{A^2}{2c_0^2r^4}\end{matrix}$$

3) $$-\sum\sum\gamma_{\alpha\beta}g_{\tau\rho}\frac{\partial\gamma_{\mu\tau}}{\partial x_\alpha}\frac{\partial\gamma_{\nu\rho}}{\partial x_\beta}$$

$$\begin{matrix}0&0&0&0\\-&-&-&-\\-&-&-&-\\0&0&0&+\frac{A^2}{c_0^2r^4}\end{matrix}$$

$$-\frac{1}{2}\sum\gamma_{\alpha\mu}\gamma_{\beta\nu}\frac{\partial g_{\tau\rho}}{\partial x_\alpha}\frac{\partial\gamma_{\tau\rho}}{\partial x_\beta}$$

$$\frac{A^2}{2}\left|\begin{matrix}\left(\frac{\partial\frac{1}{r}}{\partial x}\right)^2&\frac{\partial\frac{1}{r}}{\partial x}\frac{\partial\frac{1}{r}}{\partial y}&\cdot&0\\\cdot&\cdot&\cdot&0\\\cdot&\cdot&\cdot&0\\0&0&0&0\end{matrix}\right.$$

$$+\frac{1}{4}\gamma_{\mu\nu}\gamma_{\alpha\beta}\frac{\partial g_{<\mu\nu>\tau\rho}}{\partial x_\alpha}\frac{\partial\gamma_{\tau\rho}}{\partial x_\beta}$$

$$\frac{A^2}{4}\left|\begin{matrix}-\frac{1}{r^4}&0&0&0\\0&-\frac{1}{r^4}&0&0\\0&0&-\frac{1}{r^4}&0\\0&0&0&\frac{1}{c_0^2r^4}\end{matrix}\right.$$

[p. 7] Einstein

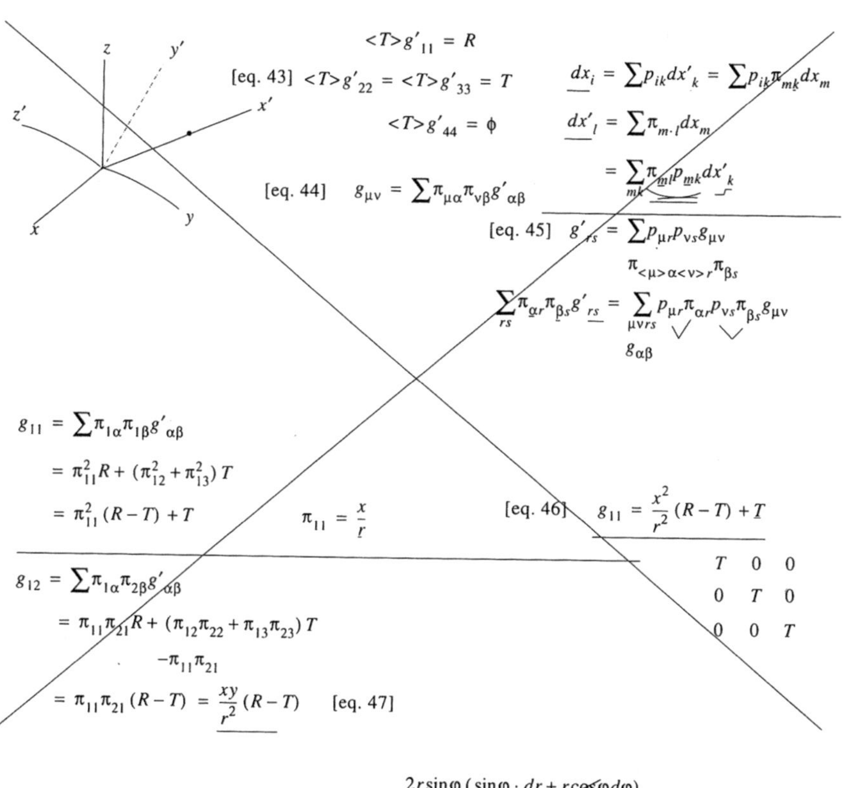

$$2r\sin\varphi\,(\underline{\sin\varphi\cdot dr} + r\cos\varphi\, d\varphi)$$

$$+2\,(\underline{r\cos\varphi}-2e)\;(\underline{\cos\varphi\cdot dr} - \underline{r\sin\varphi d\varphi}) \;=\; -2\,(2a-r)\,dr$$

$$2rdr\langle+\rangle^{0} + 2e\,(\underline{\cos\varphi dr} - r\sin\varphi d\varphi) \;=\; +2\,(2a-r)\,dr$$

$$(e\cos\varphi - 2a)\,dr \;=\; r\sin\varphi d\varphi$$

$$d\varphi \;=\; \frac{e\cos\varphi - 2a}{r\sin\varphi}\,dr$$

[p. 3] Einstein

Form der $\gamma_{\mu\nu}$ bei Radiussymmetrie.

[eq. 18] $R - T = \langle D\rangle\langle M\rangle N$ (2)

[eq. 17]
$$\begin{matrix} \frac{x^2}{r^2}N+T & \frac{xy}{r^2}N & \frac{xz}{r^2}N & 0 \\ - & \frac{y^2}{r^2}N+T & - & 0 \\ - & - & - & 0 \\ 0 & 0 & 0 & \Phi \end{matrix}$$

$|g| = \Phi T^2 (N+T)$ [eq. 19]

Diff. Gl. für die γ.

$\langle 11\rangle \mu = 1 \quad \nu = 1$

[eq. 20] $$-\Delta(\gamma^x_{11}) = \frac{A^2}{2}\left[\left(\frac{\partial\frac{1}{r}}{\partial x}\right)^2 - \frac{1}{2}\frac{1}{r^4}\right]$$

$\mu = 1 \quad \nu = 2$

[eq. 21] $$-\Delta(\gamma^x_{12}) = \frac{A^2}{2}\frac{\partial\frac{1}{r}}{\partial x}\frac{\partial\frac{1}{r}}{\partial y}$$

$\mu = 4 \quad \nu = 4$

$$-\Delta(\gamma^x_{44}) + \frac{A^2}{2c_0^2r^4} + \frac{A^2}{c_0^2r^4} = \frac{A^2}{4c_0^2r^4}$$

[eq. 22] $$\Delta(\gamma^x_{44}) = -\frac{\langle 3\rangle 5}{4}\frac{A^2}{c_0^2r^4}$$

$$\Delta T = \frac{\partial^2 T}{\partial r^2} + \frac{2}{r}\frac{\partial T}{\partial r}$$

$$\Delta\left(x^2\frac{N}{r^2}\right) = \langle 6\rangle 2\frac{N}{r^2} + \frac{\langle x(\rangle 6x^2}{r}\frac{\partial}{\partial r}\left(\frac{N}{r^2}\right) + x^2\frac{\partial^2}{\partial r^2}\left(\frac{N}{r^2}\right)$$

Diff. Gl. für $N\ T\ \Phi$.

[eq. 23] $$\overbrace{\frac{\langle\partial\rangle d^2T}{\partial r^2} + \frac{2}{r}\frac{dT}{dr}}^{\Delta T} + 2\frac{N}{r^2} + \frac{x^2}{r^2}\left(-\frac{6}{r^2}N + \frac{2}{r}\frac{dN}{dr} + \frac{d^2N}{dr^2}\right)$$
$$= -\frac{A^2}{2}\left(-\frac{1}{2r^4} + \frac{x^2}{r^6}\right)$$

[eq. 24] $$\frac{d^2T}{dr^2} + \frac{2}{r}\frac{dT}{dr} + \frac{2}{r^2}N = +\frac{A^2}{4r^4}$$

[eq. 25] $$-\frac{6}{r^2}N + \frac{2}{r}\frac{dN}{dr} + \frac{d^2N}{dr^2} = -\frac{A^2}{2r^4}$$

zweite Gleichung ergibt nichts Neues.

[eq. 26] $$\Delta\varphi = -\frac{\langle 3\rangle 5}{4}\frac{A^2}{r^4}$$

Tabelle der $\gamma^{\langle x\rangle}_{\mu\nu}$ in zweiter Näherung

$$\begin{matrix} -1+\frac{A^2}{8}\frac{x^2}{r^{\langle 2\rangle 4}} & \frac{A^2}{8}\frac{xy}{r^{\langle 2\rangle 4}} & \cdot & 0 \\ \cdot & \cdot & \cdot & 0 \\ \cdot & \cdot & \cdot & 0 \\ 0 & 0 & 0 & \frac{1}{c_0^2}\left(1+\frac{A}{r}+\frac{3}{8}\frac{A^2}{r^2}\right) \end{matrix}$$

[p. 4] Einstein

~~$\delta_x = -m$~~ 2a)

[eq. 27] $\Delta(T + x^2 \frac{N}{r^2})$

$-\frac{1}{2}$ $-6+8$

$\frac{\partial T}{\partial x} = x \langle - \rangle \frac{1}{r} \frac{\langle \partial \rangle dT}{dr}$ $\varphi = \frac{\gamma}{r^2}$

$\frac{d^2 T}{dx^2} = \frac{1}{r}\frac{dT}{dr} + x^2 \cdot \frac{1}{r}\frac{d}{dr}(\frac{1}{r}\frac{dT}{dr})$ $+\frac{3}{4}+1$

[eq. 28] $\Delta T = \frac{3}{r}\frac{dT}{dr} + r\frac{1}{r}\frac{d}{dr}(\frac{1}{r}\frac{dT}{dr})$

$\frac{\partial}{\partial x}(x^2 \frac{N}{r^2}) = 2x\frac{N}{r^2} + x^3 \cdot \frac{1}{r}\frac{d}{dr}(\frac{N}{r^2})$ $2\frac{x^2}{r}\frac{d}{dr}(\frac{N}{r^2})$

$\frac{\partial^2}{\partial x^2} = 2\frac{N}{r^2} + \langle 3 \rangle 5x^2 \cdot \frac{1}{r}\frac{d}{dr}(\frac{N}{r^2}) + x^4 \cdot \frac{1}{r}\frac{d}{dr}(\frac{1}{r}\frac{d}{dr}(\frac{N}{r^2}))$

$\frac{\partial}{\partial y} = x^2 y \cdot \frac{1}{r}\frac{d}{dr}(\frac{N}{r^2})$

$\frac{\partial^2}{\partial y^2} = x^2 \cdot \frac{1}{r}\frac{d}{dr}(\frac{N}{r^2}) + x^2 y^2 \cdot \frac{1}{r}\frac{d}{dr}(\frac{1}{r}\frac{d}{dr}\langle)\rangle(\frac{N}{r^2}))$

[eq. 29] $\Delta(x^2\frac{N}{r^2}) = 2\frac{N}{r^2} + \langle 5 \rangle 7x^2 \cdot \frac{1}{r}\frac{d}{dr}(\frac{N}{r^2}) + x^2 \cdot \frac{1}{r}\langle x \rangle r\frac{d}{dr}(\frac{1}{r}\frac{d}{dr}(\frac{N}{r^2}))$

[eq. 30] $\underbrace{\frac{3}{r}\frac{dT}{dr} + \frac{1}{r}\frac{d}{r}(\frac{1}{r}\frac{dT}{dr}) + 2\frac{N}{r^2}}_{+\frac{A^2}{4}\frac{1}{r^4}} + x^2\underbrace{\left(\langle 5 \rangle 7\frac{1}{r}\frac{d}{dr}(\frac{N}{r^2}) + r\frac{d}{dr}(\frac{1}{r}\frac{d}{dr}\frac{N}{r^2})\right)}_{-\frac{A^2}{2}\frac{1}{r^6}}$

$+\frac{A^2}{2}\left(\frac{1}{2}\frac{1}{r^4} - \frac{x^2}{r^6}\right)$

$N = \frac{1}{8}\frac{A^2}{r^2}$

$\langle (1+\alpha)(1+\beta)(1+\gamma) \rangle$

$\frac{N}{r^2} = \frac{1}{8}\frac{A^2}{r^4}$

$-\left(1 + \frac{1}{8}A^2\frac{x^2}{r^4} - \frac{1}{4}A^2\frac{1}{r^2}\right)$

$\frac{d}{dr}(\ ) = -\frac{1}{2}\frac{A^2}{r^5}$

$-\frac{3}{4}\frac{A^2}{r^2} + \frac{1}{8}\frac{A^2}{r^2} - \frac{5}{8}\frac{A^2}{r^2}$

$\frac{1}{r}(\ ) = -\frac{1}{2}\frac{A^2}{r^6}$ | $\langle 5 \rangle 7$

bcd

$abcd$

$-\frac{1}{1 + \frac{x^2}{r^4}\frac{1}{8}A^2[-] - \varepsilon} = -(1 - \quad + \varepsilon)$

$-1 + \quad -$

$\frac{d}{dr}(\quad) = 3\frac{A^2}{r^7}$

$r(\quad) = 3\frac{A^2}{r^6}$ | 1

[p. 6] Einstein

Diff. Gl. für $N\ T\ \&\ \Phi$ (3)

[eq. 36] $$<5>7\cdot\frac{1}{r}\frac{d}{dr}\left(\frac{N}{r^2}\right)+r\frac{d}{dr}\left[\frac{1}{r}\left(\frac{d}{dr}\left(\frac{N}{r^2}\right)\right)\right]=-\frac{A^2}{2}\cdot\frac{1}{r^6}\qquad N=+\frac{1}{8}\frac{A^2}{r^2}$$

[eq. 37] $$3\frac{1}{r}\frac{dT}{dr}+r\frac{1}{r}\frac{d}{dr}\left(\frac{1}{r}\frac{dT}{dr}\right)+2\frac{N}{r^2}=\frac{A^2}{4}\cdot\frac{1}{r^4}\qquad T=\frac{1}{4}\frac{A^2}{r^2}\ 0$$ [eq. 39]

[eq. 38] $$3\cdot\frac{1}{r}\frac{<\partial>d\Phi}{dr}+r\frac{1}{r}\frac{d}{dr}\left(\frac{1}{r}\frac{d\Phi}{dr}\right)=+\frac{5}{<2>4}\frac{A^2}{c_0^2r^4}\qquad \Phi=+\frac{1}{c_0^2}\frac{5}{8}\frac{A^2}{r^2}$$

$$-\frac{1}{8}+\frac{3}{4}+\frac{5}{8}$$

γ

[eq. 40] $$\begin{array}{cccc}
-1+\frac{1}{8}\frac{A^2}{r^2}\frac{x^2}{r^2}+\frac{1}{4}\frac{A^2}{r^2} & +\frac{1}{8}\frac{A^2}{r^2}\frac{xy}{r^2} & \cdot & 0\\
- & - & - & -\\
- & - & - & -\\
- & - & 0 & \frac{1}{c_0^2}\left(1+\frac{A}{r}+\frac{5}{8}\frac{A^2}{r^2}\right)
\end{array}$$

[eq. 41] $$\gamma=\frac{1}{c_0^2}\left\{-1-\frac{A}{r}-<->\frac{1}{2}\frac{A^2}{r^2}\right\}=-\frac{1}{c_0^2}\left(1+\frac{A}{r}+\frac{1}{2}\frac{A^2}{r^2}\right)$$
$$g=-c_0^2\left\{1-\frac{A}{r}+<->\frac{1}{2}\frac{A^2}{r^2}\right\}$$

Tabelle der g.

[eq. 42] $$\begin{array}{cccc}
-1-\frac{1}{8}\frac{A^2}{r^2}\frac{x^2}{r^2} & -\frac{1}{8}\frac{A^2}{r^2}\frac{xy}{r^2} & \cdot & 0\\
- & - & - & 0\\
- & - & - & 0\\
0 & 0 & 0 & c_0^2\left(1-\frac{A}{r}+\frac{3}{8}\frac{A^2}{r^2}\right)
\end{array}$$

[p. 8] Besso

<Für> <d>Die Bewegungsgleichungen des materiellen Punktes lauten: 4)

[eq. 48] 1 bis 3) $\frac{dJ_x}{dt} = \mathfrak{K}_x$ also (bei $m = 1$): [eq. 49] $-\frac{d}{dt}\frac{g_{11}\dot{x}_1 + \dots + g_{14}}{\frac{ds}{dt}} = -\frac{1}{2}\sum_{\mu\nu}\frac{\partial g_{\mu\nu}}{\partial x_1}\cdot\frac{dx_\mu}{dt}\cdot\frac{dx_\nu}{dt}\cdot\frac{dt}{ds}$ (Seite 7. Gl. 7 & 8)

[eq. 50] 4) $E = +m\left(g_{41}\frac{dx_1}{ds} + \dots + g_{44}\frac{dx_4}{ds}\right)$ also: [eq. 51] $E = -g_{44}\frac{dt}{ds}$ (Seite 7. Gl. 9)$_{II}$)

Dabei ist (<aus Gl. 5> S. <7>6 <)> auch S.7 Gl. 5), wenn man <relativ kleine Grössen,> die Grössen<ordnung von> <von der Ordnung> $\frac{A}{r}$ und <---> $\frac{q^2}{c_0^2} = \frac{\dot{x}^2 + \dot{y}^2 + \dot{z}^2}{c_0^2}$ als unendlich klein erster Ordnung betrachtet

(worin $\frac{A}{r} = \frac{M}{4\pi}\cdot\frac{8\pi}{c_0^2}\cdot\frac{K}{r} = \frac{2M}{c_0^2 r}K$, und wieder M die Sonnenmasse, c_0 die Lichtgeschwindigkeit für r – Abstand von der Sonne – unendlich), K die gew./Gravitationskonstante ist – Kraft = $K\frac{Mm}{r^2}$ – bedeutet)

und zudem/berücksichtigt dass sich am Felde mit der Zeit nichts ändert, also/$g_{\nu 4}$ für $\nu = 1$ bis 3 <also> = 0 zu setzen sind, wenn/man auch wieder <für> $x_1 = x$, $x_2 = y$, $x_3 = z$, $x_4 = t$ setzt

[eq. 52] $\frac{ds}{dt} = \sqrt{g_{11}\dot{x}^2 + <g>.. + 2g_{12}<g>\dot{x}_1\dot{x}_2 + ..2g_{14}\dot{x} + \dots + g_{44}} = \sqrt{g_{11}\dot{x}^2 + g_{22}\dot{y}^2 + g_{33}\dot{z}^2 + 2g_{12}\dot{x}\dot{y} + .. + g_{44}}$

Die $g_{\nu\nu\neq 4}$ sind aber, bis auf unendl. kl. zweiter Ordnung = – 1, die $g_{\nu\neq\mu} = 0$,

$g_{44} = c_0^2\left(1 - \frac{A}{r} + \frac{3}{8}\frac{A^2}{r^2}\right)$. Daher reducier<en>t sich <die Gleichngen> der Ausdruck für $\frac{ds}{dt}$ auf

1) <bis 3) auf> [eq. 53] $\frac{ds}{dt} = \sqrt{c_0^2 - q^2 - c_0^2\frac{A}{r} + \frac{3}{8}c_0^2\frac{A^2}{r^2}} =$ ~~$c_0\left(1 - \frac{1}{2}\frac{q^2}{c_0^2} - \frac{1}{2}\frac{A}{r}\right)$~~ ; die Gleichngen werden

1) bis 3) ~~$\frac{d}{dt}\cdot\frac{\dot{x}}{\sqrt{c_0^2 - q^2 - \frac{A}{r}}} = c_0^2\frac{d}{dt}\left[\dot{x}\left(1 + \frac{1}{2}\frac{q^2}{c_0^2} + \frac{1}{2}\frac{A}{r}\right)\right] =$~~

[eq. 54] $\frac{d}{dt}\frac{\dot{x}}{\frac{ds}{dt}} = -\frac{1}{2}\cdot\sum_{<44>}\cdot\frac{\partial g_{44}}{\partial x}\cdot\frac{dt}{dt}\cdot\frac{dt}{dt}\cdot\frac{1}{\frac{ds}{dt}} = -\frac{1}{2}\cdot\frac{<\partial>dg_{44}}{<\partial>dr}\frac{\partial r}{\partial x}\Big/\frac{ds}{dt} = -\frac{1}{2}\frac{dg_{44}}{dr}\frac{<r>x}{r}\Big/\frac{ds}{dt}$

4) $E = +g_{44}\Big/\frac{ds}{dt}$ · [eq. 55]

Aus <das> y · Gl. 1) – x · Gl. 2) ergibt sich der Flächensatz

[eq. 56] $y\frac{d}{dt}\left(\dot{x}\Big/\frac{ds}{dt}\right) - x\frac{d}{dt}\left(\dot{y}\Big/\frac{ds}{dt}\right) = \frac{d}{dt}\frac{y\dot{x} - x\dot{y}}{\frac{ds}{dt}} = 0$

Nimmt man die <--> Bahnebene als xy Ebene, so ergibt sich daraus

[eq. 57]

I) $2\dot{f} = y\dot{x} - x\dot{y} = \underline{B}\frac{ds}{dt}$ (Flächensatzkonstante = Bc_0) und
$\dot{f}$ = Flächengeschwindigkeit

II) $g_{44} = E\frac{ds}{dt}$ $(E \sim c_0)$ $E = c_0$

Da nun $\frac{ds}{dt}$ <so> eine Wurzel ist, so ist es bequemer mit den quadrierten Gleichngen (in Polarcoordinaten) zu operieren:

[p. 9] Besso

$$2\dot{f} = \dot{\varphi}r^2 = BW \quad \text{[eq. 58]}$$

$$E = \frac{g_{44}}{W} \quad \text{[eq. 59]}$$

[eq. 60] $E\dot{\varphi}r^2 = Bg_{44} = Bc_0^2\left(1 - \frac{A}{r} + \frac{3}{8}\frac{A^2}{r^2}\right)$

[eq. 61] $\dot{\varphi}r^2 = \not{Bc}\; F^{<'>} \quad (1 - \frac{A}{r})$, $F = \frac{Bc_0^2}{E}$

[eq. 62] $<E>c_0^4\left(1 - 2\frac{A}{r} + \frac{7}{4}\frac{A^2}{r^2}\right) = E^2c_0^2\left(1 - \frac{q^2}{c_0^2} - \frac{A}{r} + \frac{3}{8}\frac{A^2}{r^2}\right)$

[eq. 63] 5)

$$1 - \frac{E^2}{c_0^2} = \varepsilon$$

$$\frac{E^2}{c_0^2} = 1 - \varepsilon$$

$$\frac{c_0^2}{E^2} = 1 + \varepsilon$$

$$\frac{c_0^2}{E^2} - <E>1 = \varepsilon$$

$$\frac{1}{1-\alpha} = 1 + \alpha + \alpha^2$$

$$\frac{1}{E^2} \quad \frac{E^2 - c_0^2}{c_0^2} = \varepsilon \quad \frac{E^2 - c_0^2}{E^2} = \frac{c_0^2}{E^2}\varepsilon$$

sondern so:

$$1 - \frac{A}{r} + \frac{q^2}{c_0^2} + \frac{11}{8}\frac{A^2}{r^2} + \left(\frac{q^2}{c_0^2} + \frac{A}{r}\right)^2 = \frac{E^2}{c_0^2} - 1 = \varepsilon$$

$$\dot{\varphi}r^2 = f \quad \frac{1}{E^2}c_0^2\left(1 - 2\frac{A}{r} + \frac{7}{4}\frac{A^2}{r^2}\right) = E^2\left(1 - \frac{q^2}{c_0^2} - \frac{A}{r} + \frac{3}{8}\frac{A^2}{r^2}\right)$$

$$\frac{c_0^2}{E^2} = 1 - \frac{q^2}{c_0^2} + \frac{A}{r} + \frac{3}{8}\frac{A^2}{r^2} - \frac{14}{8}\frac{A^2}{r^2}$$

nicht so:

$$\frac{E^2}{c^2}\left(1 - \frac{q^2}{c_0^2} - \frac{A}{r} + \frac{3}{8}\frac{A^2}{r^2}\right) = 1 - 2\frac{A}{r} + \frac{7}{4}\frac{A^2}{r^2}$$

[eq. 64] $\frac{q^2}{c_0^2} = 1 - \frac{c_0^2}{E^2} + \frac{A}{r}\left(2\frac{c_0^2}{E^2} - 1\right) + \frac{A^2}{r^2}\left(\frac{3}{8} - \frac{7}{4}\frac{c_0^2}{E^2}\right)$ | $2\left(\frac{c_0^2}{E^2} - 1\right) + 1$

$$= -\varepsilon + \frac{A}{r}(2\varepsilon + 1) + \frac{A^2}{r^2}\left(-\frac{<3>11}{8}\right)$$

[eq. 65] $\frac{dr^2 + r^2d\phi^2}{dt^2} = c_0^2\left[-\varepsilon + \frac{A}{r}(1 + 2\varepsilon) - \frac{A^2}{r^2}\cdot\frac{11}{8}\right]$

[eq. 66] $r^{<2>4}\frac{d\phi^2}{dt^2} = F^2(1 - 2\frac{A}{r})$

$$\frac{1}{c_0^2}(dr^2 + r^2d\phi^2)F^2(1 - 2\frac{A}{r}) = -r^4d\phi^2 \cdot \not{c_0^2}\left(\varepsilon - (1 + 2\varepsilon)\frac{A}{r} + \frac{11}{8}\frac{A^2}{r^2}\right)$$

$$-r^2d\phi^2\frac{1}{c_0^2}F^2(1 - 2\frac{A}{r})$$

$$\frac{1}{c_0^2}dr^2F^2(1 - 2\frac{A}{r}) = d\phi^2\{-\varepsilon\not{c_0^2}\cdot r^4 + \not{c_0^2}(1 + 2\varepsilon)A\cdot r^3 - \left[\frac{11}{8}A^2 + \frac{F^2}{c_0^2}(1 - 2\frac{A}{r})\right]r^2 + 2\frac{F^2}{c_0^2}Ar\}$$

$$d\phi^{<2>} = \frac{\frac{F}{c_0}(1 - \frac{A}{r})\,dr}{\not{F(1-\frac{A}{r})}\sqrt{-\varepsilon r^4 + (1 + 2\varepsilon)\cdot A\cdot r^3 - \left(\frac{11}{8}A^2 + \frac{F^2}{c_0^2}\right)r^2 + 2\frac{F^2}{c_0^2}A^2\cdot r}} = \quad \text{[eq. 67]}$$

$$\int d\phi = \frac{F}{c_0\sqrt{\varepsilon}}\int\frac{1 - \frac{A}{r}}{\sqrt{-r^4 + (\frac{1}{\varepsilon} + 2)A\cdot r^3 - \frac{1}{\varepsilon}\left(\frac{F^2}{c_0^2} + \frac{11}{8}A^2\right)r^2 + \frac{2}{\varepsilon}\frac{F^2}{c_0^2}A\cdot r}}dr \quad \text{[eq. 68]}$$

[p. 10] Einstein

$$\int \frac{dr}{\sqrt{(r-r')(r-r'')(r-r_1)(r_2-r_{<2>})}} = \frac{2\pi}{\sqrt{r_1 r_2}}\left[1+\frac{1}{2}\left(\frac{1}{r_1}+\frac{1}{r_2}\right)\right] \quad \text{[eq. 69]}$$

$$= \frac{2\pi}{\sqrt{r_1 r_2}}\left[1+\frac{1}{<2>4}\left(\frac{1}{r_1}+\frac{1}{r_2}\right)\underbrace{(r'+r'')}_{-\frac{d}{c}}\right]$$

$$= \int \frac{dr}{r\sqrt{(r-r_1)(r_2-r_{<2>})}\sqrt{1-\frac{r'}{r}}\sqrt{1-\frac{r''}{r}}} \quad \text{[eq. 70]}$$

$$= \int \frac{\left(1+\frac{r'+r''}{2}\frac{1}{r}\right)dr}{r\sqrt{(r-r_1)(r-r_2)}} \quad \text{[eq. 71]}$$

$$\int \frac{dr}{r\sqrt{(r_{<1>}-r_1)(r-r_2}} = \frac{2\pi}{\sqrt{r_1 r_2}} \quad \text{[eq. 72]}$$

$$= -\int_{+} \frac{1}{-i\sqrt{r_1 r_2}}\frac{dr}{r} = \frac{1}{i\sqrt{r_1 r_2}}\underbrace{\int i\,d\varphi}_{2\pi i}$$

I_1

$\frac{1}{-i\sqrt{r_1 r_2}}$

-2π $-\pi$ $-\frac{3}{2}\pi$

[eq. 73]

$$\int \frac{dr}{r^2\sqrt{(r-r_1)(r_2-r)}}$$

[eq. 74]

$$\frac{1}{-i\sqrt{r_1 r_2}\sqrt{\left(1-\frac{r}{r_1}\right)\left(1-\frac{r}{r_2}\right)}} = \frac{1}{-i\sqrt{r_1 r_2}}\left(1+\frac{1}{2}\left(\frac{1}{r_1}+\frac{1}{r_2}\right)r\right)$$

$$-\int_{+0\text{-Kreis}} \frac{1}{-i\sqrt{r_1 r_2}}\frac{1}{2}\left(\frac{1}{r_1}+\frac{1}{r_2}\right)\underbrace{\int\frac{dr}{r}}_{2\pi i}$$

$$\int \frac{dr}{\sqrt{-ar^4+br^3+cr^2+dr+e}} \quad \text{[eq. 75]}$$

Unendlich kleine Wurzeln aus der Gleichung

[eq. 76] $cr^2+dr+e = 0 = c(r-r')(r-r'')$

$$d = -c(r'+r'')$$

[eq. 77] $r'+r'' = -\frac{d}{c}$

$$ar^4 <-> -br^3-cr^2-dr-e = a(r-r_1)(r-r_2)(r-r')(r-r'') \quad \text{[eq. 78]}$$

$$= ar^4$$

$$-(r_1+r_2+r'+r'')r^3$$

$$+(r_1 r_2+r_1 r'+r_2 r'+.\ +.\ +r'r'')r^2$$

Für endliche Grössen

[eq. 79] $r_1+r_2 = \frac{b}{a}$

[eq. 80] $r_1 r_2 = -\frac{c}{a}$

[p. 11] Einstein

6)

$$\int \frac{dr}{\sqrt{(r-r')\,(\;)\,(\;)\,(\;)}} = \frac{2\pi}{\sqrt{-\frac{c}{<a>1}}}\left[1+\frac{1}{4}\frac{\frac{b}{<a>1}}{-\frac{c}{<a>1}}\cdot -\frac{d}{c}\right] \quad \text{[eq. 81]}$$

$$= \frac{2\pi}{\sqrt{-<\frac{c}{a}>c}}\left[1+\frac{1}{4}\frac{bd}{c^2}\right]$$

$$\int \frac{dr\,(1+\alpha\frac{1}{r}+\beta\frac{1}{r^2})}{\sqrt{(r-r_1)\,(r_2-r)\,(r-r')\,(r-r'')}} = I \quad \text{[eq. 82]}$$

$$= \int \frac{dr}{r\sqrt{(r-r_1)\,(r_2-r)}} \;\Big|\; (1-\frac{r'}{r})^{-\frac{1}{2}}\,(1-\frac{r''}{r})^{-\frac{1}{2}}\,(1+\alpha\frac{1}{r}+<)>\beta\frac{1}{r^2})$$

Faktor auf ∞ Kl. erster Ordnung berechnet

$$\left[1+\frac{1}{2}(r'+r'')\cdot\frac{1}{r}\right](1+\alpha\frac{1}{r})$$

[eq. 83] $1+\underbrace{(\alpha+\frac{r'+r''}{2})}_{\alpha'}\frac{1}{r}$

[eq. 84]

$$I = \int \frac{dr\,(1+\alpha'\frac{1}{r})}{r\sqrt{(r-r_1)\,(r_2-r)\,\underline{(r-r')\,(r-r'')}}} = \int \frac{dr\,(1+\alpha'\frac{1}{r})}{r\sqrt{(r-r_1)\,(r_2-r)}}$$

$$= \frac{2\pi}{\sqrt{r_1r_2}}\left[1+\frac{1}{2}(\frac{1}{r_1}+\frac{1}{r_2})\,(\alpha+\frac{r'+r''}{2})\right]$$

$$I = \int \frac{1+\alpha\frac{1}{r}}{\sqrt{-r^4+ar^3+br^2+cr+d}}dr \quad \text{[eq. 85]}$$

$a = r_1+r_2+r'+r''$

$-b = r_1r_2+(r_1+r_2)\,(r'+r'')$

$c = r_1r_2(r'+r'')$

$-d = 0$

$$\text{[eq. 86]}\quad c_0B\int \frac{(dr\,(1-A\frac{1}{r})}{\sqrt{(E^2-c_0^2)\,r^4+A\,(2c_0^2-E^2)\,r^3+\left[A^2(\frac{7}{4}c_0^2-\frac{3}{8}E^2)-B^2c_0^2\right]r^2+2B^2c_0^2Ar-\frac{7}{4}B^2c_0^2A^2}}$$

$$= \frac{c_0B}{\sqrt{c_0^2-E^2}}\cdot\frac{2\pi}{\sqrt{r_1r_2}}\left[1+\frac{1}{2}(\frac{1}{r_1}+\frac{1}{r_2})\,(-A+\underbrace{-\frac{1}{2}\frac{d}{c}}_{-2A})\right]$$

$$\text{[eq. 87]}\quad \sqrt{r_1r_2} = \sqrt{-c} = \sqrt{\frac{c_0^2B^2}{c_0^2-E^2}}$$

Faktor $\frac{1}{2}$ vor dem ganzen Integral vergessen.

[p. 14] Besso

<6>7)

[eq. 98] $r' + r'' = -\dfrac{\frac{2}{\varepsilon}\frac{F^2}{c_0^2}A}{-\frac{1}{\varepsilon}\frac{F^2}{c_0^2}} = \frac{2}{\varepsilon}A + 2A$

[eq. 99] $r_1 + r_2 = +(\frac{1}{\varepsilon}+2)A - 2A = +\frac{1}{\varepsilon}A$

$$r_1 r_2 = \frac{1}{\varepsilon}\left(\frac{F^2}{c_0^2}+\frac{11}{8}A^2\right) - \{-(\frac{1}{\varepsilon}-2)A+2A\}(-2A)$$

$$= \frac{1}{\varepsilon}\frac{F^2}{c_0^2} + (\frac{11}{8}\frac{1}{\varepsilon}+4)A^2 - (\frac{1}{\varepsilon}-2)2A^2$$

$$= \frac{1}{\varepsilon}\frac{F^2}{c_0^2} + (-\frac{5}{8}\frac{1}{\varepsilon}+8)A^2$$

$$= \frac{1}{\varepsilon}\left(\frac{F^2}{c_0^2}-\frac{5}{8}A^2\right)$$

[eq. 100]

$$r_1 r_2 = \frac{1}{\varepsilon}\left(\frac{F^2}{c_0^2}<-> + \frac{<5>11}{8}A^2\right)<+> - 2A\cdot\frac{1}{\varepsilon}A$$

$$= \frac{1}{\varepsilon}\left(\frac{F^2}{c_0^2}<+> - \frac{5}{8}A^2\right)$$

[eq. 101] $\displaystyle\int d\varphi = \frac{F}{c_0\sqrt{\varepsilon}}\frac{\pi\sqrt{\varepsilon}}{\sqrt{\frac{F^2}{c_0^2}-\frac{5}{8}A^2}}\left[1+\frac{1}{2}(\frac{1}{r_1}+\frac{1}{r_2})(-A+A)\right]$

$$= \frac{F}{c_0}\frac{\pi}{\frac{F}{c_0}\sqrt{1-\frac{5}{8}\frac{A^2c_0^2}{F^2}}} = \pi\left(1+\frac{5}{16}\frac{A^2c_0^2}{F^2}\right)<=> \sim \pi(1+\frac{5}{8}\frac{A}{R^{<->}}) = \pi\left(1+\frac{5}{8}\frac{8\pi^2R^{[-]}}{c_0^2T^2}\right) \Bigg\| A = \frac{\kappa M}{4\pi} = \frac{2KM}{c_0^2}$$

$$= \pi(1+\frac{5}{8}\frac{A}{a(1-e^2)})$$

[eq. 102] $F = \frac{Bc_0^2}{E} = Bc_0 = \frac{1}{2}2f = 2\frac{\pi\cdot a^2\cdot\sqrt{1-e^2}}{2T}$

$b = a\sqrt{1-e^2}$ $\quad b^2 = a^2 - a^2e^2$

$ae = c$

$$A^2c_0^2 = 4K^2\cdot\frac{M^2}{c_0^2}$$

$$\frac{A^2c_0^2}{F^2} = 16\pi^2(\frac{r}{c_0T})^2 \quad ?$$

$$A = K\frac{8\pi}{c_0^2}\cdot\frac{M^2}{4\pi} = 2K\frac{M}{c_0^2}$$

$$\frac{A^2c_0^2}{F^2} = \frac{4K^2M^2T^2}{\pi^2c_0^2a^4(1-e^2)} = \frac{4K^2M^2}{c_0^2\cdot a^2(1-e)^2v_{max}^2}$$

$$F = a(1-e)v_{max} \qquad = \frac{4}{a^2(1-e)^2}\frac{K^2M^2}{c_0^2}\cdot\frac{1}{v_m^2/c_0^2}$$

[p. 26] Einstein

<Zwischenrechgen > 5a) I

[eq. 174] $r' + r'' = +2A$

[eq. 175] $r_1 + r_2 = (\frac{1}{\varepsilon} + 2)A - 2A = \frac{1}{\varepsilon}A$

$\left(r_1 r_2 = \left\{\frac{1}{\varepsilon}\left(\frac{F^2}{c_0^2} + \frac{11}{8}A^2\right) - \frac{1}{\varepsilon}A\right)$ [struck out]

[eq. 176] $r_1 r_2 = \frac{1}{\varepsilon}\left(\frac{F^2}{c_0^2} + \frac{11}{8}A^2\right) - \frac{1}{\varepsilon}A \cdot 2A = \frac{1}{\varepsilon}\left(\frac{F^2}{c_0^2} - \frac{5}{8}A^2\right)$

[eq. 177] $\int d\varphi = \frac{F}{c_0\sqrt{\varepsilon}} \frac{\pi c_0 \sqrt{\varepsilon}}{F} \Big/ \sqrt{1 - \frac{5}{8}\frac{A^2 c_0^2}{F^2}} \qquad = \pi\left(1 + \frac{5}{16}\frac{A^2 c_0^2}{F^2}\right)$

[eq. 178] $F = \frac{\pi a^2 \sqrt{1 - e^2} \cdot 2}{T}$

[eq. 179] $A = \frac{2KM}{c_0^2}$

$K = \frac{1}{3862^2}$

$\lg K = -2 \cdot 3.58681$

$= -7, 17362$

$\lg M = 5,51108$ | 5.6 = Dichte d. E

28, 78265
28.29373
34

$\lg V = 28.03456$
$\lg 5.6 = \quad 74819$
$\lg M_e = 28,78265$

$\lg T = \lg \; 87.97 \cdot 24 \cdot 60 \cdot 60$

= 1.94433   3
1.38021   1, 77815
3.55630
6.88084

$\lg c_0 = 10,47712$ Mittlerer Abst. = grosse Halbachse gesetzt.

$\lg \pi = 0,49715$

$\lg a = \; .58782 - 1$
13.17164
12.75946

$\lg a^2 = 25.51892$

$\lg\,(1 - e^2) = \lg = 1.2056 \qquad .08120 \quad 98124 - 1$
$0.7944 - 1 \qquad 90004 - 1$
~~000 − 1~~

$\lg \sqrt{1 - e^2} = 9906 - 1$

[eq. 180]

$\frac{Ac_0}{F} = K \cdot \frac{2M}{c_0^2} \cdot \frac{c_0 T}{\pi a^2 \sqrt{1 - e^2} \cdot 2}$

34, 2937   7.1736
6.8808   10.4771
41.1745   0.4971
43 6573   25.5189
0.5172 − 3   9906 − 1
  43.6573

lg Präzession pro halbem Umlauf.
zu mult. mit

$2 \cdot \frac{T_{\text{erde}}}{T_{\text{Merkur}}} \cdot 100$

2.30103
2.5626
4.8636
1.9443
2.9193

0.0151 − 8
2.9193
0.9344 − 6   Winkelabschnitt.
5.3145
0.2489

[p. 28] Einstein

I.

$0.5172 - 3 = \lg \frac{Ac_0}{F}$

$0,0344 - 5 = \lg (\frac{Ac_0}{F})^2$

0.6990

0.7334 − 5

1.2041

$0.5293 - 6 = \lg \frac{5}{16} (\ )^2 = 3.4 \cdot 10^{-6}$

~~5.8116~~

~~0.35~~

~~2.2~~

2.2553
1.7781
1.7782

5.8116

$2 \cdot \frac{365,2}{87,97} \cdot 100$

2.3010
2.5625

4.8635
1.9443

2.9192

Präzession pro halbem Umlauf in Bogensekunden

0.5293 − 6
5.8116

0, 3409

Präzession in 100 Jahren:

0, 3409
2.9193

3.2602

~~1898″ = 31, 5′~~
1821″ = 30′ unabhängig geprüft.

[p. 41, top half] Einstein

$$\begin{matrix} -1 & 0 & 0 & -2\alpha y \\ 0 & -1 & 0 & -2\alpha x \\ 0 & 0 & -1 & 0 \\ -2\alpha y & -2\alpha x & 0 & \text{Const.} - \alpha^2 \langle r^2 \rangle \rho^2 \end{matrix} \quad \text{[eq. 267]}$$

[eq. 268] $\gamma_{14} = -2\alpha y$, $\gamma_{24} = 2\alpha x$

$D_{\mu\nu}(g) + k\, t_{\mu\nu} = 0$ [eq. 269]

[eq. 270] $D_{44}(g) = -\Delta g_{44} - 8\alpha^2$

[eq. 271] $\kappa\, t_{\mu\nu} = \sum -\frac{1}{2} \frac{\partial g_{\tau\rho}}{\partial x_\mu} \frac{\partial \gamma_{\tau\rho}}{\partial x_\nu} + \frac{1}{4} g_{\mu\nu} \gamma_{\alpha\beta} \frac{\partial g_{\tau\rho}}{\partial x_\alpha} \frac{\partial \gamma_{\tau\rho}}{\partial x_\beta}$

$c^2 \quad -1$

$\frac{1}{c^2} \cdot 16\alpha^2$

$\tau = 1$	$\rho = 4$	$\alpha = 2$
$\tau = 2$	$\rho = 4$	$\alpha = 1$
$\tau = 4$	$\rho = 1$	$\alpha = 2$
$\tau = 4$	$\rho = 2$	$\alpha = 1$

[eq. 272] $-\Delta g_{44} - 8\alpha^2 + 4\alpha^2 = 0$

[eq. 273] $\Delta g_{44} = -4\alpha^2$

[eq. 274] $g_{44} = \text{konst} - \alpha^2 \langle r^2 \rangle \rho^2$ stimmt.

[p. 42, left half] Einstein

[p. 42]

[eq. 281] $$-\Delta\gamma_{44} <+> -\frac{8\alpha^2}{c^4} \underbrace{-\frac{1}{4}\cdot\frac{1}{c^2}(-1)\cdot 2\cdot 8\alpha^2\cdot\frac{1}{c^2}}_{+4\frac{\alpha^2}{c^4}} = 0$$

[eq. 282] $$\begin{aligned} g_{14} &= -2\alpha y \\ g_{24} &= 2\alpha x \end{aligned}$$

[eq. 283] $$\Delta\gamma_{44} = -\frac{4\alpha^2}{c^4}$$

$$\frac{\partial\gamma_{44}}{\partial x} = \frac{\partial\gamma_{44}}{\partial\rho}\frac{x}{\rho}$$

$$\frac{\partial^2\gamma_{44}}{\partial x^2} = \frac{\partial^{<2>}}{\partial x_\rho}\left(\frac{1}{\rho}\frac{\partial\gamma_{44}}{\partial\rho}\right)\frac{x^2}{\rho} + \frac{1}{\rho}\frac{\partial\gamma_{44}}{\partial\rho}$$

[eq. 284] $$\gamma_{44}^{x} = \beta\rho^2$$

[eq. 285] $$\Delta\gamma_{44} = <2>\rho\frac{\partial}{\partial\rho}\left(\frac{1}{\rho}\frac{\partial\gamma_{44}}{\partial\rho}\right) + \frac{2}{\rho}\frac{\partial\gamma_{44}}{\partial x_\rho}$$

[eq. 286] $$\Delta\gamma_{44} = 4\beta = -\frac{4\alpha^2}{c^{<2>4}}$$

[eq. 287] $$\beta = -\frac{\alpha^2}{c^4}$$

[eq. 288] $$\gamma_{44} = \frac{1}{c_0^2} \not{+} - 3\frac{\alpha^2}{c_0^4}\rho^2$$

[eq. 289] $$g_{44} = c_0^2\left(1 + <3>\frac{\alpha^2}{c_0^2}\rho^2\right)$$

[eq. 290] $$-m\frac{d}{dt}\left(-\frac{dx}{ds} + g_{14}\frac{dt}{ds}\right) = -\frac{1}{2}m\frac{\partial g_{44}}{\partial x}\cdot\frac{dt}{ds}$$

[eq. 291] $$m\ddot{x} <=> + \cdot = 3\alpha^2 x$$

[eq. 292] $$\frac{\partial g_{\mu\nu}}{\partial x_6}\left[\frac{\partial}{\partial x_\alpha}\left(\sqrt{g}\gamma_{\alpha\beta}\frac{\partial\gamma_{\mu\nu}}{\partial x_\beta}\right)\right.$$

2.3 Commentary

The Einstein-Besso manuscript is a sheaf of loose sheets with research notes documenting (for the most part) attempts by Einstein and his friend Michele Besso to account for the anomalous advance of the perihelion of Mercury on the basis of the *Entwurf* theory (cf. Pt. I, Ch. 6). The manuscript can thus be dated to the reign of that theory, which was roughly from May 1913 to October 1915. In fact, as we shall see shortly, many pages can be dated more precisely.

The bulk of the manuscript—53 pages, including the 14 pages presented here—was published in CPAE4, in facsimile as Appendix B and in transcription as Doc. 14. Three sets of numbers were added to the transcription, one numbering pages ("[p. 1]," etc.: Einstein and Besso's own numbering is erratic at best), one numbering editorial notes ("[1]," etc.), and one numbering equations ("[eq. 1]," etc.). For this volume, we used the transcription from CPAE4 without the editorial-note numbers but with the equation numbers. We will refer to those in our commentary, which is based on the editorial note, "The Einstein-Besso manuscript on the motion of the perihelion of Mercury," the annotation of the transcription in CPAE4 (pp. 344–359, pp. 360–473), and Janssen (1996, 1999).

The published part of the manuscript comes from two sources. Two of its 53 pages, [pp. 16–17], mostly in Besso's hand, are recto and verso of a letter from Charles-Eugène Guye to Einstein, May 31, 1913 (CPAE5, Doc. 443). The remaining 51 pages are on 37 sheets found in the Besso *Nachlass*. Michele's son Vero presented these to Pierre Speziali, editor and translator (into French) of the correspondence between Einstein and Besso (1972). Speziali made them available to the editors of the Einstein Papers Project in the late 1980s. A copy of these 51 pages is in the supplementary archive to the Einstein Archive, assembled and used by the editors of the Einstein Papers Project. The designation of this copy of the manuscript is EA 79 896 (cf. Pt. I, Ch. 1, notes 2 and 15). Of its 51 pages, 24 are in Einstein's hand (with an occasional entry by Besso), 24 in Besso's (most of them without any contribution from Einstein), while 3 pages, i.e., [p. 33] and [pp. 41–42], contain substantial contributions from both.

These 51 pages were sold at auction at Christie's in 1996 and again in 2002, fetching $360K and $500K, respectively (Coover 1996, 2002). One of us wrote an essay on the manuscript for the auction catalogs (Janssen 1996). In 2002, the manuscript was acquired by Aristophil in Paris. This company published, in a limited edition, a boxed set with a reprint of the auction-catalog essay (illustrated by Laurent Taudin) and a facsimile of the manuscript (Einstein and Besso 2003). Closely matching the original in physical appearance, this facsimile comes as a folder with 37 loose sheets. It contains four pages not deemed significant enough for inclusion in CPAE4: a printed letter, dated "Ende April 1913," soliciting subscriptions to a *Festschrift* for the Swiss historian Gerald Meyer von Knonau (1843–1931), which has [p. 2] on the verso; the verso of a large folded sheet with [pp. 10–11] on the recto; and the verso of [p. 20], which has a figure and parts of an equation related to rotation.

As reported in the New York Times on February 21, 2020, "the French authorities [in 1915] shut down Aristophil and arrested [its founder and president], charging him with fraud and accusing him of orchestrating what amounts to a highbrow Ponzi scheme. As he bought ... rare manuscripts and letters, he had them appraised, divided their putative value into shares and sold them as if they were stock in a corporation. Those shares were bought by 18,000 people, many of them elderly and of modest means, who collectively invested about $1 billion." The New York Times article notes that the Einstein-Besso manuscript, the first item used this way by Aristophil, was "divided into hundreds of shares and sold at a valuation of $13 million." The manuscript was sold at auction again in 2021, this time fetching a record sum of $11.5 million, about twenty times more than in 2002 and not much less than the seemingly highly inflated value assigned to it by Aristophil's pliant appraisers.

In the editorial note in CPAE4 (p. 358), the 53 pages of the Einstein-Besso manuscript are divided into two parts: Part One written in close collaboration during a well-documented visit of Besso to Einstein in Zurich in June 1913 (42 pages) [we know that on June 18 Besso went back to Gorizia, where he was living in at the time; CPAE4, p. 357, note 57]; Part Two produced by Besso alone after Einstein sent him some papers, presumed to be Part One, in early 1914 (11 pages). In his (undated) cover letter (CPAE5, Doc. 499), Einstein wrote: "Here you finally have your manuscript bundle. It is really a shame if you do not bring the matter to completion."

Since the publication of the Einstein-Besso manuscript in CPAE4 in 1995, more pages of Part Two have come to light. These pages are all in Besso's hand and do not contain a single contribution from Einstein. They were part of another sheaf of loose sheets from the Besso *Nachlass* that Laurent Besso, Vero's grandson, made available to Robert Schulmann, then director of the Einstein Papers Project, in 1998. They include 14 additional pages of the Einstein-Besso manuscript, which can all be dated to 1913–14, and 8 pages with notes on general relativity, which can all be dated to 1916 (Janssen 2007, sec. 2).

The new material strongly suggests that Einstein and Besso met again in August 1913 to resume work on their joint project (see Janssen 2007, secs. 2.1.2 and 2.2, on the content and the dating of the new pages, respectively). Part One of the Einstein-Besso manuscript should accordingly be divided into two parts, the first part written during Besso's visit in June 1913, the second part during his visit in August 1913. Some of the Einstein pages could have been written any time between June 1913 and early 1914, when Einstein presumably sent the manuscript to Besso.

The most important item in the new part of the Einstein-Besso manuscript is the so-called "Besso Memo". Unlike any other pages of the manuscript, Besso dated the first of its four pages (which are written on a folded sheet): August 28, 1913. Two pages of the memo are reproduced in facsimile in Janssen (2007, p. 786, p. 789). In this memo, Besso appears to have recorded some of his discussions with Einstein at the time about the problem of rotation (see Ch. 4 below and Janssen 2007, sec. 3) and about what would become the hole argument (see Pt. I, Sec. 4.1, above and Janssen 2007, sec. 4).

With the exception of [pp. 41–42], the 14 pages presented in this chapter can reliably be dated to June 1913 and thus belong to Part One. The Besso material on [pp. 41–42] (which is not important for our purposes here) belongs to Part Two. Most likely, the Einstein material, which again bears on the problem of rotation (Janssen 1999), was written in June 1913 but it cannot be ruled out that Einstein only added it later that year (Janssen 2007, p. 809).

The other 12 pages selected for inclusion here, 9 by Einstein and 3 by Besso, deal with the perihelion motion of a single planet in the static field of the sun. Both the *Entwurf* theory and general relativity in its final form predict that the field of the sun contributes to a planet's perihelion motion. In Newton's theory, any perihelion motion is due to perturbations of other planets (see the tables in Misner, Thorne, and Wheeler (1973, p. 1113) or Smith (2014, p. 310 and 314) for the actual numbers). In fact, Newton had shown that a planet's perihelion would be stationary as long as there is only a central inverse-square force acting on it (Harper 2011, pp. 120–121).

On [pp. 1, 3–4, 6–7], Einstein solved the *Entwurf* field equations for the static field of the sun using an iterative approximation procedure. On [pp. 8–9], Besso took over and derived a differential equation for the angle between perihelion and aphelion of a planet moving in this field. The deviation of this angle from π gives the perihelion advance in radians per half a revolution. On [pp. 10–11], Einstein performed some contour integrations to find this angle. On [p. 14] Besso expressed the formula Einstein found in terms of observable parameters. On [pp. 26 and 28], using one of the formulae Besso derived, Einstein calculated the perihelion advance of Mercury in radians per half a revolution and converted the result to seconds of arc per century. It can no longer be established from which source(s) Einstein got the various numbers he needed. It is clear, however, from several pages of the manuscript (e.g., [p. 31], not included here, and the bottom half of [p. 41]) that Besso consulted Simon Newcomb's (1895) *The Elements of the Four Inner Planets.*

Even at a superficial glance, one notices the striking difference between pages by Besso and pages by Einstein. The former have much more explanatory text to go with the calculations and many more deletions. The pages we selected are representative of the manuscript as a whole in this respect. Besso comes across as tentative and unsure of himself, Einstein as supremely confident and sure-footed. Besso nonetheless took responsibility for key parts of their joint project and corrected some errors in Einstein's calculations (see, in particular, [p. 4, eq. 28], [p. 6, eqs. 36–37]; [p. 35, eq. 211] (see Fig. 6.2 in Pt. I, Ch. 6); and [p. 14] where he corrected Einstein's overly crude approximations on [pp. 10–11]). He also raised at least one important question. On [p. 16] (not included here), in the course of summarizing Einstein's iterative procedure for solving the *Entwurf* field equations, he asked: "Is the static gravitational field [that Einstein had found for the sun] a particular solution or is it the general solution expressed in particular coordinates?" This same question comes up in the perihelion paper two years later (Einstein 1915c, p. 832; see Sec. 6.3). As one of us (JR) first suggested, Besso's original question may have been what inspired Einstein's hole argument (Janssen 2007, p. 820; cf. Pt. I, Sec. 4.1).

[P. 1] Einstein

For his iterative calculation of the metric field of the sun, Einstein used the *Entwurf* field equations in their 'contravariant' form $\Delta_{\mu\nu}(\gamma) = \kappa(\Theta_{\mu\nu} + \vartheta_{\mu\nu})$ (Einstein and Grossmann 1913, p. 17, eq. (18)). $\Theta_{\mu\nu}$ is Einstein's notation for the contravariant components $T^{\mu\nu}$ of the energy-momentum tensor of matter. $\vartheta_{\mu\nu}$ is his notation for the 'contravariant' components of a quantity representing gravitational energy-momentum. We use scare quotes to indicate that this is not a generally covariant tensor. Neither is $\Delta_{\mu\nu}(\gamma)$.

Einstein expanded the metric field as a power series in the small parameter A/r, of the order of magnitude of the Newtonian potential. In our notation,

$$g^{\mu\nu} = \overset{(0)}{g}{}^{\mu\nu} + \overset{(1)}{g}{}^{\mu\nu} + \overset{(2)}{g}{}^{\mu\nu} + \ldots$$

For the first term, Einstein simply took the standard diagonal Minkowski metric, $\eta^{\mu\nu} = \mathrm{diag}(-1, -1, -1, 1/c^2)$, in an arbitrary Lorentz frame.

To find the next term, he considered the *Entwurf* field equations for the weak static field produced by slow moving dust,

$$-\Box \overset{(1)}{g}{}^{\mu\nu} = \kappa\rho \frac{dx^\mu}{ds}\frac{dx^\nu}{ds}, \qquad \text{[eq. 1]}$$

where $\Box = \Delta - (\partial/\partial t)^2$ and the right-hand side is the product of Einstein's gravitational constant κ and the energy-momentum tensor for slow moving dust. Einstein used the notation $\gamma_{\mu\nu}$ for the contravariant metric $g^{\mu\nu}$ (see Sec. 1.3, note 1). The only non-trivial component of this equation is the 44-component:

$$-\Delta \overset{(1)}{g}{}^{44} = \frac{\kappa\rho}{c^2}, \qquad \text{[eq. 2]}$$

where we used that $dx^4/ds = (dx^4/dt)(dt/ds) \approx 1/c$ (note that $x^\mu \equiv (x, y, z, t)$). Einstein explicitly wrote c_0 for the velocity of light in vacuo. This may be a remnant of his 1912 theory for static gravitational fields in which the velocity of light played the role of the gravitational potential and thus, in general, was not a constant but a function of (x, y, z).

The solution of this equation for the special case of a spherical body of mass M, representing the sun, at some coordinate distance $r \equiv \sqrt{x^2+y^2+z^2}$ from the center, is given by

$$\overset{(1)}{g}{}^{44} = \frac{1}{c^2}\left(1+\frac{A}{r}\right),$$

where

$$A \equiv \frac{\kappa M}{4\pi}. \qquad \text{[eq. 6]}$$

Inverting the matrix for $\overset{(1)}{g}{}^{\mu\nu}$ we find:

$$\overset{(1)}{g}_{44} = c^2\left(1-\frac{A}{r}\right).$$

These are the 44-components of the matrices for $\overset{(1)}{g}_{\mu\nu}$ and $\overset{(1)}{g}{}^{\mu\nu}$ in [eqs. 4–5] under the header "table of the gravitational field for the first approximation."

To find the relation between κ and Newton's gravitational constant K (see [eqs. 3 and 6]), Einstein considered the equation of motion for a unit point mass in a static field (see [p. 13], not included here). The equation of motion can be found in the Zurich notebook (p. 5R/p. 11), in the *Entwurf* paper (Einstein and Grossmann 1913, p. 7, Eqs. 7 and 8), and on [p. 8, eqs. 48–49, 54], where Besso cites the *Entwurf* paper. For the static metric $\overset{(1)}{g}{}^{\mu\nu}$, it reduces to

$$\frac{d}{dt}\left(\frac{\dot{x}}{ds/dt}\right) = -\frac{1}{2}\frac{\partial \overset{(1)}{g}_{44}}{\partial x}\frac{dx^4}{ds}\frac{dx^4}{dt},$$

where the dot indicates a time derivative. Multiplying both sides by ds/dt, which can be taken to be constant, and using that $dx^4/dt = 1$, we arrive at

$$\ddot{x} = -\frac{1}{2}\frac{\partial \overset{(1)}{g}_{44}}{\partial x} = -\frac{\partial}{\partial x}\left(-\frac{\kappa M c^2}{8\pi r}\right).$$

Comparing this to Newton's second law for a unit point mass moving in the gravitational field of the sun,

$$\ddot{x} = -\frac{\partial \varphi}{\partial x} = -\frac{\partial}{\partial x}\left(-\frac{KM}{r}\right),$$

we conclude that:

$$\kappa = \frac{8\pi K}{c^2}. \qquad \text{[eq. 3]}$$

To find the field in second approximation, we need to include the terms quadratic in first-order derivatives of the metric in the *Entwurf* field equations that could be neglected in the first-order approximation. Inserting the zeroth- and first-order terms $\overset{(0)}{g}{}^{\mu\nu}$ and $\overset{(1)}{g}{}^{\mu\nu}$ into these quadratic terms, we arrive at the following equation for $\overset{(2)}{g}{}^{\mu\nu}$ (CPAE4, p. 349, Eq. (2)):

$$-\Delta \overset{(2)}{g}{}^{\mu\nu} + \frac{\overset{(0)}{g}{}^{\alpha\beta}}{\sqrt{-\overset{(0)}{g}}}\sqrt{-\overset{(1)}{g}}_{,\alpha}\,\overset{(1)}{g}{}^{\mu\nu}{}_{,\beta} - \overset{(0)}{g}{}^{\alpha\beta}\,\overset{(0)}{g}_{\tau\rho}\,\overset{(1)}{g}{}^{\mu\tau}{}_{,\alpha}\,\overset{(1)}{g}{}_{\nu\rho,\beta}$$
$$= -\frac{1}{2}\overset{(0)}{g}{}^{\alpha\mu}\,\overset{(0)}{g}{}^{\beta\nu}\,\overset{(1)}{g}_{\tau\rho,\alpha}\,\overset{(1)}{g}{}^{\tau\rho}{}_{,\beta} + \frac{1}{4}\overset{(0)}{g}{}^{\mu\nu}\,\overset{(0)}{g}{}^{\alpha\beta}\,\overset{(1)}{g}_{\tau\rho,\alpha}\,\overset{(1)}{g}{}^{\tau\rho}{}_{,\beta} \qquad (2.1)$$

The five terms in the equation above can be found on the bottom half of [p. 1], under the heading "Second approximation," the three terms on the left-hand side to the left of the vertical line, the two terms on the right-hand side to the right. Note that Einstein used subscripts '0' and '×' to mark the zeroth- and second-order terms $\overset{(0)}{g}{}^{\mu\nu}$ and $\overset{(2)}{g}{}^{\mu\nu}$, respectively.

Einstein substituted the expressions for $\overset{(1)}{g}{}^{\mu\nu}$ in [eqs. 4–5] into the five terms in Eq. (2.1) and recorded the results. We only present the calculations of the second term on the left-hand side. Only the 44-component of that term is non-vanishing. Using that

$$\sqrt{-\overset{(0)}{g}} = c, \quad \overset{(0)}{g}{}^{ij} = -\delta^{ij}, \quad \delta_{ij}x^i x^j = r^2, \quad \frac{\partial r}{\partial x^i} = \frac{x^i}{r},$$

$$\sqrt{-\overset{(1)}{g}}_{,i} = c\frac{\partial}{\partial x^i}\sqrt{1-\frac{A}{r}} = \frac{c}{2\sqrt{1-\frac{A}{r}}}\frac{A}{r^2}\frac{x^i}{r} \approx \frac{cA}{2r^2}\frac{x^i}{r},$$

$$\text{and } \overset{(1)}{g}{}^{44}_{,i} = \frac{1}{c^2}\frac{\partial}{\partial x^i}\left(1+\frac{A}{r}\right) = -\frac{A}{c^2r^2}\frac{x^i}{r},$$

we find that, to second order in A/r,

$$\frac{1}{\sqrt{-\overset{(0)}{g}}}\overset{(0)}{g}{}^{ij}\sqrt{-\overset{(1)}{g}}_{,i}\,\overset{(1)}{g}{}^{44}_{,j} = -\frac{1}{c}\delta_{ij}\left(\frac{cAx^i}{2r^3}\right)\left(-\frac{Ax^j}{c^2r^3}\right) = \frac{A^2}{2c^2r^4}.$$

The expressions on the bottom half of [p. 1] for the other terms in Eq. (2.1) are recovered in similar fashion.

[P. 7] Einstein

On [p. 7], the verso of [p. 6], Einstein derived the general form of a spherically symmetric metric in Cartesian coordinates (on [p. 5], not included here, Besso more laboriously went through the same derivation; cf. Droste (1915, pp. 999–1000) and Earman and Janssen (1993, p. 144)). It is unclear why Einstein crossed out this entire calculation. Maybe it is because he realized he needed the general form of the contravariant $\overset{(2)}{g}{}^{\mu\nu}$ but derived the general form of a covariant metric on this page. However, the covariant and the contravariant metrics have the same form in this case (Droste 1915, p. 1000). The material in the lower-right corner is in Besso's hand.

As the figure in the top-left corner illustrates, Einstein first determined the metric at an arbitrary point P in a Cartesian coordinate system (x', y', z') in which P lies on the x'-axis and then transformed to an arbitrary Cartesian coordinate system (x, y, z). In this primed coordinate system, the metric at P, with coordinates $(x', y', z') = (r, 0, 0)$, has a very simple form. Since it is static, $g_{i4} = g_{4i} = 0$. Since it is spherically symmetric, $g'_{ij} = 0$ whenever $i \neq j$ and $g'_{22} = g'_{33}$. Einstein set

$$g'_{11} = R, \quad g'_{22} = g'_{33} = T, \quad g'_{44} = \Phi, \qquad \text{[eq. 43]}$$

where R, T, and Φ are yet to be determined functions of r. He then wrote down the transformation law for the metric from the primed to the unprimed coordinates:

$$g_{\mu\nu} = \pi_{\mu\alpha}\pi_{\nu\beta}\,g'_{\alpha\beta}, \qquad \text{[eq. 44]}$$

where $\pi_{\mu\alpha}$ is the notation he used during this period for the transformation matrix $\partial x'^{\alpha}/\partial x^{\mu}$ (cf. Sec. 1.3, note 20). Since the transformation only affects the spatial coordinates, $g_{44} = \pi_{44}\,\pi_{44}\,g'_{44} = g'_{44}$. Einstein only explicitly computed g_{11} and g_{12}. Using [eq. 43] and $\pi_{i1} = \partial x'^1/\partial x^i = \partial r/\partial x^i = x^i/r$, he found that

$$g_{11} = \pi_{11}^2\,g'_{11} + \pi_{12}^2\,g'_{22} + \pi_{13}^2\,g'_{33} = \frac{x^2}{r^2}R + \frac{y^2}{r^2}T + \frac{z^2}{r^2}T = \frac{x^2}{r^2}(R-T) + T. \quad \text{[eq. 46]}$$

For g_{12}, [eqs. 43 and 44] give:

$$g_{12} = \pi_{11}\,\pi_{21}\,R + \pi_{12}\,\pi_{22}\,T + \pi_{13}\,\pi_{23}\,T.$$

Since π_{ij} is the matrix of a rotation, its inverse is just the transposed matrix, $\pi_{ij}^{-1} = \pi_{ij}^{\mathrm{T}} = \pi_{ji}$. Hence, $\sum_k \pi_{ik}\pi_{jk} = \sum_k \pi_{ik}\pi_{kj}^{-1} = \delta_{ij}$. Einstein could thus use that $\sum_k \pi_{1k}\,\pi_{2k} = 0$ to substitute $-\pi_{11}\,\pi_{21}$ for $\pi_{12}\,\pi_{22} + \pi_{13}\,\pi_{23}$ and rewrite g_{12} as

$$g_{12} = \pi_{11}\,\pi_{21}\,(R-T) = \frac{xy}{r^2}(R-T). \quad \text{[eq. 47]}$$

The other components of $g_{\mu\nu}$ can be found in the same way. The end result is given in [eq. 17] at the top of [p. 3].

[P. 3] Einstein

At the top of [p. 3], under the header "Form of $\gamma_{\mu\nu}$ [i.e., $g^{\mu\nu}$] in the case of spherical symmetry," Einstein wrote down the matrix for a spherically symmetric (contravariant) metric $g^{\mu\nu}$ in Cartesian coordinates. A derivation of (the covariant form of) this metric can be found on [p. 7]. At this point, Einstein introduced the function $N \equiv R - T$ ([eq. 18]). With the help of this function, the spherically symmetric metric can be written as:

$$g^{ij} = \frac{x^i x^j}{r^2}N(r) + \delta^{ij}T(r), \quad g^{44} = \Phi(r), \quad g^{i4} = g^{4i} = 0. \qquad \text{[eq. 17]}$$

Einstein noted that the determinant $|g^{\mu\nu}|$—which in his notation is γ and not $g \equiv |g_{\mu\nu}|$ (cf. [p. 6, eq. 41])—is given by

$$|g^{\mu\nu}| = \Phi T^2(N+T). \qquad \text{[eq. 19]}$$

This follows directly from the diagonal form in which the metric is introduced on [p. 7] in the primed coordinate system (see [p. 7, eq. 43]): $|g^{\mu\nu}| = |g'^{\mu\nu}| = RT^2\Phi$. Substituting $R = N + T$, we recover [eq. 19].

On [p. 1], Einstein already determined that $\overset{(0)}{g}{}^{\mu\nu} = \eta^{\mu\nu}$ and that

$$\overset{(1)}{g}{}^{\mu\nu} = \mathrm{diag}\left(0,0,0,\frac{1}{c^2}\frac{A}{r}\right).$$

To first order in A/r, the three functions determining the metric are therefore given by

$$\overset{(0)}{N} = \overset{(1)}{N} = 0, \quad \overset{(0)}{T} = -1, \quad \overset{(1)}{T} = 0, \quad \overset{(0)}{\Phi} = \frac{1}{c^2}, \quad \overset{(1)}{\Phi} = \frac{1}{c^2}\frac{A}{r} \qquad (2.2)$$

To find the second-order contributions to these functions, Einstein substituted [eq. 17] for $\overset{(2)}{g}{}^{\mu\nu}$ in the first term on the left-hand side of Eq. (2.1). But first he substituted the zeroth- and first-order solutions $\overset{(0)}{g}{}^{\mu\nu}$ and $\overset{(1)}{g}{}^{\mu\nu}$ in the remaining terms of Eq. (2.1). Under the header "Differential equations for γ" (i.e., $\overset{(2)}{g}{}^{\mu\nu}$), he wrote down the result for a few components.

$$-\Delta \overset{(2)}{g}^{11} = \frac{A^2}{2}\left(\left(\frac{\partial}{\partial x}\left(\frac{1}{r}\right)\right)^2 - \frac{1}{2r^4}\right) = -\frac{A^2}{4r^4} + x^2\frac{A^2}{2r^6}. \qquad \text{[eq. 20]}$$

$$-\Delta \overset{(2)}{g}^{12} = \frac{A^2}{2}\frac{\partial}{\partial x}\left(\frac{1}{r}\right)\frac{\partial}{\partial y}\left(\frac{1}{r}\right) = xy\frac{A^2}{2r^6} \qquad \text{[eq. 21]}$$

(the material to the right of [eq. 21] is in Besso's hand). For the 44-component, finally, he found (modulo an erroneous minus sign):

$$\Delta \overset{(2)}{g}^{44} = \frac{5}{4}\frac{A^2}{c^2 r^4}. \qquad \text{[eq. 22] (corrected)}$$

On [p. 4], Einstein derived an expression for the action of Δ on the 11-component of the metric in [eq. 17]. On the bottom half of [p. 3], the resulting equations for $\overset{(2)}{N}$ and $\overset{(2)}{T}$ are stated without derivation (see [eqs. 23–25]). [Eq. 26] for $\overset{(2)}{\Phi}$ follows directly from [eq. 22] and inherits the sign error from it. Moreover, the factor $1/c^2$ got lost.

Einstein crossed out the calculation on the bottom half of [p. 3] and started over at the top of [p. 6]. We will briefly return to the crossed-out material on [p. 3] after we have covered the derivation on [p. 4].

[P. 4] Einstein

On [p. 4] Einstein derived the equations for the contributions of second order in A/r to the functions $N(r)$, $T(r)$, and $\Phi(r)$ introduced in [p. 3, eq. 17]. Rewriting the left-hand side of [p. 3, eq. 20] in terms of these functions, we find:

$$\Delta \overset{(2)}{g}^{11} = \Delta\left(\frac{x^2}{r^2}\overset{(2)}{N} + \overset{(2)}{T}\right). \qquad \text{[eq. 27]}$$

Einstein now derived equations for $\Delta\overset{(2)}{T}$ and $\Delta(x^2\overset{(2)}{N}/r^2)$ (see [eqs. 28–29]). In our reconstruction of the calculations on [p. 4], we will suppress these superscripts, as they would further clutter the already tedious though straightforward algebra, but all N's and T's below should be read as $\overset{(2)}{N}$'s and $\overset{(2)}{T}$'s.

Einstein began by computing $\partial^2 T/\partial x^2$:

$$\frac{\partial^2 T}{\partial x^2} = \frac{\partial}{\partial x}\left(\frac{dT}{dr}\frac{x}{r}\right) = \frac{d^2T}{dr^2}\frac{x^2}{r^2} + \frac{1}{r}\frac{dT}{dr} - \frac{dT}{dr}\frac{x}{r^2}\frac{x}{r}.$$

The first and the third term on the right-hand side combine to form

$$\frac{x^2}{r}\frac{d}{dr}\left(\frac{1}{r}\frac{dT}{dr}\right).$$

Using similar expressions for $\partial^2 T/\partial y^2$ and $\partial^2 T/\partial z^2$, we find that

$$\Delta T = \frac{3}{r}\frac{dT}{dr} + r\frac{d}{dr}\left(\frac{1}{r}\frac{dT}{dr}\right) \qquad \text{[eq. 28]}$$

(the second term on the right-hand side originally had $1/r$ instead of r: the correction is in Besso's hand).

Next, Einstein turned his attention to $\dfrac{\partial^2}{\partial x^2}\left(x^2\dfrac{N}{r^2}\right)$. The first derivative is given by

$$\frac{\partial}{\partial x}\left(x^2\frac{N}{r^2}\right) = 2x\frac{N}{r^2} + \frac{x^2}{r^2}\frac{dN}{dr}\frac{x}{r} - \frac{2x^2}{r^3}\frac{x}{r}N.$$

The last two terms combine to form

$$\frac{x^3}{r}\frac{d}{dr}\left(\frac{N}{r^2}\right).$$

The second derivative is thus given by

$$\frac{\partial^2}{\partial x^2}\left(x^2\frac{N}{r^2}\right) = \frac{\partial}{\partial x}\left(\frac{2xN}{r^2} + \frac{x^3}{r}\frac{d}{dr}\left(\frac{N}{r^2}\right)\right).$$

Working out the first term on the right-hand side, we get

$$\frac{\partial}{\partial x}\left(\frac{2xN}{r^2}\right) = \frac{2N}{r^2} + \frac{2x}{r^2}\frac{dN}{dr}\frac{x}{r} - \frac{4xN}{r^3}\frac{x}{r} = \frac{2N}{r^2} + \frac{2x^2}{r}\frac{d}{dr}\left(\frac{N}{r^2}\right)$$

(where in the last step we combined two terms same way we did above); working out the second term, we get

$$\frac{\partial}{\partial x}\left(\frac{x^3}{r}\frac{d}{dr}\left(\frac{N}{r^2}\right)\right) = \frac{3x^2}{r}\frac{d}{dr}\left(\frac{N}{r^2}\right) - \frac{x^3}{r^2}\frac{x}{r}\frac{d}{dr}\left(\frac{N}{r^2}\right) + \frac{x^3}{r}\frac{\partial}{\partial x}\left(\frac{d}{dr}\left(\frac{N}{r^2}\right)\right).$$

We examine the third term on the right-hand side separately:

$$\begin{aligned}\frac{\partial}{\partial x}\left(\frac{d}{dr}\left(\frac{N}{r^2}\right)\right) &= \frac{\partial}{\partial x}\left(\frac{1}{r^2}\frac{dN}{dr} - \frac{2N}{r^3}\right)\\ &= \frac{1}{r^2}\frac{d^2N}{dr^2}\frac{x}{r} - \frac{2}{r^3}\frac{x}{r}\frac{dN}{dr} - \frac{2}{r^3}\frac{dN}{dr}\frac{x}{r} + \frac{6}{r^4}\frac{x}{r}N.\end{aligned}$$

Collecting all terms proportional to x^2 in $\dfrac{\partial^2}{\partial x^2}\left(x^2\dfrac{N}{r^2}\right)$, we find $\dfrac{5x^2}{r}\dfrac{d}{dr}\left(\dfrac{N}{r^2}\right)$; collecting all terms proportional to x^4, we find

$$-\frac{x^4}{r^3}\left(\frac{1}{r^2}\frac{dN}{dr} - \frac{2N}{r^3}\right) + \frac{x^4}{r}\left(\frac{1}{r^3}\frac{d^2N}{dr^2} - \frac{4}{r^4}\frac{dN}{dr} + \frac{6N}{r^5}\right),$$

which reduces to

$$\frac{x^4}{r}\left(\frac{1}{r^3}\frac{d^2N}{dr^2} - \frac{5}{r^4}\frac{dN}{dr} + \frac{8N}{r^5}\right).$$

One readily verifies that the factor in parentheses can be written more compactly:

$$\begin{aligned}\frac{d}{dr}\left(\frac{1}{r}\frac{d}{dr}\left(\frac{N}{r^2}\right)\right) &= \frac{d}{dr}\left(\frac{1}{r^3}\frac{dN}{dr} - \frac{2N}{r^4}\right)\\ &= \frac{1}{r^3}\frac{d^2N}{dr^2} - \frac{3}{r^4}\frac{dN}{dr} - \frac{2}{r^4}\frac{dN}{dr} + \frac{8N}{r^5}.\end{aligned}$$

Einstein thus arrived at the following intermediate result:

$$\frac{\partial^2}{\partial x^2}\left(x^2\frac{N}{r^2}\right) = \frac{2N}{r^2} + \frac{5x^2}{r}\frac{d}{dr}\left(\frac{N}{r^2}\right) + \frac{x^4}{r}\frac{d}{dr}\left(\frac{1}{r}\frac{d}{dr}\left(\frac{N}{r^2}\right)\right).$$

He similarly found that

$$x^2\frac{\partial^2}{\partial y^2}\left(\frac{N}{r^2}\right)=\frac{x^2}{r}\frac{d}{dr}\left(\frac{N}{r^2}\right)+\frac{x^2y^2}{r}\frac{d}{dr}\left(\frac{1}{r}\frac{d}{dr}\left(\frac{N}{r^2}\right)\right).$$

If y is replaced by z this turns into the expression for $\partial^2/\partial z^2$ acting on x^2N/r^2.

Adding the expressions for $\partial^2/\partial x^2$, $\partial^2/\partial y^2$, and $\partial^2/\partial z^2$ acting on x^2N/r^2, we obtain:

$$\Delta\left(x^2\frac{N}{r^2}\right)=\frac{2N}{r^2}+\frac{7x^2}{r}\frac{d}{dr}\left(\frac{N}{r^2}\right)+x^2r\frac{d}{dr}\left(\frac{1}{r}\frac{d}{dr}\left(\frac{N}{r^2}\right)\right). \qquad \text{[eq. 29]}$$

Adding [eq. 28] to [eq. 29] and substituting the resulting expression for $\Delta\left(\frac{x^2}{r^2}N+T\right)$ (cf. [eq. 27]) into [p. 3, eq. 20], Einstein arrived at

$$\underbrace{\frac{3}{r}\frac{dT}{dr}+r\frac{d}{dr}\left(\frac{1}{r}\frac{dT}{dr}\right)+\frac{2N}{r^2}}_{=\frac{A^2}{4r^4}}+x^2\underbrace{\left(\frac{7}{r}\frac{d}{dr}\left(\frac{N}{r^2}\right)+r\frac{d}{dr}\left(\frac{1}{r}\frac{d}{dr}\left(\frac{N}{r^2}\right)\right)\right)}_{=-\frac{A^2}{2r^6}}. \qquad \text{[eq. 30] (a), (b)}$$

As indicated by the upbrackets, the equation for $\overset{(2)}{g}{}^{11}$ splits into two equations, one for the terms proportional to x^2 and one for the remaining terms (the material in the lower-left corner, under [eq. 30] (a), is in Besso's hand).

[Eq. 30] is fully equivalent to [p. 3, eq. 23]. Like [eq. 30], [eq. 23] splits into two parts:

$$\frac{d^2T}{dr^2}+\frac{2}{r}\frac{dT}{dr}+\frac{2}{r^2}N=\frac{A^2}{4r^4}, \qquad \text{[p. 3, eq. 24]}$$

$$-\frac{6}{r^2}N+\frac{2}{r}\frac{dN}{dr}+\frac{d^2N}{dr^2}=-\frac{A^2}{2r^4}. \qquad \text{[p. 3, eq. 25]}$$

One readily verifies the equivalence of [eq. 30] (a)–(b) and [p. 3, eqs. 24–25].

As Einstein noted on [p. 3] ("second equation gives nothing new"), [eq. 21] for $\overset{(2)}{g}{}^{12}$ leads to equations for $\overset{(2)}{T}$ and $\overset{(2)}{N}$ that are equivalent to [eqs. 24–25]/[eqs. 30] (a)–(b).

The equation for $\overset{(2)}{\Phi}$, the second-order contribution to the third function determining the metric, is not given on [p. 4] but can be found (with a spurious minus sign and a factor $1/c^2$ missing) in the crossed out part at the bottom of [p. 3]:

$$\Delta\overset{(2)}{\Phi}=\frac{5}{4}\frac{A^2}{c^2r^4}. \qquad \text{[p. 3, eq. 26]}$$

[P. 6] Einstein

On [p. 6], Einstein gave the end result of his derivation of the spherically symmetric metric field of the sun in Cartesian coordinates to second order in the parameter A/r proportional to the Newtonian gravitational potential.

At the top of [p. 6], we once again find the equations derived on [pp. 3–4] for the second-order contributions to the functions $N(r)$, $T(r)$, and $\Phi(r)$. These equations, [eqs. 36 and 37], are identical to [p. 4, eqs. 30] (b) and (a), respectively, and equivalent to [p. 3, eqs. 25 and 24], respectively. [Eq. 38] is (the corrected version of) [p. 3, eq. 26] for $\overset{(2)}{\Phi}$ with the Laplacian Δ expressed in spherical coordinates (cf. [p. 4, eq. 28] for $\Delta\overset{(2)}{T}$):

$$\frac{3}{r}\frac{d\overset{(2)}{\Phi}}{dr}+r\frac{d}{dr}\left(\frac{1}{r}\frac{d\overset{(2)}{\Phi}}{dr}\right)=\frac{5}{4}\frac{A^2}{c^2r^4}. \qquad \text{[eq. 38]}$$

As in [p. 4, eq. 28], Besso corrected a factor $1/r$ to r in the expressions for $\Delta\overset{(2)}{T}$ and $\Delta\overset{(2)}{\Phi}$ in [eqs. 37–38].

To the right of the vertical line drawn next to [eqs. 36–38] for $\overset{(2)}{N}$, $\overset{(2)}{T}$, and $\overset{(2)}{\Phi}$, Einstein listed the solutions of these equations:

$$\overset{(2)}{N}=\frac{1}{8}\frac{A^2}{r^2}, \quad \overset{(2)}{T}=0, \quad \overset{(2)}{\Phi}=\frac{1}{c^2}\frac{5}{8}\frac{A^2}{r^2}. \qquad \text{[eq. 39]}$$

As expected, these solutions result in contributions of second order in A/r to the metric. One readily verifies that these are indeed solutions by inserting them into [eq. 38] and [eqs. 36–37].

Adding these second-order contributions to N, T, and Φ to the zeroth- and first-order contributions found earlier (see Eq. 2.2), we find

$$N=\frac{1}{8}\frac{A^2}{r^2}, \quad T=-1, \quad \Phi=\frac{1}{c^2}\left(1+\frac{A}{r}+\frac{5}{8}\frac{A^2}{r^2}\right).$$

Substituting these expressions into [p. 3, eq. 17], Einstein arrived at the following result for the contravariant components of the metric field of the sun to second order in A/r:

$$g^{ij}=-\delta^{ij}+\frac{1}{8}\frac{A^2}{r^2}\frac{x^ix^j}{r^2}, \quad g^{44}=\frac{1}{c^2}\left(1+\frac{A}{r}+\frac{5}{8}\frac{A^2}{r^2}\right), \quad g^{i4}=g^{4i}=0. \qquad \text{[eq. 40]}$$

Inverting this matrix, Einstein found the covariant components of this metric field to second order in A/r:

$$g_{ij}=-\delta^{ij}-\frac{1}{8}\frac{A^2}{r^2}\frac{x^ix^j}{r^2}, \quad g_{44}=c^2\left(1-\frac{A}{r}+\frac{3}{8}\frac{A^2}{r^2}\right), \quad g_{i4}=g_{4i}=0. \qquad \text{[eq. 42]}$$

As becomes clear on [pp. 8–9], only g_{44} is needed to second order in A/r to calculate the motion of a planet's perihelion in the field of the sun. On [p. 2], not included here, Einstein went through a calculation similar to the one on [pp. 1, 3–4, 6–7] but starting from the 'covariant' form of the *Entwurf* field equations, $-D_{\mu\nu}=\kappa(t_{\mu\nu}+T_{\mu\nu})$ (Einstein and Grossmann 1913, p. 17, eq. (21)) and deriving only the expression for g_{44} to order A^2/r^2. [P. 2] is the verso of a printed letter dated "Ende April 1913." Einstein, in all likelihood, did the calculations on [pp. 1, 3–4, 6–7] before those on [p. 2]. It cannot be determined, of course, exactly when Einstein cannibalized this piece of paper for his calculations. But it is plausible that this would have been around the time of Besso's visit in June 1913.

[P. 8] Besso

On [pp. 8–9], Besso used the equations of motion of a test particle in a metric field to derive a differential equation for the angle between perihelion and aphelion of the orbit of a single planet in the field of the sun given by [p. 6, eq. 42].

He started from the Lagrangian for a test particle of rest mass m in an arbitrary metric field. Following Einstein and Grossmann (1913, p. 7, eq. 5), Besso used H instead of L for the Lagrangian:

$$H = -m\frac{ds}{dt} = -m\sqrt{\frac{ds^2}{dt^2}} = -m\sqrt{g_{\mu\nu}\dot{x}^\mu\dot{x}^\nu}.$$

The Euler-Lagrange equations are (p. 7, eq. 6):

$$\frac{d}{dt}\left(\frac{\partial H}{\partial \dot{x}^\mu}\right) - \frac{\partial H}{\partial x^\mu} = 0.$$

If the generalized momentum and the gravitational four-force are defined as $J^\mu \equiv \partial H/\partial \dot{x}^\mu$ and $\mathfrak{K}^\mu \equiv \partial H/\partial x^\mu$, respectively (p. 7, eqs. 7–8), the x-component of the Euler-Lagrange equations can be written as $dJ_x/dt = \mathfrak{K}_x$ ([eq. 48]), where

$$J_x = \frac{\partial H}{\partial \dot{x}^1} = -\frac{m}{2\sqrt{ds^2/dt^2}}2g_{1\nu}\dot{x}^\nu = -\frac{mg_{1\nu}dx^\nu}{ds},$$

and

$$\mathfrak{K}_x = -\frac{\partial H}{\partial x^1} = -\frac{m}{2\sqrt{ds^2/dt^2}}\frac{\partial g_{\mu\nu}}{\partial x^1}\dot{x}^\mu\dot{x}^\nu = -\frac{1}{2}m\frac{\partial g_{\mu\nu}}{\partial x^1}\frac{dx^\mu}{ds}\frac{dx^\nu}{dt}.$$

Under the header, "The equations of motion for the material point are," Besso could thus write down the x-component of the equations of motion as

$$\frac{d}{dt}\left(\frac{g_{1\nu}\dot{x}^\nu}{ds/dt}\right) = \frac{1}{2}\frac{\partial g_{\mu\nu}}{\partial x^1}\frac{dx^\mu}{dt}\frac{dx^\nu}{dt}\frac{dt}{ds}. \qquad \text{[eq. 49]}$$

The energy E of a test particle is given by the Hamiltonian, i.e., the Legendre transform $\mathbf{J}\cdot\dot{\mathbf{x}} - H$ of the Lagrangian H (p. 7, eq. 9):

$$E = -\frac{mg_{i\nu}dx^\nu}{ds}\dot{x}^i + m\frac{ds}{dt} = m\frac{-g_{i\nu}dx^i dx^\nu + ds^2}{dsdt} = m\frac{g_{4\nu}dx^\nu}{ds}, \qquad \text{[eq. 50]}$$

where in the last step we used that $ds^2 = g_{\mu\nu}dx^\mu dx^\nu$ and that

$$g_{\mu\nu}dx^\mu dx^\nu - g_{i\nu}dx^i dx^\nu = g_{4\nu}dtdx^\nu.$$

It follows that, for a stationary metric ($g_{i4} = g_{4i} = 0$),

$$E = m g_{44}\frac{dt}{ds} \qquad \text{[eq. 51] (corrected)}$$

(the manuscript has a minus sign on the right-hand side and omits the factor m).

Besso was interested in a slow-moving planet in the spherically symmetric weak static field of the sun. The virial theorem told Besso that potential and kinetic energy, and hence the quantities A/r and q^2/c^2 (with $q \equiv \dot{x}^2 + \dot{y}^2 + \dot{z}^2$), are of the same order of magnitude. This allowed him to write down [eqs. 52–53] for ds/dt to second order in A/r. The convoluted sentence, replete with deletions, introducing these equations reads: "Here [i.e., in [eq. 51]] is, if one considers the quantities A/r and q^2/c^2 as infinitesimally small quantities of the first order ... and in addition takes into account that the field is constant in time, so that $g_{\nu 4}$ for $\nu = 1$ to 3 should be set $= 0$, and also once again sets $x_1 = x$, $x_2 = y$, $x_3 = z$, $x_4 = t$:"

$$\begin{aligned}\frac{ds}{dt} &= \sqrt{\left(\overset{(0)}{g}_{ij} + \overset{(1)}{g}_{ij}\right)\dot{x}^i\dot{x}^j + \overset{(0)}{g}_{44} + \overset{(1)}{g}_{44} + \overset{(2)}{g}_{44}} \\ &= c\sqrt{1 - \frac{q^2}{c^2} - \frac{A}{r} + \frac{3}{8}\frac{A^2}{r^2}}, \qquad \text{[eqs. 52,53]}\end{aligned}$$

where we used [p. 6, eq. 42] to set $\overset{(1)}{g}_{ij} = 0$ and

$$g_{44} = c^2\left(1 - \frac{A}{r} + \frac{3}{8}\frac{A^2}{r^2}\right). \qquad \text{[p. 6, eq. 42]}$$

In general relativity in its final November 1915 form, it is no longer true that $\overset{(1)}{g}_{ij} = 0$ (see Einstein 1915c, p. 834; cf. Sec. 6.3.3, note 8, and Pt. I, Ch. 6).

Again neglecting terms smaller than second order in A/r, Besso could write [eq. 49] and the corresponding y- and z-components of the equations of motion for this special case as

$$\frac{d}{dt}\left(\frac{\dot{x}^i}{ds/dt}\right) = -\frac{1}{2}\frac{\partial g_{44}}{\partial x^i}\frac{dt}{ds} = -\frac{1}{2}\frac{dg_{44}}{dr}\frac{x^i}{r}\frac{dt}{ds}, \qquad \text{[eq. 54]}$$

and, silently correcting the sign error in [eq. 51], the energy E per unit rest mass as

$$E = \frac{g_{44}}{ds/dt}. \qquad \text{[eq. 55]}$$

Besso now derived the analogue in the *Entwurf* theory of Kepler's area law. In Newtonian theory, the areal velocity, the area swept out in unit time by the vector $\mathbf{x}$ giving the position of the test particle in some arbitrary Cartesian coordinate system, is given by $\frac{1}{2}(\mathbf{x}\times\dot{\mathbf{x}})$. The angular momentum, $\mathbf{L} \equiv m(\mathbf{x}\times\dot{\mathbf{x}})$, is thus equal to $2m$ times the areal velocity. For a central force, the angular momentum and hence the areal velocity are conserved:

$$\frac{d\mathbf{L}}{dt} = m\frac{d}{dt}(\mathbf{x}\times\dot{\mathbf{x}}) = \dot{\mathbf{x}}\times\dot{\mathbf{x}} + \mathbf{x}\times\ddot{\mathbf{x}} = 0,$$

where the second term vanishes on account of the equations of motion, $\ddot{\mathbf{x}} = -(d\varphi/dr)(\mathbf{x}/r)$. It follows that the plane of the orbit of a particle moving under the influence of a central force remains fixed and that the areal velocity is a constant (i.e., Kepler's area law).

Besso picked Cartesian coordinates in which the planetary orbit lies in the xy-plane. The angular momentum and the areal velocity vector are thus in the z-direction. Besso introduced the notation $\dot{f}$ for the magnitude of the areal velocity (*Flächengeschwindigkeit*). He set $2\dot{f}$ equal to $y\dot{x} - x\dot{y}$, which is actually *minus* the z-component of $\mathbf{x}\times\dot{\mathbf{x}}$, but this sign error does not materially affect the rest of his argument (at the bottom of [p. 13], not included here, Einstein went through the same argument using the correct sign). Besso noted that

$$\frac{d}{dt}\left(\frac{y\dot{x} - x\dot{y}}{ds/dt}\right) = y\frac{d}{dt}\left(\frac{\dot{x}}{ds/dt}\right) - x\frac{d}{dt}\left(\frac{\dot{y}}{ds/dt}\right) = 0, \qquad \text{[eq. 56]}$$

where in the last step we used [eq. 54]. Besso concluded that $(y\dot{x} - x\dot{y})(dt/ds)$ should be some constant, B. He called the product of B and c, the leading term in ds/dt (if $g_{\mu\nu} = \eta_{\mu\nu}$, $ds = cdt\sqrt{1 - q^2/c^2}$), the "area law constant" (*Flächensatzkonstante*). Correcting the sign error in Besso's definition of the areal velocity, we can thus write:

$$2\dot{f} = x\dot{y} - y\dot{x} = B\frac{ds}{dt}. \qquad \text{[eq. 57] (corrected)}$$

At the bottom of the page, Besso noted: "Since ds/dt is a square root, it is more convenient to work with the squared equations in polar coordinates." This is what he proceeded to do at the top of [p. 9].

Note that the two equations for the perihelion motion, [eqs. 55 and 57], only involve the g_{44} component to second order in A/r (cf. [eq. 53] for ds/dt). This may be why Einstein rederived the metric field for the sun on [p. 2], calculating only g_{44} to order A^2/r^2 (cf. the observation at the end of our commentary for [p. 6]).

[P. 9] Besso

On [p. 9], Besso finished the derivation he began on [p. 8] of a differential equation for the angle between perihelion and aphelion. First, he rewrote [p. 8, eqs. 57 and 55]—expressing essentially conservation of angular momentum and energy, respectively—as

$$\dot{\varphi} r^2 = BW \qquad \text{[eq. 58]}$$

(where the left-hand side is $x\dot{y} - y\dot{x}$ in polar coordinates and $W \equiv ds/dt$), and

$$E = \frac{g_{44}}{W}. \qquad \text{[eq. 59]}$$

Using [eq. 59] to eliminate W from [eq. 58] and using [p. 6, eq. 42] for g_{44}, Besso found

$$E\dot{\varphi} r^2 = Bg_{44} = Bc^2\left(1 - \frac{A}{r} + \frac{3}{8}\frac{A^2}{r^2}\right). \qquad \text{[eq. 60]}$$

Besso neglected the last term. Although this is not clear at this point, we will see later that he was justified in doing so (see our comments following Eq. (2.10) below). Introducing $F \equiv Bc^2/E$, he wrote [eq. 60] as

$$\dot{\varphi} r^2 = F\left(1 - \frac{A}{r}\right). \qquad \text{[eq. 61]}$$

On the next line, Besso inserted [p. 6, eq. 42] for g_{44} and [p. 8, eq. 53] for $W = ds/dt$ into the equation $g_{44}^2 = E^2W^2$ (i.e., [eq. 59] squared). Neglecting terms smaller than of order A^2/r^2, he found:

$$c^4\left(1 - \frac{2A}{r} + \frac{A^2}{r^2} + \frac{6}{8}\frac{A^2}{r^2}\right) = E^2c^2\left(1 - \frac{q^2}{c^2} - \frac{A}{r} + \frac{3}{8}\frac{A^2}{r^2}\right). \qquad \text{[eq. 62]}$$

Dividing both sides by E^2c^2, putting q^2/c^2 on the left- and everything else on the right-hand side, he rewrote this as:

$$\frac{q^2}{c^2} = 1 - \frac{c^2}{E^2} + \frac{A}{r}\left(\frac{2c^2}{E^2} - 1\right) + \frac{A^2}{r^2}\left(\frac{3}{8} - \frac{7}{4}\frac{c^2}{E^2}\right). \qquad \text{[eq. 64]}$$

[Eq. 59] shows that c^2/E^2 is of the form $1+\varepsilon$, where ε is of order A/r (see the calculation in the upper-right corner of the page). Besso could thus substitute $1+\varepsilon$ for c^2/E^2 in the term with A/r on the right-hand side of [eq. 64] and 1 in the term with A^2/r^2. Expressing q^2 on the left-hand side in polar coordinates, he arrived at:

$$\frac{dr^2 + r^2d\varphi^2}{c^2dt^2} = -\varepsilon + \frac{A}{r}(1+2\varepsilon) - \frac{11}{8}\frac{A^2}{r^2}. \qquad \text{[eq. 65]}$$

Besso now used [eq. 61] squared (to order A/r),

$$r^4\frac{d\varphi^2}{dt^2} = F^2\left(1 - \frac{2A}{r}\right), \qquad \text{[eq. 66]}$$

to eliminate dt^2 from [eq. 65]:

$$\frac{1}{c^2}\left(dr^2 + r^2d\varphi^2\right)F^2\left(1 - \frac{2A}{r}\right) = r^4d\varphi^2\left(-\varepsilon + \frac{A}{r}(1+2\varepsilon) - \frac{11}{8}\frac{A^2}{r^2}\right).$$

Collecting all terms with $d\varphi^2$ on the left- and all terms with dr^2 on the right-hand side, we can rewrite this equation as

$$\frac{r^2F^2}{c^2}\left(1-\frac{2A}{r}\right)d\varphi^2 - r^4\left(-\varepsilon+\frac{A}{r}(1+2\varepsilon)-\frac{11}{8}\frac{A^2}{r^2}\right)d\varphi^2 = -\frac{F^2}{c^2}\left(1-\frac{2A}{r}\right)dr^2.$$

Rearranging terms on the left-hand side, grouping them by powers of r, we find:

$$-\varepsilon\left(-r^4+A\left(\frac{1}{\varepsilon}+2\right)r^3-\frac{1}{\varepsilon}\left(\frac{F^2}{c^2}+\frac{11}{8}A^2\right)r^2+\frac{2AF^2}{\varepsilon c^2}r\right)d\varphi^2.$$

Dividing both sides of the equation above by the expression multiplying $d\varphi^2$ and taking the square root of the resulting equation, we find the differential equation for the angle φ as a function of r that Besso was after:

$$d\varphi = \frac{F}{c\sqrt{\varepsilon}}\frac{1-\frac{A}{r}}{\sqrt{-r^4+A\left(\frac{1}{\varepsilon}+2\right)r^3-\frac{1}{\varepsilon}\left(\frac{F^2}{c^2}+\frac{11}{8}A^2\right)r^2+\frac{2AF^2}{\varepsilon c^2}r}}\,dr. \quad \text{[eq. 67]}$$

Integrating the right-hand side from the minimum value r_1 at perihelion to the maximum value r_2 at aphelion, we find the angle between perihelion and aphelion (cf. [eq. 68]).

[P. 10] Einstein

On [pp. 10–11], Einstein performed two contour integrations to evaluate $\int_{r_1}^{r_2} d\varphi$ (see [p. 9, eq. 68] or, equivalently, [p. 11, eq. 86]) to find the angle between perihelion and aphelion. He made two mistakes. First, as he noticed at the bottom of [p. 11], he dropped a "factor $\frac{1}{2}$ in front of the whole integral." He noticed this because he found an angle between perihelion and aphelion close to 2π, whereas this angle clearly should be close to π. Second, the approximation he used in the final steps of his derivation of the formula for this angle was so crude that he did not find any deviation from π and hence no perihelion motion. On [p. 14] Besso repeated these steps more carefully and did find a small deviation from π.

The fourth-order polynomial under the square-root sign in [p. 9, eq. 68] can be factorized as (cf. [eq. 69])

$$(r-r_1)(r_2-r)(r-r')(r-r''). \tag{2.3}$$

The roots r_1 and r_2 are the values of r at perihelion and aphelion, respectively. They are both much larger than the roots r' and r''. It is therefore convenient to expand the integrand of [p. 9, eq. 68] in r/r' and r/r'', with the help of (cf. [eq. 70])

$$\frac{1}{\sqrt{(r-r')(r-r'')}} = \frac{1}{r\sqrt{\left(1-\frac{r'}{r}\right)\left(1-\frac{r''}{r}\right)}} = \frac{1}{r}\left(1+\frac{r'+r''}{2r}\right).$$

The angle between perihelion and aphelion can then be written as the sum of two integrals:

$$\int_{r_1}^{r_2} d\varphi = C I_f + C\left(\tfrac{1}{2}(r'+r'')-A\right) I_g \tag{2.4}$$

where

$$I_f \equiv \int_{r_1}^{r_2} dr\, f(r), \quad \text{with} \quad f(r) \equiv \frac{1}{r\sqrt{(r-r_1)(r_2-r)}},$$

$$I_g \equiv \int_{r_1}^{r_2} dr\, g(r), \quad \text{with} \quad g(r) \equiv \frac{1}{r^2\sqrt{(r-r_1)(r_2-r)}}.$$

The factor C is our abbreviation for $F/c_0\sqrt{\varepsilon}$ in [p. 9, eq. 68]. In our discussion of [pp. 10–11, 13] we will follow Einstein and Besso in writing the velocity of light in vacuo as c_0 to avoid confusion with the coefficient c of r^2 in the polynomial under the square-root sign in [p. 9, eq. 68].

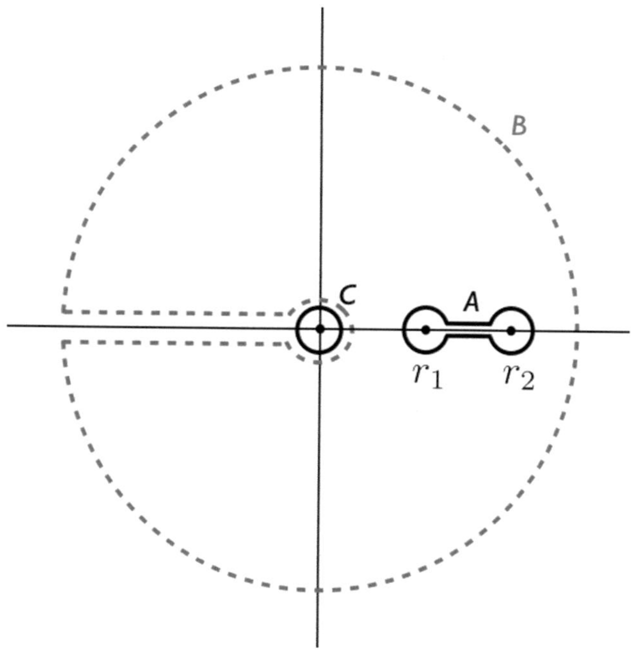

Figure 2.1 The contour A in the complex plane can continuously be deformed into the contour B without passing through the poles 0, r_1, and r_2 on the real axis of the functions $f(z)$ and $g(z)$ being integrated on [p. 10]. If the radius of B goes to infinity, the only contributions to these integrals come from the small circle C around the pole $z = 0$ (the integral from $-\infty$ to 0 just below the real axis cancels the integral from 0 to $-\infty$ just above the real axis). Einstein's own drawing on [p. 10] shows A and C. With the help of B one readily sees that the integrals over A (clockwise) and C (counterclockwise) give the same result.

Below the first horizontal line on [p. 10] (see [eq. 72]), Einstein evaluated I_f; below the second horizontal line (see [eq. 73]), he evaluated I_g. He used that $2I_f$ and $2I_g$ are equal to integrals in the complex plane over the contour A with negative (i.e., clockwise) orientation (the integrals just above and just below the real axis are over different sheets of the functions $f(z)$ and $g(z)$ and therefore have opposite signs). As illustrated in Fig. 2.1, an enhanced version of Einstein's own figure to the right of [eq. 72], these integrals can be replaced by integrals over the contour C, a small circle around $z = 0$ in the complex plane, with a positive (i.e., counterclockwise) orientation.

Although Einstein did not appeal to it in the manuscript, Cauchy's residue theorem tells us that

$$2I_f = 2\pi i \operatorname*{Res}_{z=0} f(z), \qquad 2I_g = 2\pi i \operatorname*{Res}_{z=0} g(z),$$

where the residues are

$$\operatorname*{Res}_{z=0} f(z) = \left(z f(z)\right)\Big|_{z=0} = \frac{1}{i\sqrt{r_1 r_2}},$$

$$\operatorname*{Res}_{z=0} g(z) = \frac{d}{dz}\left(z^2 g(z)\right)\Big|_{z=0} = \frac{1}{2i\sqrt{r_1 r_2}}\left(\frac{1}{r_1} + \frac{1}{r_2}\right).$$

We thus find that

$$2I_f = \frac{2\pi}{\sqrt{r_1 r_2}}, \qquad 2I_g = \frac{\pi}{\sqrt{r_1 r_2}}\left(\frac{1}{r_1} + \frac{1}{r_2}\right). \tag{2.5}$$

Einstein omitted the factors of 2 on the left-hand side, which is the source of the error he noted at the bottom of [p. 11].

Substituting the results for I_f and I_g into Eq. (2.4), we find:

$$\int_{r_1}^{r_2} d\varphi = \frac{C\pi}{\sqrt{r_1 r_2}}\left(1 + \frac{1}{2}\left(\frac{1}{r_1} + \frac{1}{r_2}\right)\left(\frac{r' + r''}{2} - A\right)\right). \tag{2.6}$$

The next step is to express the roots of the polynomial in Eq. (2.3) in terms of its coefficients (cf. [eq. 75]):

$$-a r^4 + b r^3 + c r^2 + d r + e. \tag{2.7}$$

The roots r' and r'' are so small that their third and fourth powers are negligible. These roots should thus satisfy (or, as Einstein wrote, "infinitesimally small roots from the equation"):

$$c\left(r^2 + \frac{d}{c} r + \frac{e}{c}\right) = c(r - r')(r - r'') = c(r^2 - (r' + r'')r + r'r'').$$

It follows that

$$r' + r'' = -\frac{d}{c}. \qquad \text{[eq. 77]}$$

Writing the polynomial in Eq. (2.7) (with $a = -1$) as (cf. [eq. 78])

$$\begin{aligned} -r^4 + (r_1 + r_2 + r' + r'')r^3 - (r_1 r_2 + (r_1 + r_2)(r' + r''))r^2 \\ + r_1 r_2 (r' + r'')r - r_1 r_2 r' r'' \end{aligned} \tag{2.8}$$

and neglecting any and all terms containing r' or r'', Einstein found

$$-r^4 + (r_1 + r_2)r^3 - r_1 r_2 r^2,$$

from which he concluded that:

$$r_1 + r_2 = b, \quad r_1 r_2 = -c. \qquad \text{[eq. 79, 80]}$$

It follows that

$$\frac{1}{r_1} + \frac{1}{r_2} = \frac{r_1 + r_2}{r_1 r_2} = -\frac{b}{c}. \tag{2.9}$$

On [p. 14] Besso revisited this identification of roots and coefficients neglecting only terms of second order in r' and r''.

[P. 11] Einstein

Inserting [p. 10, eq. 77] and Eq. (2.9) into Eq. (2.6), we arrive at

$$\int_{r_1}^{r_2} d\varphi = \frac{C\pi}{\sqrt{-c}}\left(1+\frac{1}{2}\frac{b}{c}\left(\frac{d}{2c}+A\right)\right). \qquad \text{[eq. 81] (corrected)}$$

Besides the spurious factor 2, Einstein omitted the factor C (our abbreviation for $F/c_0\sqrt{\varepsilon}$; cf. [p. 9, eq. 67]) and the term with A on the right-hand side.

The part of [p. 11] between the double horizontal line under [eq. 81] and the horizontal line above [eq. 86] mostly repeats parts of the argument on [p. 10]. If we set $C = 1$ and $A = -\alpha$ in Eq. (2.6), the right-hand side turns into the final expression for the integral I in [eq. 84]. [Eq. 85] and the four lines below it are in Besso's hand (cf. Eq. (2.8)). Besso may also have crossed out the factor 2 in [eq. 84].

[Eq. 86] is Einstein's version of the right-hand side of [p. 9, eq. 68] derived by Besso for the angle between perihelion and aphelion. On [pp. 25, 27, and 33], not included here, Besso checked whether these two equations are equivalent. No pages with Einstein's derivation of [eq. 86] survive. Einstein probably discarded them before mailing the manuscript to Besso in early 1914. Using Eq. (2.7), with $a = -1$, for the polynomial under the square-root sign, we can write [eq. 86] as:

$$\int_{r_1}^{r_2} d\varphi = \frac{c_0 B}{\sqrt{c_0^2 - E^2}}\int_{r_1}^{r_2} \frac{1 - A/r}{\sqrt{-r^4 + br^3 + cr^2 + dr + e}}dr, \tag{2.10}$$

where the coefficients b, c, d, and e that we read off of Einstein's [eq. 86] are

$$b = A\left(\frac{2c_0^2 - E^2}{c_0^2 - E^2}\right), \qquad c = \frac{A^2\left(\frac{7}{4}c_0^2 - \frac{3}{8}E^2\right) - B^2c_0^2}{c_0^2 - E^2},$$

$$d = \frac{2B^2c_0^2A}{c_0^2 - E^2}, \qquad e = -\frac{7}{4}\frac{B^2c_0^2A^2}{c_0^2 - E^2}.$$

The zeroth-order term, e, is absent from Besso's result, [p. 9, eq. 68]. This is because Besso neglected the A^2/r^2-term in [p. 9, eq. 60]. Since $e = -r_1 r_2 r' r''$ (see Eq. (2.8)), Besso was justified in setting $e = 0$.

The same cannot be said for the approximations in [p. 10, eqs. 79–80] on which Einstein relied in [eq. 81]. Setting $c = -\frac{B^2c_0^2}{c_0^2 - E^2}$ (he was justified in neglecting the term with A^2), he found $\frac{d}{c} = -2A$ and $\sqrt{-c} = \frac{Bc_0}{\sqrt{c_0^2 - E^2}}$ ([eq. 87]). When these expressions along with $C = \frac{c_0 B}{\sqrt{c_0^2 - E^2}}$ are inserted into [eq. 81], the result is that the angle between aphelion and perihelion is equal to π, which would mean that the *Entwurf* theory predicts no perihelion motion at all in the case of a single planet orbiting the sun.

Because of the factor 2 he dropped on [p. 10] (see Eq. (2.5)), Einstein found 2π instead of π. He immediately recognized he was off by a factor 2. As he wrote directly below [eq. 87]: "Factor $\frac{1}{2}$ forgotten in front of the whole integral." He left it to Besso to find the deviation from π.

[P. 14] Besso

On [p. 14], Besso went through a more careful version of Einstein's derivation at the bottom of [p. 10] of the relation between roots and coefficients of the fourth-order polynomial under the square-root sign in the integral giving the angle between perihelion and aphelion. Besso used [p. 9, eq. 67] for this integral, which he himself derived, rather than [p. 11, eq. 86], which Einstein had found. Writing the polynomial in the form of Eq. (2.7), we thus arrive at the following identification of its coefficients:

$$a=-1, \quad b=\left(\frac{1}{\varepsilon}+2\right)A, \quad c=-\frac{1}{\varepsilon}\left(\frac{F^2}{c_0^2}+\frac{11}{8}A^2\right), \quad d=\frac{2F^2A}{\varepsilon c_0^2}, \quad e=0. \tag{2.11}$$

Using [p. 10, eq. 77] and neglecting the term with A^2 in c, we find:

$$r'+r''=-\frac{d}{c}=2A. \qquad \text{[eq. 98]}$$

Comparing Eq. (2.7) and Eq. (2.8), we read off the following relations between roots and coefficients of the polynomial:

$$r_1+r_2+r'+r''=b, \tag{2.12}$$

$$r_1r_2+(r_1+r_2)(r'+r'')=-c. \tag{2.13}$$

In [p. 10, eqs. 79–80], Einstein neglected the terms with r' and r'' in these relations.

Using Eq. (2.11) for b and [eq. 98] for $r'+r''$ in Eq. (2.12), we find:

$$r_1+r_2=\left(\frac{1}{\varepsilon}+2\right)A-2A=\frac{A}{\varepsilon}. \qquad \text{[eq. 99]}$$

Similarly, using Eq. (2.11) for c (now without neglecting the term with A^2), [eq. 98] for $r'+r''$, and [eq. 99] for r_1+r_2 in Eq. (2.13), we find:

$$r_1r_2=\frac{1}{\varepsilon}\left(\frac{F^2}{c_0^2}+\frac{11}{8}A^2\right)-\frac{2A^2}{\varepsilon}=\frac{1}{\varepsilon}\left(\frac{F^2}{c_0^2}-\frac{5}{8}A^2\right). \qquad \text{[eq. 100]}$$

Substituting [eq. 98] into Eq. (2.6), we see that $\frac{1}{2}(r'+r'')-A=0$. Eq. (2.6) thus reduces to $\int_{r_1}^{r_2} d\varphi=\frac{C\pi}{\sqrt{r_1r_2}}$. Inserting $C=\frac{F}{c_0\sqrt{\varepsilon}}$ (cf. [p. 9, eq. 67]) and [eq. 100] for r_1r_2, we find:

$$\int_{r_1}^{r_2} d\varphi=\frac{\frac{F}{c_0\sqrt{\varepsilon}}\pi}{\sqrt{\frac{1}{\varepsilon}\left(\frac{F^2}{c_0^2}-\frac{5}{8}A^2\right)}}=\frac{\pi}{\sqrt{1-\frac{5}{8}\frac{A^2c_0^2}{F^2}}}\approx\pi\left(1+\frac{5}{16}\frac{A^2c_0^2}{F^2}\right). \qquad \text{[eq. 101]}$$

For a single planet orbiting the sun, the *Entwurf* theory thus predicts a perihelion advance of

$$\frac{5}{16}\pi\left(\frac{Ac_0}{F}\right)^2 \tag{2.14}$$

radians per half a revolution.

On the remainder of [p. 14], Besso expressed F and A in terms of parameters for which one could easily find numerical values.

In [p. 9, eq. 61], F was defined as Bc_0^2/E. Using the definition of ε in [p. 9, eq. 63] to set $c_0/E = 1$ (as is certainly allowed when evaluating F in expression (2.14) for the perihelion advance), $F = Bc_0$, which is the leading term in $2\dot{f}$ in [p. 8, eq. 57]. F is thus twice the areal velocity which we can take to be a constant when evaluating expression (2.14):

$$F = 2\dot{f} = 2\frac{\pi ab}{T} = \frac{2\pi a^2\sqrt{1-e^2}}{T}, \qquad \text{[eq. 102]}$$

where a, b, T, and e are the planetary orbit's semi-major axis, semi-minor axis, period, and eccentricity, respectively. Besso used the triangle he drew next to [eq. 102] to help him reconstruct the formula $\pi ab = \pi a^2\sqrt{1-e^2}$ for the area of an ellipse.

In [p. 1, eqs. 3 and 6], A was defined as $2KM/c_0^2$. Setting the centripetal force $m\omega^2 r$, with $\omega = 2\pi/T$, equal to the gravitational force KMm/r^2 and setting r equal to the major axis a we find $4\pi^2a^3/T^2 = KM$, which is just Kepler's third law (the ratio a^3/T^2 is the same for all planets in the solar system). It follows that

$$A = \frac{8\pi^2a^3}{c_0^2T^2}. \tag{2.15}$$

Using this relation along with [eq. 102], Besso could rewrite expression (2.14) for the perihelion advance in radians per half a revolution as

$$\frac{5}{16}\pi\frac{A^2c_0^2}{F^2} = \frac{5}{16}\pi A\left(\frac{8\pi^2a^3}{T^2}\right)\left(\frac{T^2}{4\pi^2a^4(1-e^2)}\right) = \frac{5}{8}\pi\frac{A}{a(1-e^2)}. \tag{2.16}$$

In the 1915 perihelion paper, Einstein (1915c, p. 838, Eq. (13)) found that his new theory predicted a perihelion advance of

$$3\pi\frac{\alpha}{a(1-e^2)}$$

radians per revolution, where α is just a different notation for A. The *Entwurf* theory thus predicts a perihelion advance $\frac{5/8}{3/2} = \frac{5}{12}$ the size of that predicted by general relativity, which in the case of Mercury amounts to 18 seconds of arc per century.

In the 1915 perihelion paper, Einstein switches from r to $x = \frac{1}{r}$ and integrates an equation for $\frac{d\varphi}{dx}$ between the minimum and maximum values α_1 and α_2 of x instead of integrating an equation for $\frac{d\varphi}{dr}$ of the form of [eq. 67] on [p. 9] between the minimum and maximum values r_1 and r_2 of r. This greatly simplifies the calculation. Einstein now only has to evaluate an integral of the form

$$\int_{\alpha_1}^{\alpha_2}\frac{dx}{\sqrt{-(x-\alpha_1)(x-\alpha_2)}}$$

(see Sec. 6.3.3, note 21). This integral does not have a pole at $x = 0$. The contour A in the complex plane in Fig. 2.1 can thus be continuously deformed to a large circle with radius $R \gg \alpha_2$ and the integral above can be replaced by twice the integral of $\frac{1}{iz}$ over this circle. Writing $z = Re^{i\varphi}$ and $dz = iRe^{i\varphi}\,d\varphi$, we have

$$\int_{\alpha_1}^{\alpha_2}\frac{dx}{\sqrt{-(x-\alpha_1)(x-\alpha_2)}} = \frac{1}{2}\int_0^{2\pi}\frac{iRe^{i\varphi}\,d\varphi}{iRe^{i\varphi}} = \frac{1}{2}\int_0^{2\pi}d\varphi = \pi. \tag{2.17}$$

[P. 26] Einstein

On [pp. 26 and 28], Einstein computed the numerical value of expression (2.14) derived by Besso on [p. 14] for the perihelion advance in radians per half a revolution for Mercury and converted the result to seconds of arc per century.

The top half of [p. 26] repeats equations derived by Besso on [p. 14]. [Eqs. 174–175] are the same as [p. 14, eqs. 98–100] for the relation between roots and coefficients of the fourth-order polynomial under the square-root sign in [p. 9, eq. 68] for the angle between perihelion and aphelion. [Eq. 177] for this angle is the same as [p. 14, eq. 101]. [Eq. 178] is the same as [p. 14, eq. 102]. [Eq. 179], originally introduced in [p. 1, eq. 6], was also used by Besso on [p. 14] (see Eq. (2.15)).

With the help of [eqs. 178 and 179], Einstein rewrote part of Besso's expression (2.14) as

$$\frac{Ac}{F} = c\frac{2KM}{c^2}\frac{T}{2\pi a^2\sqrt{1-e^2}} = \frac{KMT}{c\pi a^2\sqrt{1-e^2}} \qquad \text{[eq. 180]}$$

(where we dropped the subscript '0' in the velocity of light c_0).

The two columns immediately below [eq. 180] contain the logarithms of the quantities on the right-hand side of this equation. The first two numbers in the left column give the logarithms of M and T, respectively; the first five numbers in the right column give the logarithms of K^{-1}, c, π, a^2, and $\sqrt{1-e^2}$, respectively. Subtracting the sum of these five numbers from the sum of $\log M$ and $\log T$, one arrives at $\log(Ac/F) = .5172 - 3$, which is the starting point of the calculation on [p. 28]. The calculations in the three lines immediately above [eq. 180] and the calculations to the left of the vertical line on [p. 26] show how the values for the various factors in Ac/F were found that were used in the calculation to the right of the vertical line. With the exception of the value for M, these values are reasonably close to the modern values.

The value of K is given as $1/3862^2$, the value of T as 87.97 days. From the calculation here and from $a = .387 \cdot 1.48 \cdot 10^{13} = 5.72 \cdot 10^{12}$ cm ([eq. 266] on [p. 40], not included here), it can be inferred that a was computed via the relation $a = (\bar{r}_m/\bar{r}_e)\bar{r}_e$, where $\bar{r}_m$ and $\bar{r}_e$ are the mean distances from the sun of Mercury and the earth, respectively. As Einstein jotted down next to the result for $\log c$: "Average distance set equal to semi-major axis"). Before calculating the logarithm of $1-e^2$, he wrote it as $(1-e)(1+e)$. He used the value .2056 for e.

As we already noted in Pt. I, Ch. 6, the value Einstein inserted for M is off by a factor 10. As can be inferred from the calculation of $\log M$ on [p. 26], an almost identical calculation on [p. 34] (by Einstein, not included here), and the equation

$$M = 3.24 \cdot 10^5 \cdot 5.6 \cdot 1.08 \cdot 10^{2\langle 8\rangle 7} = 1.96 \cdot 10^{3\langle 4\rangle 3},$$

which occurs several times in the manuscript—[p. 30, eq. 185, Einstein] (see Pt. I, Fig. 6.3), [p. 35, eq. 211, Einstein with corrections by Besso] (see Pt. I, Fig. 6.2), and, without the corrections (and not included here), [p. 40, Einstein, eq. 264]—Einstein found the value for M as the product of three factors, the ratio of the sun's mass to the earth's mass, the earth's density and the earth's volume. It is the value for the earth's volume that is a factor 10 too large.

Due to the error in M, the value Einstein found for Ac/F is a factor 10 too large. Since the expression (2.14) involves the square of this quantity, this leads to a value for the motion of Mercury's perihelion that is a factor 100 too large.

In the crossed-out passage in the lower-right corner, Einstein began to convert the perihelion advance from radians per half a revolution to seconds of arc per century. He made a fresh start with this conversion on [p. 28].

[P. 28] Einstein

Starting from the value for $\log(Ac/F)$ he had computed on [p. 26], which is off by a factor 10, Einstein found that

$$\frac{5}{16}\left(\frac{Ac}{F}\right)^2 = 3.4\cdot 10^{-6},$$

which is off by a factor 100. Comparison with expression (2.14) shows that this should give the perihelion motion of Mercury in fractions of π per half a revolution. Einstein recorded this same result on [p. 29] and at the bottom of [p. 40] (pages not included here). On [p. 30] (also not included here), Einstein once again computed the perihelion advance in fractions of π per half a revolution. He made several mistakes in this calculation and arrived at a value of $1.65\cdot 10^{-6}$. Right underneath this result, however, he wrote the correct result, $3.4\cdot 10^{-8}$. This suggests that he had come to realize at that point that the result on [p. 28] was off by a factor of 100. [P. 28] gives no indication that Einstein realized his mistake. The mistake was first found, it seems, by Besso (see [p. 35, eq. 211] shown in Fig. 6.2 in Pt. I, Ch. 6).

To convert his result to seconds of arc per century, Einstein computed the logarithms of $180\cdot 60\cdot 60$, the number of seconds of arc in π radians, and $2(365.2/87.97)\,100$, the number of half revolutions of Mercury in a century. He thus found the "precession per half a revolution in seconds of arc" and, finally, the "precession in 100 years" in seconds of arc: "$1821'' = 30'$." Next to this bizarre result, he wrote: "independently checked" (see the discussion in Pt. I, Ch. 6). The correct result of $18''$ is not stated anywhere in the manuscript, although, as we have seen, there are clear indications that Einstein and Besso discovered the source of the erroneous factor 100 in the result stated on [p. 28].

On an undated page of his so-called Scratch Notebook (CPAE3, Appendix A, [p. 61]), Einstein gave, without derivation, the following expression for the "advance per revolution":

$$10\pi^3\left(\frac{a}{cT'}\right)^2.$$

One arrives at this expression by substituting Eq. (2.15) for A and [p. 14, eq. 102] for F into expression (2.14) for the perihelion advance in radians per half a revolution and multiplying the result by 2:

$$2\cdot\frac{5}{16}\pi\left(\frac{Ac}{F}\right)^2 = \frac{5}{8}\pi\left(\frac{8\pi^2 a^3}{cT^2}\,\frac{T}{2\pi a^2\sqrt{1-e^2}}\right)^2 = 10\pi^3\left(\frac{a}{cT'}\right)^2,$$

where $T' \equiv T\sqrt{1-e^2}$. Two years later, Einstein (1915c, p. 839, Eq. (14)) gave the final formula for the perihelion advance in the new theory in this same form. The only difference is that the new theory replaces the factor 10 by 24. On [p. 61] of the Scratch Notebook, Einstein used the formula of the *Entwurf* theory to compute the perihelion advance of Mercury and converted the result to seconds of arc per century. This time he arrived at essentially the correct (though disappointing) result: $17''$.

[P. 41, top half] Einstein

On [pp. 41–42], Einstein used the same iterative approximation procedure used on [pp. 1–6] to find the metric field of the sun to check whether the rotation metric, as we called it (Pt. I, Ch. 3), the Minkowski metric in a slowly rotating Cartesian coordinate system, is a vacuum solution of the *Entwurf* equations.

On the final page of the Scratch Notebook mentioned above (CPAE3, Appendix A, [p. 66]), Einstein wrote three components of the rotation metric (see Sec. 1.3, Eqs. (1.25) and (1.26)),

$$g_{44} = 1 - \omega^2 r^2, \quad g_{14} = \omega y, \quad g_{24} = -\omega x,$$

where ω is the angular frequency of the rotation and $r \equiv \sqrt{x^2+y^2}$ (the term 1 in g_{44} should be c^2). Right next to these components he wrote: "Is the first equation a consequence of the last two on the basis of the theory?" (Janssen 1999, pp. 137–138).

On [pp. 41–42] of the Einstein-Besso manuscript, Einstein tried to answer this question for the *Entwurf* theory by substituting the components of the rotation metric of first order in ω in the 44-component of the *Entwurf* field equations and solving for the ω^2-term in g_{44}. The solution he found was

$$-\alpha^2\rho^2, \qquad \text{cf. [eq. 274]}$$

which is just the ω^2-term in the 44-component of the rotation metric. (We will follow the notation of [pp. 41–42] in using ρ rather than r for $\sqrt{x^2+y^2}$ but the notation of the Scratch Notebook in using ω rather than α for the angular frequency.) Next to [eq. 274], he thus wrote: "Is correct" [*stimmt*].

Einstein, however, made several mistakes in this calculation. As can be inferred from [eq. 267] (in Besso's hand), [eq. 268] and [p. 42, eq. 282], he incorrectly read off the components of the metric from the line element, resulting in spurious factors of 2 (cf. Sec. 1.3, Eqs. (1.24)–(1.25)). He also lost a minus sign at some point. When these mistakes are corrected, the ω^2-term in g_{44} changes from $-\omega^2\rho^2$ to $-\frac{3}{4}\omega^2\rho^2$.

On [p. 42], Einstein similarly calculated the ω^2-term in the 44-component of the contravariant metric. He discovered that the expressions he found for the 44-components of the covariant and the contravariant metric are incompatible with one another. However, he still did not correct the expression for g_{44} he had found on [p. 41]. Instead, it appears, he convinced himself that he must have made an error in the calculation of g^{44} on [p. 42]. All this strongly suggests that, when he did these calculations (probably in June 1913), he was firmly convinced that the rotation metric is a solution of the *Entwurf* field equations. Despite warnings recorded in the Besso memo of August 1913 (Janssen 2007, p. 785), it was only in September 1915 that Einstein finally accepted that it is not (see the letter to Freundlich presented in Ch. 4).

We reconstruct the derivations on [pp. 41–42], modernizing and enhancing Einstein's notation the same way we did in our commentary on [p. 1] (cf. Eq. (2.1)). Our reconstruction follows Janssen (1999, pp. 139–148).

On [p. 41], Einstein used the vacuum *Entwurf* field equations in their 'covariant' form:

$$D_{\mu\nu}(g) + \kappa t_{\mu\nu} = 0 \qquad \text{[eq. 269]} \tag{2.18}$$

(square quotes because neither the left-hand side as a whole nor the two terms it consists of are generally covariant tensors). In modernized notation, the first term is defined as

$$D_{\mu\nu}(g) \equiv \frac{1}{\sqrt{-g}}\left(g^{\alpha\beta}\sqrt{-g}\,g_{\mu\nu,\beta}\right)_{,\alpha} - g^{\alpha\beta}g^{\tau\rho}g_{\mu\tau,\alpha}\,g_{\nu\rho,\beta}, \tag{2.19}$$

while the second, representing (κ times) the energy-momentum density of the gravitational field, is defined as

$$\kappa t_{\mu\nu} \equiv -\frac{1}{2}g_{\tau\rho,\mu}\,g^{\tau\rho}{}_{,\nu} + \frac{1}{4}g_{\mu\nu}\,g^{\alpha\beta}\,g_{\tau\rho,\alpha}\,g^{\tau\rho}{}_{,\beta} \qquad \text{[eq. 271]} \tag{2.20}$$

(Einstein and Grossmann 1913, pp. 16–17, Eqs. (21), (16), and (14), respectively).

It is easily seen that, to first order in ω, the rotation metric is a solution of [eq. 269]. The spatial derivatives of the ω-terms in the rotation metric are all ω or ω/c^2. In first-order approximation, the one term in the *Entwurf* field equations with second-order derivatives of the metric thus vanishes and terms with products of first-order derivatives either vanish or are of order ω^2 and can be neglected.

Einstein considered the 44-component of the field equations to second order,

$$\overset{(2)}{D}_{44}(g) + \kappa \overset{(2)}{t}_{44} = 0, \tag{2.21}$$

and inserted the ω-terms of the rotation metric,

$$g^{14} = g^{41} = -\frac{\omega}{c^2} y, \quad g^{24} = g^{42} = \frac{\omega}{c^2} x, \qquad \text{[eq. 268] (corrected)}$$

$$g_{14} = g_{41} = -\omega y, \quad g_{24} = g_{42} = \omega x, \qquad \text{[p. 42, eq. 282] (corrected)}$$

into this equation. Einstein's expressions for these components are all off by a factor 2. Note that the signs are the opposite of those in Eqs. (1.24)–(1.25) in Sec. 1.3. This just means that the rotation is clockwise rather than counterclockwise.

With the help of Eq. (2.19) and using that $\overset{(1)}{g}_{44} = 0$ and $g_{\mu\nu,4} = 0$ for the rotation metric, we find that

$$\overset{(2)}{D}_{44}(g) = \overset{(0)}{g}{}^{ij} \overset{(2)}{g}_{44,ij} - \overset{(0)}{g}{}^{\alpha\beta} \overset{(0)}{g}{}^{\tau\rho} \overset{(1)}{g}_{4\tau,\alpha} \overset{(1)}{g}_{4\rho,\beta}.$$

The first term on the right-hand side is just $-\Delta \overset{(2)}{g}_{44}$; the second gives two identical contributions, one for the index combination $(\alpha = \beta = 1, \tau = \rho = 2)$ and one for $(\alpha = \beta = 2, \tau = \rho = 1)$:

$$\left(\overset{(1)}{g}_{42,1}\right)^2 = \left(\overset{(1)}{g}_{41,2}\right)^2 = \omega^2.$$

Hence:

$$\overset{(2)}{D}_{44}(g) = -\Delta \overset{(2)}{g}_{44} - 2\omega^2. \qquad \text{[eq. 270] (corrected)}$$

Because of the erroneous factors 2 in [eq. 268] and [p. 42, eq. 282], [eq. 270] has $-8\omega^2$ instead of $-2\omega^2$.

With the help of Eq. (2.20), we find that

$$\kappa \overset{(2)}{t}_{44} = -\frac{1}{2} \overset{(1)}{g}_{\tau\rho,4} \overset{(1)}{g}{}^{\tau\rho}{}_{,4} + \frac{1}{4} \overset{(0)}{g}_{44} \overset{(0)}{g}{}^{\alpha\beta} \overset{(1)}{g}_{\tau\rho,\alpha} \overset{(1)}{g}{}^{\tau\rho}{}_{,\beta}.$$

The first term on the right-hand side vanishes; the second gives four identical contributions for the four index combinations written next to [eqs. 270–271]:

$$\begin{aligned} &\tau = 1,\ \rho = 4,\ \alpha = \beta = 2; \\ &\tau = 2,\ \rho = 4,\ \alpha = \beta = 1; \\ &\tau = 4,\ \rho = 1,\ \alpha = \beta = 2; \\ &\tau = 4,\ \rho = 2,\ \alpha = \beta = 1. \end{aligned}$$

For the first of these index combinations, for instance, we find

$$\frac{1}{4} \overset{(0)}{g}_{44} \overset{(0)}{g}{}^{22} \overset{(1)}{g}_{14,2} \overset{(1)}{g}{}^{14}{}_{,2} = -\frac{c^2}{4} \frac{\omega^2}{c^2}.$$

Hence

$$\kappa \overset{(2)}{t}_{44} = -\omega^2.$$

The evaluation of the various factors in the second term of [eq. 271], written underneath this term, shows that Einstein, because of the erroneous factors 2 in [eq. 268] and [p. 42, eq. 282], set $\kappa \overset{(2)}{t}_{44}$ equal to

$$\frac{1}{4} \cdot c^2 \cdot -1 \cdot \frac{1}{c^2} \cdot 16\alpha^2 = -4\alpha^2$$

rather than $-\alpha^2$ (or, in our notation, $-\omega^2$).

Combining the correct expressions for $\overset{(2)}{D}_{44}(g)$ and $\kappa \overset{(2)}{t}_{44}$ for the rotation metric, we find

$$-\Delta \overset{(2)}{g}_{44} - 2\,\omega^2 - \omega^2 = 0 \qquad \text{[eq. 272] (corrected)}$$

Because of the spurious factors of 2 and a sign error, the last term on left-hand side of [eq. 272] is $+4\omega^2$ rather than $-\omega^2$. The sign error was probably the result of wishful thinking on Einstein's part. It ensures, as he probably foresaw at this point, that $\overset{(2)}{g}_{44}$ is just the ω^2-term in the 44-component of the rotation metric.

The equation for $\overset{(2)}{g}_{44}$ thus reduces to

$$\Delta \overset{(2)}{g}_{44} = -3\,\omega^2. \qquad \text{[eq. 273] (corrected)}$$

To solve this equation, we need a formula for the action of the Laplacian Δ on a function of $\rho \equiv \sqrt{x^2+y^2}$. On [p. 42], in the two lines leading up to [eq. 285], we see Einstein derive this formula in the same way that he derived one for the action of Δ on a function of $r \equiv \sqrt{x^2+y^2+z^2}$ in [p. 4, eq. 28]. For an arbitrary function $f(\rho)$ we have:

$$\frac{\partial^2 f}{\partial x^2} = \frac{x^2}{\rho}\frac{d}{d\rho}\left(\frac{1}{\rho}\frac{df}{d\rho}\right) + \frac{1}{\rho}\frac{df}{d\rho}.$$

Adding a similar expression for $\partial^2 f/\partial y^2$, we find that

$$\Delta f = \rho \frac{d}{d\rho}\left(\frac{1}{\rho}\frac{df}{d\rho}\right) + \frac{2}{\rho}\frac{df}{d\rho}. \qquad \text{[p. 42, eq. 285] with } f = \gamma_{44}$$

With the help of this formula, one immediately sees that

$$\overset{(2)}{g}_{44} = -\frac{3}{4}\omega^2\rho^2 \tag{2.22}$$

is the solution of (the corrected version of) [eq. 273]. This means that the rotation metric is *not* a vacuum solution of the *Entwurf* field equations. Since Einstein's version of [eq. 273] has $-4\,\omega^2$ rather than $-3\,\omega^2$, he concluded that $\overset{(2)}{g}_{44} = -\omega^2\rho^2$ and that the rotation metric *is* a vacuum solution.

[P. 42, top half, left part] Einstein

On [p. 42], Einstein used the vacuum *Entwurf* field equations in their 'contravariant' form (cf. [p. 41, eq. 269]):

$$\Delta_{\mu\nu}(\gamma) - \kappa\,\vartheta_{\mu\nu} = 0,$$

where

$$\Delta_{\mu\nu}(\gamma) \equiv \frac{1}{\sqrt{-g}}\left(g^{\alpha\beta}\sqrt{-g}\,g^{\mu\nu}{}_{,\beta}\right)_{,\alpha} - g^{\alpha\beta}g_{\tau\rho}\,g^{\mu\tau}{}_{,\alpha}\,g^{\nu\rho}{}_{,\beta},$$

and

$$\kappa\vartheta_{\mu\nu} \equiv -\frac{1}{2}g^{\alpha\mu}g^{\beta\nu}g_{\tau\rho,\alpha}g^{\tau\rho}{}_{,\beta} + \frac{1}{4}g^{\mu\nu}g^{\alpha\beta}g_{\tau\rho,\alpha}g^{\tau\rho}{}_{,\beta}$$

(Einstein and Grossmann 1913, pp. 15–17, Eqs. (18), (15), and (13), respectively).

Einstein considered the 44-component of the field equations to second order,

$$\overset{(2)}{\Delta}_{44}(\gamma) - \kappa\overset{(2)}{\vartheta}_{44} = 0, \tag{2.23}$$

inserted the components of the rotation metric of first order in ω into this equation, and solved for the second-order contribution to g^{44}. He used [eq. 282] and [p. 41, eq. 268] for the first-order components, which are off by a factor of 2. Using the correct expressions for these ω-terms, we find:

$$\overset{(2)}{\Delta}_{44}(\gamma) = -\Delta\overset{(2)}{g}{}^{44} - \overset{(0)}{g}{}^{\alpha\beta}\overset{(0)}{g}_{\tau\rho}\overset{(1)}{g}{}^{4\tau}{}_{,\alpha}\overset{(1)}{g}{}^{4\rho}{}_{,\beta} = -\Delta\overset{(2)}{g}{}^{44} - 2\frac{\omega^2}{c^4},$$

$$\kappa\overset{(2)}{\vartheta}_{44} = \frac{1}{4}\overset{(0)}{g}{}^{44}\overset{(0)}{g}{}^{\alpha\beta}\overset{(1)}{g}_{\tau\rho,\alpha}\overset{(1)}{g}{}^{\tau\rho}{}_{,\beta} = -\frac{\omega^2}{c^4}.$$

The equation for $\overset{(2)}{\vartheta}{}^{44}$ thus becomes:

$$-\Delta\overset{(2)}{g}{}^{44} - 2\frac{\omega^2}{c^4} + \frac{\omega^2}{c^4} = 0, \qquad \text{[eq. 281] (corrected)}$$

or

$$\Delta\overset{(2)}{g}{}^{44} = -\frac{\omega^2}{c^4}. \qquad \text{[eq. 283] (corrected)}$$

In the manuscript, the coefficients of the ω^2-terms in [eq. 281] and [eq. 283] are a factor 4 too large because of the erroneous factors of 2 in [eq. 282] and [p. 41, eq. 268].

With the help of [eq. 285], we see that

$$\overset{(2)}{g}{}^{44} = -\frac{1}{4}\frac{\omega^2}{c^4}\rho^2 \tag{2.24}$$

is the solution of (the corrected version of) [eq. 283]. The rotation metric has $\overset{(2)}{g}{}^{44} = 0$ (see Eq. (1.26) in Sec. 1.3). So this calculation shows, once again, that the rotation metric is not a vacuum solution of the *Entwurf* field equations.

Einstein found $\overset{(2)}{g}{}^{44} = -\omega^2\rho^2/c^4$. In [eq. 284], he wrote $\gamma^{\times}_{44}$, his notation for $\overset{(2)}{g}{}^{44}$, as $\beta\rho^2$. From [eqs. 283, 285–286] he concluded that $\beta = -\alpha^2/c^4$, where α is his notation for ω ([eq. 287]). For g^{44}, Einstein would then have arrived at:

$$g^{44} = \frac{1}{c^2} - \frac{\omega^2}{c^4} \qquad \text{[eq. 288] without extra factor 3}$$

We conjecture that the factor 3 was only added to the last term of [eq. 188] after Einstein had found [eq. 289] for g_{44}, which must have been puzzling and disappointing to him. The reasoning behind this conjecture is as follows (Janssen 1999, pp. 147–148).

It looks as if Einstein used his result for $\overset{(2)}{g}{}^{44}$ and the condition $g_{\mu\alpha}g^{\alpha\nu} = \delta^{\nu}_{\mu}$ to derive an expression for $\overset{(2)}{g}_{44}$. Collecting terms of second order in ω in $g_{4\alpha}g^{\alpha 4} = 1$, we find:

$$\overset{(1)}{g}_{41}\overset{(1)}{g}{}^{14} + \overset{(1)}{g}_{42}\overset{(1)}{g}{}^{24} + \overset{(0)}{g}_{44}\overset{(2)}{g}{}^{44} + \overset{(2)}{g}_{44}\overset{(0)}{g}{}^{44} = \frac{\omega^2\rho^2}{c^2} + c^2\overset{(2)}{g}{}^{44} + \frac{1}{c^2}\overset{(2)}{g}_{44} = 0. \tag{2.25}$$

Inserting Eq. (2.24) for $\overset{(2)}{g}{}^{44}$, we simply recover Eq. (2.22) for $\overset{(2)}{g}_{44}$.

For Einstein, because of the extra factors of 2, Eq. (2.25) would have taken the form:

$$c^2 \overset{(2)}{g}{}^{44} + \frac{\overset{(2)}{g}_{44}}{c^2} = -\frac{4\omega^2\rho^2}{c^2}.$$

Inserting $-\omega^2\rho^2/c^4$ for $\overset{(2)}{g}{}^{44}$ into this equation, one finds

$$\overset{(2)}{g}_{44} = -3\omega^2\rho^2.$$

This explains, we submit, why Einstein wrote

$$g_{44} = c^2\left(1 + \langle 3\rangle \frac{\omega^2}{c^2}\rho^2\right) \qquad \text{[eq. 289]} \tag{2.26}$$

on the line following [eq. 288] for g^{44} (the plus sign in [eq. 289] should clearly have been a minus sign). This expression for g_{44} differs from the one in [p. 41, eq. 274], next to which Einstein had written "is correct." In fact, the factor 3 in [eq. 289] is correct: it is what Einstein would have found had he not made the sign error in [p. 41, eq. 272]. We conjecture that Einstein crossed out the factor 3 in [eq. 289] for g_{44} and added it in [eq. 288] for g^{44}, thereby reassuring himself that the rotation metric really is a vacuum solution of the *Entwurf* field equations. It apparently did not bother him that he got the wrong result for g^{44}.

In [Eq. 290] (cf. [p. 8, eq. 49]), Einstein started to evaluate the equation of motion for a test particle in the metric field he found on [pp. 41–42] but this calculation breaks off after one line. Besso took up this same calculation on the bottom half of [p. 41]. He took it a few steps further than Einstein but it remains unclear what the purpose of the calculation was.

Chapter 3
The Formal Foundation of the General Theory of Relativity. The Differential Equations of the Gravitational Field (Part D, §§ 12–15)

3.1 Facsimile

D. Die Differentialgesetze des Gravitationsfeldes.

Im letzten Abschnitt wurden die Koeffizienten $g_{\mu\nu}$, welche physikalisch als Komponenten des Gravitationspotentials aufzufassen sind, als gegebene Funktionen der x_ν betrachtet. Es sind noch die Differentialgesetze aufzusuchen, welchen diese Größen genügen. Das erkenntnistheoretisch Befriedigende der bisher entwickelten Theorie liegt darin, daß dieselbe dem Relativitätsprinzip in dessen weitgehendster Bedeutung Genüge leistet. Dies beruht, formal betrachtet, darauf, daß die Gleichungssysteme allgemein, d. h. beliebigen Substitutionen der x_ν gegenüber, kovariant sind. [1]

Es scheint hiernach die Forderung geboten, daß auch die Differentialgesetze für die $g_{\mu\nu}$ allgemein kovariant sein müssen. Wir wollen aber zeigen, daß wir diese Forderung einschränken müssen, wenn wir dem Kausalgesetz vollständig Genüge leisten wollen. Wir beweisen nämlich, daß Gesetze, welche den Ablauf des Geschehens im Gravitationsfelde bestimmen, unmöglich allgemein kovariant sein können. [2]

M. Janssen, J. Renn, *How Einstein Found His Field Equations*, Classic Texts in the Sciences, https://doi.org/10.1007/978-3-030-97955-3_10

§ 12. Beweis von der Notwendigkeit einer Einschränkung der Koordinatenwahl. [3]

Wir betrachten einen endlichen Teil Σ des Kontinuums, in welchem ein materieller Vorgang nicht stattfindet. Das physikalische Geschehen in Σ ist dann vollständig bestimmt, wenn in bezug auf ein zur Beschreibung benutztes Koordinatensystem K die Größen $g_{\mu\nu}$ als Funktion der x_ν gegeben werden. Die Gesamtheit dieser Funktionen werde symbolisch durch $G(x)$ bezeichnet. [4]

Es werde ein neues Koordinatensystem K' eingeführt, welches außerhalb Σ mit K übereinstimme, innerhalb Σ aber von K abweiche, derart, daß die auf K' bezogenen $g'_{\mu\nu}$ wie die $g_{\mu\nu}$ (nebst ihren Ableitungen) überall stetig sind. Die Gesamtheit der $g'_{\mu\nu}$ bezeichnen wir symbolisch durch $G'(x')$. $G'(x')$ und $G(x)$ beschreiben das nämliche Gravitationsfeld. Ersetzen wir in den Funktionen $g'_{\mu\nu}$ die Koordinaten x'_ν durch die Koordinaten x_ν, d. h. bilden wir $G'(x)$, so beschreibt $G'(x)$ ebenfalls ein Gravitationsfeld bezüglich K, welches aber nicht übereinstimmt mit dem tatsächlichen (bzw. ursprünglich gegebenen) Gravitationsfelde.

Setzen wir nun voraus, daß die Differentialgleichungen des Gravitationsfeldes allgemein kovariant sind, so sind sie für $G'(x')$ erfüllt (bezüglich K'), wenn sie bezüglich K für $G(x)$ erfüllt sind. Sie sind dann also auch bezüglich K für $G'(x)$ erfüllt. Bezüglich K existierten dann die voneinander verschiedenen Lösungen $G(x)$ und $G'(x)$, trotzdem an den Gebietsgrenzen beide Lösungen übereinstimmten, d. h. [5] durch allgemein kovariante Differentialgleichungen für das Gravitationsfeld kann das Geschehen in demselben nicht eindeutig festgelegt werden.

Verlangen wir daher, daß der Ablauf des Geschehens im Gravitationsfelde durch die aufzustellenden Gesetze vollständig bestimmt sei, so sind wir genötigt, die Wahl des Koordinatensystems derart einzuschränken, daß es ohne Verletzung der einschränkenden Bedingungen unmöglich ist, ein neues Koordinatensystem K' von der vorhin charakterisierten Art einzuführen. Die Fortsetzung des Koordinatensystems ins Innere eines Gebietes Σ hinein darf nicht willkürlich sein.

§ 13. Kovarianz bezüglich linearer Transformationen. Angepaßte Koordinatensysteme.

Nachdem wir gesehen haben, daß das Koordinatensystem Bedingungen zu unterwerfen ist, müssen wir einige Arten der Spezialisierung der Koordinatenwahl ins Auge fassen. Eine sehr weitgehende Spezialisierung erhält man, wenn man nur lineare Transformationen zuläßt. Würden wir von den Gleichungen der Physik nur verlangen, daß sie linearen Transformationen gegenüber kovariant sein müssen, so würde unsere Theorie ihre Hauptstütze einbüßen. Denn eine Transformation auf ein beschleunigtes oder rotierendes System würde dann keine berechtigte Transformation sein, und die in § 1 hervorgehobene physikalische Gleichwertigkeit des »Zentrifugalfeldes« und Schwerefeldes würde durch die Theorie nicht auf eine Wesensgleichheit zurückgeführt. [6] Anderseits aber ist es (wie sich im folgenden zeigen wird) vorteilhaft, zu fordern, daß zu den berechtigten Transformationen auch die linearen gehören. Es sei daher zunächst kurz einiges gesagt über die Modifikation, welche die im Absatz B dargelegte Kovariantentheorie erfährt, wenn statt beliebiger nur lineare Transformationen als berechtigte zugelassen werden.

Kovarianten bezüglich linearer Transformationen. Die in § 3 bis § 8 dargestellten algebraischen Eigenschaften der Tensoren werden dadurch, daß man nur lineare Transformationen zuläßt, nicht vereinfacht; hingegen vereinfachen sich die Regeln für die Bildung der Tensoren durch Differentiation (§ 9) bedeutend.

Es ist nämlich allgemein

$$\frac{\partial}{\partial x'_\varrho} = \sum_\delta \frac{\partial x_\delta}{\partial x'_\varrho} \frac{\partial}{\partial x_\delta}.$$

Also ist z. B. für einen kovarianten Tensor zweiten Ranges gemäß (§ 5 a)

$$\frac{\partial A'_{\mu\nu}}{\partial x'_\varrho} = \sum_{\alpha\beta\delta} \frac{\partial x_\delta}{\partial x'_\varrho} \frac{\partial}{\partial x_\delta}\left(\frac{\partial x_\alpha}{\partial x'_\mu} \frac{\partial x_\beta}{\partial x'_\nu} A_{\alpha\beta}\right).$$

Für lineare Substitutionen sind die Ableitungen $\frac{\partial x_\alpha}{\partial x'_\mu}$ usw. von den x_δ unabhängig, so daß man hat

$$\frac{\partial A'_{\mu\nu}}{\partial x'_\varrho} = \sum_{\alpha\beta\delta} \frac{\partial x_\alpha}{\partial x'_\mu} \frac{\partial x_\beta}{\partial x'_\nu} \frac{\partial x_\delta}{\partial x'_\varrho} \frac{\partial A_{\alpha\beta}}{\partial x_\delta}.$$

$\left(\frac{\partial A_{\alpha\beta}}{\partial x_\delta}\right)$ ist also ein kovarianter Tensor dritten Ranges.

Allgemein kann gezeigt werden, daß man durch Differentiation der Komponenten eines beliebigen Tensors nach den Koordinaten wieder einen Tensor erhält, dessen Rang um 1 erhöht ist, wobei der hinzutretende Index kovarianten Charakter hat. Dies ist also die Operation der Erweiterung bei Beschränkung auf lineare Transformationen. Da die Erweiterung in Verbindung mit den algebraischen Operationen die Grundlage für die Kovariantenbildung überhaupt bildet, beherrschen wir damit das System der Kovarianten bezüglich linearer Transformationen. Wir wenden uns nun zu einer Überlegung, die zu einer viel weniger weitgehenden Beschränkung der Koordinatenwahl hinführt.

Transformationsgesetz des Integrals I. Es sei H eine Funktion der $g^{\mu\nu}$ und ihrer ersten Ableitungen $\frac{\partial g^{\mu\nu}}{\partial x_\sigma}$, die wir zur Abkürzung auch $g^{\mu\nu}_\sigma$ nennen. I bezeichne das über einen endlichen Teil Σ des Kontinuums erstreckte Integral

$$J = \int H\sqrt{-g}\,d\tau \tag{61}$$

Das zunächst benutzte Koordinatensystem sei K_1. Wir fragen nach der Änderung ΔJ, welche J erfährt, wenn man vom System K_1 auf das unendlich wenig verschiedene Koordinatensystem K_2 übergeht. Bezeichnet man mit $\Delta\phi$ den Zuwachs, welchen die beliebige, auf einen Punkt des Kontinuums sich beziehende Größe ϕ bei der Transformation erleidet, so hat man zunächst gemäß (17)

$$\Delta(\sqrt{-g}\,d\tau) = 0 \tag{62}$$

und ferner

$$\Delta H = \sum_{\mu\nu\sigma}\left(\frac{\partial H}{\partial g^{\mu\nu}}\Delta g^{\mu\nu} + \frac{\partial H}{\partial g^{\mu\nu}_\sigma}\Delta g^{\mu\nu}_\sigma\right). \tag{62a}$$

Die $\Delta g^{\mu\nu}$ lassen sich vermöge (8) durch die Δx_μ ausdrücken, indem man die Beziehungen

$$\begin{aligned}\Delta g^{\mu\nu} &= g^{\mu\nu\prime} - g^{\mu\nu}\\ \Delta x_\mu &= x'_\mu - x_\mu\end{aligned}$$

berücksichtigt. Man erhält

$$\Delta g^{\mu\nu} = \sum_\alpha\left(g^{\mu\alpha}\frac{\partial\Delta x_\nu}{\partial x_\alpha} + g^{\nu\alpha}\frac{\partial\Delta x_\mu}{\partial x_\alpha}\right) \tag{63}$$

$$\Delta g^{\mu\nu}_\sigma = \sum_\alpha\left\{\frac{\partial}{\partial x_\sigma}\left(g^{\mu\alpha}\frac{\partial\Delta x_\nu}{\partial x_\alpha} + g^{\nu\alpha}\frac{\partial\Delta x_\mu}{\partial x_\alpha}\right) - \frac{\partial g^{\mu\nu}}{\partial x_\alpha}\frac{\partial\Delta x_\alpha}{\partial x_\sigma}\right\}. \tag{63a}$$

Die Gleichungen (62a), (63), (63a) liefern ΔH als lineare homogene Funktion der ersten und zweiten Ableitungen der Δx_μ nach den Koordinaten.

Bisher haben wir über die Art, wie H von den $g^{\mu\nu}$ und $g^{\mu\nu}_\sigma$ abhängen soll, noch keine Festsetzung getroffen. Wir nehmen nun an, daß H bezüglich linearer Transformationen eine Invariante sei; d. h. ΔH soll verschwinden, falls die $\frac{\partial^2\Delta x_\mu}{\partial x_\alpha\partial x_\sigma}$ verschwinden. Unter dieser Voraussetzung erhalten wir [7]

$$\frac{1}{2}\Delta H = \sum_{\mu\nu\sigma\alpha} g^{\nu\alpha}\frac{\partial H}{\partial g^{\mu\nu}_\sigma}\frac{\partial^2\Delta x_\mu}{\partial x_\sigma\partial x_\alpha}. \tag{64}$$

Mit Hilfe von (64) und (62) erhält man

$$\frac{1}{2}\Delta J = \int d\tau \sum_{\mu\nu\sigma\alpha} g^{\nu\alpha} \frac{\partial H\sqrt{-g}}{\partial g^{\mu\nu}_{\sigma}} \frac{\partial^2 \Delta x_\mu}{\partial x_\alpha \partial x_\sigma},$$

und hieraus durch partielle Integration:

$$\frac{1}{2}\Delta J = \int d\tau \sum_{\mu} (\Delta x_\mu B_\mu) + F, \quad (65)$$

wobei gesetzt ist [8]

$$B_\mu = \sum_{\alpha\sigma\nu} \frac{\partial^2}{\partial x_\sigma \partial x_\alpha}\left(g^{\nu\alpha} \frac{\partial H\sqrt{-g}}{\partial g^{\mu\nu}_{\sigma}}\right) \quad (65\text{a})$$

$$F = \int d\tau \sum_{\alpha\sigma\nu} \frac{\partial}{\partial x_\alpha}\left[g^{\nu\alpha} \frac{\partial H\sqrt{-g}}{\partial g^{\mu\nu}_{\sigma}} \frac{\partial \Delta x_\mu}{\partial x_\sigma} - \frac{\partial}{\partial x_\sigma}\left(g^{\nu\sigma} \frac{\partial H\sqrt{-g}}{\partial g^{\mu\nu}_{\alpha}}\right)\Delta x_\mu\right]. \quad (65\text{b})$$

F läßt sich in ein Oberflächenintegral verwandeln. Es verschwindet, wenn an der Begrenzung die Δx_μ und $\frac{\partial \Delta x_\mu}{\partial x_\sigma}$ verschwinden.

Angepaßte Koordinatensysteme. Wir betrachten wieder den nach allen Koordinaten endlichen Teil Σ unseres Kontinuums, der zunächst auf das Koordinatensystem K bezogen sei. Von diesem Koordinatensystem K ausgehend, denke man sich sukzessive, einander unendlich benachbarte Koordinatensysteme K', K'' usw. eingeführt, derart, daß für den Übergang von jedem System zu dem folgenden die Δx_μ und $\frac{\partial \Delta x_\mu}{\partial x_\alpha}$ an der Begrenzung verschwinden. Wir nennen alle diese Systeme »Koordinatensystem mit übereinstimmenden Begrenzungskoordinaten«. Für jede infinitesimale Koordinatentransformation zwischen benachbarten Koordinatensystemen der Gesamtheit K, K', $K'' \cdots$ ist [9]

$$F = 0,$$

so daß hier statt (65) die Gleichung

$$\frac{1}{2}\Delta J = -\int d\tau \, \Delta x_\mu B_\mu \quad (66)$$

tritt. Unter allen Systemen mit übereinstimmenden Begrenzungskoordinaten wird es solche geben, für welche J ein Extrenuum ist gegenüber den J-Werten aller benachbarten Systeme mit übereinstimmenden Begrenzungskoordinaten; solche Koordinatensysteme nennen wir »dem Gravitationsfeld angepaßte Koordinntensysteme«. Für angepaßte Systeme gelten nach (66), weil die Δx_μ im Innern von Σ frei wählbar sind, die Gleichungen

$$B_\mu = 0. \quad (67)$$

Umgekehrt ist (67) hinreichende Bedingung dafür, daß das Koordinatensystem ein dem Gravitationsfeld angepaßtes ist. [10]

Indem wir im folgenden Differenzialgleichungen des Gravitationsfeldes aufstellen, welche nur für angepaßte Koordinatensysteme Gültigkeit beanspruchen, vermeiden wir die im § 13 dargelegte Schwierigkeit. In der Tat ist es bei Beschränkung auf angepaßte Koordinatensysteme nicht gestattet, ein außerhalb Σ gegebenes Koordinatensystem ins Innere von Σ in beliebiger Weise stetig fortzusetzen. [11]

§ 14. Der H-Tensor. [12]

Die Gleichung (65) führt uns zu einem Satze, der für die ganze Theorie von fundamentaler Bedeutung ist. Wenn wir das Gravitationsfeld der $g_{\mu\nu}$ unendlich wenig variieren, d. h. die $g_{\mu\nu}$ durch $g^{\mu\nu}+\delta g^{\mu\nu}$ ersetzen, wobei die $\delta g^{\mu\nu}$ in einer endlich breiten, der Begrenzung von Σ anliegenden Zone verschwinden mögen, so wird H in $H+\delta H$ und das Integral J in $J+\delta J$ übergehen. Wir behaupten nun, daß stets die Gleichung

$$\Delta\{\delta J\} = 0 \tag{68}$$

gilt, wie die $\delta g_{\mu\nu}$ auch gewählt werden mögen, falls nur die Koordinatensysteme (K_1 und K_2) bezüglich des unvariierten Gravitationsfeldes angepaßte Koordinatensysteme sind; d. h. bei Beschränkung auf angepaßte Koordinatensysteme ist δJ eine Invariante.

Zum Beweise denken wir uns die Variationen $\delta g^{\mu\nu}$ aus zwei Teilen zusammengesetzt; wir schreiben also

$$\delta g^{\mu\nu} = \delta_1 g^{\mu\nu} + \delta_2 g^{\mu\nu}, \tag{69}$$

welche Teilvariationen in folgender Weise gewählt werden:

a) Die $\delta_1 g^{\mu\nu}$ seien so gewählt, daß das Koordinatensystem K_1 nicht nur dem (wirklichen) Gravitationsfelde der $g^{\mu\nu}$, sondern auch dem (variierten) Gravitationsfelde der $g^{\mu\nu}+\delta g^{\mu\nu}$ angepaßt sei. Es bedeutet dies, daß nicht nur die Gleichung

$$B_\mu = 0,$$

sondern auch die Gleichungen

$$\delta_1 B_\mu = 0 \tag{70}$$

gelten soll. Die $\delta_1 g^{\mu\nu}$ sind also nicht voneinander unabhängig, sondern es bestehen zwischen ihnen 4 Differentialgleichungen.

b) Die $\delta_2 g^{\mu\nu}$ seien so gewählt, wie sie ohne Änderung des Gravitationsfeldes durch bloße Variation des Koordinatensystems erzielt werden könnten, und zwar durch eine Variation in demjenigen Teilgebiete von Σ, in welchem die $\delta g^{\mu\nu}$ von Null verschieden sind. Eine derartige Variation ist durch vier voneinander unabhängige Funktionen (Variationen der Koordinaten) bestimmt. Es ist klar, daß im allgemeinen $\delta_2 B_\mu \neq 0$ ist.

Die Superposition dieser beiden Variationen ist also durch

$$(10-4)+4=10$$

voneinander unabhängige Funktionen bestimmt; sie wird also einer beliebigen Variation der $\delta g^{\mu\nu}$ äquivalent sein. Der Beweis unseres Satzes ist also geleistet, wenn Gleichung (68) für beide Teilvariationen bewiesen ist.

Beweis für die Variation δ_1: Durch δ_1-Variation von (65) erhält man unmittelbar

$$\frac{1}{2}\Delta(\delta_1 J) = \int d\tau \sum_\mu (\Delta x_\mu \, \delta_1 B_\mu) + \delta_1 F. \tag{65a}$$

Da an der Begrenzung von Σ die δ_1-Variationen der $g^{\mu\nu}$ und ihrer sämtlichen Ableitungen verschwinden, so verschwindet gemäß (65b) die in ein Oberflächenintegral verwandelbare Größe $\delta_1 F$. Hiernach und nach (70) geht (65a) über in die behauptete Beziehung

$$\Delta(\delta_1 J) = 0. \tag{68a}$$

Beweis für die Variation δ_2: Die Variation $\delta_2 J$ entspricht einer infinitesimalen Koordinatentransformation bei festgehaltenen Begrenzungskoordinaten. Da das Koordinatensystem bezüglich des unvariierten Gravitationsfeldes ein angepaßtes sein soll, so ist also gemäß der Definition des angepaßten Koordinatensystems

$$\delta_2 J = 0.$$

Es werde zunächst angenommen, daß die betrachtete Variation des Gravitationsfeldes bezüglich des Koordinatensystems K_1 als eine δ_2-Variation gewählt sei; dann ist also zunächst

$$\delta_2 (J_1) = 0.$$

Ist diese Variation dann auch bezüglich K_2 eine δ_2-Variation, was nachher bewiesen werden wird, so gilt bezüglich K_2 die analoge Gleichung

$$\delta_2 (J_2) = 0.$$

Durch Subtraktion folgt dann die zu beweisende Gleichung

$$\delta_2(\Delta J) = \Delta(\delta_2 J) = 0. \tag{68b}$$

Es ist noch der Nachweis zu erbringen, daß die betrachtete Variation auch bezüglich K_2 eine δ_2-Variation ist. Wir bezeichnen symbolisch mit G_1 bzw. G_2 die unvariierten, auf K_1 bzw. K_2 bezogenen Tensoren der $g^{\mu\nu}$, mit G_1^* bzw. G_2^* die variierten, auf K_1 bzw. K_2 bezogenen Tensoren $g^{\mu\nu}$. Von G_1 zu G_2 bzw. von G_1^* zu G_2^* gelangt man durch die Koordinatentransformation T; die inverse Substitution sei T^{-1}. Ferner gelange man von G_1 zu G_1^* durch die Koordinatentransformation t. Dann erhält man G_2^* aus G_2 durch die Aufeinanderfolge

$$T^{-1}-t-T$$

von Transformationen, also wieder durch eine Koordinatentransformation. Damit ist gezeigt, daß die betrachtete Variation der $g^{\mu\nu}$ auch bezüglich K_2 eine δ_2-Variation ist.

Aus (68a) und (68b) folgt endlich die zu beweisende Gleichung (68).

Aus dem bewiesenen Satze leiten wir die Existenz eines aus 10 Komponenten bestehenden Komplexes ab, der bei Beschränkung auf angepaßte Koordinatensysteme Tensorcharakter besitzt. Nach (61) hat man

$$\delta J = \delta\left\{\int H\sqrt{-g}\,d\tau\right\}$$
$$= \int d\tau \sum_{\mu\nu\sigma}\left\{\frac{\partial(H\sqrt{-g})}{\partial g^{\mu\nu}}\delta g^{\mu\nu} + \frac{\partial(H\sqrt{-g})}{\partial g^{\mu\nu}_\sigma}\delta g^{\mu\nu}_\sigma\right\}$$

oder, da $\delta g^{\mu\nu}_\sigma = \frac{\partial}{\partial x_\sigma}(\delta g^{\mu\nu})$, nach partieller Integration und mit Rücksicht darauf, daß die $\delta(g^{\mu\nu})$ an der Begrenzung verschwinden.

$$\delta J = \int d\tau \sum_{\mu\nu}\delta g^{\mu\nu}\left\{\frac{\partial H\sqrt{-g}}{\partial g^{\mu\nu}} - \sum_\sigma \frac{\partial}{\partial x_\sigma}\left(\frac{\partial H\sqrt{-g}}{\partial g^{\mu\nu}_\sigma}\right)\right\} \tag{71}$$

Wir haben nun bewiesen, daß δJ bei Beschränkung auf angepaßte Koordinatensysteme eine Invariante ist. Da die $\delta g^{\mu\nu}$ nur in einem unendlich kleinen Gebiete von Null verschieden zu sein brauchen und $\sqrt{-g}\,d\tau$ ein Skalar ist, so ist auch der durch $\sqrt{-g}$ dividierte Integrand eine Invariante, d. h. die Größe

$$\frac{1}{\sqrt{-g}}\sum \delta g^{\mu\nu}\mathfrak{E}_{\mu\nu}, \tag{72}$$

wobei

$$\mathfrak{E}_{\mu\nu} = \frac{\partial H\sqrt{-g}}{\partial g^{\mu\nu}} - \sum_\sigma \frac{\partial}{\partial x_\sigma}\left(\frac{\partial H\sqrt{-g}}{\partial g^{\mu\nu}_\sigma}\right) \tag{73}$$

gesetzt ist. Nun ist aber $\delta g^{\mu\nu}$ ebenso wie $g^{\mu\nu}$ ein kontravarianter Tensor, und es sind die Verhältnisse der $\delta g^{\mu\nu}$ frei wählbar. Daraus folgt, daß

$$\frac{\mathfrak{E}_{\mu\nu}}{\sqrt{-g}}$$

bei Beschränkung auf angepaßte Koordinatensysteme und Substitutionen zwischen solchen ein kovarianter Tensor, $\mathfrak{E}_{\mu\nu}$ selbst der entsprechende kovariante V-Tensor ist, und zwar nach (73) ein symmetrischer Tensor.

§ 15. Ableitung der Feldgleichungen.

Es liegt die Annahme nahe, daß in den gesuchten Feldgleichungen der Gravitation, welche an die Stelle der Poissonschen Gleichung der Newtonschen Theorie zu treten haben, der Tensor $\mathfrak{E}_{\mu\nu}$ eine fundamentale Rolle spiele. Denn wir haben nach den Überlegungen der §§ 13 und 14 zu fordern, daß die gesuchten Gleichungen — ebenso wie der Tensor $\mathfrak{E}_{\mu\nu}$ — nur bezüglich angepaßter Koordinatensysteme kovariant seien. Da wir ferner im Anschluß an (42a) gesehen haben, daß für [13]
die Einwirkung des Gravitationsfeldes auf die Materie der Energietensor $\mathfrak{T}_\sigma^\nu$ maßgebend ist, so werden die gesuchten Gleichungen in einer Verknüpfung der Tensoren $\mathfrak{E}_{\mu\nu}$ und $\mathfrak{T}_\sigma^\nu$ bestehen. Es liegt also nahe, die gesuchten Gleichungen so anzusetzen:

$$\mathfrak{E}_{\sigma\tau} = \varkappa \mathfrak{T}_{\sigma\tau} \tag{74}$$

Dabei ist $\varkappa$ eine universelle Konstante und $\mathfrak{T}_{\sigma\tau}$ der symmetrische kovariante V-Tensor, der zu dem gemischten Energietensor $\mathfrak{T}_\sigma^\nu$ gehört, gemäß der Relation

$$\left.\begin{aligned} \mathfrak{T}_{\sigma\tau} &= \sum_\nu g_{\nu\tau} \mathfrak{T}_\sigma^\nu \\ \text{bzw.}\quad \mathfrak{T}_\sigma^\nu &= \sum_\tau \mathfrak{T}_{\sigma\tau} g^{\nu\tau} \end{aligned}\right\} \tag{75}$$

Bestimmung der Funktion H. Damit sind die gesuchten Gleichungen insofern noch nicht vollständig gegeben, als wir die Funktion H noch nicht festgelegt haben. Wir wissen bisher nur, daß H von den $g^{\mu\nu}$ und $g_\sigma^{\mu\nu}$ allein abhängt und bezüglich linearer Transformationen ein Skalar ist[1]. Eine weitere Bedingung, welcher H genügen muß, erhalten wir auf folgendem Wege.

Ist $\mathfrak{T}_\sigma^\nu$ der Energietensor des gesamten, in dem betrachteten Gebiete vorhandenen materiellen Vorganges, so verschwindet in (42a) der V-Vierervektor $(\mathfrak{K}_\sigma)$ der Kraftdichte. (42a) sagt dann aus, daß die Divergenz des Energietensors $\mathfrak{T}_\sigma^\nu$ des materiellen Vorganges verschwindet; gleiches gilt dann gemäß (74) für den Tensor $\mathfrak{E}_{\sigma\tau}$, bzw. für den aus

[1] Ohne die letztere, in § 14 eingeführte Beschränkung hätten wir für B_μ nicht den in (65a) gegebenen Ausdruck gefunden; die im folgenden im Texte angegebene Betrachtung zur Bestimmung von H scheitert, wenn man jene Beschränkung fallen läßt. Hierin liegt ihre Rechtfertigung.

demselben zu bildenden gemischten V-Tensor $\mathfrak{E}_\sigma^\nu$. Es muß also für jedes Gravitationsfeld die Beziehung erfüllt sein (vgl. (41b) und (34)). [14]

$$\sum_{\nu\tau} \frac{\partial}{\partial x_\nu}(g^{\tau\nu} \mathfrak{E}_{\sigma\tau}) + \frac{1}{2} \sum_{\mu\nu} \frac{\partial g^{\mu\nu}}{\partial x_\sigma} \mathfrak{E}_{\mu\nu} = 0 .$$

Diese Beziehung läßt sich auf Grund von (73) und (65a) in die Form bringen: [15]

$$\sum_\nu \frac{\partial S_\sigma^\nu}{\partial x_\nu} - B_\sigma = 0, \tag{76}$$

wobei

$$S_\sigma^\nu = \sum_{\mu\tau} \left(g^{\nu\tau} \frac{\partial H\sqrt{-g}}{\partial g^{\sigma\tau}} + g_\mu^{\nu\tau} \frac{\partial H\sqrt{-g}}{\partial g_\mu^{\sigma\tau}} + \frac{1}{2} \delta_\sigma^\nu H \sqrt{-g} - \frac{1}{2} g_\sigma^{\mu\tau} \frac{\partial H\sqrt{-g}}{\partial g_\nu^{\mu\tau}} \right) \tag{76a}$$

gesetzt ist ($\delta_\sigma^\nu = 1$ bzw. 0, je nachdem $\sigma = \nu$ oder $\sigma \pm \nu$ ist).

Wenn die $\mathfrak{T}_{\sigma\tau}$ gegeben sind, so können die 10 Gleichungen (74) dazu dienen, die 10 Funktionen $g^{\mu\nu}$ zu bestimmen. Außerdem müssen die $g^{\mu\nu}$ aber noch die vier Gleichungen (67) erfüllen, da das Koordinatensystem ein angepaßtes sein soll. Wir haben also mehr Gleichungen als zu suchende Funktionen. Dies geht nur dann an, wenn die Gleichungen nicht alle voneinander unabhängig sind. Es wird gefordert werden müssen, daß die Erfüllung der Gleichungen (74) zur Folge hat, daß auch die Gleichungen (67) erfüllt sind. Ein Blick auf (76) und (76a) zeigt, daß dies dann erreicht ist, wenn S_σ^ν (welche Größe wie H eine Funktion der $g^{\mu\nu}$ und $g_\sigma^{\mu\nu}$ ist) identisch verschwindet für jede Kombination der Indizes. H muß also gemäß den Bedingungen

$$S_\sigma^\nu \equiv 0 \tag{77}$$

gewählt werden.

Ohne einen formalen Grund dafür angeben zu können, fordere ich ferner, daß H eine ganze homogene Funktion zweiten Grades in den $g_\sigma^{\mu\nu}$ sei. Dann ist H bis auf einen konstanten Faktor vollkommen bestimmt. Denn da es bezüglich linearer Transformationen ein Skalar sein soll, muß[1] es mit Rücksicht auf die eben angegebene Festsetzung eine lineare Kombination der folgenden fünf Größen sein:

$$\sum g_{\mu\nu} \frac{\partial g^{\mu\nu}}{\partial x_\sigma} \frac{\partial g^{\sigma\tau}}{\partial x_\tau} ; \quad \sum g^{\sigma\sigma'} g_{\mu\nu} \frac{\partial g^{\mu\nu}}{\partial x_\sigma} g_{\mu'\nu'} \frac{\partial g^{\mu'\nu'}}{\partial x'_\sigma} ; \quad \sum g^{\sigma\sigma'} \frac{\partial g^{\tau\mu}}{\partial x_\mu} \frac{\partial g^{\tau'\nu}}{\partial x_{\nu'}} ;$$

$$\sum g_{\mu\mu'} g_{\nu\nu'} g^{\tau\tau'} \frac{\partial g^{\mu\nu}}{\partial x_\tau} \frac{\partial g^{\mu'\nu'}}{\partial x'_\sigma} ; \quad \sum g_{\alpha\beta} \frac{\partial g^{\alpha\tau}}{\partial x_\tau} \frac{\partial g^{\beta\tau}}{\partial x_\sigma} .$$

Die Bedingungen (77) führen endlich dazu, die Funktion H, abgesehen von einem konstanten Faktor der vierten dieser Größen gleich-

[1] Der Beweis hierfür ist einfach, aber weitläufig, deshalb lasse ich ihn weg.

zusetzen. Wir setzen daher[1] mit Rücksicht auf (35), indem wir über [16]
die Konstante willkürlich verfügen: [17]

$$H = \frac{1}{4} \sum_{\alpha\beta\tau\varrho} g^{\alpha\beta} \frac{\partial g_{\tau\varrho}}{\partial x_\alpha} \frac{\partial g^{\tau\varrho}}{\partial x_\beta} . \qquad (78)$$

Wir beschränken uns darauf, zu zeigen, daß bei dieser Wahl von H Gleichung (77) wirklich erfüllt ist. Mit Hilfe der Relationen

$$dg = g \sum_{\sigma\tau} g^{\sigma\tau} dg_{\sigma\tau} = -g \sum_{\sigma\tau} g_{\sigma\tau} dg^{\sigma\tau}$$

$$dg_{\alpha\beta} = -\sum_{\mu\nu} g_{\alpha\mu} g_{\beta\nu} dg^{\mu\nu}$$

erhält man aus (78)

$$\left.\begin{aligned} \sum_\tau g^{\nu\tau} \frac{\partial H\sqrt{-g}}{\partial g^{\sigma\tau}} &= \frac{1}{2} H\sqrt{-g}\,\delta^\nu_\sigma + \frac{1}{4}\sqrt{-g} \sum_{\mu\mu'\tau} g^{\nu\tau} \frac{\partial g_{\mu\mu'}}{\partial x_\sigma} \frac{\partial g^{\mu\mu'}}{\partial x_\tau} \\ &\quad - \frac{1}{2}\sqrt{-g} \sum_{\varrho\varrho'\varkappa} g^{\varrho\varrho'} \frac{\partial g_{\sigma\varkappa}}{\partial x_\varrho} \frac{\partial g^{\nu\varkappa}}{\partial x_{\varrho'}} \\ \sum_{\mu\tau} g^{\nu\tau}_\mu \frac{\partial H\sqrt{-g}}{\partial g^{\sigma\tau}_\mu} &= \frac{1}{2}\sqrt{-g} \sum_{\varrho\varrho'\varkappa} g^{\varrho\varrho'} \frac{\partial g_{\sigma\varkappa}}{\partial x_\varrho} \frac{\partial g^{\nu\varkappa}}{\partial x_{\varrho'}} \\ \frac{1}{2} \sum_{\mu\tau} g^{\mu\tau}_\sigma \frac{\partial H\sqrt{-g}}{\partial g^{\mu\tau}_\nu} &= \frac{1}{4}\sqrt{-g} \sum_{\mu\mu'\tau} g^{\mu\tau} \frac{\partial g_{\mu\mu'}}{\partial x_\sigma} \frac{\partial g^{\mu\mu'}}{\partial x_\tau} . \end{aligned}\right\} \qquad (79)$$

Hieraus folgt die Behauptung. [18]

Wir sind nun auf rein formalem Wege, d. h. ohne direkte Heranziehung unserer physikalischen Kenntnisse von der Gravitation, zu ganz
bestimmten Feldgleichungen gelangt. Um dieselben in ausführlicher [19]
Schreibweise zu erhalten, multiplizieren wir (74) mit $g^{\nu\tau}$ und summieren über den Index τ; wir erhalten so mit Rücksicht auf (73)

$$\varkappa\mathfrak{T}^\nu_\sigma = \sum_{\tau\alpha} g^{\nu\tau} \left(\frac{\partial H\sqrt{-g}}{\partial g^{\sigma\tau}} \frac{\partial}{\partial x_\alpha} \left[\frac{\partial H\sqrt{-g}}{\partial g^{\sigma\tau}_\alpha} \right] \right) \qquad (80)$$

oder

$$-\sum_{\alpha\tau} \frac{\partial}{\partial x_\alpha} \left(g^{\nu\tau} \frac{\partial H\sqrt{-g}}{\partial g^{\sigma\tau}_\alpha} \right) = \varkappa\mathfrak{T}^\nu_\sigma + \sum_{\alpha\tau} \left(-g^{\nu\tau} \frac{\partial H\sqrt{-g}}{\partial g^{\sigma\tau}} + g^{\nu\tau}_\alpha \frac{\partial H\sqrt{-g}}{\partial g^{\sigma\tau}_\alpha} \right) . \qquad (80a)$$

Dabei gilt, weil unser Koordinatensystem ein angepaßtes ist, gemäß (67) und (65 a) die Gleichung

---

[1] Drückt man H durch die Komponenten $\Gamma^\nu_{\sigma\tau}$ des Gravitationsfeldes aus (vgl. (46)), so erhält man $H = -\sum_{\mu\varrho\tau\tau'} g^{\tau\tau'} \Gamma^\varrho_{\mu\tau} \Gamma^\mu_{\varrho\tau'}$.

$$\sum_{\alpha\tau\nu} \frac{\partial}{\partial x_\nu}\frac{\partial}{\partial x_\alpha}\left(g^{\nu\tau}\frac{\partial H\sqrt{-g}}{\partial g^{\sigma\tau}_\alpha}\right)=0,$$

also mit Rücksicht auf (80) die Gleichung

$$\sum_\nu \frac{\partial}{\partial x_\nu}\left\{\mathfrak{T}_\sigma + \frac{1}{\varkappa}\sum_{\alpha\tau}\left(-g^{\nu\tau}\frac{\partial H\sqrt{-g}}{\partial g^{\sigma\tau}} + g^{\nu\tau}_\alpha \frac{\partial H\sqrt{-g}}{\partial g^{\sigma\tau}_\alpha}\right)\right\}=0. \quad (80\text{b})$$

Vermöge (78), (79) und (46) können wir an die Stelle der Gleichungen (80a) und (80b) die folgenden setzen [20]

$$\sum_{\alpha\beta} \frac{\partial}{\partial x^\alpha}(\sqrt{-g}\, g^{\alpha\beta}\, \Gamma^\nu_{\sigma\beta}) = -\varkappa(\mathfrak{T}^\nu_\sigma + \mathfrak{t}^\nu_\sigma), \quad (81)$$

$$\sum_\nu \frac{\partial}{\partial x_\nu}(\mathfrak{T}^\nu_\sigma + \mathfrak{t}^\nu_\sigma) = 0, \quad (42\text{c})$$

wobei [21]

$$\Gamma^\nu_{\sigma\beta} = \frac{1}{2}\sum_\tau g^{\nu\tau}\frac{\partial g_{\sigma\tau}}{\partial x_\beta}, \quad (81\text{a})$$

$$\left.\begin{aligned}\mathfrak{t}^\nu_\sigma &= -\frac{\sqrt{-g}}{4\varkappa}\sum_{\mu\mu'\varrho\tau}\left(g^{\nu\tau}\frac{\partial g_{\mu\mu'}}{\partial x_\sigma}\frac{\partial g^{\mu\mu'}}{\partial x_\tau} - \frac{1}{2}\delta^\nu_\sigma g^{\varrho\tau}\frac{\partial g_{\mu\mu'}}{\partial x_\varrho}\frac{\partial g^{\mu\mu'}}{\partial x_\tau}\right)\\ &= \frac{\sqrt{-g}}{\varkappa}\sum_{\mu\varrho\tau\tau'}\left(g^{\nu\tau}\Gamma^\varrho_{\mu\sigma}\Gamma^\mu_{\varrho\tau} - \frac{1}{2}\delta^\nu_\sigma g^{\tau\tau'}\Gamma^\varrho_{\mu\tau}\Gamma^\mu_{\varrho\tau'}\right)\end{aligned}\right\} \quad (81\text{b})$$

Die Gleichungen (81) in Verbindung mit (81a) und (81b) sind die Differentialgleichungen des Gravitationsfeldes. Die Gleichungen (42c) drücken nach den in § 10 gegebenen Überlegungen die Erhaltungssätze des Impulses und der Energie für Materie und Gravitationsfeld zusammen aus. $\mathfrak{t}^\nu_\sigma$ sind diejenigen auf das Gravitationsfeld bezüglichen Größen, welche den Komponenten $\mathfrak{T}^\nu_\sigma$ des Energietensors (V-Tensors) der physikalischen Bedeutung nach analog sind. Es sei hervorgehoben, daß die $\mathfrak{t}^\nu_\sigma$ nicht beliebigen berechtigten, sondern nur linearen Transformationen gegenüber Tensorkovarianz besitzen; trotzdem nennen wir ($\mathfrak{t}^\nu_\sigma$) den Energietensor des Gravitationsfeldes. Analoges gilt für die Komponenten $\Gamma^\nu_{\sigma\beta}$ der Feldstärke des Gravitationsfeldes.

Das Gleichungssystem (81) läßt trotz seiner Kompliziertheit eine einfache physikalische Interpretation zu. Die linke Seite drückt eine Art Divergenz des Gravitationsfeldes aus. Diese wird — wie die rechte Seite zeigt — bedingt durch die Komponenten des totalen Energietensors. Sehr wichtig ist dabei das Ergebnis, daß der Energietensor des Gravitationsfeldes selbst in gleicher Weise felderregend wirksam ist wie der Energietensor der Materie. . . . . [22]

3.2 Translation

The Formal Foundation of the General Theory of Relativity.

D. The Differential Laws of the Gravitational Field

In the previous section we considered the coefficients $g_{\mu\nu}$, which are to be interpreted physically as the components of the gravitational potential, as known functions of the [coordinates] x_ν. It remains to find the differential laws these quantities satisfy. What is epistemologically satisfactory about the theory presented so far is that the principle of relativity is satisfied in its broadest possible sense. From a formal point of view, this is because the systems of equations are *generally* covariant, i.e., covariant under arbitrary substitutions of the x_ν.[1]

It thus seems natural to demand that the differential equations for the $g_{\mu\nu}$ also be generally covariant. However, we want to show that we have to restrict this demand if we want to satisfy the law of cause and effect. In fact, we shall prove that the laws that characterize the course of events in a gravitational field cannot possibly be *generally* covariant.[2]

§ 12. Proof of a Necessary Restriction in the Choice of Coordinates[3]

We consider a finite portion Σ of the continuum, in which no material process occurs.[4] What happens physically in Σ is then completely determined if the quantities $g_{\mu\nu}$ are given as functions of the coordinates x_ν, with respect to a coordinate system K used for the description. The totality of these functions will be symbolically denoted by $G(x)$.

Let a new coordinate system K' be introduced that coincides with K outside of Σ but deviates from K within Σ in such a way that the $g'_{\mu\nu}$ referred to K', like the $g_{\mu\nu}$ (and their derivatives), are everywhere continuous. The totality of the $g'_{\mu\nu}$ will be symbolically designated by $G'(x')$. $G'(x')$ and $G(x)$ describe the same gravitational field. If we replace the coordinates x'_ν by the coordinates x_ν in the functions $g'_{\mu\nu}$, i.e., if we form $G'(x)$, then $G'(x)$ also describes a gravitational field with respect to K, which however, does not correspond to the actual (i.e., originally given) gravitational field.

Now if we assume that the differential equations of the gravitational field are generally covariant, then they are satisfied for $G'(x')$ (with respect to K') if they are satisfied for $G(x)$ with respect to K. They are then also satisfied with respect to K for $G'(x)$. Relative to K there then exist two solutions $G(x)$ and $G'(x)$, which are different from each other, even though at the boundary of the region they coincide.[5] That means that *what happens in the gravitational field cannot be uniquely determined by generally covariant differential equations for this field.*

Therefore, if we demand that the course of events in the gravitational field be completely determined, we are forced to restrict the choice of the coordinate system in such a way that it is impossible, without a violation of these restrictions, to introduce a new coordinate system K' of the type characterized above. The continuation of the coordinate system into the interior of a region Σ cannot be arbitrary.

§ 13. Covariance under Linear Transformations Adapted Coordinate Systems

Now that we have seen that the coordinate system has to be subjected to some conditions, we must take a look at several kinds of specializations in the choice of coordinates. A very far-reaching specialization is obtained if one only allows linear transformations. If we were to demand that the equations of physics be covariant merely under *linear* transformations, our theory would be deprived of its main pillar of support. In that case, a transformation to an accelerated or rotating system would no longer be a justified [*berechtigte*] transformation and the theory would not reduce the physical equivalence of a "centrifugal field" and a gravitational field—emphasized in § 1—to an equality of essence [*Wesensgleichheit*].[6] By contrast, however, it is (as will become clear in what follows) advantageous to demand that linear transformations are among the justified transformations. We therefore need to briefly address the modification of the theory of covariants, set out in section B, when only linear rather than arbitrary transformations are allowed as justified transformations.

Quantities covariant under linear transformations. The algebraic properties of tensors presented in § 3 to § 8 are not simplified if one only allows linear transformations. By contrast, the rules for the formation of tensors by differentiation (§ 9) become significantly simpler.

In all generality we have

$$\frac{\partial}{\partial x'_\rho} = \sum_\delta \frac{\partial x_\delta}{\partial x'_\rho} \frac{\partial}{\partial x_\delta}.$$

Hence, one has, e.g., for a covariant tensor of rank two, according to (5a) $\left[\text{i.e., } A'_{\mu\nu} = \Sigma_{\alpha\beta} \frac{\partial x_\alpha}{\partial x'_\mu} \frac{\partial x_\beta}{\partial x'_\nu} A_{\alpha\beta}\right]$

$$\frac{\partial A'_{\mu\nu}}{\partial x'_\rho} = \sum_{\alpha\beta\delta} \frac{\partial x_\delta}{\partial x'_\rho} \frac{\partial}{\partial x_\delta} \left(\frac{\partial x_\alpha}{\partial x'_\mu} \frac{\partial x_\beta}{\partial x'_\nu} A_{\alpha\beta} \right).$$

For linear transformations, the derivatives $\frac{\partial x_\alpha}{\partial x'_\mu}$, etc., are independent of the x_δ and one has

$$\frac{\partial A'_{\mu\nu}}{\partial x'_\rho} = \sum_{\alpha\beta\delta} \frac{\partial x_\alpha}{\partial x'_\mu} \frac{\partial x_\beta}{\partial x'_\nu} \frac{\partial x_\delta}{\partial x'_\rho} \frac{\partial A_{\alpha\beta}}{\partial x_\delta}.$$

Therefore, $\left(\frac{\partial A_{\alpha\beta}}{\partial x_\delta}\right)$ is a covariant tensor of rank three.

It can be shown in full generality that differentiation of the components of any tensor [under linear transformations] with respect to its coordinates produces a tensor [under linear transformations] with a rank raised by one, where the additional index has covariant character. So this is the operation of taking the *exterior derivative* [*Erweiterung*] when we restrict ourselves to linear transformations. And since taking exterior derivatives combined with algebraic operations forms the basis for forming any covariants, this gives us mastery over the entire system of covariants under linear transformations. We now turn to a consideration that leads to a much less restrictive choice of coordinates.

The transformation law of the integral J. Let H be a function of the $g^{\mu\nu}$ and their first derivatives $\frac{\partial g^{\mu\nu}}{\partial x_\sigma}$, where the latter are called $g^{\mu\nu}_\sigma$ for short. Let J be an integral extended over a finite part Σ of the continuum

$$J = \int H\sqrt{-g}d\tau. \tag{61}$$

Let K_1 be the coordinate system initially used. We ask for the change ΔJ in J when we go from system K_1 to a different coordinate system K_2 infinitesimally close to K_1. If $\Delta\phi$ denotes the increase—due to transformation—of an arbitrary quantity ϕ at some point in the continuum, then one has, according to (17) [i.e., $\sqrt{g}d\tau = d\tau^*_0$], first

$$\Delta(\sqrt{-g}d\tau) = 0 \tag{62}$$

and furthermore

$$\Delta H = \sum_{\mu\nu\sigma}\left(\frac{\partial H}{\partial g^{\mu\nu}}\Delta g^{\mu\nu} + \frac{\partial H}{\partial g^{\mu\nu}_\sigma}\Delta g^{\mu\nu}_\sigma\right). \tag{62a}$$

The $\Delta g^{\mu\nu}$ can be expressed by means of (8) $\left[\text{i.e., } A^{\mu\nu\prime} = \sum_{\alpha\beta} \frac{\partial x'_\mu}{\partial x_\alpha}\frac{\partial x'_\nu}{\partial x_\beta} A^{\alpha\beta}\right]$ in terms of the Δx_μ if one takes the relations

$$\begin{aligned} \Delta g^{\mu\nu} &= g^{\mu\nu\prime} - g^{\mu\nu} \\ \Delta x_\mu &= x'_\mu - x_\mu \end{aligned}$$

into account. One obtains

$$\Delta g^{\mu\nu} = \sum_{\alpha} \left(g^{\mu\alpha} \frac{\partial \Delta x_\nu}{\partial x_\alpha} + g^{\nu\alpha} \frac{\partial \Delta x_\mu}{\partial x_\alpha} \right) \tag{63}$$

$$\Delta g_\sigma^{\mu\nu} = \sum_{\alpha} \left\{ \frac{\partial}{\partial x_\sigma} \left(g^{\mu\alpha} \frac{\partial \Delta x_\nu}{\partial x_\alpha} + g^{\nu\alpha} \frac{\partial \Delta x_\mu}{\partial x_\alpha} \right) - \frac{\partial g^{\mu\nu}}{\partial x_\alpha} \frac{\partial \Delta x_\alpha}{\partial x_\sigma} \right\}. \tag{63a}$$

The equations (62a), (63), (63a), give ΔH as a linear homogeneous function of the first and second derivatives of the Δx_μ with respect to the coordinates.

So far, we have not made any assumptions of how H depends on the $g^{\mu\nu}$ and the $g_\sigma^{\mu\nu}$. We shall now assume that H is invariant under linear transformations, i.e., that ΔH vanishes whenever the $\frac{\partial^2 \Delta x_\mu}{\partial x_\alpha \partial x_\sigma}$ do. Under this assumption we obtain[7]

$$\frac{1}{2} \Delta H = \sum_{\mu\nu\sigma\alpha} g^{\nu\alpha} \frac{\partial H}{\partial g_\sigma^{\mu\nu}} \frac{\partial^2 \Delta x_\mu}{\partial x_\sigma \partial x_\alpha}. \tag{64}$$

With the help of (64) and (62) one obtains

$$\frac{1}{2} \Delta J = \int d\tau \sum_{\mu\nu\sigma\alpha} g^{\nu\alpha} \frac{\partial H \sqrt{-g}}{\partial g_\sigma^{\mu\nu}} \frac{\partial^2 \Delta x_\mu}{\partial x_\sigma \partial x_\alpha}$$

and from this, by partial integration,

$$\frac{1}{2} \Delta J = \int d\tau \sum_{\mu} (\Delta x_\mu B_\mu) + F, \tag{65}$$

where we used the abbreviations[8]

$$B_\mu = \sum_{\alpha\sigma\nu} \frac{\partial^2}{\partial x_\sigma \partial x_\alpha} \left(g^{\nu\alpha} \frac{\partial H \sqrt{-g}}{\partial g_\sigma^{\mu\nu}} \right) \tag{65a}$$

$$F = \int d\tau \sum_{\alpha\sigma\nu\mu} \frac{\partial}{\partial x_\alpha} \left[g^{\nu\alpha} \frac{\partial H \sqrt{-g}}{\partial g_\sigma^{\mu\nu}} \frac{\partial \Delta x_\mu}{\partial x_\sigma} - \frac{\partial}{\partial x_\sigma} \left(g^{\nu\sigma} \frac{\partial H \sqrt{-g}}{\partial g_\alpha^{\mu\nu}} \right) \Delta x_\mu \right]. \tag{65b}$$

F can be transformed into a surface integral. It vanishes when the Δx_μ and the $\frac{\partial \Delta x_\mu}{\partial x_\sigma}$ vanish at the boundary.

Adapted coordinate systems.[9] We once again consider the region Σ of our continuum, finite in all its [four] coordinates, initially given in the coordinate system K. We now imagine that, starting from K, we successively introduce coordinate systems K', K'' etc., all infinitely close to each other, such that the Δx_μ and the $\frac{\partial \Delta x_\mu}{\partial \Delta x_\alpha}$ vanish at the boundaries. We call all such systems "coordinate systems with coinciding boundary coordinates." For any infinitesimal coordinate transformation between neighboring coordinate systems in the set $K, K', K'' \ldots$ we have

$$F = 0,$$

so that instead of (65) we have the equation

$$\frac{1}{2}\Delta J = \sum_{\mu} \int d\tau \Delta x_{\mu} B_{\mu}. \tag{66}$$

Among all systems with coinciding boundary coordinates will be some for which J attains an extremum compared with the J-values of neighboring systems with coinciding boundary values. Such coordinate systems we call "coordinate systems adapted [*angepaßt*] to the gravitational field." According to (66), the equations

$$B_{\mu} = 0 \tag{67}$$

hold for these adapted systems because the Δx_{μ} can be chosen freely inside of Σ.

Inversely, (67) is a sufficient condition that the coordinate system is adapted to the gravitational field.[10]

By constructing, in what follows, differential equations for the gravitational field that claim validity only in adapted coordinate systems, we avoid the difficulty mentioned in § 13. With the restriction to adapted coordinate systems, it is no longer allowed to continuously extend in an arbitrary manner a coordinate system given for the exterior of Σ into the interior of Σ.[11]

§ 14. The H-Tensor[12]

Equation (65) leads us to a theorem that is of fundamental importance to the entire theory. If we vary the gravitational field of the $g_{\mu\nu}$ by an infinitely small amount, i.e., replace the $g_{\mu\nu}$ by $g^{\mu\nu} + \delta g^{\mu\nu}$, where the $\delta g^{\mu\nu}$ shall vanish in a zone of finite width adjacent to the boundary of Σ, the H becomes $H + \delta H$ and J becomes $J + \delta J$. We now claim that the equation

$$\Delta\{\delta J\} = 0 \tag{68}$$

always holds whichever way the $\delta g_{\mu\nu}$ might be chosen, provided the coordinate systems (K_1 and K_2) are *adapted* coordinate systems relative to the unvaried gravitational field. This means that under the restriction to adapted coordinate systems, δJ is an invariant.

In order to prove this, we imagine the variations $\delta g^{\mu\nu}$ to be composed of two parts, and we therefore write

$$\delta g^{\mu\nu} = \delta_1 g^{\mu\nu} + \delta_2 g^{\mu\nu}, \tag{69}$$

where the two parts of the variation are chosen such that

a. The $\delta_1 g^{\mu\nu}$ are taken in a manner that the coordinate system K_1 is not only *adapted* to the (true) gravitational field of the $g^{\mu\nu}$ but also to the (varied) gravitational field of the $g^{\mu\nu} + \delta g^{\mu\nu}$. This means that not only the equations

$$B_\mu = 0$$

but also the equations

$$\delta_1 B_\mu = 0 \tag{70}$$

are valid. In other words, the $\delta_1 g^{\mu\nu}$ are not independent of each other, there are rather 4 differential equations between them.

b. The $\delta_2 g^{\mu\nu}$ are taken just as one would get them without changing the gravitational field, by mere variation of the coordinate system, specifically, by variation in that subregion of Σ in which the $\delta g^{\mu\nu}$ differ from zero. Such a variation is determined by four mutually independent functions (variations of coordinates). Obviously, in general, $\delta_2 B_\mu \neq 0$.

The superposition of the two variations is therefore determined by

$$(10-4)+4=10$$

mutually independent functions, and thus will be equivalent to an *arbitrary* variation of the $\delta g^{\mu\nu}$. Hence, the proof of our theorem will be completed when equation (68) is proven for the two partial variations.

Proof for the variation of δ_1: By δ_1-variation of (65) one obtains in a straightforward manner

$$\frac{1}{2}\Delta(\delta_1 J) = \int d\tau \sum_\mu (\Delta x_\mu \delta_1 B_\mu) + \delta_1 F. \tag{65a}$$

Since the δ_1-variations of the $g^{\mu\nu}$ and all their derivatives vanish at the boundary of Σ, the quantity $\delta_1 F$ (which can be transformed into a surface integral) also vanishes according to (65b). After this and with (70), our equation (65a) turns into the relation

$$\Delta(\delta_1 J) = 0. \tag{68a}$$

Proof for the variation of δ_2: The variation $\delta_2 J$ is equivalent to an infinitesimal coordinate transformation under fixed coordinates of the boundary. Since the coordinate system is an adapted one relative to the unvaried gravitational field, it follows from the definition of adapted coordinate systems that

$$\delta_2 J = 0.$$

Next, we assume the variation of the gravitational field relative to the coordinate system K_1 to be chosen as a δ_2-variation; then we have

$$\delta_2(J_1) = 0.$$

If this variation is a δ_2-variation also relative to K_2—as we shall prove later—then we have an analogous equation relative to K_2, i.e.,

$$\delta_2(J_2) = 0.$$

The equation to be proven follows then by subtraction

$$\delta_2(\Delta J) = \Delta(\delta_2 J) = 0. \tag{68b}$$

We still have to show that the variation under consideration is a δ_2-variation also relative to K_2. The unvaried tensors $g^{\mu\nu}$ relative to K_1 and K_2 be denoted symbolically by G_1 and G_2, respectively; similarly, the varied tensors $g^{\mu\nu}$ relative to K_1 and K_2 are G_1^* and G_2^*, respectively. The coordinate transformation T brings us from G_1 to G_2, also from G_1^* and G_2^* resp.; and the inverse transformation will be T^{-1}. Furthermore,the coordinate transformation t brings G_1 to G_1^*. Consequently, G_2^* is obtained from G_2 by the sequence of transformations

$$T^{-1} - t - T,$$

which is again a coordinate transformation. In this manner it is shown that the variation of the $g^{\mu\nu}$ which we have under consideration here is also a δ_2-variation relative to K_2.

The equation (68), which is to be proven, follows finally from (68a) and (68b).

From the proven theorem we deduce the existence of a complex of 10 components, which has tensorial character if we limit ourselves to adapted coordinate systems. According to (61) one has

$$\delta J = \delta\left\{\int H\sqrt{-g}d\tau\right\}$$
$$= \int d\tau \sum_{\mu\nu\sigma}\left\{\frac{\partial H\sqrt{-g}}{\partial g^{\mu\nu}}\delta g^{\mu\nu} + \frac{\partial H\sqrt{-g}}{\partial g_\sigma^{\mu\nu}}\delta g_\sigma^{\mu\nu}\right\}$$

or, since $\delta g_\sigma^{\mu\nu} = \dfrac{\partial}{\partial x_\sigma}(\delta g^{\mu\nu})$, after partial integration and considering the vanishing of the $\delta(g^{\mu\nu})$ at the boundary

$$\delta J = \int d\tau \sum_{\mu\nu}\delta g^{\mu\nu}\left\{\frac{\partial H\sqrt{-g}}{\partial g^{\mu\nu}} - \sum_\sigma \frac{\partial}{\partial x_\sigma}\left(\frac{\partial H\sqrt{-g}}{\partial g_\sigma^{\mu\nu}}\right)\right\}. \tag{71}$$

We have now proven that under limitation to adapted coordinate systems δJ is an invariant. Since the $\delta g^{\mu\nu}$ need differ from zero only in an infinitely small region, and since $\sqrt{-g}d\tau$ is a scalar, the integral divided by $\sqrt{-g}$ is also an invariant, i.e., the quantity

$$\frac{1}{\sqrt{-g}}\sum \delta g^{\mu\nu}\mathfrak{E}_{\mu\nu}, \tag{72}$$

where

$$\mathfrak{E}_{\mu\nu} = \frac{\partial H\sqrt{-g}}{\partial g^{\mu\nu}} - \sum_{\sigma} \frac{\partial}{\partial x_\sigma}\left(\frac{\partial H\sqrt{-g}}{\partial g_\sigma^{\mu\nu}}\right). \tag{73}$$

Now, however, $\delta g^{\mu\nu}$ is a contravariant tensor just as $g^{\mu\nu}$ is, and the ratios of the $\delta g^{\mu\nu}$ can be chosen freely. From this follows that under limitation to adapted coordinate systems and substitutions between them,

$$\frac{\mathfrak{E}_{\mu\nu}}{\sqrt{-g}}$$

is a covariant tensor and $\mathfrak{E}_{\mu\nu}$ *itself is the corresponding covariant V-tensor and according to (73) a symmetric tensor.*

§ 15. Derivation of the Field Equations

One may expect the tensor $\mathfrak{E}_{\mu\nu}$ to play a fundamental role in the field equations of gravitation that we want to find, and that those equations have to take the place of the Poisson equation has in Newtonian theory. After the considerations in §§ 13 and 14 we have to demand that the desired equations—as well as the tensor $\mathfrak{E}_{\mu\nu}$—are only covariant with respect to adapted coordinate systems. Since we already saw, following (42a),[13] that the energy tensor $\mathfrak{T}_\sigma^\nu$ is decisive for the action of the gravitational field upon matter, the equations we are looking for will tie together the tensors $\mathfrak{E}_{\mu\nu}$ and $\mathfrak{T}_\sigma^\nu$. It is therefore natural to assume that desired equations are

$$\mathfrak{E}_{\sigma\tau} = \kappa \mathfrak{T}_{\sigma\tau}. \tag{74}$$

Here κ is a universal constant and $\mathfrak{T}_{\sigma\tau}$ is the symmetric covariant V-tensor, associated with the mixed energy tensor $\mathfrak{T}_\sigma^\nu$ by the relation

$$\left.\begin{aligned} \mathfrak{T}_{\sigma\tau} &= \sum_\nu g_{\nu\tau}\mathfrak{T}_\sigma^\nu \\ \text{and}\quad \mathfrak{T}_\sigma^\nu &= \sum_\tau \mathfrak{T}_{\sigma\tau} g^{\nu\tau} \text{resp.} \end{aligned}\right\}. \tag{75}$$

The determination of the function H. The equations we are looking for are not yet completely given insofar as we have not yet determined the function H. Presently, we only know H to depend solely upon the $g^{\mu\nu}$ and the $g_\sigma^{\mu\nu}$, and to be a scalar under *linear* transformations.[1] A further condition that H must satisfy is found in the following manner.

The V-four-vector $(\mathfrak{K}_\sigma)$ of the force density vanishes in (42a) if $\mathfrak{T}_\sigma^\nu$ is the energy tensor of all the material processes in the region under consideration. Equation (42a) then states that the divergence of the energy tensor $\mathfrak{T}_\sigma^\nu$ of the material processes vanishes; and the same applies—according to (74)—to the tensor $\mathfrak{E}_{\sigma\tau}$ or resp. for the mixed V-tensor $\mathfrak{E}_\sigma^\nu$ to be formed from it.

[1] We would not have found the expression (65a) for B_μ without the latter limitation, which we introduced in § 14. The consideration given in the following text would fail if we dropped this limitation. This fact is the justification for its introduction.

Consequently, every gravitational field must satisfy the relation (see (41b) and (34)):[14]

$$\sum_{\nu\tau}\frac{\partial}{\partial x_\nu}(g^{\tau\nu}\mathfrak{E}_{\sigma\tau})+\frac{1}{2}\sum_{\mu\nu}\frac{\partial g^{\mu\nu}}{\partial x_\sigma}\mathfrak{E}_{\mu\nu}=0.$$

By means of (73) and (65a), this relation can be brought into the form[15]

$$\sum_\nu\frac{\partial S_\sigma^\nu}{\partial x_\nu}-B_\sigma=0, \tag{76}$$

where

$$S_\sigma^\nu=\sum_{\mu\tau}\left(g^{\nu\tau}\frac{\partial H\sqrt{-g}}{\partial g^{\sigma\tau}}+g_\mu^{\nu\tau}\frac{\partial H\sqrt{-g}}{\partial g_\mu^{\sigma\tau}}+\frac{1}{2}\delta_\sigma^\nu H\sqrt{-g}-\frac{1}{2}g_\sigma^{\mu\tau}\frac{\partial H\sqrt{-g}}{\partial g_\nu^{\mu\tau}}\right). \tag{76a}$$

and $\delta_\sigma^\nu=1$ or 0 depending upon $\sigma=\nu$ or $\sigma\neq\nu$.

The ten equations (74) can be used to determine the ten functions $g^{\mu\nu}$ if the $\mathfrak{T}_{\sigma\tau}$ are given. Furthermore, the $g^{\mu\nu}$ must also satisfy the four equations (67) because the coordinate system is to be an adapted one; We have, therefore, more equations than we have functions to be found. This is only possible if the equations are not all mutually independent of each other. It must be demanded that satisfying equations (74) implies that equations (67) are also satisfied. A glance at (76) and (76a) shows that this is achieved if S_σ^ν (a quantity which is a function of the $g^{\mu\nu}$ and the $g_\sigma^{\mu\nu}$ just as H is) vanishes identically for every combination of indices. H then has to be chosen in agreement with the conditions

$$S_\sigma^\nu\equiv 0. \tag{77}$$

Without being able to state a formal reason, I demand furthermore that H is an integral homogeneous function of the second degree in the $g_\sigma^{\mu\nu}$. In this case H is completely determined up to a constant factor. Since H shall be a scalar under linear transformations, it must[1] (considering what we just postulated) be a linear combination of the following five quantities:

$$\sum_{\mu\nu\sigma\tau}g_{\mu\nu}\frac{\partial g^{\mu\nu}}{\partial x_\sigma}\frac{\partial g^{\sigma\tau}}{\partial x_\tau};\quad \sum_{\mu\nu\mu'\nu'\sigma\sigma'}g^{\sigma\sigma'}g_{\mu\nu}\frac{\partial g^{\mu\nu}}{\partial x_\sigma}g_{\mu'\nu'}\frac{\partial g^{\mu'\nu'}}{\partial x_{\sigma'}};\quad \sum_{\sigma\sigma'\mu\nu}g_{\sigma\sigma'}\frac{\partial g^{\sigma\mu}}{\partial x_\mu}\frac{\partial g^{\sigma'\nu}}{\partial x_\nu};$$

$$\sum_{\mu\mu'\nu\nu'\sigma\sigma'}g_{\mu\mu'}g_{\nu\nu'}g^{\sigma\sigma'}\frac{\partial g^{\mu\nu}}{\partial x_\sigma}\frac{\partial g^{\mu'\nu'}}{\partial x_{\sigma'}};\quad \sum_{\alpha\beta\sigma\tau}g_{\alpha\beta}\frac{\partial g^{\alpha\sigma}}{\partial x_\tau}\frac{\partial g^{\beta\tau}}{\partial x_\sigma}.$$

[1] The proof is simple but involved, and for this reason I delete it.

Conditions (77), finally, lead us to equate H—aside from a constant factor—to the fourth one of these quantities.[16] We therefore set[1] under consideration of (35) and making free use of the constant,[17]

$$H = \frac{1}{4} \sum_{\alpha\beta\tau\rho} g^{\alpha\beta} \frac{\partial g_{\tau\rho}}{\partial x_\alpha} \frac{\partial g^{\tau\rho}}{\partial x_\beta}. \tag{78}$$

We limit ourselves to show that this choice of H actually satisfies (77). Using the relations

$$dg = g \sum_{\sigma\tau} g^{\sigma\tau} dg_{\sigma\tau} = -g \sum_{\sigma\tau} g_{\sigma\tau} dg^{\sigma\tau}$$
$$dg_{\alpha\beta} = -\sum_{\mu\nu} g_{\alpha\mu} g_{\beta\nu} dg^{\mu\nu},$$

one obtains from (78)

$$\left.\begin{aligned}
\sum_{\tau} g^{\nu\tau} \frac{\partial H\sqrt{-g}}{\partial g^{\sigma\tau}} &= -\frac{1}{2} H\sqrt{-g}\,\delta_\sigma^\nu + \frac{1}{4}\sqrt{-g} \sum_{\mu\mu'\tau} g^{\nu\tau} \frac{\partial g_{\mu\mu'}}{\partial x_\sigma} \frac{\partial g^{\mu\mu'}}{\partial x_\tau} \\
&\quad - \frac{1}{2}\sqrt{-g} \sum_{\rho\rho'\kappa} g^{\rho\rho'} \frac{\partial g_{\sigma\kappa}}{\partial x_\rho} \frac{\partial g^{\nu\kappa}}{\partial x_{\rho'}} \\
\sum_{\mu\tau} g_\mu^{\nu\tau} \frac{\partial H\sqrt{-g}}{\partial g_\mu^{\sigma\tau}} &= \frac{1}{2}\sqrt{-g} \sum_{\rho\rho'\kappa} g^{\rho\rho'} \frac{\partial g_{\sigma\kappa}}{\partial x_\rho} \frac{\partial g^{\nu\kappa}}{\partial x_{\rho'}} \\
\frac{1}{2} \sum_{\mu\tau} g_\sigma^{\mu\tau} \frac{\partial H\sqrt{-g}}{\partial g_\nu^{\mu\tau}} &= \frac{1}{4}\sqrt{-g} \sum_{\mu\mu'\tau} g^{\nu\tau} \frac{\partial g_{\mu\mu'}}{\partial x_\sigma} \frac{\partial g^{\mu\mu'}}{\partial x_\tau}.
\end{aligned}\right\} \tag{79}$$

From these relations follows the claim made above.[18]

Without using our physical knowledge of gravitation, we have arrived in a purely formal manner at quite distinct field equations.[19] In order to get them into a more explicit notation, we multiply (74) by $g^{\nu\tau}$ and sum over the index τ. Considering (73), we thus get

$$\kappa \mathfrak{T}_\sigma^\nu = \sum_{\tau\alpha} g^{\nu\tau} \left(\frac{\partial H\sqrt{-g}}{\partial g^{\sigma\tau}} - \frac{\partial}{\partial x_\alpha} \left[\frac{\partial H\sqrt{-g}}{\partial g_\alpha^{\sigma\tau}} \right] \right) \tag{80}$$

or

$$-\sum_{\alpha\tau} \frac{\partial}{\partial x_\alpha} \left(g^{\nu\tau} \frac{\partial H\sqrt{-g}}{\partial g_\alpha^{\sigma\tau}} \right) = \kappa \mathfrak{T}_\sigma^\nu + \sum_{\alpha\tau} \left(-g^{\nu\tau} \frac{\partial H\sqrt{-g}}{\partial g^{\sigma\tau}} - g_\alpha^{\nu\tau} \frac{\partial H\sqrt{-g}}{\partial g_\alpha^{\sigma\tau}} \right). \tag{80a}$$

[1] Expressing H by the components $\Gamma_{\sigma\tau}^\nu$ of the gravitational field (see (46)), one obtains

$$H = -\sum_{\mu\rho\tau\tau'} g^{\tau\tau'} \Gamma_{\mu\tau}^\rho \Gamma_{\rho\tau'}^\mu.$$

Since our coordinate system is an adapted one, the equation

$$\sum_{\alpha\tau\nu} \frac{\partial}{\partial x_\nu} \frac{\partial}{\partial x_\alpha} \left(g^{\nu\tau} \frac{\partial H\sqrt{-g}}{\partial g_\alpha^{\sigma\tau}} \right) = 0$$

also holds due to (67) and (65a) and, therefore, considering (80), also the equation

$$\sum_\nu \frac{\partial}{\partial x_\nu} \left\{ \mathfrak{T}_\sigma^\nu + \frac{1}{\kappa} \sum_{\alpha\tau} \left(-g^{\nu\tau} \frac{\partial H\sqrt{-g}}{\partial g^{\sigma\tau}} - g_\alpha^{\nu\tau} \frac{\partial H\sqrt{-g}}{\partial g_\alpha^{\sigma\tau}} \right) \right\} = 0. \tag{80b}$$

Using (78), (79), and (46), we can replace equations (80a) and (80b) by the following ones:[20]

$$\sum_{\alpha\beta} \frac{\partial}{\partial x_\alpha} (\sqrt{-g} g^{\alpha\beta} \Gamma_{\sigma\beta}^\nu) = -\kappa (\mathfrak{T}_\sigma^\nu + \mathfrak{t}_\sigma^\nu), \tag{81}$$

$$\sum_\nu \frac{\partial}{\partial x_\nu} (\mathfrak{T}_\sigma^\nu + \mathfrak{t}_\sigma^\nu) = 0, \tag{42c}$$

where[21]

$$\Gamma_{\sigma\beta}^\nu = \frac{1}{2} \sum_\tau g^{\nu\tau} \frac{\partial g_{\sigma\tau}}{\partial x_\beta}, \tag{81a}$$

$$\left. \begin{aligned} \mathfrak{t}_\sigma^\nu &= -\frac{\sqrt{-g}}{4\kappa} \sum_{\mu\mu'\rho\tau} \left(g^{\nu\tau} \frac{\partial g_{\mu\mu'}}{\partial x_\sigma} \frac{\partial g^{\mu\mu'}}{\partial x_\tau} - \frac{1}{2} \delta_\sigma^\nu g^{\rho\tau} \frac{\partial g_{\mu\mu'}}{\partial x_\rho} \frac{\partial g^{\mu\mu'}}{\partial x_\tau} \right) \\ &= \frac{\sqrt{-g}}{\kappa} \sum_{\mu\rho\tau\tau'} \left(g^{\nu\tau} \Gamma_{\mu\sigma}^\rho \Gamma_{\rho\tau}^\mu - \frac{1}{2} \delta_\sigma^\nu g^{\tau\tau'} \Gamma_{\mu\tau}^\rho \Gamma_{\rho\tau'}^\mu \right) \end{aligned} \right\}. \tag{81b}$$

The equations (81) together with (81a) and (81b) are the differential equations of the gravitational field. Following the deliberations of §10, the equations (42c) represent the conservation laws of momentum and energy for matter and gravitational field combined. The $\mathfrak{t}_\sigma^\nu$ are those quantities, related to the gravitational field, which are in physical analogy to the components $\mathfrak{T}_\sigma^\nu$ of the energy tensor (V-tensor). It is to be emphasized that the $\mathfrak{t}_\sigma^\nu$ do not have tensorial covariance under arbitrary admissible transformations but only under linear transformations. Nevertheless, we call $(\mathfrak{t}_\sigma^\nu)$ the energy tensor of the gravitational field. A similar analogy applies to the components $\Gamma_{\sigma\beta}^\nu$ of the field strength of the gravitational field.

The system of equations (81) allows for a simple physical interpretation in spite of its complicated form. The left-hand side represents a kind of divergence of the gravitational field. As the right-hand side shows, this is caused by the components of the total energy tensor. A very important aspect of this is the result that the energy tensor of the gravitational field itself acts as field-generating, just as the energy tensor of matter.[22]

3.3 Commentary

Presented in this chapter is the section on field equations of a lengthy paper with the final version of the *Entwurf* theory (Einstein 1914c, Sec. D, §§ 12–15, pp. 1066–1077). The paper, submitted to the Berlin Academy October 29, 1914 and published in its Proceedings November 26, 1914, is written as a review article, including a largely self-contained exposition of the necessary mathematics (Sec. B, "From the theory of covariants," §§ 3–8, pp. 1034–1054). It was replaced by a new review article after the transition from the *Entwurf* theory to the theory of November 1915 (Einstein 1916b, CPAE6, Doc. 30, see Ch. 9). This 1914 article can be found in facsimile as Doc. 9 in CPAE6. Part of our commentary is based on sec. 3 of "Untying the knot" (Janssen and Renn 2007).

Introduction to section D

1. Throughout the period covered by this volume, i.e., from late 1912 to late 1916, Einstein conflated general *covariance* and general *relativity*, i.e., the extension of the relativity of uniform (non-accelerated) motion of special relativity to arbitrary non-uniform (accelerated) motion. See Pt. I, Ch. 1, especially notes 3 and 4, and Janssen (2012; 2014, and references therein, such as, in particular, Norton 1999).
2. This argument against generally covariant field equations is known as the "hole argument," after the German *Lochbetrachtung*, the term Einstein used in a letter to Besso of January 3, 1916 (CPAE8, Doc. 178), in which he used the so-called "point coincidence argument" to explain why the hole argument fails (see Pt. I, Ch. 1, Sec. 4.1).

§ 12

3. The translation of this section is taken from Stachel (1989, pp. 72–73).
4. This is the fourth time Einstein published the hole argument. In two earlier versions (Einstein 1914a, p. 260; Einstein and Grossmann 1914, p. 218), Einstein used L rather than Σ, as he does here, or Φ, as in Einstein (1914b, p. 178), to refer to the matter-free region from which the hole (*Loch*) argument derives its name. Despite the suggestive use of L, the term *Lochbetrachtung* (see note 2 above) is not used in any of these four publications.
5. In the 1980s, Stachel (1989, p. 72) could not rule out that this fourth version represents "a significant evolution in Einstein's thinking about the 'hole' argument, as Earman and Glymour (1978) argue." The three earlier versions of the argument left room for accusing Einstein of the elementary mistake of taking (in the notation Einstein uses here) $G(x)$ and $G'(x')$ to be two different fields, whereas they are, as any modern relativist would immediately recognize, just one and the same field expressed in two different coordinate systems. Reexamining the issue, Norton (1984, p. 131) concluded that there is enough textual evidence to "suggest that all four versions of the argument were understood by Einstein to have the same content as the fourth and that his greatest mistake was only to present the first three versions in too compact a form to be readily understood." The discovery in 1998 of a manuscript in Besso's hand dated August 28, 1913, settled the issue. This "Besso memo" contains an embryonic version of the hole argument, which shows that Einstein from the very beginning thought of $G'(x)$ rather than $G'(x')$ as the field that is different from $G(x)$ (Janssen 2007, pp. 821–823). As Besso put it (and it is safe to assume that he was just recording what Einstein told him): "If in coordinate system 1, there is a solution K_1, then this same construct is also a solution in 2, K_2; K_2, however, also a solution in 1" (*Ist im Coordinatensystem 1 eine Lösung* K_1, *so ist dieses selbe Gebilde auch eine Lösung in 2,* K_2*;* K_2 *aber auch eine Lösung in 1*). See Janssen (2007, p. 789) for a facsimile of this page of the Besso memo.

§ 13

6. The "equality of essence" (*Wesensgleichheit*) of gravitational forces and the inertial forces of rotation requires that the rotation metric, the metric of Minkowski space-time in rotating coordinates, be a vacuum solution of the gravitational field equations (Janssen 1999, 2007). Between 1913 and 1915, Einstein changed his mind several times as to whether or not the *Entwurf* field equations meet this requirement. A calculation in the Einstein-Besso manuscript in which he made several mistakes convinced him that they did (see our commentary on [pp. 41–42] of the manuscript in Ch. 2). Later in 1913 he started to have his doubts, which he dispelled again in early 1914. In the introduction of this paper, Einstein (1914c, pp. 1031–1032) confidently asserted that the rotation metric is a vacuum solution. A renewed calculation without any mistakes in September 1915 finally drove home the point that it is not. Troubled by this realization, he asked his protégé Erwin Freundlich to look into the matter (see the letter presented in Ch. 4).
7. We need to take a closer at the derivation of Eq. (64). We start from

$$\Delta H = \frac{\partial H}{\partial g^{\mu\nu}} \Delta g^{\mu\nu} + \frac{\partial H}{\partial g_\sigma^{\mu\nu}} \Delta g_\sigma^{\mu\nu}. \tag{62a}$$

From

$$\Delta g^{\mu\nu} \equiv g'^{\mu\nu} - g^{\mu\nu} = \frac{\partial x'^\mu}{\partial x^\alpha}\frac{\partial x'^\nu}{\partial x^\beta} g^{\alpha\beta} - g^{\mu\nu}$$

$$= \left(\delta_\alpha^\mu + \frac{\partial \Delta x^\mu}{\partial x^\alpha}\right)\left(\delta_\beta^\nu + \frac{\partial \Delta x^\nu}{\partial x^\beta}\right) g^{\alpha\beta} - g^{\mu\nu},$$

it follows that

$$\Delta g^{\mu\nu} = g^{\mu\alpha}\frac{\partial \Delta x^\nu}{\partial x^\alpha} + g^{\nu\alpha}\frac{\partial \Delta x^\mu}{\partial x^\alpha}, \tag{63}$$

which is Eq. (63) except that we follow the modern practice of writing the indices of the coordinates "upstairs" rather than "downstairs" as Einstein did (and continued to do in the period covered by this volume). Similarly, from

$$\Delta g_\sigma^{\mu\nu} \equiv g_\sigma'^{\mu\nu} - g_\sigma^{\mu\nu} = \frac{\partial x^\rho}{\partial x'^\sigma}\frac{\partial}{\partial x^\rho}\left(\frac{\partial x'^\mu}{\partial x^\alpha}\frac{\partial x'^\nu}{\partial x^\beta} g^{\alpha\beta}\right) - g_\sigma^{\mu\nu}$$

$$= \left(\delta_\sigma^\rho - \frac{\partial \Delta x^\rho}{\partial x^\sigma}\right)\frac{\partial}{\partial x^\rho}\left(\left(\delta_\alpha^\mu - \frac{\partial \Delta x^\mu}{\partial x^\alpha}\right)\left(\delta_\beta^\nu + \frac{\partial \Delta x^\nu}{\partial x^\beta}\right) g^{\alpha\beta}\right) - g_\sigma^{\mu\nu},$$

it follows that

$$\Delta g_\sigma^{\mu\nu} = \frac{\partial}{\partial x^\sigma}(\Delta g^{\mu\nu}) - \frac{\partial \Delta x^\alpha}{\partial x^\sigma} g_\alpha^{\mu\nu},$$

which is equivalent to Eq. (63a). Inserting these expressions for $\Delta g^{\mu\nu}$ and $\Delta g_\sigma^{\mu\nu}$ into Eq. (62a) for ΔH, we arrive at

$$\Delta H = \frac{\partial H}{\partial g^{\mu\nu}}\left(g^{\mu\alpha}\frac{\partial \Delta x^\nu}{\partial x^\alpha} + g^{\nu\alpha}\frac{\partial \Delta x^\mu}{\partial x^\alpha}\right) + \frac{\partial H}{\partial g_\alpha^{\mu\nu}}\left(\frac{\partial}{\partial x^\alpha}(\Delta g^{\mu\nu}) - \frac{\partial \Delta x^\beta}{\partial x^\alpha} g_\beta^{\mu\nu}\right),$$

which can be rewritten as

$$\Delta H = 2\frac{\partial H}{\partial g^{\mu\nu}} g^{\mu\alpha}\frac{\partial \Delta x^\nu}{\partial x^\alpha} + \frac{\partial H}{\partial g^{\mu\nu}_\alpha}\left(\frac{\partial}{\partial x^\alpha}\left(2g^{\mu\beta}\frac{\partial \Delta x^\nu}{\partial x^\beta}\right) - \frac{\partial \Delta x^\beta}{\partial x^\alpha} g^{\mu\nu}_\beta\right)$$

$$= 2\frac{\partial H}{\partial g^{\mu\nu}} g^{\mu\alpha}\frac{\partial \Delta x^\nu}{\partial x^\alpha} + 2\frac{\partial H}{\partial g^{\mu\nu}_\alpha} g^{\mu\beta}_\alpha \frac{\partial \Delta x^\nu}{\partial x^\beta}$$

$$+ 2\frac{\partial H}{\partial g^{\mu\nu}_\alpha} g^{\mu\beta}\frac{\partial^2 \Delta x^\nu}{\partial x^\alpha \partial x^\beta} - \frac{\partial H}{\partial g^{\mu\nu}_\alpha} g^{\mu\nu}_\beta \frac{\partial \Delta x^\beta}{\partial x^\alpha}.$$

Relabeling summation indices and grouping terms with $\frac{\partial \Delta x^\sigma}{\partial x^\nu}$ and $\frac{\partial^2 \Delta x^\mu}{\partial x^\rho \partial x^\sigma}$, we arrive at

$$\Delta H = \left\{2g^{\alpha\nu}\frac{\partial H}{\partial g^{\alpha\sigma}} + 2g^{\beta\nu}_\alpha \frac{\partial H}{\partial g^{\beta\sigma}_\alpha} - g^{\alpha\beta}_\sigma \frac{\partial H}{\partial g^{\alpha\beta}_\nu}\right\}\frac{\partial \Delta x^\sigma}{\partial x_\nu} + 2\frac{\partial H}{\partial g^{\mu\tau}_\rho} g^{\sigma\tau}\frac{\partial^2 \Delta x^\mu}{\partial x^\rho \partial x^\sigma}.$$

For linear transformations, all second-order derivatives of Δx^μ vanish and only the first term on the right-hand side of this equation survives. It follows that, for H to be covariant under linear transformations, the expression in curly brackets in this equation must vanish. On the assumption that H *is* covariant under linear transformations, the equation above thus reduces to

$$\frac{1}{2}\Delta H = \frac{\partial H}{\partial g^{\mu\tau}_\rho} g^{\sigma\tau}\frac{\partial^2 \Delta x^\mu}{\partial x^\rho \partial x^\sigma}. \tag{64}$$

Eventually, Einstein was interested in the covariance of the field equations under linear transformations and hence in the covariance of the action $J = \int H\sqrt{-g}d\tau$ (Eq. (61)) under linear transformations. This leads us to write the condition for covariance under linear transformations in a slightly different form. Einstein, it seems, did not bother to evaluate the specific form of this condition at this point, something that would come back to haunt him. Consider the change in J under the infinitesimal coordinate transformation $x'^\mu = x^\mu + \Delta x^\mu$. Since $\sqrt{-g}d\tau$ is generally covariant (see Eq. (62)), it follows that $\Delta J = \int \Delta H\sqrt{-g}d\tau$. Inserting the expression for ΔH above, we find that

$$\frac{1}{2}\Delta J = \int \left\{g^{\alpha\nu}\frac{\partial H}{\partial g^{\alpha\sigma}}\sqrt{-g} + g^{\beta\nu}_\alpha \frac{\partial H\sqrt{-g}}{\partial g^{\beta\sigma}_\alpha} - \frac{1}{2}g^{\alpha\beta}_\sigma \frac{\partial H\sqrt{-g}}{\partial g^{\alpha\beta}_\nu}\right\}\frac{\partial \Delta x^\sigma}{\partial x_\nu} d\tau$$

$$+ \int \frac{\partial H\sqrt{-g}}{\partial g^{\mu\tau}_\rho} g^{\sigma\tau}\frac{\partial^2 \Delta x^\mu}{\partial x^\rho \partial x^\sigma} d\tau$$

Using that $\frac{\partial \sqrt{-g}}{\partial g^{\alpha\sigma}} = -\frac{1}{2}\sqrt{-g}\, g_{\alpha\sigma}$ and that $g^{\alpha\nu} g_{\alpha\sigma} = \delta^\nu_\sigma$, we can write the first term in the expression in curly brackets as

$$g^{\alpha\nu}\frac{\partial H}{\partial g^{\alpha\sigma}}\sqrt{-g} = g^{\alpha\nu}\frac{\partial H\sqrt{-g}}{\partial g^{\alpha\sigma}} + \frac{1}{2}H\sqrt{-g}\,\delta^\nu_\sigma.$$

The expression in curly brackets can thus be rewritten as:

$$g^{\alpha\nu}\frac{\partial H\sqrt{-g}}{\partial g^{\alpha\sigma}} + \frac{1}{2}H\sqrt{-g}\,\delta^\nu_\sigma + g^{\beta\nu}_\alpha \frac{\partial H\sqrt{-g}}{\partial g^{\beta\sigma}_\alpha} - \frac{1}{2}g^{\alpha\beta}_\sigma \frac{\partial H\sqrt{-g}}{\partial g^{\alpha\beta}_\nu}.$$

In § 15, in a different context, we will encounter this same expression, now called S^ν_σ (see Eq. (76a) on p. 1075).

8. Einstein and Grossmann (1914, p. 224, Eq. (II)) had already introduced the condition $B_\mu = 0$ for the special case that H is chosen so as to give the *Entwurf* field equations (see p. 1076, Eq. (78)). Note that Einstein goes to the trouble here of evaluating the surface terms F in Eq. (65) (see Eq. (65b)) but did not bother to evaluate the coefficient of $\partial\Delta x^\mu/\partial x^\nu$ in the expression for ΔH, which turned out to be much more important.
9. Einstein and Grossmann (1914, sec. 4) first introduced the notions of "adapted" (*angepaßte*) coordinates and "justified" (*berechtigte*) transformations between them. See Pt. I, Secs. 3.1 (especially note 8) and 4.2 for discussion of these notions and the closely related notions of (what we call) coordinate restrictions and (what Einstein called) "non-autonomous" (*unselbständige*) transformations.
10. In a letter to Einstein of March 28, 1915 (CPAE8, Doc. 67), Levi-Civita constructed a counter-example to Einstein's claim that $B_\mu = 0$ is a sufficient condition for coordinates to be adapted to the metric field. In our notation (see Pt. I, Sec. 3.1) and with most indices suppressed, Levi-Civita found a non-autonomous transformation $x \to x'(x, g(x))$ that satisfies the condition for justified transformations between adapted coordinates (i.e., $B_\mu(g(x)) = B_\mu(g'(x')) = 0$ with the H in B_μ being the Lagrangian for the *Entwurf* field equations) but under which the *Entwurf* field equations are nonetheless not invariant (i.e., $g(x)$ is a vacuum solution but $g'(x')$ is not). In Levi-Civita's example, $g(x)$ is just the standard diagonal Minkowski metric, $\eta^{\mu\nu} = \mathrm{diag}(-1,-1,-1,1)$ (Janssen and Renn 2007, p. 868, note 58).
11. As he made clear at the beginning of § 13, Einstein was looking for a restriction on coordinate systems that would steer clear of the hole argument against generally covariant field equations but would still allow non-linear transformations such as those to a rotating frame in Minkowski space-time. It turns out, however, that for the choice of H that leads to the *Entwurf* field equations, $B_\mu \neq 0$ for the rotation metric (Janssen 1999, pp. 150–151, note 47; cf. Pt. I, Sec. 4.2). Einstein never seems to have bothered to carefully calculate B_μ for this important special case.

§ 14

12. In this section, Einstein tries to prove that if the variation δJ of the action integral $J = \int H\sqrt{-g}\,d\tau$ (where H can be any function of $g^{\mu\nu}$ and $g^{\mu\nu}_\sigma$ that is invariant under linear transformations) transforms as a scalar under justified transformations between adapted coordinates, a quantity $\mathfrak{E}_{\mu\nu}/\sqrt{-g}$ (the "H-tensor" of the title of this section), which when set equal to 0 gives the Euler-Lagrange equations corresponding to H, transforms as a tensor under such transformations (for discussion, see Norton 1984, pp. 134–136). $\mathfrak{E}_{\mu\nu}$ itself transforms as a "V-tensor," Einstein's term for tensor densities.

    Einstein's proof was criticized by Tullio Levi-Civita (for discussion, see Cattani and De Maria 1989, especially pp. 185–193). Einstein staunchly defended his proof in a series of letters to Levi-Civita in March–May 1915 (only one of Levi-Civita's letters is still extant). In the first of these, dated March 5, 1915, he called it "the most important proof of theory, purchased with streams of perspiration" (CPAE8, Doc. 60). Two months later, as can be inferred from the last of these letters, dated May 5, 1915, the debate ended without a clear resolution (CPAE8, Doc. 80).

§ 15

13. Eq. (42a) on p. 1056 of this paper gives the energy-momentum balance law, setting $\sqrt{-g}$ times the covariant divergence of the energy-momentum tensor for matter ($T^{\nu}_{\sigma;\nu}$) equal to a four-force density ($\mathfrak{K}_\nu$). Eq. (42a) states this law in terms of the mixed tensor density $\mathfrak{T}^\sigma_\nu = \sqrt{-g}\,T^\sigma_\nu$ (cf. Pt. I, Sec. 3.2, note 15, which we followed in labeling the indices):

$$\frac{\partial \mathfrak{T}^\alpha_\mu}{\partial x_\alpha} - \frac{1}{2} g^{\beta\rho} \frac{\partial g_{\rho\alpha}}{\partial x_\mu} \mathfrak{T}^\alpha_\beta = \mathfrak{K}_\mu$$

If $T_{\mu\nu}$ is the energy-momentum tensor for all matter, $\mathfrak{K}_\mu$ vanishes and the second term on the left-hand side describes the force exerted by the gravitational field on matter.

14. In Eq. (74), Einstein introduced field equations of the form $\mathfrak{E}_{\mu\nu} = \kappa\,\mathfrak{T}_{\mu\nu}$, where $\mathfrak{E}_{\mu\nu}$ is defined as

$$\mathfrak{E}_{\mu\nu} \equiv \frac{\partial H\sqrt{-g}}{\partial g^{\mu\nu}} - \frac{\partial}{\partial x^\alpha}\left(\frac{\partial H\sqrt{-g}}{\partial g_\alpha^{\mu\nu}}\right) \tag{73}$$

(so the vacuum field equations are just the Euler-Lagrange equations for the Lagrangian $H\sqrt{-g}$). Setting $\mathfrak{K}_\mu = 0$ in the energy-momentum balance law (see note 13 above), substituting $g^{\alpha\nu}\mathfrak{E}_{\mu\nu}/\kappa$ for $\mathfrak{T}_\mu^\alpha$ in the resulting equation and using that

$$-g^{\beta\rho}\frac{\partial g_{\rho\alpha}}{\partial x^\mu}g^{\alpha\nu} = g^{\beta\rho}g_{\rho\alpha}\frac{\partial g^{\alpha\nu}}{\partial x^\mu} = \delta_\alpha^\beta\frac{\partial g^{\alpha\nu}}{\partial x^\mu} = \frac{\partial g^{\beta\nu}}{\partial x^\mu} \equiv g_\mu^{\beta\nu}.$$

we arrive at the equation at the top of p. 1075:

$$\frac{\partial}{\partial x^\alpha}\left(g^{\alpha\nu}\mathfrak{E}_{\mu\nu}\right) + \frac{1}{2}g_\mu^{\beta\nu}\mathfrak{E}_{\beta\nu} = 0.$$

15. Substituting the expression for $\mathfrak{E}_{\mu\nu}$ in Eq. (73) into the equation at the top of p. 1075 (see note 14 above), we arrive at

$$\begin{aligned}\frac{\partial}{\partial x^\alpha}&\left(g^{\alpha\nu}\frac{\partial H\sqrt{-g}}{\partial g^{\mu\nu}} - g^{\alpha\nu}\frac{\partial}{\partial x^\beta}\left(\frac{\partial H\sqrt{-g}}{\partial g_\beta^{\mu\nu}}\right)\right)\\ &+\frac{1}{2}g_\mu^{\beta\nu}\left(\frac{\partial H\sqrt{-g}}{\partial g^{\beta\nu}} - \frac{\partial}{\partial x^\alpha}\left(\frac{\partial H\sqrt{-g}}{\partial g_\alpha^{\beta\nu}}\right)\right) = 0.\end{aligned} \tag{3.1}$$

Using the product rule of differentiation, we replace the second term on the first line by

$$-\frac{\partial^2}{\partial x^\alpha\partial x^\beta}\left(g^{\alpha\nu}\frac{\partial H\sqrt{-g}}{\partial g_\beta^{\mu\nu}}\right) + \frac{\partial}{\partial x^\alpha}\left(g_\beta^{\alpha\nu}\frac{\partial H\sqrt{-g}}{\partial g_\beta^{\mu\nu}}\right). \tag{3.2}$$

In the first of these two terms we recognize the quantity B_μ introduced in Eq. (65a) in § 13 (cf. note 8 above). The second term on the first line of Eq. (3.1) and the two terms on the second line can be written as the divergence of the quantity we derived in our commentary on § 13 (see note 7) but which Einstein only introduced and called S_ν^μ in Eq. (76a) in § 15. On account of

$$\frac{\partial H\sqrt{-g}}{\partial x^\mu} = \frac{\partial H\sqrt{-g}}{\partial g^{\alpha\beta}}g_\mu^{\alpha\beta} + \frac{\partial H\sqrt{-g}}{\partial g_\rho^{\alpha\beta}}g_{\rho\mu}^{\alpha\beta},$$

we can write the second line of Eq. (3.1) as

$$\frac{1}{2}\frac{\partial H\sqrt{-g}}{\partial x^\mu} - \frac{1}{2}\frac{\partial H\sqrt{-g}}{\partial g_\alpha^{\beta\nu}}g_{\alpha\mu}^{\beta\nu} - \frac{1}{2}g_\mu^{\beta\nu}\frac{\partial}{\partial x^\alpha}\left(\frac{\partial H\sqrt{-g}}{\partial g_\alpha^{\beta\nu}}\right),$$

which in turn can be rewritten as

$$\frac{\partial}{\partial x^\alpha}\left(\frac{1}{2}\delta_\mu^\alpha H\sqrt{-g} - \frac{1}{2}g_\mu^{\beta\nu}\frac{\partial H\sqrt{-g}}{\partial g_\alpha^{\beta\nu}}\right). \tag{3.3}$$

Inserting the expressions in Eqs. (3.2) and (3.3) into Eq. (3.1), we conclude that the latter can be written as

$$\frac{\partial S^{\alpha}_{\mu}}{\partial x^{\alpha}} - B_{\mu} = 0, \tag{76}$$

with

$$S^{\alpha}_{\mu} \equiv g^{\alpha\nu}\frac{\partial H\sqrt{-g}}{\partial g^{\mu\nu}} + g^{\alpha\nu}_{\beta}\frac{\partial H\sqrt{-g}}{\partial g^{\mu\nu}_{\beta}} + \frac{1}{2}\delta^{\alpha}_{\mu}H\sqrt{-g} - \frac{1}{2}g^{\beta\nu}_{\mu}\frac{\partial H\sqrt{-g}}{\partial g^{\beta\nu}_{\alpha}}. \tag{76a}$$

16. Einstein's choice for H leads to the *Entwurf* field equations (modulo a factor 2: see note 17 below). The condition $S^{\nu}_{\sigma} = 0$, however, is satisfied for any function H that is some linear combination of the expressions given at the bottom of p. 1075. This is because any such H, as we showed in note 7, would give rise to an action $J = \int H\sqrt{-g}d\tau$ invariant under linear transformations. Einstein, however, derived the condition $S^{\nu}_{\sigma} = 0$ from considerations of energy-momentum conservation (see note 15) and only came to recognize in 1915 (see Ch. 5) that this same condition also follows from the invariance of J under linear transformations.
17. The expression for H given in Eq. (78) needs to be multiplied by 2 to get the *Entwurf* equations as given in previous publications of Einstein (and Grossmann) (Janssen and Renn 2007, p. 869, note 68). The expression for H in terms of the components $\Gamma^{\nu}_{\sigma\beta}$ of the gravitational field given in footnote 1 returns (without the minus sign and now called $\mathfrak{L}$) in Einstein (1915a, p. 784, Eq. (27)) but with a new definition of $\Gamma^{\nu}_{\sigma\beta}$.
18. The left-hand sides of the three equations in Eq. (79) correspond to the first, the second and minus the fourth term in Eq. (76a) for S^{ν}_{σ}. The first term on the right-hand side of the first of these three equations corresponds to minus the third term in Eq. (76a). Adding the first and the second of these three equations and subtracting the third, we see that the four remaining terms cancel each other.
19. This uniqueness argument for the *Entwurf* field equations is, in fact, fallacious (see note 16 above). Einstein would only come to recognize this in the Fall of 1915 (see the letter to Lorentz of October 12, 1915 [CPAE8, Doc. 129] presented in Ch. 5). The correspondence between Einstein and Levi-Civita mentioned in notes 10 and 12 was triggered by a letter from Max Abraham to Levi-Civita of February 23, 1915, in which Abraham put his finger on the problem: "Among all the possible invariants that could be used to construct the function H he chooses very arbitrarily the one that yields his field equations" (quoted and discussed by Cattani and De Maria 1989, pp. 184–185). Oddly, in the extant correspondence between Einstein and Levi-Civita of early 1915, the uniqueness argument does not come up (but recall that all but one of Levi-Civita's letters appear to be missing).
20. Einstein first presented the *Entwurf* field equations in this form—setting the divergence of what is essentially the gravitational field, involving a gradient of the gravitational potential ($g_{\mu\nu}$), equal to the sum of the energy-momentum densities for matter and the gravitational field—in his lecture on gravity at the *Naturforscherversammlung* in Vienna in September the year before (Einstein 1913, p. 1258, Eq. (7b)). This is also the form in which the equations are given in Einstein and Grossmann (1914, p. 217, Eq. (II)). Because of the factor $\frac{1}{4}$ rather than $\frac{1}{2}$ in the definition of H in Eq. (78) (see note 17), the left-hand side of Eq. (81) differs by a factor $\frac{1}{2}$ from the equations in these earlier publications. The notation for $\mathfrak{T}^{\nu}_{\sigma}$ and $\mathfrak{t}^{\nu}_{\sigma}$ in these earlier publications is $\mathfrak{T}_{\sigma\nu}$ and $\mathfrak{t}_{\sigma\nu}$, respectively, and the expression for $\mathfrak{t}^{\nu}_{\sigma}$ in Eq. (81b) is $\frac{1}{2}$ times the corresponding expression in Einstein (1913, p. 1259, first column, first equation) and Einstein and Grossmann (1914, p. 217, Eq. (2a)).
21. The expression below for "components of the gravitational field" $\Gamma^{\nu}_{\sigma\beta}$, introduced in Eq. (46) of § 9 of this paper, would be replaced by the Christoffel symbols in Einstein (1915a, p. 782; see Sec. 6.1 and Pt. I, Sec. 5.1)
22. The requirement that the energy(-momentum) of the gravitational field enters the field equations the exact same way as the energy(-momentum) of matter goes back

to Einstein’s 1912 theory for the static field, the gradient of a gravitational potential represented by a variable speed of light. The source term in the original version of the field equations of this theory only had the energy density of matter (Einstein 1912a, p. 360). To avoid a conflict between these field equations and energy-momentum conservation, he added a term with the energy density of the gravitational field, noting that “if every energy density ... generates a (negative) divergence of gravitational lines of force, this must also be true for the energy density of the gravitational field itself” (Einstein 1912b, p. 457). This requirement served as one of the constraints Einstein used in his search for field equations for his metric theory of gravity in the Zurich notebook (Renn 2007a, Vol. 2, p. 498, pp. 550–551). It also played a key role in the transition from the field equations of the first November 1915 paper to those of the fourth (see Ch. 6 and Pt. I, Sec. 5.4).

Chapter 4
Einstein to Erwin Freundlich, September 30, 1915

4.1 Transcription

[Berlin,] 30. IX. [1915]

Lieber Freundlich!

Ich will Naumann gerne schreiben, und zwar schon in den nächsten Tagen. Morgen Früh sehe ich Planck, mit dem ich auch darüber spreche.[1] Ich schreibe Ihnen jetzt in einer wissenschaftlichen Angelegenheit, die mich ungeheuer elektrisiert. Ich bin nämlich in der Gravitationstheorie auf einen logischen Widerspruch quantitativer Art gestossen, der mir beweist, dass in meinem Gebäude irgendwo eine rechnerische Unrichtigkeit stecken muss.

Denken Sie ein unendlich langsam rotierendes Koordinatensystem (Rotationsgeschwindigkeit ω). In diesem ist, wie sich durch einfache Transformation leicht zeigen lässt, das Gravitationsfeld durch das $g_{\mu\nu}$-System[2]

$$\begin{matrix} -1 & 0 & 0 & \omega y \\ 0 & -1 & 0 & -\omega x \\ 0 & 0 & -1 & 0 \\ \omega y & -\omega x & 0 & 1 \end{matrix}$$

gegeben. Ich kann nun mittelst der Gleichungen die nächste Approximation berechnen (Glieder proportional ω^2) und finde aus der letzten Feldgleichung der Gravitation[3]

$$g_{44} = 1 - \frac{3}{4}\omega^2(x^2 + y^2),$$

während die unmittelbare Transformation aus dem galileischen Fall ergibt

$$g_{44} = 1 - \omega^2(x^2 + y^2).$$

Dies ist ein flagranter Widerspruch.[4] Ich zweifle daher nicht daran, dass auch die Theorie der Perihelbewegung an dem gleichen Fehler krankt.[5] Entweder sind die Gleichungen schon numerisch unrichtig (Zahlenkoeffizienten) oder ich wende die Gleichungen prinzipiell falsch an. Ich glaube nicht, dass ich selbst imstande bin, den Fehler zu finden, da mein Geist in dieser Sache zu ausgefahrene Geleise hat. Ich muss mich vielmehr darauf verlassen, dass ein

M. Janssen, J. Renn, *How Einstein Found His Field Equations*, Classic Texts in the Sciences, https://doi.org/10.1007/978-3-030-97955-3_11

Nebenmensch mit unverdorbener Gehirnmasse den Fehler findet. Versäumen Sie nicht, wenn Sie Zeit haben, sich mit dem Gegenstande zu beschäftigen.[6]

Mit bestem Gruss Ihr

A. Einstein.

4.2 Translation

[Berlin,] 30 September [1915]

Dear Freundlich,

I shall be glad to write to Naumann, as soon as in the next day or so. Tomorrow morning I am going to see Planck, with whom I shall also discuss it.[1] I am writing you now about a scientific matter that electrifies me enormously. For I have come upon a logical contradiction of a quantitative nature in the theory of gravitation, which proves to me that there must be a calculational error somewhere within my framework.

Imagine an infinitely slowly rotating coordinate system (rotation velocity ω). In this one, as can easily be shown by simple transformation, the gravitational field is given by the $g_{\mu\nu}$ system[2]

$$\begin{matrix} -1 & 0 & 0 & \omega y \\ 0 & -1 & 0 & -\omega x \\ 0 & 0 & -1 & 0 \\ \omega y & -\omega x & 0 & 1 \end{matrix}$$

Using the formulas, I can now calculate the closest approximation (terms proportional to ω^2) and find from the last gravitational field equation[3]

$$g_{44} = 1 - \frac{3}{4}\omega^2(x^2 + y^2),$$

while the direct transformation from the Galilean case yields

$$g_{44} = 1 - \omega^2(x^2 + y^2).$$

This is a blatant contradiction.[4] I do not doubt, therefore, that the theory covering perihelion motion suffers from the same fault.[5] Either the equations are numerically incorrect from the start (numerical coefficients), or I am applying the equations in a principally incorrect way. I do not believe that I myself am in the position to find the error, because my mind follows the same old rut too much in this matter. Rather, I must depend on a fellow human being with unspoiled brain matter to find the error. If you have time, do not fail to study the topic.[6]

With best regards, yours,

A. Einstein.

4.3 Commentary

The source for this letter is a photocopy in the supplementary archive of the Einstein Papers Project. Its designation in the Einstein Archive is EA 80 061 (cf. Pt. I, Ch. 1, notes 2 and 15). The transcription presented here comes from CPAE8, Doc. 123. See Janssen (1999) for detailed discussion of both content and context of this letter (which make it clear that the year must be 1915). Following the editors of CPAE8, we omitted unrelated algebraic calculations, probably in Freundlich's hand, appended to this document.

1. Otto Naumann was ministerial director for university matters in the Prussian Ministry of Education (*Kultusministerium*). Freundlich, who had led the ill-fated eclipse expedition to the Crimea in 1914 in hopes to observe the bending of light (see Pt. I, Ch. 5, note 1), sought to be relieved of his routine duties at the Neubabelsberg Observatory outside Potsdam so that he could spend more time on astronomical tests of Einstein's new theory of gravity.

   Hermann Struve, his superior at the Observatory, strongly opposed this idea, especially after Freundlich incurred the wrath of Struve's Munich colleague, the prominent astronomer Hugo von Seeliger, by criticizing a paper of his (Seeliger 1915) on the anomalous advance of Mercury's perihelion (Freundlich 1915a). Von Seeliger did not waste time getting back at Freundlich. Going over Freundlich's head, he wrote to Struve in June 1915, drawing Struve's attention to an elementary error in a subsequent paper in which Freundlich (1915b) claimed that the gravitational redshift predicted by Einstein's theory could explain why stars with redshifted spectra seem to outnumber those with blueshifted ones. Freundlich had to promise Struve to publish a correction. When he finally got around to doing so the following year, he failed to acknowledge Von Seeliger's intervention and tried to salvage his conclusion by arbitrarily adjusting some parameters. Freundlich thus incensed his detractor in Munich even more, resulting in more angry letters to Babelsberg.

   Even by the summer of 1915, though, tensions with Struve had run so high that Freundlich decided to go over his superior's head and take his plight to the *Kultusministerium*. Einstein agreed to support Freundlich in this quest. In February 1915, Einstein had already enlisted the help of Planck in a first failed attempt to pressure Struve into letting Freundlich pursue his interest in testing Einstein's theory (see CPAE8, Docs. 53 and 54).

   In early October 1915, as he promised Freundlich in this letter, Einstein drafted a letter to Naumann, which starts:

   > A day or so ago Dr. Freundlich from the N[eubabelsberg] Observatory called on me. He told me that during an official meeting you alluded to the possibility that he be released from his duties as assistant at the Observatory for a few years, without losing his salary. I was extremely pleased to hear this and so was my colleague Planck, who recently encouraged me to ask you in writing not to drop this liberating idea (CPAE8, Doc. 124).

   It is not clear whether Einstein actually sent this letter. Nearly two months later, however, he did pay Naumann a visit to advocate on Freundlich's behalf. In a letter that can reliably be dated to late November 1915, Einstein debriefed Freundlich on this meeting with Naumann right after it happened (CPAE8, Doc. 151; the letter starts: "I just come from Naumann, who responded to me at last," suggesting he had tried to contact him earlier). He followed up with a letter to Naumann dated December 7, 1915, which starts:

   > I have just been informed by my colleague Planck by telephone that he spoke with you about the Freundlich matter. It gives me great pleasure to see that

you have not lost sight of this matter, on account of which I recently took the liberty to call on you (CPAE8, Doc. 160).

A few days earlier Einstein had met with Struve at the Prussian Academy to discuss the "Freundlich matter," without letting on that he, Freundlich and Planck were also pursuing the matter with Naumann. This Struve learned from a letter from Naumann dated December 16, 1915, who enclosed a copy of Einstein's letter December 7 to him with his own. That Freundlich and Einstein had gone to Naumann behind his back may well have strengthened Struve's resolve not to grant the request for a more independent position for Freundlich. In his official reply to Naumann of December 20, 1915, Struve sharply criticized Freundlich, complained that he had neglected his official duties, and suggested he might want to look for another job (see CPAE8, Doc. 160, note 8).

For discussion of these and subsequent developments in the "Freundlich affair" see Ch. 7, note 17; CPAE8, introduction, sec. II; and Hentschel (1992, 1994).

2. This is the *rotation metric*, the metric of Minkowski space-time in a rotating Cartesian coordinate system, to first order in the angular velocity ω. For a derivation of the form of the rotation metric, see our commentary on [p. 22L] of the Zurich Notebook (Ch. 1, note 20, Eqs. (1.21)–(1.26)).
3. Einstein had already done the calculation described here two years earlier. It can be found on [pp. 41–42] of the Einstein-Besso manuscript. For a reconstruction of this calculation, see our commentary on [p. 41] of the manuscript (Ch. 2, Eqs. (2.18)–(2.22)). As we saw there, Einstein made several errors in that calculation, leading him to conclude that the *Entwurf* field equations do reproduce the 44-component of the rotation metric correctly.

   These errors illustrate the dangers of thinking one knows the answer before doing a calculation. In a memo dated August 28, 1913, Besso unequivocally states that the *Entwurf* field equations do *not* reproduce the 44-component of the rotation metric correctly. In a letter of August 10, 1913, Ehrenfest told Lorentz that Einstein had done this calculation "five or six times" finding "a different result almost every time" (Janssen 2007, pp. 833–834).

   In late 1913, Einstein seems to have made his peace with the verdict of the Besso memo on the basis of a specious and short-lived argument from energy-momentum conservation that convinced him that the *Entwurf* field equations only were (and only had to be) invariant under linear transformations (see Pt. I, Ch. 4, Introduction).

   By March 1914, he had changed his mind again and convinced himself on the basis of another specious argument that the rotation metric was (and had to be) a vacuum solution of the *Entwurf* field equations. He confidently asserted this in a letter to Besso of March 10, 1914 (CPAE5, Doc. 514) (illustrating the asymmetry in their relationship Besso did not question his friend's about-face) and in a letter to the philosopher Joseph Petzoldt of April 14, 1914 (CPAE8, Doc. 5). In fact, he hailed it as a signature success in the detailed exposition on the *Entwurf* theory published in November that year (Einstein 1914c, pp. 1031–1032).

   That Einstein decided to redo this calculation yet again the following year can be construed as evidence that he was starting to lose confidence in the *Entwurf* theory. A version of the calculation free of errors, which leads to the result reported in this letter to Freundlich, can be found on a sheet of paper that was subsequently used for the draft of a letter to Naumann mentioned in note 1. For a facsimile, a transcription and an analysis of the calculation in this Naumann draft, see Janssen (1999, pp. 149–150; see also CPAE8, Doc. 124, note 5).
4. Einstein wanted the rotation metric to be a vacuum solution of the gravitational field equations to ensure that inertial forces in a rotating frame can be interpreted as gravitational forces in a frame at rest (Janssen 2012, sec. 3, pp. 169–172). The contradiction arrived at here shows that the *Entwurf* field equations do not satisfy this requirement.

5. The iterative approximation procedure used to check whether the rotation metric is a solution of the *Entwurf* field equations is the same as the one Einstein used to find the perihelion motion of Mercury, both in the Einstein-Besso manuscript (see Ch. 2) and in the 1915 perihelion paper (Einstein 1915c, see Sec. 6.3). Within a few weeks, he would come to realize that there is nothing wrong with the approximation procedure and that the problem lies with the *Entwurf* field equations.
6. Einstein did not breathe a word about this "blatant contradiction" in a letter to Lorentz two weeks later (see Ch. 5). Einstein only mentioned the problem to Lorentz once the new field equations of November 1915 had eliminated the problem (see Einstein to Lorentz, January 1, 1916 [CPAE8, Doc. 177]; see note 3 in our commentary on a similar letter to Sommerfeld in Ch. 7 for the relevant passage of this letter to Lorentz). It seems that Einstein, understandably, preferred to run this problem by a protégé rather than a senior colleague.

Chapter 5
Einstein to Hendrik A. Lorentz, October 12, 1915

5.1 Transcription

[Berlin,] 12. X. 15.

Hoch verehrter, lieber Herr Kollege!

Nachträgliche Überlegungen zu dem letzten Briefe, den ich an Sie richtete, haben gezeigt, dass ich in diesem Briefe Unrichtiges behauptete.[1] Thatsächlich liefert die invariantentheoretische Methode nicht mehr als das Hamilton'sche Prinzip wenn es sich um die Bestimmung der Ihrer Funktion $Q(=H\sqrt{-g})$ handelt.[2] Dass ich dies letztes Jahr nicht merkte liegt daran, dass ich auf Seite 1069 meiner Abhandlung leichtsinnig die Voraussetzung einführte, H sei eine Invariante bezüglich *linearer* Transformationen. Unterlässt man diese Voraussetzung, so erhält man folgendes Resultat.

Wie auch Q gewählt werden möge, wenn man das Koordinatensystem so wählt, das J durch Koordinatenwahl bei gegebenem Gravitationsfelde zu einem Extremum wird, oder dass[3]

$$C_\mu = B_\mu - \sum_\lambda \frac{\partial S_\mu^\lambda}{\partial x_\lambda} = 0$$

wobei $S_\mu^\lambda = \sum_{\sigma\nu}\left(g^{\nu\lambda}\frac{\partial Q}{\partial g^{\mu\nu}} + g_\sigma^{\nu\lambda}\frac{\partial Q}{\partial g_\sigma^{\mu\nu}} - \frac{1}{2}g_\mu^{\sigma\nu}\frac{\partial Q}{\partial g_\lambda^{\sigma\nu}} + \frac{1}{2}Q\delta_\mu^\lambda\right)$,

so ist[4]

$$\frac{\partial Q}{\partial g^{\mu\nu}} - \sum \frac{\partial}{\partial x_\sigma}\left(\frac{\partial Q}{\partial g_\sigma^{\mu\nu}}\right)$$

stets ein Tensor bezüglich solcher Koordinatensysteme. Das Postulat der Kovarianz bezw. Relativität kann also nicht zur Bestimmung der Funktion Q dienen.

Diese Bestimmung gründet man am besten auf folgendes physikalische Postulat.[5]

Die Feldgleichungen lauten in gemischter Form

$$-\sum \frac{\partial}{\partial x_\sigma}\left(g^{\nu\lambda}\frac{\partial Q}{\partial g_\sigma^{\mu\nu}}\right) = \kappa\mathfrak{T}_\mu^\lambda + \left(-\sum_\nu g^{\nu\lambda}\frac{\partial Q}{\partial g^{\mu\nu}} - \sum_{\nu\sigma} g_\sigma^{\nu\lambda}\frac{\partial Q}{\partial g_\sigma^{\mu\nu}}\right),$$

M. Janssen, J. Renn, *How Einstein Found His Field Equations*, Classic Texts in the Sciences, https://doi.org/10.1007/978-3-030-97955-3_12

die Erhaltungsgleichungen

$$\left.\begin{aligned} \sum_{\lambda} \frac{\partial}{\partial x_{\lambda}} (\mathfrak{T}_{\mu}^{\lambda} + t_{\mu}^{\lambda}) = 0, \\ \text{wobei } t_{\mu}^{\lambda} = \frac{1}{2\kappa} \sum_{\sigma\nu} \left(-g_{\mu}^{\nu\sigma} \frac{\partial Q}{\partial g_{\lambda}^{\nu\sigma}} + Q\delta_{\mu}^{\lambda} \right) \end{aligned}\right\}$$

Die Divergenz des Gravitationsfeldes muss vermöge der Feldgleichungen durch die Summe der gravitierenden Massen (Energien) *der Materie und des Gravitationsfeldes zusammen* bestimmt sein. Dies trifft bei unserer Feldgleichung nur dann zu, wenn das zweite Glied der rechten Seite [d]em mit κ multiplizierten Energietensor t_{μ}^{λ} des Gravitationsfeldes gleichgesetzt wird.

Man kommt so auf die Bedingung

$$S_{\mu}^{\lambda} \equiv 0.$$

Dies ist gleichzeitig die Bedingung dafür, dass QdV eine Invariante bezüglich *linearer* Substitutionen ist.[6] Letzterer Umstand wäre an sich gleichgültig. Aber er erleichtert das Aufsuchen von Q. Es ergibt sich nämlich aus dieser Invarianz unmittelbar, dass $\frac{Q}{\sqrt{-g}}$ eine lineare homog. Funktion der fünf auf Seite 1075 unten in meiner Abhandlung angegebenen Ausdrücke sein muss.[7] Dass $\frac{Q}{\sqrt{-g}}$ von mir gleich dem vierten der dort angegebenen Ausdrücke gesetzt wurde, lässt sich dadurch rechtfertigen, dass die Theorie nur bei dieser Wahl die Newton'sche als Näherung enthält. Dass ich glaubte, diese Auswahl auf die Gleichung S_{μ}^{λ} stützen zu können, beruhte auf Irrtum.[8]

Es grüsst Sie herzlich Ihr

A. Einstein.

5.2 Translation

[Berlin,] 12 October 1915

Highly esteemed, dear Colleague,

Subsequent reflections on the last letter I sent you have shown that I made erroneous assertions in that letter.[1]

As a matter of fact, the invariant theory method does not yield more than Hamilton's principle when determining your function $Q(= H\sqrt{-g})$.[2] The reason I did not notice this last year is that on page 1069 of my article I carelessly introduced the condition that H be invariant under *linear* transformations. If one drops this condition, one obtains the following result.

No matter how Q is chosen, if one chooses the coordinate system such that J through the choice of coordinates for a given gravitational field has an extremum, or that[3]

$$C_\mu = B_\mu - \sum_\lambda \frac{\partial S_\mu^\lambda}{\partial x_\lambda} = 0$$

where $S_\mu^\lambda = \sum_{\sigma\nu} \left(g^{\nu\lambda} \frac{\partial Q}{\partial g^{\mu\nu}} + g_\sigma^{\nu\lambda} \frac{\partial Q}{\partial g_\sigma^{\mu\nu}} - \frac{1}{2} g_\mu^{\sigma\nu} \frac{\partial Q}{\partial g_\lambda^{\sigma\nu}} + \frac{1}{2} Q \delta_\mu^\lambda \right)$,

then[4]

$$\frac{\partial Q}{\partial g^{\mu\nu}} - \sum \frac{\partial}{\partial x_\sigma} \left(\frac{\partial Q}{\partial g_\sigma^{\mu\nu}} \right)$$

is always a tensor with respect to such coordinate systems. The postulate of covariance or relativity thus cannot serve to determine function Q.

This determination is best based upon the following physical postulate.[5] The field equations in mixed form are

$$-\sum \frac{\partial}{\partial x_\sigma} \left(g^{\nu\lambda} \frac{\partial Q}{\partial g_\sigma^{\mu\nu}} \right) = \kappa \mathfrak{T}_\mu^\lambda + \left(-\sum_\nu g^{\nu\lambda} \frac{\partial Q}{\partial g^{\mu\nu}} - \sum_{\nu\sigma} g_\sigma^{\nu\lambda} \frac{\partial Q}{\partial g_\sigma^{\mu\nu}} \right),$$

the conservation equations,

$$\left. \begin{aligned} & \sum_\lambda \frac{\partial}{\partial x_\lambda} (\mathfrak{T}_\mu^\lambda + t_\mu^\lambda) = 0, \\ \text{where } t_\mu^\lambda &= \frac{1}{2\kappa} \sum_{\sigma\nu} \left(-g_\mu^{\nu\sigma} \frac{\partial Q}{\partial g_\lambda^{\nu\sigma}} + Q \delta_\mu^\lambda \right) \end{aligned} \right\}.$$

The divergence of the gravitational field must be determined according to the field equations by the sum of the gravitational masses (energies) *of both the matter and the gravitational field.* This is true for our field equation only if the second term on the right-hand side is set equal to the energy tensor t_μ^λ of the gravitational field multiplied by κ.

One thus arrives at the condition

$$S_\mu^\lambda \equiv 0.$$

This is simultaneously the condition for QdV being an invariant under *linear* substitutions.[6] In and of itself, the latter circumstance would be uninteresting. However, it makes the search Q easier. For it follows directly from this invariance that $\frac{Q}{\sqrt{-g}}$ must be a linear homogeneous function of the five expressions on page 1075 below of my article.[7] That I set $\frac{Q}{\sqrt{-g}}$ equal to the fourth expression given there can be justified by the fact that it is only with this choice that the theory contains Newton's as an approximation. That I believed I could base this choice on the S_μ^λ equation was based on an error.[8]

Cordial greetings from your

A. Einstein.

5.3 Commentary

The original, a signed autograph letter, is in the *Archief H. A. Lorentz* at the Boerhaave Museum in Leiden. There is a copy in the Einstein Archive with the designation EA 16 442 (cf. Pt. I, Ch. 1, notes 2 and 15). The transcription presented here comes from CPAE8, Doc. 129. The letter is also presented in transcription in Kox (2018, Doc. 301). Our commentary is based on "Untying the knot" (Janssen and Renn 2007, pp. 845, 860, 865–866).

1. In a letter to Lorentz of September 23, 1915 (CPAE8, Doc. 122), Einstein had made some brief comments on a paper in which Lorentz (1914–15) gave a variational derivation of the *Entwurf* field equations, as Einstein had done in 1914 in his second paper with Grossmann and in his definitive exposition of the *Entwurf* theory later that year (Einstein and Grossmann 1914; Einstein 1914c).
2. Lorentz (1914–15, p. 1086; p. 763 in the English translation) used Q for the Lagrangian of the gravitational field whereas Einstein (1914c, p. 1069, Eq. (61)) used $H\sqrt{-g}$, writing the corresponding action as $J = \int H\sqrt{-g}d\tau$. Einstein (1914c, pp. 1075–1076) claimed that covariance considerations lead uniquely to the Lagrangian for the *Entwurf* field equations (see Sec. 3.3, note 16 and 19, and Pt. I, Sec. 4.2)
3. The condition $C_\mu = 0$ and the definition of S_μ^λ below correspond to Eqs. (76) and (76a) in §15 of Einstein (1914c, p. 1075; see Sec. 3.3, note 14 and 15, for discussion).
4. The expression below is called $\mathfrak{E}_{\mu\nu}$ in Einstein (1914c, p. 1073, Eq. (73)). In §14 of this paper, Einstein had argued that $\mathfrak{E}_{\mu\nu}/\sqrt{-g}$ transforms as a tensor (the "*H*-tensor") under transformations between coordinate systems adapted to the metric field under which the variation δJ of the action transforms as a scalar (see Sec. 3.3, note 12, for discussion). Both in §14 and in §15 of the paper, he used the condition $B_\mu = 0$, introduced in Eq. (65a) in §13, to pick out such adapted coordinates. In §15, after deriving the condition $C_\mu = 0$, he set $S_\mu^\lambda = 0$, in which case $C_\mu = 0$ reduces to $B_\mu = 0$. This paragraph in this letter suggests that Einstein felt that the proof that $\mathfrak{E}_{\mu\nu}/\sqrt{-g}$ transforms as a tensor would still go through if $B_\mu = 0$ is replaced by $C_\mu = 0$ as the condition for adapted coordinates.
5. The argument for setting $S_\mu^\lambda = 0$ in Einstein (1914c, p. 1075) is replaced by a new argument based on considerations of energy-momentum conservation. Writing the field equations in mixed covariant/contravariant form, one can identify

$$\kappa \mathrm{t}(Q, \mathrm{source})_\mu^\lambda \equiv -\sum_\nu g^{\nu\lambda} \frac{\partial Q}{\partial g^{\mu\nu}} - \sum_{\nu\sigma} g_\sigma^{\nu\lambda} \frac{\partial Q}{\partial g_\sigma^{\mu\nu}}$$

as the quantity representing the energy-momentum density of the gravitational field in the field equations, where it acts as its own source.

This quantity is not automatically the same as $\mathfrak{t}(Q,\text{conservation})^{\lambda}_{\mu}$, the quantity representing gravitational energy-momentum density in the conservation law

$$\frac{\partial}{\partial x_\lambda}\left(\mathfrak{T}^{\lambda}_{\mu}+\mathfrak{t}^{\lambda}_{\mu}\right)=0.$$

To derive an expression for $\mathfrak{t}(Q,\text{conservation})^{\lambda}_{\mu}$, we substitute the left-hand side of the covariant field equations for $\mathfrak{T}_{\alpha\beta}$, divided by κ, in the energy-momentum balance law (cf. Pt. I, Sec. 3.2, note 15),

$$\frac{\partial \mathfrak{T}^{\lambda}_{\mu}}{\partial x_\lambda}+\frac{1}{2}\frac{\partial g^{\alpha\beta}}{\partial x_\mu}\mathfrak{T}_{\alpha\beta}=0.$$

The second term of this balance law then turns into:

$$\frac{1}{2\kappa}\frac{\partial g^{\alpha\beta}}{\partial x_\mu}\left(\frac{\partial Q}{\partial g^{\alpha\beta}}-\frac{\partial}{\partial x^\lambda}\left(\frac{\partial Q}{\partial g^{\alpha\beta}_{\lambda}}\right)\right).$$

Using that

$$\frac{\partial Q}{\partial x^\mu}=\frac{\partial Q}{\partial g^{\alpha\beta}}g^{\alpha\beta}_{\mu}+\frac{\partial Q}{\partial g^{\nu\sigma}_{\lambda}}g^{\alpha\beta}_{\lambda\mu},$$

we can rewrite the first term in this expression (modulo a factor $1/2\kappa$) as

$$\frac{\partial g^{\alpha\beta}}{\partial x_\mu}\frac{\partial Q}{\partial g^{\alpha\beta}}=\frac{\partial Q}{\partial x^\mu}-\frac{\partial Q}{\partial g^{\nu\sigma}_{\lambda}}g^{\alpha\beta}_{\lambda\mu}.$$

Combining this with the second term, we can rewrite the whole expression as $1/2\kappa$ times

$$\frac{\partial Q}{\partial x^\mu}-g^{\alpha\beta}_{\mu\lambda}\frac{\partial Q}{\partial g^{\nu\sigma}_{\lambda}}-g^{\alpha\beta}_{\mu}\frac{\partial}{\partial x^\lambda}\left(\frac{\partial Q}{\partial g^{\alpha\beta}_{\lambda}}\right)=\frac{\partial}{\partial x^\lambda}\left(\delta^{\lambda}_{\mu}Q-g^{\alpha\beta}_{\mu}\frac{\partial Q}{\partial g^{\alpha\beta}_{\lambda}}\right).$$

This then allows us to identify

$$\kappa\mathfrak{t}(Q,\text{conservation})^{\lambda}_{\mu}\equiv\frac{1}{2}\left(\delta^{\lambda}_{\mu}Q-g^{\alpha\beta}_{\mu}\frac{\partial Q}{\partial g^{\alpha\beta}_{\lambda}}\right)$$

as the gravitational energy-momentum density as it appears in the conservation law. The condition Einstein was after then follows from the demand that the gravitational energy-momentum density is represented by the same quantity in the conservation law and in the field equations:

$$S^{\lambda}_{\mu}=\kappa\mathfrak{t}(Q,\text{conservation})^{\lambda}_{\mu}-\kappa\mathfrak{t}(Q,\text{source})^{\lambda}_{\mu}=0.$$

6. This is what Einstein (1914c, p. 1069) missed when he "carelessly" assumed the Lagrangian H to be invariant under linear transformations. See Sec. 3.3, note 7, for a derivation of the condition $S^{\lambda}_{\mu}=0$ from this assumption.
7. Since all five candidates for H listed in Einstein (1914c, p. 1075) are invariant under linear transformations, the condition $S^{\lambda}_{\mu}=0$ is satisfied for any linear combination of them.
8. Einstein (1914c, p. 1076) had claimed that one is driven to the choice of H giving the *Entwurf* field equations "in a purely formal manner, without using our physical knowledge of gravitation." He now realized that he did need some of this physical knowledge to motivate the choice of H.

Chapter 6
The November 1915 Papers

M. Janssen, J. Renn, *How Einstein Found His Field Equations*, Classic Texts in the Sciences, https://doi.org/10.1007/978-3-030-97955-3_13

6.1 On the general theory of relativity

6.1.1 Facsimile

Zur allgemeinen Relativitätstheorie.

Von A. Einstein.

In den letzten Jahren war ich bemüht, auf die Voraussetzung der Relativität auch nicht gleichförmiger Bewegungen eine allgemeine Relativitätstheorie zu gründen. Ich glaubte in der Tat, das einzige Gravitationsgesetz gefunden zu haben, das dem sinngemäß gefaßten, allgemeinen Relativitätspostulate entspricht, und suchte die Notwendigkeit gerade dieser Lösung in einer im vorigen Jahre in diesen Sitzungsberichten erschienenen Arbeit[1] darzutun. [1]

Eine erneute Kritik zeigte mir, daß sich jene Notwendigkeit auf dem dort eingeschlagenen Wege absolut nicht erweisen läßt; daß dies doch der Fall zu sein schien, beruhte auf Irrtum. Das Postulat der Relativität, soweit ich es dort gefordert habe, ist stets erfüllt, wenn man das Hamiltonsche Prinzip zugrunde legt; es liefert aber in Wahrheit keine Handhabe für eine Ermittelung der Hamiltonschen Funktion H des Gravitationsfeldes. In der Tat drückt die die Wahl von H einschränkende Gleichung (77) a. a. O. nichts anderes aus, als daß H eine Invariante bezüglich linearer Transformationen sein soll, welche Forderung mit der der Relativität der Beschleunigung nichts zu schaffen hat. Ferner wird die durch Gleichung (78) a. a. O. getroffene Wahl durch Gleichung (77) keineswegs festgelegt. [2]

Aus diesen Gründen verlor ich das Vertrauen zu den von mir aufgestellten Feldgleichungen vollständig und suchte nach einem Wege, der die Möglichkeiten in einer natürlichen Weise einschränkte. So gelangte ich zu der Forderung einer allgemeineren Kovarianz der Feldgleichungen zurück, von der ich vor drei Jahren, als ich zusammen mit meinem Freunde Grossmann arbeitete, nur mit schwerem Herzen abgegangen war. In der Tat waren wir damals der im nachfolgenden gegebenen Lösung des Problems bereits ganz nahe gekommen. [3]

Wie die spezielle Relativitätstheorie auf das Postulat gegründet ist, daß ihre Gleichungen bezüglich linearer, orthogonaler Transfor-

[1] Die formale Grundlage der allgemeinen Relativitätstheorie. Sitzungsberichte XLI, 1914, S. 1066—1077. Im folgenden werden Gleichungen dieser Abhandlungen beim Zitieren durch den Zusatz »a. a. O.« von solchen der vorliegenden Arbeit unterschieden.

mationen kovariant sein sollen, so ruht die hier darzulegende Theorie auf dem Postulat der Kovarianz aller Gleichungssysteme bezüglich Transformationen von der Substitutionsdeterminante 1.

Dem Zauber dieser Theorie wird sich kaum jemand entziehen können, der sie wirklich erfaßt hat; sie bedeutet einen wahren Triumph der durch Gauss, Riemann, Christoffel, Ricci und Levi-Civiter begründeten Methode des allgemeinen Differentialkalküls. [4]

§ 1. Bildungsgesetze der Kovarianten.

Da ich in meiner Arbeit vom letzten Jahre eine ausführliche Darlegung der Methoden des absoluten Differentialkalküls gegeben habe, kann ich mich hier bei der Darlegung der hier zu benutzenden Bildungsgesetze der Kovarianten kurz fassen; wir brauchen nur zu untersuchen, was sich an der Kovariantentheorie dadurch verändert, daß nur Substitutionen von der Determinante 1 zugelassen werden. [5]

Die für beliebige Substitutionen gültige Gleichung

$$d\tau' = \frac{\partial(x_1' \cdots x_4')}{\partial(x_1 \cdots x_4)} d\tau$$

geht zufolge der Prämisse unsrer Theorie

$$\frac{\partial(x_1' \cdots x_4')}{\partial(x_1 \cdots x_4)} = 1 \qquad (1)$$

über in

$$d\tau' = d\tau; \qquad (2)$$

das vierdimensionale Volumelement $d\tau$ ist also eine Invariante. Da ferner (Gleichung (17) a. a. O.) $\sqrt{-g}\,d\tau$ eine Invariante bezüglich beliebiger Substitutionen ist, so ist für die uns interessierende Gruppe auch

$$\sqrt{-g'} = \sqrt{-g} \qquad (3)$$

Die Determinante aus den $g_{\mu\nu}$ ist also eine Invariante. Vermöge des Skalarcharakters von $\sqrt{-g}$ lassen die Grundformeln der Kovariantenbildung gegenüber den bei allgemeiner Kovarianz gültigen eine Vereinfachung zu, die kurz gesagt darin beruht, daß in den Grundformeln die Faktoren $\sqrt{-g}$ und $\frac{1}{\sqrt{-g}}$ nicht mehr auftreten, und der Unterschied zwischen Tensoren und V-Tensoren wegfällt. Im einzelnen ergibt sich folgendes: [6]

1. An Stelle der Tensoren $G_{iklm} = \sqrt{-g}\,\delta_{iklm}$

und $G^{iklm} = \frac{1}{\sqrt{-g}}\,\delta_{iklm}$

((19) und (21a) a. a. O.) treten die einfacher gebauten Tensoren

$$G_{iklm} = G^{iklm} = \delta_{iklm} \tag{4}$$

2. Die Grundformeln (29) a. a. O. und (30) a. a. O. für die Erweiterung der Tensoren lassen sich auf Grund unserer Prämisse nicht durch einfachere ersetzen, wohl aber die Definitionsgleichung der Divergenz, welche in der Kombination der Gleichungen (30) a. a. O. und (31) a. a. O. besteht. Sie läßt sich so schreiben [7]

$$A^{\alpha_1\cdots\alpha_l} = \sum_s \frac{\partial A^{\alpha_1\cdots\alpha_l s}}{\partial x_s} + \sum_{s\tau}\left[\begin{Bmatrix} s\tau \\ \alpha_1 \end{Bmatrix} A^{\tau\alpha_2\cdots\alpha_l s} + \cdots\cdots \begin{Bmatrix} s\tau \\ \alpha_l \end{Bmatrix} A^{\alpha_1\cdots\alpha_{l-1}\tau s}\right] + \sum_{s\tau}\begin{Bmatrix} s\tau \\ s \end{Bmatrix} A^{\alpha_1\cdots\alpha_l\tau}. \tag{5}$$

Nun ist aber gemäß (24) a. a. O. und (24a) a. a. O.

$$\sum_\tau \begin{Bmatrix} s\tau \\ s \end{Bmatrix} = \frac{1}{2}\sum_{\alpha s} g^{s\alpha}\left(\frac{\partial g_{s\alpha}}{\partial x_\tau} + \frac{\partial g_{\tau\alpha}}{\partial x_s} - \frac{\partial g_{s\tau}}{\partial x_\alpha}\right) = \frac{1}{2}\sum g^{s\alpha}\frac{\partial g_{s\alpha}}{\partial x_\tau} = \frac{\partial(lg\sqrt{-g})}{\partial x_\tau}. \tag{6}$$

Es hat also diese Größe wegen (3) Vektorcharakter. Folglich ist das letzte Glied der rechten Seite von (5) selbst ein kontravarianter Tensor vom Range l. Wir sind daher berechtigt, an Stelle von (5) die einfachere Definition der Divergenz

$$A^{\alpha_1\cdots\alpha_l} = \sum \frac{\partial A^{\alpha_1\cdots\alpha_l s}}{\partial x_s} + \sum_{\sigma\tau}\left[\begin{Bmatrix} s\tau \\ \alpha_1 \end{Bmatrix} A^{\tau\alpha_2\cdots\alpha_l s} + \cdots\cdot \begin{Bmatrix} s\tau \\ \alpha_l \end{Bmatrix} A^{\alpha_1-\alpha_{l-1}\tau s}\right] \tag{5a}$$

zu setzen, was wir konsequent tun wollen.

So wäre z. B. die Definition (37) a. a. O.

$$\Phi = \frac{1}{\sqrt{-g}}\sum_\mu \frac{\partial}{\partial x_\mu}(\sqrt{-g}\,A^\mu)$$

durch die einfachere Definition

$$\Phi = \sum_\mu \frac{\partial A^\mu}{\partial x_\mu} \tag{7}$$

zu ersetzen, die Gleichung (40) a. a. O. für die Divergenz des kontravarianten Sechservektors durch die einfachere

$$A^\mu = \sum_\nu \frac{\partial A^{\mu\nu}}{\partial x_\nu}. \tag{8}$$

An Stelle von (41a) a. a. O. tritt infolge unserer Festsetzung

$$A_\sigma = \sum_\nu \frac{\partial A_\sigma^\nu}{\partial x_\nu} - \frac{1}{2}\sum_{\mu\nu\tau} g^{\tau\mu}\frac{\partial g_{\mu\nu}}{\partial x_\sigma} A_\tau^\nu. \tag{9}$$

Ein Vergleich mit (41b) zeigt, daß bei unserer Festsetzung das Gesetz für die Divergenz dasselbe ist, wie gemäß dem allgemeinen Differentialkalkül das Gesetz für die Divergenz des V-Tensors. Daß diese Bemerkung für beliebige Tensordivergenzen gilt, läßt sich aus (5) und (5a) leicht ableiten.

3. Die tiefgreifendste Vereinfachung bringt unsere Beschränkung auf Transformationen von der Determinante 1 hervor für diejenigen Kovarianten, die aus den $g_{\mu\nu}$ und ihren Ableitungen allein gebildet werden können. Die Mathematik lehrt, daß diese Kovarianten alle von dem RIEMANN-CHRISTOFFELschen Tensor vierten Ranges abgeleitet werden können, welcher (in seiner kovarianten Form) lautet: [8]

$$\left.\begin{aligned}(ik, lm) = \frac{1}{2}\left(\frac{\partial^2 g_{im}}{\partial x_k \partial x_l} + \frac{\partial^2 g_{kl}}{\partial x_i \partial x_m} - \frac{\partial^2 g_{il}}{\partial x_k \partial x_m} - \frac{\partial^2 g_{mk}}{\partial x_l \partial x_i}\right) \\ + \sum_{\rho\sigma} g^{\rho\sigma}\left(\begin{bmatrix} im \\ \rho \end{bmatrix}\begin{bmatrix} kl \\ \sigma \end{bmatrix} - \begin{bmatrix} il \\ \rho \end{bmatrix}\begin{bmatrix} km \\ \sigma \end{bmatrix}\right)\end{aligned}\right\}. \tag{10}$$

Das Problem der Gravitation bringt es mit sich, daß wir uns besonders für die Tensoren zweiten Ranges interessieren, welche aus diesem Tensor vierten Ranges und den $g_{\mu\nu}$ durch innere Multiplikation gebildet werden können. Infolge der aus (10) ersichtlichen Symmetrie-Eigenschaften des RIEMANNschen Tensors

$$\left.\begin{aligned}(ik, lm) &= (lm, ik) \\ (ik, lm) &= -(ki, lm)\end{aligned}\right\} \tag{11}$$

kann eine solche Bildung nur auf eine Weise vorgenommen werden; es ergibt sich der Tensor

$$G_{im} = \sum_{kl} g^{kl}(ik, lm). \tag{12}$$

Wir leiten diesen Tensor für unsere Zwecke jedoch vorteilhafter aus einer zweiten, von CHRISTOFFEL angegebenen Form des Tensors (10) ab, nämlich aus[1]

$$\{ik, lm\} = \sum_\rho g^{k\rho}(i\rho, lm) = \frac{\partial \left\{ {il \atop k} \right\}}{\partial x_m} - \frac{\partial \left\{ {im \atop k} \right\}}{\partial x_l} + \sum_\rho \left[\begin{Bmatrix} il \\ \rho \end{Bmatrix}\begin{Bmatrix} \rho m \\ k \end{Bmatrix} - \begin{Bmatrix} im \\ \rho \end{Bmatrix}\begin{Bmatrix} \rho l \\ k \end{Bmatrix}\right]. \tag{13}$$

Aus diesem ergibt sich der Tensor G_{im}, indem man ihn mit dem Tensor

$$\delta_k^l = \sum_\alpha g_{k\alpha} g^{\alpha l}$$

multipliziert (innere Multiplikation):

[1] Einen einfachen Beweis für den Tensorcharakter dieses Ausdrucks findet man auf S. 1053 meiner mehrfach zitierten Arbeit.

$$G_{im} = \{il, lm\} = R_{im} + S_{im} \tag{13}$$

$$R_{im} = -\frac{\partial \left\{ {im \atop l} \right\}}{\partial x_l} + \sum_\rho \left\{ {il \atop \rho} \right\} \left\{ {\rho m \atop l} \right\} \tag{13a}$$

$$S_{im} = \frac{\partial \left\{ {il \atop l} \right\}}{\partial x_m} - \left\{ {im \atop \rho} \right\} \left\{ {\rho l \atop l} \right\}. \tag{13b}$$

Beschränkt man sich auf Transformationen von der Determinante 1, so ist nicht nur (G_{im}) ein Tensor, sondern es besitzen auch (R_{im}) und (S_{im}) Tensorcharakter. In der Tat folgt aus dem Umstande, daß $\sqrt{-g}$ ein Skalar ist, wegen (6), daß $\left\{ {il \atop l} \right\}$ ein kovarianter Vierervektor ist. (S_{im}) ist aber gemäß (29) a. a. O. nichts anderes als die Erweiterung dieses Vierervektors, also auch ein Tensor. Aus dem Tensorcharakter von (G_{im}) und (S_{im}) folgt nach (13) auch der Tensorcharakter von (R_{im}). Dieser letztere Tensor ist für die Theorie der Gravitation von größter Bedeutung. [9]

§ 2. Bemerkungen zu den Differentialgesetzen der »materiellen« Vorgänge.

1. Impuls-Energie-Satz für die Materie (einschließlich der elektromagnetischen Vorgänge im Vakuum.

An die Stelle der Gleichung (42a) a. a. O. hat nach den allgemeinen Betrachtungen des vorigen Paragraphen die Gleichung [10]

$$\sum_\nu \frac{\partial T_\sigma^\nu}{\partial x_\nu} = \frac{1}{2} \sum_{\mu\tau\nu} g^{\tau\mu} \frac{\partial g_{\mu\nu}}{\partial x_\sigma} T_\tau^\nu + K_\sigma \tag{14}$$

zu treten; dabei ist T_σ^ν ein gewöhnlicher Tensor, K_σ ein gewöhnlicher kovarianter Vierervektor (kein V-Tensor bzw. V-Vektor). An diese Gleichung haben wir eine für das Folgende wichtige Bemerkung zu knüpfen. Diese Erhaltungsgleichung hat mich früher dazu verleitet, die Größen

$$\frac{1}{2} \sum_\mu g^{\tau\mu} \frac{\partial g_{\mu\nu}}{\partial x_\sigma}$$

als den natürlichen Ausdruck für die Komponenten des Gravitationsfeldes anzusehen, obwohl es im Hinblick auf die Formeln des absoluten Differentialkalküls näher liegt, die Christoffelschen Symbole

$$\left\{ {\nu\sigma \atop \tau} \right\}$$

statt jener Größen einzuführen. Dies war ein verhängnisvolles Vorurteil. Eine Bevorzugung des Christoffelschen Symbols rechtfertigt [11] sich insbesondere wegen der Symmetrie bezüglich seiner beiden Indices kovarianten Charakters (hier ν und σ) und deswegen, weil dasselbe in den fundamental wichtigen Gleichungen der geodätischen Linie (23b) a. a. O. auftritt, welche, vom physikalischen Gesichtspunkte aus betrachtet, die Bewegungsgleichung des materiellen Punktes in einem Gravitationsfelde sind. Gleichung (14) bildet ebenfalls kein Gegenargument, denn das erste Glied ihrer rechten Seite kann in die Form [12]

$$\sum_{\nu\tau} \begin{Bmatrix} \sigma\nu \\ \tau \end{Bmatrix} T_\tau^\nu$$

gebracht werden.

Wir bezeichnen daher im folgenden als Komponenten des Gravitationsfeldes die Größen

$$\Gamma_{\mu\nu}^\sigma = -\begin{Bmatrix} \mu\nu \\ \sigma \end{Bmatrix} = -\sum_\alpha g^{\sigma\alpha} \begin{bmatrix} \mu\nu \\ \alpha \end{bmatrix} = -\frac{1}{2} \sum_\alpha g^{\sigma\alpha} \left(\frac{\partial g_{\mu\alpha}}{\partial x_\nu} + \frac{\partial g_{\nu\alpha}}{\partial x_\mu} - \frac{\partial g_{\mu\nu}}{\partial x_\alpha} \right). \quad (15)$$

Bezeichnet T_σ^ν den Energietensor des gesamten »materiellen« Geschehens, so verschwindet K_σ; der Erhaltungssatz (14) nimmt dann die Form an

$$\sum_\alpha \frac{\partial T_\sigma^\alpha}{\partial x_\alpha} = -\sum_{\alpha\beta} \Gamma_{\sigma\beta}^\alpha T_\alpha^\beta. \quad (14a)$$

Wir merken an, daß die Bewegungsgleichungen (23b) a. a. O. des materiellen Punktes im Schwerefelde die Form annehmen

$$\frac{d^2 x_\tau}{ds^2} = \sum_{\mu\nu} \Gamma_{\mu\nu}^\tau \frac{dx_\mu}{ds} \frac{dx_\nu}{ds}. \quad (15)$$

2. An den Betrachtungen der Paragraphen 10 und 11 der zitierten Abhandlung ändert sich nichts, nur haben nun die dort als *V*-Skalare und *V*-Tensoren bezeichneten Gebilde den Charakter gewöhnlicher Skalare bzw. Tensoren.

§ 3. Die Feldgleichungen der Gravitation.

Nach dem bisher Gesagten liegt es nahe, die Feldgleichungen der Gravitation in der Form

$$R_{\mu\nu} = -\varkappa T_{\mu\nu} \quad (16)$$

anzusetzen, da wir bereits wissen, daß diese Gleichungen gegenüber beliebigen Transformationen von der Determinante 1 kovariant sind. In der Tat genügen diese Gleichungen allen Bedingungen, die wir an sie zu stellen haben. Ausführlicher geschrieben lauten sie gemäß (13a) und (15)

$$\sum_\alpha \frac{\partial \Gamma_{\mu\nu}^\alpha}{\partial x_\alpha} + \sum_{\alpha\beta} \Gamma_{\mu\beta}^\alpha \Gamma_{\nu\alpha}^\beta = -\varkappa T_{\mu\nu}. \quad (16a)$$

Wir wollen nun zeigen, daß diese Feldgleichungen in die Hamiltonsche Form [13]

$$\left.\begin{aligned}\delta\left\{\int\left(\mathfrak{L}-\varkappa\sum_{\mu\nu}g^{\mu\nu}T_{\mu\nu}\right)d\tau\right\}\\ \mathfrak{L}=\sum_{\sigma\tau\alpha\beta}g^{\sigma\tau}\Gamma^{\alpha}_{\sigma\beta}\Gamma^{\beta}_{\tau\alpha}\end{aligned}\right\}\tag{17}$$

gebracht werden können, wobei die $g^{\mu\nu}$ zu variieren, die $T_{\mu\nu}$ als Konstante zu behandeln sind. Es ist nämlich (17) gleichbedeutend mit den Gleichungen

$$\sum_\alpha\frac{\partial}{\partial x_\alpha}\left(\frac{\partial\mathfrak{L}}{\partial g^{\mu\nu}_\alpha}\right)-\frac{\partial\mathfrak{L}}{\partial g^{\mu\nu}}=-\varkappa T_{\mu\nu},\tag{18}$$

wobei $\mathfrak{L}$ als Funktion der $g^{\mu\nu}$ und $\frac{\partial g^{\mu\nu}}{\partial x_\sigma}$ $(=g^{\mu\nu}_\sigma)$ zu denken ist. Anderseits ergeben sich durch eine längere, aber ohne Schwierigkeiten durchzuführende Rechnung die Beziehungen [14]

$$\frac{\partial\mathfrak{L}}{\partial g^{\mu\nu}}=-\sum_{\alpha\beta}\Gamma^{\alpha}_{\mu\beta}\Gamma^{\beta}_{\nu\alpha}\tag{19}$$

$$\frac{\partial\mathfrak{L}}{\partial g^{\mu\nu}_\alpha}=\Gamma^{\alpha}_{\mu\nu}.\tag{19a}$$

Diese ergeben zusammen mit (18) die Feldgleichungen (16a).

Nun läßt sich auch leicht zeigen, daß dem Prinzip von der Erhaltung der Energie und des Impulses Genüge geleistet wird. Multipliziert man (18) mit $g^{\mu\nu}_\sigma$ und summiert man über die Indices μ und ν, so erhält man nach geläufiger Umformung [15]

$$\sum_{\alpha\mu\nu}\frac{\partial}{\partial x_\alpha}\left(g^{\mu\nu}_\sigma\frac{\partial\mathfrak{L}}{\partial g^{\mu\nu}_\alpha}\right)-\frac{\partial\mathfrak{L}}{\partial x_\sigma}=-\varkappa\sum_{\mu\nu}T_{\mu\nu}g^{\mu\nu}_\sigma.$$

Anderseits ist nach (14) für den gesamten Energietensor der Materie

$$\sum_\lambda\frac{\partial T^\lambda_\sigma}{\partial x_\lambda}=-\frac{1}{2}\sum_{\mu\nu}\frac{\partial g^{\mu\nu}}{\partial x_\sigma}T_{\mu\nu}.$$

Aus den beiden letzten Gleichungen folgt

$$\sum_\lambda\frac{\partial}{\partial x_\lambda}(T^\lambda_\sigma+t^\lambda_\sigma)=0,\tag{20}$$

wobei [16]

$$t^\lambda_\sigma=\frac{1}{2\varkappa}\left(\mathfrak{L}\delta^\lambda_\sigma-\sum_{\mu\nu}g^{\mu\nu}_\sigma\frac{\partial\mathfrak{L}}{\partial g^{\mu\nu}_\lambda}\right)\tag{20a}$$

den »Energietensor« des Gravitationsfeldes bezeichnet, der übrigens nur linearen Transformationen gegenüber Tensorcharakter hat. Aus (20a) und (19a) erhält man nach einfacher Umformung

$$t_\sigma^\lambda = \frac{1}{2}\delta_\sigma^\lambda \sum_{\mu\nu\alpha\beta} g^{\mu\nu}\Gamma_{\mu\beta}^\alpha\Gamma_{\nu\alpha}^\beta - \sum_{\mu\nu\alpha} g^{\mu\nu}\Gamma_{\mu\sigma}^\alpha\Gamma_{\nu\alpha}^\lambda \qquad (20\text{b})$$

Endlich ist es noch von Interesse, zwei skalare Gleichungen abzuleiten, die aus den Feldgleichungen hervorgehen. Multiplizieren wir (16a) mit $g^{\mu\nu}$ und summieren wir über μ und ν, so erhalten wir nach einfacher Umformung [17]

$$\sum_{\alpha\beta}\frac{\partial^2 g^{\alpha\beta}}{\partial x_\alpha \partial x_\beta} - \sum_{\sigma\tau\alpha\beta} g^{\sigma\tau}\Gamma_{\sigma\beta}^\alpha\Gamma_{\tau\alpha}^\beta + \sum_{\alpha\beta}\frac{\partial}{\partial x_\alpha}\left(g^{\alpha\beta}\frac{\partial \lg\sqrt{-g}}{\partial x_\beta}\right) = -\varkappa\sum_\sigma T_\sigma^\sigma\,. \qquad (21)$$

Multiplizieren wir anderseits (16a) mit $g^{\nu\lambda}$ und summieren über ν, so erhalten wir

$$\sum_{\alpha\nu}\frac{\partial}{\partial x_\alpha}(g^{\nu\lambda}\Gamma_{\mu\nu}^\alpha) - \sum_{\alpha\beta\nu} g^{\nu\beta}\Gamma_{\nu\mu}^\alpha\Gamma_{\beta\alpha}^\lambda = -\varkappa T_\mu^\lambda\,,$$

oder mit Rücksicht auf (20b)

$$\sum_{\alpha\nu}\frac{\partial}{\partial x_\alpha}(g^{\nu\lambda}\Gamma_{\mu\nu}^\alpha) - \frac{1}{2}\delta_\mu^\lambda \sum_{\mu\nu\alpha\beta} g^{\mu\nu}\Gamma_{\mu\beta}^\alpha\Gamma_{\nu\alpha}^\beta = -\varkappa(T_\mu^\lambda + t_\mu^\lambda)\,.$$

Hieraus folgt weiter mit Rücksicht auf (20) nach einfacher Umformung die Gleichung

$$\frac{\partial}{\partial x_\mu}\left[\sum_{\alpha\beta}\frac{\partial^2 g^{\alpha\beta}}{\partial x_\alpha \partial x_\beta} - \sum_{\sigma\tau\alpha\beta} g^{\sigma\tau}\Gamma_{\sigma\beta}^\alpha\Gamma_{\tau\alpha}^\beta\right] = 0\,. \qquad (22)$$

Wir aber fordern etwas weitergehend:

$$\sum_{\alpha\beta}\frac{\partial^2 g^{\alpha\beta}}{\partial x_\alpha \partial x_\beta} - \sum_{\sigma\tau\alpha\beta} g^{\sigma\tau}\Gamma_{\sigma\beta}^\alpha\Gamma_{\tau\alpha}^\beta = 0\,, \qquad (22\text{a})$$

so daß (21) übergeht in

$$\sum_{\alpha\beta}\frac{\partial}{\partial x_\alpha}\left(g^{\alpha\beta}\frac{\partial \lg\sqrt{-g}}{\partial x_\beta}\right) = -\varkappa\sum_\sigma T_\sigma^\sigma \qquad (21\text{a})$$

Aus Gleichung (21a) geht hervor, daß es unmöglich ist, das Koordinatensystem so zu wählen, daß $\sqrt{-g}$ gleich 1 wird; denn der Skalar des Energietensors kann nicht zu null gemacht werden. [18]

Die Gleichung (22a) ist eine Beziehung, der die $g_{\mu\nu}$ allein unterworfen sind und die in einem neuen Koordinatensystem nicht mehr gelten würde, das durch eine unerlaubte Transformation aus dem ursprünglich benutzten Koordinatensystem hervorginge. Diese Gleichung sagt also aus, wie das Koordinatensystem der Mannigfaltigkeit angepaßt werden muß. [19]

§ 4. Einige Bemerkungen über die physikalischen Qualitäten der Theorie.

Die Gleichungen (22a) geben in erster Näherung [20]

$$\sum_{\alpha\beta} \frac{\partial^2 g^{\alpha\beta}}{\partial x_\alpha \partial x_\beta} = 0 .$$

Hierdurch ist das Koordinatensystem noch nicht festgelegt, indem zur Bestimmung desselben 4 Gleichungen nötig sind. Wir dürfen deshalb für die erste Näherung willkürlich festsetzen

$$\sum_{\beta} \frac{\partial g^{\alpha\beta}}{\partial x_\beta} = 0 . \qquad (22)$$

Ferner wollen wir zur Vereinfachung der Darstellung die imaginäre Zeit als vierte Variable einführen. Dann nehmen die Feldgleichungen (16a) in erster Näherung die Form an

$$\frac{1}{2} \sum_{\alpha} \frac{\partial^2 g_{\mu\nu}}{\partial x_\alpha^2} = \varkappa T_{\mu\nu} , \qquad (16\text{b})$$

von welcher sogleich ersichtlich ist, daß sie das Newtonsche Gesetz als Näherung enthält. —

Daß die Relativität der Bewegung gemäß der neuen Theorie wirklich gewahrt ist, geht daraus hervor, daß unter den erlaubten Transformationen solche sind, die einer Drehung des neuen Systems gegen das alte mit beliebig veränderlicher Winkelgeschwindigkeit entsprechen, [21] sowie solche Transformationen, bei welchen der Anfangspunkt des neuen Systems im alten System eine beliebig vorgeschriebene Bewegung ausführt.

In der Tat sind die Substitutionen

$$\begin{aligned} x' &= x \cos\tau + y \sin\tau \\ y' &= -x \sin\tau + y \cos\tau \\ z' &= z \\ t' &= t \end{aligned}$$

und

$$\begin{aligned} x' &= x - \tau_1 \\ y' &= y - \tau_2 \\ z' &= z - \tau_3 \\ t' &= t , \end{aligned}$$

wobei τ bzw. τ_1, τ_2, τ_3 beliebige Funktionen von t sind, Substitutionen von der Determinante 1.

Ausgegeben am 11. November.

Berlin, gedruckt in der Reichsdruckerei.

6.1.2 Translation

On the General Theory of Relativity

In recent years, I have tried to develop a general theory of relativity based on the assumption that even non-uniform motion is relative. In fact, I believed to have found the only law of gravity corresponding to a sensibly conceived postulate of general relativity and I tried to demonstrate the necessity of this particular solution in a paper[1] that appeared last year in these proceedings.[1]

A new analysis has convinced me that it is absolutely impossible to prove this necessity following the path I chose there. That this did seem possible was based on an error. The postulate of relativity—*to the extent I demanded it there*—is always satisfied if one starts from the Hamiltonian principle; but it turns out that this does not give one a handle on finding the Hamiltonian function H for the gravitational field. Equation (77) l.c. which constrains the choice of H actually only says that H has to be invariant under linear transformations, a requirement that has nothing to do with the relativity of acceleration. Moreover, the choice given in equation (78) l.c. is not determined by equation (77) at all.[2]

For these reasons I completely lost confidence in the field equations I had constructed and looked for a way to restrict the possibilities in a natural manner. I thus found my way back to the demand for a more general covariance of the field equations, which I had abandoned with a heavy heart three years ago when I was collaborating with my friend GROSSMANN. In fact, back then we already came very close to the solution of the problem given below.[3]

Just as the special theory of relativity is based on the postulate that all equations are covariant under linear orthogonal transformations, the theory presented here is based on the postulate of the *covariance of all systems of equations under transformations with determinant 1.*

Hardly anybody who has truly grasped theory will be able to avoid coming under its spell. It is a real triumph of the general differential calculus founded by GAUSS, RIEMANN, CHRISTOFFEL, RICCI, and LEVI-CIVITA.[4]

§ 1. Laws of Forming Covariants

I can be brief about the laws for forming covariants since I gave an elaborate description of the methods of absolute differential calculus in my paper of last year. We thus only need to investigate what changes in the theory of covariants if only transformations with determinant 1 are allowed.[5]

---

[1] "Die formale Grundlage der Relativitätstheorie," *Proceedings* 41 (1914), pp. 1066–1077. Equations in this earlier paper are referred to with the additional note "l.c." to distinguish them from those in the present paper.

The equation

$$d\tau' = \frac{\partial(x_1' \ldots x_4')}{\partial(x_1 \ldots x_4)x} d\tau,$$

which is valid for arbitrary transformations, turns, as a result of the premise of our theory, i.e.,

$$\frac{\partial(x_1' \ldots x_4')}{\partial(x_1 \ldots x_4)} = 1, \tag{1}$$

into

$$d\tau' = d\tau. \tag{2}$$

The four-dimensional volume element $d\tau$ is thus an invariant. Since furthermore (equation (17) l.c.) $\sqrt{-g}d\tau$ is an invariant under arbitrary transformations, it follows for the group that interests us that

$$\sqrt{-g'} = \sqrt{-g}. \tag{3}$$

Hence, the determinant of $g_{\mu\nu}$ is an invariant. Because of the scalar character of $\sqrt{-g}$ one can simplify the basic formulas of the formation of covariants compared to those for general covariance. What it comes down to is that the factors $\sqrt{-g}$ and $1/\sqrt{-g}$ no longer occur in the basic formulas and that the distinction between tensors and V-tensors disappears.[6] Specifically one gets:

1. The tensors $G_{iklm} = \sqrt{-g}\,\delta_{iklm}$ and $G^{iklm} = \frac{1}{\sqrt{-g}}\delta_{iklm}$ ((19) and (21a) l.c.) are replaced by tensors of a simpler structure,

$$G_{iklm} = G^{iklm} = \delta_{iklm}. \tag{4}$$

2. The basic formulas (29) and (30) l.c. for the extension of tensors can, under our premise, not be replaced by simpler ones, but the equations that define divergence, a combination of the equations (30) and (31) l.c., can be simplified.[7] It can be written as

$$A^{\alpha_1 \ldots \alpha_l} = \sum_s \frac{\partial A^{\alpha_1 \ldots \alpha_l s}}{\partial x_s} + \sum_{s\tau}\left[\begin{Bmatrix} s\tau \\ \alpha_1 \end{Bmatrix} A^{\tau\alpha_2 \ldots \alpha_l s} + \ldots \begin{Bmatrix} s\tau \\ \alpha_l \end{Bmatrix} A^{\alpha_1 \ldots \alpha_{l-1}\tau s}\right] + \sum_{s\tau}\begin{Bmatrix} s\tau \\ s \end{Bmatrix} A^{\alpha_1 \ldots \alpha_l \tau}. \tag{5}$$

But according to (24) and (24a) l.c.

$$\sum_s \begin{Bmatrix} s\tau \\ s \end{Bmatrix} = \frac{1}{2}\sum_{\alpha s} g^{s\alpha}\left(\frac{\partial g_{s\alpha}}{\partial x_\tau} + \frac{\partial g_{\tau\alpha}}{\partial x_s} - \frac{\partial g_{s\tau}}{\partial x_\alpha}\right) = \frac{1}{2}\sum_{\alpha s} g^{s\alpha}\frac{\partial g_{s\alpha}}{\partial x_\tau} = \frac{\partial(\lg\sqrt{-g})}{\partial x_\tau}. \tag{6}$$

Because of (3) this quantity is a vector. Consequently, the last term on the right-hand side of (5) is, by itself, a contravariant tensor of rank l. We are therefore justified in replacing (5) by a simpler definition of the divergence,

$$A^{\alpha_1\ldots\alpha_l} = \sum_s \frac{\partial A^{\alpha_1\ldots\alpha_l s}}{\partial x_s} + \sum_{s\tau}\left[\begin{Bmatrix} s\tau \\ \alpha_1 \end{Bmatrix} A^{\tau\alpha_2\ldots\alpha_l s} + \ldots . \begin{Bmatrix} s\tau \\ \alpha_l \end{Bmatrix} A^{\alpha_1\cdot\alpha_{l-1}\tau s}\right], \tag{5a}$$

which we will use throughout.

For example, we will replace the definition (37) l.c.

$$\Phi = \frac{1}{\sqrt{-g}}\sum_\mu \frac{\partial}{\partial x_\mu}(\sqrt{-g}A^\mu)$$

by the simpler definition

$$\Phi = \sum_\mu \frac{\partial A^\mu}{\partial x_\mu}, \tag{7}$$

and equation (40) l.c. for the divergence of the contravariant six-vector by the simpler

$$A^\mu = \sum_\nu \frac{\partial A^{\mu\nu}}{\partial x_\nu}. \tag{8}$$

Instead of (41a) l.c. we get, following this rule,

$$A_\sigma = \sum_\nu \frac{\partial A_\sigma^\nu}{\partial x_\nu} - \frac{1}{2}\sum_{\mu\nu\tau} g^{\tau\mu}\frac{\partial g_{\mu\nu}}{\partial x_\sigma}A_\tau^\nu. \tag{9}$$

A comparison with (41b) reveals that, with this rule, the law for the divergence is the same as that for the divergence of V-tensors in the general differential calculus. This remark applies to the divergence of any tensor, as can be derived from (5) and (5a).

3. The most far-reaching simplification of our restriction to transformations with determinant 1 affects those covariants constructed purely out of $g_{\mu\nu}$ and its derivatives. Mathematics teaches us that such covariants can all be obtained from the RIEMANN-CHRISTOFFEL tensor of rank four, which (in its covariant form) is given by:[8]

$$\left.\begin{aligned} (ik,lm) = \frac{1}{2}\left(\frac{\partial^2 g_{im}}{\partial x_k \partial x_l} + \frac{\partial^2 g_{kl}}{\partial x_i \partial x_m} - \frac{\partial^2 g_{il}}{\partial x_k \partial x_m} - \frac{\partial^2 g_{mk}}{\partial x_l \partial x_i}\right) \\ + \sum_{\rho\sigma} g^{\rho\sigma}\left(\begin{bmatrix} im \\ \rho \end{bmatrix}\begin{bmatrix} kl \\ \sigma \end{bmatrix} - \begin{bmatrix} il \\ \rho \end{bmatrix}\begin{bmatrix} km \\ \sigma \end{bmatrix}\right) \end{aligned}\right\}. \tag{10}$$

For the problem of gravitation we are especially interested in tensors of rank two, which can be formed by inner multiplication of this tensor of rank four with $g_{\mu\nu}$. Because of the symmetry properties of the RIEMANNIAN tensor, apparent from (10),

$$\begin{aligned} (ik,lm) &= (lm,ik) \\ (ik,lm) &= -(ki,lm), \end{aligned} \tag{11}$$

this multiplication can be formed only in one way. In this way one obtains the tensor

$$G_{im} = \sum_{kl} g^{kl} (ik, lm). \tag{12}$$

For our purposes, it is more convenient to derive this tensor from a different form of (10), due to CHRISTOFFEL,[1] i.e.,

$$\{ik, lm\} = \sum_{\rho} g^{k\rho} (i\rho, lm) = \frac{\partial \left\{ {il \atop k} \right\}}{\partial x_m} - \frac{\partial \left\{ {im \atop k} \right\}}{\partial x_l} + \sum_{\rho} \left[\left\{ {il \atop \rho} \right\} \left\{ {\rho m \atop k} \right\} - \left\{ {im \atop \rho} \right\} \left\{ {\rho l \atop k} \right\} \right]. \tag{13}$$

One arrives at the tensor G_{im} if the tensor

$$\delta_k^l = \sum_{\alpha} g_{k\alpha} g^{\alpha l}$$

is multiplied (inner multiplication) with this tensor:

$$G_{im} = \{il, lm\} = R_{im} + S_{im} \tag{13}$$

$$R_{im} = -\frac{\partial \left\{ {im \atop l} \right\}}{\partial x_l} + \sum_{\rho} \left\{ {il \atop \rho} \right\} \left\{ {\rho m \atop l} \right\} \tag{13a}$$

$$S_{im} = \frac{\partial \left\{ {il \atop l} \right\}}{\partial x_m} - \left\{ {im \atop \rho} \right\} \left\{ {\rho l \atop l} \right\}. \tag{13b}$$

Under the restriction to transformations with determinant 1, not only is (G_{im}) a tensor, but so are (R_{im}) and (S_{im}). Because $\sqrt{-g}$ is a scalar and because of (6), $\left\{ {il \atop l} \right\}$ is a covariant four-vector. According to (29) l.c., however, (S_{im}) is nothing but the extension of this four-vector, which means that it too is a tensor. From the tensor character of (G_{im}) and (S_{im}) and given (13) the tensor character of (R_{im}) follows. The tensor (R_{im}) is of the utmost importance for the theory of gravitation.[9]

§ 2. Notes on the Differential Laws of "Material" Processes

1. The energy-momentum theorem for matter (including electromagnetic processes in the vacuum).

According to the general considerations of the previous section, equation (42a) l.c. has to be replaced by[10]

---

[1] A simple proof of the tensor character of this expression can be found on page 1053 of the paper I cited several times already.

$$\sum_{\nu} \frac{\partial T_{\sigma}^{\nu}}{\partial x_{\nu}} = \frac{1}{2} \sum_{\mu\tau\nu} g^{\tau\mu} \frac{\partial g_{\mu\nu}}{\partial x_{\sigma}} T_{\tau}^{\nu} + K_{\sigma}, \tag{14}$$

where T_{σ}^{ν} is an ordinary tensor, K_{σ} an ordinary four-vector (not a V-tensor, V-vector, resp.). We need to add a comment to this equation that is important for what foilows. This conservation law has seduced me on the past to look upon the quantities

$$\frac{1}{2} \sum_{\mu} g^{\tau\mu} \frac{\partial g_{\mu\nu}}{\partial x_{\sigma}}$$

as the natural expressions for the components of the gravitational field, even though the formulas of the absolute differential calculus suggest that it would be more natural to introduce the CHRISTOFFEL symbols

$$\begin{Bmatrix} \nu\sigma \\ \tau \end{Bmatrix}$$

for those quantities instead. This was a fateful prejudice.[11] The preference for the CHRISTOFFEL symbols is especially justified because of the symmetry in their covariant indices (here ν and σ) and because they occur in the fundamentally important geodesic equation (23b) l.c., which, from a physical point of view, are the equations of motion of a material point in a gravitational field. Equation (14) does not form a counterargument, since the first term on its right-hand side can be brought into the form[12]

$$\sum_{\nu\tau} \begin{Bmatrix} \sigma\nu \\ \tau \end{Bmatrix} T_{\tau}^{\nu}.$$

From now on we shall therefore call the quantities

$$\Gamma_{\mu\nu}^{\sigma} = -\begin{Bmatrix} \mu\nu \\ \sigma \end{Bmatrix} = -\sum_{\alpha} g^{\sigma\alpha} \left(\frac{\partial g_{\mu\alpha}}{\partial x_{\nu}} + \frac{\partial g_{\nu\alpha}}{\partial x_{\mu}} - \frac{\partial g_{\mu\nu}}{\partial x_{\alpha}} \right) \tag{15}$$

the components of the gravitational field. If T_{σ}^{ν} denotes the energy tensor of all "material" processes, K_{ν} vanishes and the conservation law (14) takes the form

$$\sum_{\alpha} \frac{\partial T_{\sigma}^{\alpha}}{\partial x_{\alpha}} = -\sum_{\alpha\beta} \Gamma_{\sigma\beta}^{\alpha} T_{\alpha}^{\beta}. \tag{14a}$$

We note that the equations of motion (23b) l.c. of a material point in a gravitational field take the form

$$\frac{d^2 x_{\tau}}{ds^2} = \sum_{\mu\nu} \Gamma_{\mu\nu}^{\tau} \frac{dx_{\mu}}{ds} \frac{dx_{\nu}}{ds}. \tag{15}$$

2. The considerations in sections 10 and 11 of the cited paper remain unchanged, except that the quantities which were V-scalars and V-tensors there are now ordinary scalars and tensors, respectively.

§ 3. The Field Equations of Gravitation

From what has been said so far, it is only natural to posit field equations of gravitation of the form

$$R_{\mu\nu} = -\kappa T_{\mu\nu}, \tag{16}$$

as we already know that these equations are covariant under arbitrary transformations with determinant 1. In fact, these equations satisfy all conditions we want to impose on them. Written out in more detail with the help of (13a) and (15), they are

$$\sum_{\alpha} \frac{\partial \Gamma^{\alpha}_{\mu\nu}}{\partial x_{\alpha}} + \sum_{\alpha\beta} \Gamma^{\alpha}_{\mu\beta} \Gamma^{\beta}_{\nu\alpha} = -\kappa T_{\mu\nu}. \tag{16a}$$

We now show that these field equations can be brought into HAMILTONIAN form[13]

$$\left.\begin{aligned} &\delta\left\{\int\left(\mathfrak{L} - \kappa \sum_{\mu\nu} g^{\mu\nu} T_{\mu\nu}\right) d\tau\right\} \\ &\mathfrak{L} = \sum_{\sigma\tau\alpha\beta} g^{\sigma\tau} \Gamma^{\alpha}_{\sigma\beta} \Gamma^{\beta}_{\tau\alpha} \end{aligned}\right\}, \tag{17}$$

where the $g^{\mu\nu}$ have to be varied while the $T_{\mu\nu}$ are to be treated as constants. (17) is equivalent to the equations

$$\sum_{\alpha} \frac{\partial}{\partial x_{\alpha}} \left(\frac{\partial \mathfrak{L}}{\partial g^{\mu\nu}_{\alpha}}\right) - \frac{\partial \mathfrak{L}}{\partial g^{\mu\nu}} = -\kappa T_{\mu\nu}, \tag{18}$$

where $\mathfrak{L}$ has to be thought of as a function of $g^{\mu\nu}$ and $\dfrac{\partial g^{\mu\nu}}{\partial x_{\sigma}} (= g^{\mu\nu}_{\sigma})$. A somewhat lengthier but straightforward calculation leads to the relations[14]

$$\frac{\partial \mathfrak{L}}{\partial g^{\mu\nu}} = -\sum_{\alpha\beta} \Gamma^{\alpha}_{\mu\beta} \Gamma^{\beta}_{\nu\alpha} \tag{19}$$

$$\frac{\partial \mathfrak{L}}{\partial g^{\mu\nu}_{\alpha}} = \Gamma^{\alpha}_{\mu\nu}. \tag{19a}$$

These together with (18) yield the field equations (16a).

It can now also easily be shown that the principle of the conservation of energy and momentum is satisfied.[15] Multiplying (18) by $g^{\mu\nu}_{\sigma}$ and summing over the indices μ and ν, one obtains after the usual kind of rearrangement

$$\sum_{\alpha\mu\nu} \frac{\partial}{\partial x_{\alpha}} \left(g^{\mu\nu}_{\sigma} \frac{\partial \mathfrak{L}}{\partial g^{\mu\nu}_{\alpha}}\right) - \frac{\partial \mathfrak{L}}{\partial x_{\sigma}} = -\kappa \sum_{\mu\nu} T_{\mu\nu} g^{\mu\nu}_{\sigma}.$$

According to (14), however, the *total* energy tensor of matter satisfies

$$\sum_{\lambda}\frac{\partial T_{\sigma}^{\lambda}}{\partial x_{\lambda}} = -\frac{1}{2}\sum_{\mu\nu}\frac{\partial g^{\mu\nu}}{\partial x_{\sigma}}T_{\mu\nu}.$$

It follows from the last two equations that

$$\sum_{\lambda}\frac{\partial}{\partial x_{\lambda}}(T_{\sigma}^{\lambda}+t_{\sigma}^{\lambda}) = 0, \tag{20}$$

where[16]

$$t_{\sigma}^{\lambda} = \frac{1}{2\kappa}\left(\mathfrak{L}\delta_{\sigma}^{\lambda} - \sum_{\mu\nu} g_{\sigma}^{\mu\nu}\frac{\partial\mathfrak{L}}{\partial g_{\lambda}^{\mu\nu}}\right) \tag{20a}$$

denotes the "energy tensor" of the gravitational field, which, by the way, has tensor character only for linear transformations. Upon simple rearrangement, (20a) and (19a) give:

$$t_{\sigma}^{\lambda} = \frac{1}{2}\delta_{\sigma}^{\lambda}\sum_{\mu\nu\alpha\beta} g^{\mu\nu}\Gamma_{\mu\beta}^{\alpha}\Gamma_{\nu\alpha}^{\beta} - \sum_{\mu\nu\alpha} g^{\mu\nu}\Gamma_{\mu\sigma}^{\alpha}\Gamma_{\nu\alpha}^{\lambda}. \tag{20b}$$

Finally, it is of interest to derive two scalar equations that follow from the field equations.[17] After multiplying (16a) by $g^{\mu\nu}$ with summation over μ and ν, we find after simple rearrangement

$$\sum_{\alpha\beta}\frac{\partial^2 g^{\alpha\beta}}{\partial x_{\alpha}\partial x_{\beta}} - \sum_{\sigma\tau\alpha\beta} g^{\sigma\tau}\Gamma_{\sigma\beta}^{\alpha}\Gamma_{\tau\alpha}^{\beta} + \sum_{\alpha\beta}\frac{\partial}{\partial x_{\alpha}}\left(g^{\alpha\beta}\frac{\partial \lg\sqrt{-g}}{\partial x_{\beta}}\right) = -\kappa\sum_{\sigma}T_{\sigma}^{\sigma}. \tag{21}$$

However, multiplying (16a) by $g^{\nu\lambda}$ and summing over ν, we find

$$\sum_{\alpha\nu}\frac{\partial}{\partial x_{\alpha}}(g^{\nu\lambda}\Gamma_{\mu\nu}^{\alpha}) - \sum_{\alpha\beta\nu} g^{\nu\beta}\Gamma_{\nu\mu}^{\alpha}\Gamma_{\beta\alpha}^{\lambda} = -\kappa T_{\mu}^{\tau},$$

or, with the help of (20b),

$$\sum_{\alpha\nu}\frac{\partial}{\partial x_{\alpha}}(g^{\nu\lambda}\Gamma_{\mu\nu}^{\alpha}) - \frac{1}{2}\delta_{\mu}^{\lambda}\sum_{\mu\nu\alpha\beta} g^{\mu\nu}\Gamma_{\mu\beta}^{\alpha}\Gamma_{\nu\alpha}^{\beta} = -\kappa(T_{\mu}^{\lambda}+t_{\mu}^{\lambda}).$$

Taking (20) into account, we find after some simple rearrangement

$$\frac{\partial}{\partial x_{\mu}}\left[\sum_{\alpha\beta}\frac{\partial^2 g^{\alpha\beta}}{\partial x_{\alpha}\partial x_{\beta}} - \sum_{\sigma\tau\alpha\beta} g^{\sigma\tau}\Gamma_{\sigma\beta}^{\alpha}\Gamma_{\tau\alpha}^{\beta}\right] = 0. \tag{22}$$

However, we impose the somewhat stronger demand:

$$\sum_{\alpha\beta} \frac{\partial^2 g^{\alpha\beta}}{\partial x_\alpha \partial x_\beta} - \sum_{\sigma\tau\alpha\beta} g^{\sigma\tau} \Gamma^\alpha_{\sigma\beta} \Gamma^\beta_{\tau\alpha} = 0, \tag{22a}$$

in which case (21) turns into

$$\sum_{\alpha\beta} \frac{\partial}{\partial x_\alpha} \left(g^{\alpha\beta} \frac{\partial \lg \sqrt{-g}}{\partial x_\beta} \right) = -\kappa \sum_\sigma T^\sigma_\sigma. \tag{21a}$$

It follows from equation (21a) that it is impossible to choose the coordinate system such that $\sqrt{-g}$ equals 1, since the scalar of the energy tensor cannot be set equal to zero.[18]

Equation (22a) is a relation that only involves $g_{\mu\nu}$. It would not hold in a new coordinate system resulting from the original one by a transformation that is not allowed. The equation thus tells us how the coordinate system has to be adapted to the manifold.[19]

§ 4. Some Remarks about the Physical Features of the Theory

In first approximation, the equations (22a) give[20]

$$\sum_{\alpha\beta} \frac{\partial^2 g^{\alpha\beta}}{\partial x_\alpha \partial x_\beta} = 0.$$

This does not yet fix the coordinate system, as that would require 4 equations. We are therefore entitled to arbitrarily set

$$\sum_\beta \frac{\partial g^{\alpha\beta}}{\partial x_\beta} = 0 \tag{22}$$

in first approximation. For further simplification we introduce the imaginary time as a fourth variable. The field equations (16a) then take the form, in first approximation,

$$\frac{1}{2} \sum_\alpha \frac{\partial^2 g_{\mu\nu}}{\partial x_\alpha^2} = \kappa T_{\mu\nu}, \tag{16b}$$

from which one immediately sees that they contain NEWTON's law as an approximation.—

That the new theory really implements the relativity of motion follows from the fact that among the permissible transformations there are those that correspond to a rotation of the new system with respect to the old one (with arbitrarily variable angular velocity),[21] and also those in which the origin of the new system has an arbitrarily prescribed motion with respect to that of the old one.

In fact, the transformations

$$\begin{aligned} x' &= \quad x\cos\tau + y\sin\tau \\ y' &= \quad -x\sin\tau + y\cos\tau \\ z' &= z \\ t' &= t \end{aligned}$$

and

$$\begin{aligned} x' &= x - \tau_1 \\ y' &= y - \tau_2 \\ z' &= z - \tau_3 \\ t' &= t, \end{aligned}$$

where τ and τ_1, τ_2, τ_3 are arbitrary functions of t, are indeed transformations with determinant 1.

6.1.3 Commentary

Presented in this chapter are the four communications on general relativity to the Berlin Academy of November 1915 (Einstein 1915a, 1915b, 1915c, 1915d). The first one, presented in this section, was submitted November 4, 1915 and published November 11, 1915. It can be found in facsimile as Doc. 21 in CPAE6. See Ch. 3 for the part dealing with field equations of Einstein (1914c), a paper referred to in many places in the three November 1915 papers dealing with field equations (Einstein 1915a, 1915c, 1915d). Our commentary on the first paper is based on sec. 6 of "Untying the knot" (Janssen and Renn 2007). Tilman Sauer recently brought to light the page proofs of this article that Einstein sent to Hilbert (see Pt. I, Sec. 2.3, note 10). These are identical to the published paper but Hilbert corrected several typos (see, e.g., note 16) and added some marginalia (see, e.g., note 11).

Introduction

1. For discussion of how Einstein tried to generalize the principle of relativity from uniform to non-uniform motion by extending the covariance of the sought-after theory to non-linear transformations, see Pt. I, Ch. 1, notes 3 and 4 and Janssen (2012, 2014).
2. In a letter to Lorentz a few weeks earlier (see Ch. 5), Einstein explained the flaw in his uniqueness argument for the *Entwurf* field equations in Einstein (1914c, pp. 1075–1076; cf. Sec. 3.3, note 16).
3. The field equations proposed in this paper can be found on p. 22R of the Zurich notebook (see Ch. 1 and Pt. I, Fig. 1.3). Note that Einstein's formulations—"thus I found my way back" (*so gelangte ich zurück*) to "a more general" (*allgemeinere*) covariance—suggests a gradual rather than an abrupt change of direction (Janssen and Renn 2007, p. 912, note 158; cf. note 4 below).
4. The final sentence of the introduction strongly suggests that Einstein found his new field equations by switching back from the "physical strategy" that had led him to the *Entwurf* theory to the "mathematical strategy" he had tried but abandoned in the Zurich notebook. This impression, we argue, is misleading (see Pt. I, Secs. 2.3 and 5.3).

§ 1

5. Given Einstein's remarks in the introduction, the question arises why Einstein only demanded covariance under unimodular transformations rather than general covariance. Our best answer to this question is that tweaking the variational formalism in Einstein (1914c), after discovering that the *Entwurf* field equations do not allow the rotation metric and are not unique (see Chs. 4 and 5), he hit upon field equations based on half the Ricci tensor, itself a tensor only under unimodular transformations, rather than the full Ricci tensor. For further discussion, see Pt. I, Sec. 5.3.
6. Einstein uses the term "V-tensor" for what is now called a tensor density (cf. Sec. 3.3, note 12).
7. The simplified formula (9) for the covariant divergence derived in the entry numbered 2 in § 1 is put to good use in the discussion of energy-momentum conservation in § 2 (see note 10). Note that Einstein uses the term "extension" (*Erweiterung*) for what is now called a covariant derivative.
8. At the top of p. 14L of the Zurich notebook, Einstein first wrote down the fully covariant form of the Riemann tensor (ik,lm) (modern notation: $R_{\rho\mu\sigma\nu}$) in Eq. (10), writing Grossmann's name right next to it. He then went on to form the Ricci tensor $G_{im} = \{il,lm\} = g^{kl}(ik,lm)$ (modern notation: $R^{\rho}{}_{\mu\rho\nu} = R_{\mu\nu}$) in Eq. (12) (cf. Sec. 1.3, notes 1 and 2).
9. This way of splitting the Ricci tensor into two parts, both transforming as tensors under unimodular transformations, can already be found on p. 22R of the Zurich notebook. The expression for the Ricci tensor obtained by setting $k = l$ in the first of the two equations numbered (13) for $\{ik,lm\}$ is given at the top of p. 22R, again (see note 8) with Grossmann's name written next to it (see Sec. 1.1 and Pt. I, Fig. 1.3). This corroborates Einstein's claim in the introduction of this paper that he and Grossmann had already come close to these field equations: in fact, they had already considered these exact same field equations. See Pt. I, Sec. 3.1 and Sec. 1.3, note 18, for further discussion.

§ 2

10. Eq. (42a) in Einstein (1914c, p. 1056) gives the energy-momentum balance between matter and gravitational field:

$$\mathfrak{T}^{\nu}_{\sigma,\nu} = \frac{1}{2} g^{\tau\mu} g_{\mu\nu,\sigma} \mathfrak{T}^{\nu}_{\tau} + \mathfrak{K}_{\nu}.$$

If $T_{\mu\nu}$ includes the energy-momentum of all matter, $\mathfrak{K}_{\nu} = 0$ (cf. Sec. 3.3, note 13). Since $\mathfrak{T}^{\nu}_{\sigma} = \sqrt{-g}\, T^{\nu}_{\sigma}$ and $\mathfrak{K}_{\nu} = \sqrt{-g}\, K_{\nu}$ (i.e., these quantities are a tensor density and a vector density, respectively, a "V-tensor" and a "V-vector" in Einstein's terminology), the equation would reduce to Eq. (14) in this paper *if* $\sqrt{-g}$ were a constant. But $\sqrt{-g}$ cannot be a constant: even though the theory is invariant under unimodular transformations, it does not allow unimodular coordinates, defined by the condition $\sqrt{-g} = 1$ (see note 18). Einstein got from Eq. (42a) in his earlier paper to Eq. (14) in this one along the lines of the calculation under (2) in § 1.

As we showed in Pt. I, Sec. 3.2, note 15, the equation above with $\mathfrak{K}_{\nu} = 0$ is equivalent to the vanishing of the covariant divergence (defined in § 1, Eq. (5)):

$$T^{\nu}_{\sigma;\nu} = T^{\nu}_{\sigma,\nu} + \begin{Bmatrix} \nu \\ \tau\nu \end{Bmatrix} T^{\tau}_{\sigma} - \begin{Bmatrix} \tau \\ \sigma\nu \end{Bmatrix} T^{\nu}_{\tau}.$$

Note that $\begin{Bmatrix} \nu \\ \tau\nu \end{Bmatrix} = (\log\sqrt{-g})_{,\tau}$ and that

$$\begin{Bmatrix} \tau \\ \sigma\nu \end{Bmatrix} T^{\nu}_{\tau} = \frac{1}{2} g^{\tau\mu} \left(g_{\mu\sigma,\nu} + g_{\mu\nu,\sigma} - g_{\sigma\nu,\mu} \right) T^{\nu}_{\tau} = \frac{1}{2} g^{\tau\mu} g_{\mu\nu,\sigma} T^{\nu}_{\tau},$$

where in the last step we used that $(g_{\mu\sigma,\nu} - g_{\nu\sigma,\mu}) T^{\mu\nu} = 0$ as it is the contraction of expressions anti-symmetric and symmetric in μ and ν. We thus arrive at

$$T^{\nu}_{\sigma;\nu} = T^{\nu}_{\sigma,\nu} - \frac{1}{2} g^{\tau\mu} g_{\mu\nu,\sigma} T^{\nu}_{\tau} + (\log\sqrt{-g})_{,\tau} T^{\tau}_{\sigma}.$$

Since $T^{\nu}_{\sigma;\nu}$ is a generally-covariant vector and $(\log\sqrt{-g})_{,\tau} T^{\tau}_{\sigma}$ is a vector under unimodular transformations, it follows that

$$T^{\nu}_{\sigma,\nu} - \frac{1}{2} g^{\tau\mu} g_{\mu\nu,\sigma} T^{\nu}_{\tau}$$

is also a vector under unimodular transformations. Setting this vector equal to the force vector K_σ, Einstein obtained the energy-momentum balance law for a theory invariant under unimodular transformations.

11. As Hilbert noted in the margin of page proofs of this paper, referring to Einstein (1914c): "The Γs are *not* the same as the ones in Eq. (46) on p. 1058 of the first paper" (*Die Γ sind* nicht *dieselben wie in Formel (46) Seite 1058 der ersten Arbeit*). In the *Entwurf* theory, the expression for the gravitational field $\Gamma^{\alpha}_{\beta\mu}$ had just one term with the gradient of the metric, the potential of the gravitational field in the theory. In the new theory, it is a sum of three terms with gradients of the metric, i.e., minus the Christoffel symbols:

$$\Gamma^{\alpha}_{\beta\mu} \equiv -\begin{Bmatrix} \alpha \\ \beta\mu \end{Bmatrix} = -\frac{1}{2} g^{\alpha\rho} (g_{\rho\beta,\mu} + g_{\rho\mu,\beta} - g_{\beta\mu,\rho}).$$

See Pt. I, Sec. 5.1 and note 13 below for discussion of why Einstein called his earlier choice "a fateful prejudice" and why he called his new choice "the key to the solution" in a letter to Sommerfeld a few weeks later (see Ch. 7).

12. As we verified in note 10, $\begin{Bmatrix} \tau \\ \sigma\nu \end{Bmatrix} T^{\nu}_{\tau} = \frac{1}{2} g^{\tau\mu} \frac{\partial g_{\mu\nu}}{\partial x_\sigma} T^{\nu}_{\tau}$.

§ 3

13. The first line of Eq. (17) should be $\delta\{\ldots\} = 0$. The Lagrangian for Einstein's new theory is the same function of the gravitational field $\Gamma^{\alpha}_{\beta\mu}$ as the Lagrangian for the *Entwurf* field equations given in Einstein (1914c, p. 1076, note 1; cf. Sec. 3.3, notes 17 and 20). If the variational formalism developed in this earlier paper is retained but the gravitational field is redefined as minus the Christoffel symbols (see note 11), this Lagrangian gives field equations similar but not identical to those extracted from the Ricci tensor in the Zurich notebook under the restriction to unimodular transformations (Janssen and Renn 2007, pp. 875–877). Imposing that same restriction on his variational formalism, however, as Einstein does in this paper, this Lagrangian reproduces precisely these field equations considered but rejected in the Zurich notebook. That Einstein could derive them from a variational principle made it easy to show that they satisfy energy-momentum conservation.
14. Einstein gave a self-contained derivation of Eqs. (19) and (19a) in a letter to Ehrenfest early the following year (under "(1) Lagrangian form of the field equations;" CPAE8, Doc. 185; see Ch. 8).
15. Contracting the field equations in Eq. (18) with $g^{\mu\nu}_{\sigma}$, we get

$$g^{\mu\nu}_{\sigma} \left(\frac{\partial}{\partial x_\alpha} \left(\frac{\partial \mathfrak{L}}{\partial g^{\mu\nu}_{\alpha}} \right) - \frac{\partial \mathfrak{L}}{\partial g^{\mu\nu}} \right) = -\kappa g^{\mu\nu}_{\sigma} T_{\mu\nu},$$

The energy-momentum balance law (see note 10) allows us to rewrite $g^{\mu\nu}_{\sigma} T_{\mu\nu}$ on the right-hand side as

$$g^{\mu\nu}_{\sigma} T_{\mu\nu} = g^{\mu\nu}_{\sigma} g_{\mu\lambda} T^{\lambda}_{\nu} = -g^{\mu\nu} g_{\mu\lambda,\sigma} T^{\lambda}_{\nu} = -2T^{\lambda}_{\sigma,\lambda}.$$

The left-hand side can be rewritten as

$$\frac{\partial}{\partial x_{\alpha}}\left(g^{\mu\nu}_{\sigma}\frac{\partial \mathfrak{L}}{\partial g^{\mu\nu}_{\alpha}}\right) - g^{\mu\nu}_{\sigma\alpha}\frac{\partial \mathfrak{L}}{\partial g^{\mu\nu}_{\alpha}} - g^{\mu\nu}_{\sigma}\frac{\partial \mathfrak{L}}{\partial g^{\mu\nu}}.$$

The last two terms add up to $-\partial\mathfrak{L}/\partial x_{\sigma}$. Substituting these results into the equation we started from and dividing both sides by 2κ, we find:

$$\frac{1}{2\kappa}\left(\frac{\partial}{\partial x_{\alpha}}\left(g^{\mu\nu}_{\sigma}\frac{\partial \mathfrak{L}}{\partial g^{\mu\nu}_{\alpha}}\right) - \frac{\partial \mathfrak{L}}{\partial x_{\sigma}}\right) = \frac{\partial T^{\lambda}_{\sigma}}{\partial x_{\lambda}}.$$

Rewriting the left-hand side as

$$-\frac{1}{2\kappa}\frac{\partial}{\partial x_{\lambda}}\left(\delta^{\lambda}_{\sigma}\mathfrak{L} - g^{\mu\nu}_{\sigma}\frac{\partial \mathfrak{L}}{\partial g^{\mu\nu}_{\lambda}}\right),$$

we can read off Eq. (20a) for the gravitational energy-momentum pseudo-tensor t^{λ}_{σ}. It has the same form as the expression for the pseudo-tensor density $\mathfrak{t}^{\lambda}_{\sigma}$ in the *Entwurf* theory that Einstein derived in a letter to Lorentz of October 12, 1915 (cf. Sec. 5.3, note 5).

16. Substituting the Lagrangian $\mathfrak{L}$ in Eq. (17) into Eq. (20a) for t^{λ}_{σ} and using Eq. (19a), we arrive at

$$\kappa t^{\lambda}_{\sigma} = \frac{1}{2}\left(\delta^{\lambda}_{\sigma}[g^{\mu\nu}\Gamma^{\alpha}_{\beta\mu}\Gamma^{\beta}_{\alpha\nu}] - g^{\mu\nu}_{\sigma}\Gamma^{\lambda}_{\mu\nu}\right).$$

Since the covariant derivative of the metric vanishes (Einstein gave a simple proof of this in a letter to Ehrenfest in early 1916 [see Sec. 8.3, note 14])

$$0 = g^{\mu\nu}{}_{;\sigma} = g^{\mu\nu}_{\sigma} + \left\{\begin{matrix}\mu\\ \alpha\sigma\end{matrix}\right\}g^{\alpha\nu} + \left\{\begin{matrix}\nu\\ \alpha\sigma\end{matrix}\right\}g^{\mu\alpha} = g^{\mu\nu}_{\sigma} - \Gamma^{\mu}_{\alpha\sigma}g^{\alpha\nu} - \Gamma^{\nu}_{\alpha\sigma}g^{\mu\alpha},$$

we can write the second term as

$$-\frac{1}{2}g^{\mu\nu}_{\sigma}\Gamma^{\lambda}_{\mu\nu} = -\frac{1}{2}(\Gamma^{\mu}_{\alpha\sigma}g^{\alpha\nu} + \Gamma^{\nu}_{\alpha\sigma}g^{\mu\alpha})\Gamma^{\lambda}_{\mu\nu} = -g^{\alpha\nu}\Gamma^{\mu}_{\alpha\sigma}\Gamma^{\lambda}_{\mu\nu}.$$

We thus arrive at Eq. (20b) for t^{λ}_{σ}:

$$\kappa t^{\lambda}_{\sigma} = \frac{1}{2}\delta^{\lambda}_{\sigma}g^{\mu\nu}\Gamma^{\alpha}_{\beta\mu}\Gamma^{\beta}_{\alpha\nu} - g^{\mu\nu}\Gamma^{\alpha}_{\mu\sigma}\Gamma^{\lambda}_{\alpha\nu}.$$

There is a factor κ missing on the left-hand side of this equation, as was noted by Hilbert: "Where did the κ go?" (*Wo ist das κ hingekommen*). This expression for t^{λ}_{σ} is virtually identical to the one for $\mathfrak{t}^{\lambda}_{\sigma}$ in terms of $\Gamma^{\alpha}_{\beta\mu}$ in the *Entwurf* theory (Einstein 1914c, p. 1077, Eq. (81b); they have opposite signs because the Lagrangians have opposite signs; cf. Sec. 3.3, note 17). For a careful comparison of these two expressions, see Janssen and Renn (2007, Appendix).

17. In the passage starting four lines above Eq. (21) and ending three lines below Eq. (21a), Einstein probes the relation between energy-momentum conservation and the transformation properties of the field equations in his new theory.

Before going over the mathematical manipulations in this passage, we reconstruct the rationale behind them (cf. Pt. I, Sec. 5.2). In the *Entwurf* theory, four conditions, written as $B_{\mu} = 0$, both guarantee energy-momentum conservation and determine the class of "justified" (*berechtigte*) transformations between coordinate systems "adapted" (*angepaßt*) to a given a metric field (Einstein and Grossmann 1914, p. 224,

Eq. (IV); Einstein 1914c, p. 1070, Eq. (67); cf. Pt. I, Sec. 4.2). Even though Einstein does not introduce the notation B_μ for its left-hand side, Eq. (22) is the analogue in the new theory of these conditions as far as energy-momentum conservation is concerned. In the new theory, however, these conditions no longer characterize the transformation properties of the field equations, which, by construction, are invariant under what would seem to be the much broader class of arbitrary unimodular transformations. This then, in all likelihood, is why Einstein—in moving from Eq. (22) to Eq. (22a) to Eq. (21a)—chose to satisfy the four conditions $B_\mu = 0$ in his new theory by imposing one (stronger) condition, which he then rewrote as a condition on $\sqrt{-g}$ over and above the one coming directly from the restriction to unimodular transformations, namely that $\sqrt{-g}$ transforms as a scalar. We now examine this argument in detail.

Fully contracting the field equations in the form of Eq. (16a) and slightly rewriting the result, we find

$$\left(g^{\mu\nu}\Gamma^\alpha_{\mu\nu}\right)_{,\alpha} - g^{\mu\nu}_\alpha \Gamma^\alpha_{\mu\nu} + g^{\mu\nu}\Gamma^\alpha_{\beta\mu}\Gamma^\beta_{\alpha\nu} = -\kappa T.$$

If we insert the definition of $\Gamma^\alpha_{\mu\nu}$ (see Eq. (15)), the first term on the left-hand side becomes

$$-\frac{1}{2}\left(g^{\mu\nu}g^{\alpha\rho}\left(2g_{\rho\mu,\nu} - g_{\mu\nu,\rho}\right)\right)_{,\alpha}.$$

The first term in this expression is easily seen to reduce to $g^{\alpha\nu}_{\nu\alpha}$; the second term is equal to

$$\frac{1}{2}\left(g^{\alpha\rho}g^{\mu\nu}g_{\mu\nu,\rho}\right)_{,\alpha} = \left(g^{\alpha\rho}\left(\log\sqrt{-g}\right)_{,\rho}\right)_{,\alpha}$$

(see, e.g., Einstein 1914c, p. 1051, Eq. (32)). Since $g^{\mu\nu}_\alpha = \Gamma^\mu_{\alpha\rho}g^{\rho\nu} + \Gamma^\nu_{\alpha\rho}g^{\mu\rho}$ (see note 16 above), the other two terms on the left-hand side of the fully contracted field equations reduce to:

$$-\left(\Gamma^\mu_{\alpha\rho}g^{\rho\nu} + \Gamma^\nu_{\alpha\rho}g^{\mu\rho}\right)\Gamma^\alpha_{\mu\nu} + g^{\mu\nu}\Gamma^\alpha_{\beta\mu}\Gamma^\beta_{\alpha\nu} = -g^{\mu\nu}\Gamma^\alpha_{\beta\mu}\Gamma^\beta_{\alpha\nu}.$$

Combining these results, we can write the fully contracted field equations in the form of Eq. (21):

$$g^{\alpha\beta}_{\alpha\beta} - g^{\mu\nu}\Gamma^\alpha_{\beta\mu}\Gamma^\beta_{\alpha\nu} + \left(g^{\alpha\rho}\left(\log\sqrt{-g}\right)_{,\rho}\right)_{,\alpha} = -\kappa T. \tag{21}$$

Note that the second term on the left-hand side is equal to $-\kappa t$, where t is the trace of the pseudo-tensor for gravitational energy-momentum:

$$\kappa t \equiv \kappa t^\lambda_\lambda = \frac{1}{2}\delta^\lambda_\lambda g^{\mu\nu}\Gamma^\alpha_{\beta\mu}\Gamma^\beta_{\alpha\nu} - g^{\mu\nu}\Gamma^\alpha_{\mu\lambda}\Gamma^\lambda_{\alpha\nu} = g^{\mu\nu}\Gamma^\alpha_{\beta\mu}\Gamma^\beta_{\alpha\nu}$$

(where we used that $\delta^\lambda_\lambda = 4$). The fully contracted field equations can thus be written more compactly as:

$$g^{\alpha\beta}_{\alpha\beta} - \kappa t + \left(g^{\alpha\rho}\left(\log\sqrt{-g}\right)_{,\rho}\right)_{,\alpha} = -\kappa T. \tag{A}$$

Contracting the field equations with $g^{\nu\lambda}$ and slightly rewriting the result, we find:

$$(g^{\nu\lambda}\Gamma^\alpha_{\mu\nu})_{,\alpha} - g^{\nu\lambda}_\alpha\Gamma^\alpha_{\mu\nu} + g^{\nu\lambda}\Gamma^\alpha_{\beta\mu}\Gamma^\beta_{\alpha\nu} = -\kappa T^\lambda_\mu.$$

Upon substitution of $\Gamma^\nu_{\alpha\rho}g^{\rho\lambda} + \Gamma^\lambda_{\alpha\rho}g^{\nu\rho}$ for $g^{\nu\lambda}_\alpha$, which gives

$$(g^{\nu\lambda}\Gamma^{\alpha}_{\mu\nu})_{,\alpha} - g^{\rho\lambda}\Gamma^{\nu}_{\alpha\rho}\Gamma^{\alpha}_{\mu\nu} - g^{\nu\rho}\Gamma^{\lambda}_{\alpha\rho}\Gamma^{\alpha}_{\mu\nu} + g^{\nu\lambda}\Gamma^{\alpha}_{\beta\mu}\Gamma^{\beta}_{\alpha\nu} = -\kappa T^{\lambda}_{\mu},$$

two terms cancel and the equation reduces to:

$$(g^{\nu\lambda}\Gamma^{\alpha}_{\mu\nu})_{,\alpha} - g^{\nu\rho}\Gamma^{\lambda}_{\alpha\rho}\Gamma^{\alpha}_{\mu\nu} = -\kappa T^{\lambda}_{\mu}.$$

In the second term on the right-hand side, we recognize one of the terms in the expression for κt^{λ}_{μ} in Eq. (20b) (see note 16 above). Using that

$$-g^{\nu\rho}\Gamma^{\lambda}_{\alpha\rho}\Gamma^{\alpha}_{\mu\nu} = \kappa t^{\lambda}_{\mu} - \frac{1}{2}\delta^{\lambda}_{\mu}g^{\rho\sigma}\Gamma^{\alpha}_{\beta\rho}\Gamma^{\beta}_{\alpha\sigma},$$

we arrive at the unnumbered equation above Eq. (22):

$$(g^{\nu\lambda}\Gamma^{\alpha}_{\mu\nu})_{,\alpha} - \frac{1}{2}\delta^{\lambda}_{\mu}g^{\rho\sigma}\Gamma^{\alpha}_{\beta\rho}\Gamma^{\beta}_{\alpha\sigma} = -\kappa(T^{\lambda}_{\mu} + t^{\lambda}_{\mu}).$$

To guarantee energy-momentum conservation, i.e., the vanishing of the four-divergence of $T^{\lambda}_{\mu} + t^{\lambda}_{\mu}$, we demand that the four-divergence of the left-hand side also vanishes:

$$\left((g^{\nu\lambda}\Gamma^{\alpha}_{\mu\nu})_{,\alpha} - \frac{1}{2}\delta^{\lambda}_{\mu}g^{\rho\sigma}\Gamma^{\alpha}_{\beta\rho}\Gamma^{\beta}_{\alpha\sigma}\right)_{,\lambda} = 0.$$

This is the counterpart of the condition $B_{\mu} = 0$ in the *Entwurf* theory. Einstein, however, does not introduce the notation B_{μ} (or any special notation) in this case. He also suppresses the details of how he rewrote the equation above in the form of Eq. (22).

To arrive at Eq. (22), we first rewrite the equation above as

$$(g^{\nu\lambda}\Gamma^{\alpha}_{\mu\nu})_{,\lambda\alpha} - \frac{1}{2}(g^{\rho\sigma}\Gamma^{\alpha}_{\beta\rho}\Gamma^{\beta}_{\alpha\sigma})_{,\mu} = 0.$$

Using the definition of $\Gamma^{\alpha}_{\mu\nu}$, we write the first term on the left-hand side as

$$-\frac{1}{2}\left(g^{\nu\lambda}g^{\alpha\rho}(g_{\rho\mu,\nu} + g_{\rho\nu,\mu} - g_{\mu\nu,\rho})\right)_{,\lambda\alpha}$$

or

$$-\frac{1}{2}\left(-g^{\nu\lambda}g^{\alpha\rho}_{\nu}g_{\rho\mu} - g^{\nu\lambda}g^{\alpha\rho}_{\mu}g_{\rho\nu} + g^{\nu\lambda}_{\rho}g^{\alpha\rho}g_{\mu\nu}\right)_{,\lambda\alpha}.$$

The first and the third term in parentheses form a quantity anti-symmetric in λ and α which vanishes upon contraction with $\partial/\partial x^{\lambda}$ and $\partial/\partial x^{\alpha}$. The second term can be rewritten as $\frac{1}{2}\delta^{\lambda}_{\rho}g^{\alpha\rho}_{\mu} = \frac{1}{2}g^{\alpha\lambda}_{\mu}$. Finally, $(g^{\alpha\lambda}_{\mu})_{,\lambda\alpha} = (g^{\alpha\lambda}_{\alpha\lambda})_{,\mu}$. This then is how we obtain Eq. (22),

$$(g^{\alpha\beta}_{\alpha\beta} - g^{\rho\sigma}\Gamma^{\alpha}_{\beta\rho}\Gamma^{\beta}_{\alpha\sigma})_{,\mu} = 0. \tag{22}$$

In Eq. (22a), Einstein imposes the stronger condition,

$$g^{\alpha\beta}_{\alpha\beta} - g^{\rho\sigma}\Gamma^{\alpha}_{\beta\rho}\Gamma^{\beta}_{\alpha\sigma} = 0, \tag{22a}$$

which can be written more compactly as

$$g^{\alpha\beta}_{\alpha\beta} - \kappa t = 0 \tag{B}$$

The combination of the conditions labeled (A) and (B) above shows that the field equations guarantee energy-momentum conservation as long as the condition in Eq.

(21a) holds (cf. Pt. I, Ch. 5, notes 15 and 16):

$$\left(g^{\alpha\beta}\left(\log\sqrt{-g}\right)_{,\beta}\right)_{,\alpha}=-\kappa T.$$

18. To recapitulate (see note 17), his experience with the *Entwurf* theory had led Einstein to expect that the same four conditions that ensure that the field equations entail energy-momentum conservation also determine the covariance properties of the field equations. In his new theory, however, there seems to be a mismatch between the two. The only restriction on the covariance of the field equations is that they only retain their form under transformations that leave $\sqrt{-g}$ unchanged. This is why, we conjectured, Einstein replaced the four conditions for energy-momentum conservation by one condition on $\sqrt{-g}$ (see Eq. (21a)). This condition, however, is an odd one. Since the energy-momentum tensor for ordinary matter has a non-vanishing trace (or "scalar" as Einstein calls it), it implies that $\sqrt{-g}$ cannot be a constant. So we have a theory invariant under unimodular transformations in which unimodular coordinates (defined by $\sqrt{-g}=1$) are not allowed. As we shall see, the second and the fourth of Einstein's November 1915 communications to the Berlin Academy are essentially two different proposals for getting around this peculiar prohibition against unimodular coordinates.
19. Even though Einstein's new theory is invariant under arbitrary *autonomous* (*selbständige*) unimodular coordinate transformation, the extra condition required by energy-momentum conservation (whether in the form of Eq. (22a) or Eq, (21a)) means that *non-autonomous* (*unselbständige*) transformations between "adapted" (*angepaßt*) coordinates continue to play a role. For further discussion, see our commentary on the Zurich notebook (Sec. 1.3, note 20) and Pt. I, Sec. 3.1, especially note 8.

§ 4

20. As he had already done in the Zurich notebook (see p. 22R and Sec. 1.3, note 19), Einstein used the condition ${g^{\mu\nu}}_{,\nu}=0$ (which we called the *Hertz condition/restriction*) to rewrite the term $\Gamma^{\alpha}_{\mu\nu,\alpha}$ in the field equations in Eq. (16a) as the sum of one term with second-order derivatives of the metric that reduces to the Poisson equation of Newtonian theory for weak static fields and two terms that are products of first-order derivatives, which can be neglected for such fields. Given the definition of $\Gamma^{\alpha}_{\mu\nu}$ (see note 11), we have:

$$\Gamma^{\alpha}_{\mu\nu,\alpha}=-\frac{1}{2}\left(g^{\alpha\rho}\left(g_{\rho\mu,\nu}+g_{\rho\nu,\mu}-g_{\mu\nu,\rho}\right)\right)_{,\alpha}.$$

The last term gives the left-hand side of the field equations in first approximation in Eq. (16b):

$$\frac{1}{2}g^{\alpha\rho}g_{\mu\nu,\rho\alpha}=\frac{1}{2}\partial^{\alpha}\partial_{\alpha}g_{\mu\nu}.$$

The other two terms can be rewritten as

$$\frac{1}{2}\left({g^{\alpha\rho}}_{,\nu}g_{\rho\mu}+{g^{\alpha\rho}}_{,\mu}g_{\rho\nu}\right)_{,\alpha}.$$

Since ${g^{\alpha\rho}}_{,\nu\alpha}$ and ${g^{\alpha\rho}}_{,\mu\alpha}$ vanish on account of the Hertz restriction, both terms reduce to products of first-order derivatives of the metric:

$$\frac{1}{2}\left({g^{\alpha\rho}}_{,\nu}g_{\rho\mu,\alpha}+{g^{\alpha\rho}}_{,\mu}g_{\rho\nu,\alpha}\right).$$

Taken in isolation, it looks as if $g^{\mu\nu}{}_{,\nu} = 0$ has the status of a modern coordinate condition. Einstein, however, introduces it to satisfy Eq. (22a) in first-order approximation and Eq. (22a), as the final paragraph of § 3 makes clear, is still a coordinate restriction.

Finally, we note that a metric of the form $g_{\mu\nu} = \mathrm{diag}(-1,-1,-1,f(x,y,z))$, which Einstein still expected to represent weak static fields at this point (cf. Sec. 6.2.3, note 4), satisfies the condition $g^{\mu\nu}{}_{,\nu} = 0$ (see Pt. I, Sec. 3.1).

21. The rotation metric, the metric of Minkowski space time in a rotating Cartesian coordinate system, is a solution of Einstein's new field equations because the standard diagonal metric with constant components of Minkowski space-time in an inertial Cartesian coordinate system trivially satisfies these field equations (this is true for all candidate field equations Einstein considered) and, as Einstein notes below, the transformation from the inertial to the rotating coordinate system is unimodular.

However, $g^{\mu\nu}_{\mathrm{rot},\nu} \neq 0$ (see Sec. 1.3, note 20). The rotation metric thus does not satisfy the coordinate restriction Einstein added when he first considered the field equations proposed in this paper in the Zurich notebook. This, in all likelihood, is why he originally rejected these equations (see Norton 1984, p. 119 and Pt. I, Sec. 3.1). That he no longer saw this as a problem at this point may suggest that the status of $g^{\mu\nu}{}_{,\nu} = 0$ was now that of a coordinate condition, like a gauge condition in electrodynamics, rather a coordinate restriction, i.e., that Einstein had come to realize that he could use it to show that his theory was compatible with Newton's without being forced to then use it in all other applications of his theory as well. Note, however, that Einstein only demanded that $g^{\mu\nu}{}_{,\nu} \neq 0$ (Eq. (22)) be satisfied *to first order* and that the only non-zero terms in $g^{\mu\nu}_{\mathrm{rot},\nu}$ are of second order in the angular frequency ω. Hence, at least for small values of ω, $g^{\mu\nu}_{\mathrm{rot},\nu} = 0$. For further discussion, see Pt. I, Sec. 5.2.

6.2 On the general theory of relativity (addendum)

6.2.1 *Facsimile*

Zur allgemeinen Relativitätstheorie (Nachtrag).

Von A. Einstein.

In einer neulich erschienenen Untersuchung[1] habe ich gezeigt, wie auf Riemanns Kovariantentheorie mehrdimensionaler Mannigfaltigkeiten eine Theorie des Gravitationsfeldes gegründet werden kann. Hier soll nun dargetan werden, daß durch Einführung einer allerdings kühnen zusätzlichen Hypothese über die Struktur der Materie ein noch strafferer logischer Aufbau der Theorie erzielt werden kann. [1]

Die Hypothese, deren Berechtigung in Erwägung gezogen werden soll, betrifft folgenden Gegenstand. Der Energietensor der »Materie« T^{λ}_{μ} besitzt einen Skalar $\sum_{\mu} T^{\mu}_{\mu}$. Es ist wohlbekannt, daß dieser für das elektromagnetische Feld verschwindet. Dagegen scheint er für die eigentliche Materie von Null verschieden zu sein. Betrachten wir nämlich als einfachsten Spezialfall die »inkohärente« kontinuierliche Flüssigkeit (Druck vernachlässigt), so pflegen wir ja für sie zu setzen

$$T^{\mu\nu} = \sqrt{-g}\,\rho_0 \frac{dx_\mu}{ds}\frac{dx_\nu}{ds},$$

so daß wir haben

$$\sum_{\mu} T^{\mu}_{\mu} = \sum_{\mu\nu} g_{\mu\nu} T^{\mu\nu} = \rho_0 \sqrt{-g}.$$

Hier verschwindet also nach dem Ansatz der Skalar des Energietensors nicht.

Es ist nun daran zu erinnern, daß nach unseren Kenntnissen die »Materie« nicht als ein primitiv Gegebenes, physikalisch Einfaches aufzufassen ist. Es gibt sogar nicht wenige, die hoffen, die Materie auf rein elektromagnetische Vorgänge reduzieren zu können, die allerdings einer gegenüber Maxwells Elektrodynamik vervollständigten Theorie gemäß vor sich gehen würden. [2] Nehmen wir nun einmal an, daß in einer so vervollständigten Elektrodynamik der Skalar des Energietensors ebenfalls verschwinden würde! Würde dann das soeben aufgezeigte Resultat beweisen, daß die Materie mit Hilfe dieser Theorie nicht konstruiert werden könnte? Ich glaube diese Frage verneinen

[1] Diese Sitzungsberichte S. 778.

zu können. Denn es wäre sehr wohl möglich, daß in der »Materie«, auf die sich der eben angegebene Ausdruck bezieht, Gravitationsfelder einen wesentlichen Bestandteil ausmachen. Dann kann $\sum T_\mu^\mu$ für das ganze Gebilde scheinbar positiv sein, während in Wirklichkeit nur $\sum_\mu (T_\mu^\mu + t_\mu^\mu)$ positiv ist, während $\sum_\mu T_\mu^\mu$ überall verschwindet. Wir setzen im folgenden voraus, daß die Bedingung $\sum T_\mu^\mu = 0$ tatsächlich allgemein erfüllt sei.

Wer die Hypothese, daß molekulare Gravitationsfelder einen wesentlichen Bestandteil der Materie ausmachen, nicht von vornherein ablehnt, wird in dem Folgenden eine kräftige Stütze dieser Auffassung sehen[1]. [3]

Ableitung der Feldgleichungen.

Unsere Hypothese erlaubt es, den letzten Schritt zu tun, welchen der allgemeine Relativitätsgedanke als wünschbar erscheinen läßt. Sie ermöglicht nämlich, auch die Feldgleichungen der Gravitation in allgemein kovarianter Form anzugeben. In der früheren Mitteilung habe ich gezeigt (Gleichung (13)), daß

$$G_{im} = \sum_l \{il, lm\} = R_{im} + S_{im} \tag{13}$$

ein kovarianter Tensor bezüglich beliebiger Substitutionen ist. Dabei ist gesetzt

$$R_{im} = -\sum_l \frac{\partial \begin{Bmatrix} im \\ l \end{Bmatrix}}{\partial x_l} + \sum_{\varrho l} \begin{Bmatrix} il \\ \varrho \end{Bmatrix} \begin{Bmatrix} \varrho m \\ l \end{Bmatrix} \tag{13a}$$

$$S_{im} = \sum_l \frac{\partial \begin{Bmatrix} il \\ l \end{Bmatrix}}{\partial x_m} - \sum_{\varrho l} \begin{Bmatrix} im \\ \varrho \end{Bmatrix} \begin{Bmatrix} \varrho l \\ l \end{Bmatrix} \tag{13b}$$

Dieser Tensor G_{im} ist der einzige Tensor, der für die Aufstellung allgemein kovarianter Gravitationsgleichungen zur Verfügung steht.

Setzen wir nun fest, daß die Feldgleichungen der Gravitation lauten sollen

$$G_{\mu\nu} = -\varkappa T_{\mu\nu}, \tag{16b}$$

so haben wir damit allgemein kovariante Feldgleichungen gewonnen. Diese drücken zusammen mit den vom absoluten Differentialkalkül gelieferten allgemein kovarianten Gesetzen für das »materielle« Geschehen die Kausalzusammenhänge in der Natur so aus, daß irgendwelche besondere Wahl des Koordinatensystems, welche ja logisch mit den zu

[1] Bei Niederschrift der früheren Mitteilung war mir die prinzipielle Zulässigkeit der Hypothese $\sum T_\mu^\mu = 0$ noch nicht zu Bewußtsein gekommen.

beschreibenden Gesetzmäßigkeiten nichts zu tun hat, auch bei deren Formulierung nicht verwendet wird.

Von diesem System aus kann man durch nachträgliche Koordinatenwahl leicht zu dem System von Gesetzmäßigkeiten zurückgelangen, welches ich in meiner letzten Mitteilung aufgestellt habe, und zwar ohne an den Gesetzen tatsächlich etwas zu ändern. Es ist nämlich klar, daß wir ein neues Koordinatensystem einführen können, derart, daß mit Bezug auf dieses überall

$$\sqrt{-g} = 1$$

ist. Dann verschwindet S_{im}, so daß man zu dem System der Feldgleichungen [4]

$$R_{\mu\nu} = -\varkappa T_{\mu\nu} \tag{16}$$

der letzten Mitteilung zurückgelangt. Die vom absoluten Differentialkalkül gelieferten Formeln degenerieren dabei genau in der in der letzten Mitteilung angegebenen Weise. Auch jetzt läßt ferner unsere Koordinatenwahl nur Transformationen von der Determinante 1 zu. [5]

Der Unterschied zwischen dem Inhalte unserer aus den allgemein kovarianten gewonnenen Feldgleichungen und dem Inhalte der Feldgleichungen unserer letzten Mitteilung liegt nur darin, daß in der letzten Mitteilung der Wert für $\sqrt{-g}$ nicht vorgeschrieben werden konnte. Derselbe war vielmehr durch die Gleichung

$$\sum_{\alpha\beta} \frac{\partial}{\partial x_\alpha}\left(g^{\alpha\beta} \frac{\partial \lg\sqrt{-g}}{\partial x_\alpha}\right) = -\varkappa \sum_\sigma T_\sigma^\sigma \tag{21a}$$

bestimmt. Aus dieser Gleichung sieht man, daß dort $\sqrt{-g}$ nur dann konstant sein kann, wenn der Skalar des Energietensors verschwindet.

Bei unserer jetzigen Ableitung ist vermöge unserer willkürlich getroffenen Koordinatenwahl $\sqrt{-g} = 1$. Statt der Gleichung (21a) folgt daher jetzt aus unsern Feldgleichungen das Verschwinden des Skalars des Energietensors der »Materie«. Die unsern Ausgangspunkt bildenden allgemein kovarianten Feldgleichungen (16b) führen also nur dann zu keinem Widerspruch, wenn die in der Einleitung dargelegte Hypothese zutrifft. Dann aber sind wir gleichzeitig berechtigt, unseren früheren Feldgleichungen die beschränkende Bedingung

$$\sqrt{-g} = 1 \tag{21b}$$

zuzufügen.

Ausgegeben am 18. November.

Berlin, gedruckt in der Reichsdruckerei.

6.2.2 Translation

On the General Theory of Relativity (Addendum)

In a recent investigation,[1] I have shown how a theory of the gravitational field can be based on RIEMANN's theory of covariants of multidimensional manifolds. Here I want to show that an even tighter logical structure of the theory can be achieved by introducing an admittedly bold additional hypothesis about the structure of matter.[1]

The hypothesis the justification of which is to be taken under consideration here relates to the following object. The energy tensor of "matter" T_μ^λ has a scalar $\sum_\mu T_\mu^\mu$. It is well known that this scalar vanishes for the electromagnetic field. By contrast, it appears to differ from zero for matter *proper*. After all, we consider an "incoherent" continuous fluid (pressure neglected) as the simplest special case, we are used to setting

$$T^{\mu\nu} = \sqrt{-g}\,\rho_0 \frac{dx_\mu}{ds}\frac{dx_\nu}{ds},$$

and we have

$$\sum_\mu T_\nu^\mu = \sum_{\mu\nu} g_{\mu\nu} T^{\mu\nu} = \rho_0 \sqrt{-g}.$$

The scalar of the energy tensor does not vanish in this case.

It should be recalled, however, that to our knowledge "matter" is not to be conceived as something directly given and physically simple. There actually are more than a few who hope to reduce matter to purely electrodynamic processes, which would admittedly have to happen in accordance with a completed version of MAXWELL's electrodynamics.[2] Now let us assume that in such a completed electrodynamics the scalar of the energy tensor would still vanish! In that case, would the result above prove that matter cannot be constructed in this theory? I believe the answer is no. It could very well be that gravitational fields are an important part of the "matter" to which the expression above pertains. In that case, $\sum T_\mu^\mu$ can seem to be positive for the entire construct while in reality only $\sum(T_\mu^\mu + t_\mu^\mu)$ is positive and $\sum T_\mu^\mu$ vanishes everywhere. *In the following we assume that the condition* $\sum T_\mu^\mu = 0$ *actually holds in all generality.*

---

[1] These *Proceedings*, p. 778.

Whoever does not a priori reject the hypothesis that gravitational fields make up an essential part of matter will find strong support for this conception in the following.[1], [3]

Derivation of the Field Equations

Our hypothesis allows us to take the final step congenial to the idea of general relativity. It allows us also to write even the gravitational field equations in *generally* covariant form. In the earlier paper I showed (equation (13)) that

$$G_{im} = \sum_l \{il, lm\} = R_{im} + S_{im} \tag{13}$$

is a covariant tensor for arbitrary transformations. Here

$$R_{im} = -\sum_l \frac{\partial \begin{Bmatrix} im \\ l \end{Bmatrix}}{\partial x_l} + \sum_{\rho l} \begin{Bmatrix} il \\ \rho \end{Bmatrix} \begin{Bmatrix} \rho m \\ l \end{Bmatrix} \tag{13a}$$

$$S_{im} = \sum_l \frac{\partial \begin{Bmatrix} il \\ l \end{Bmatrix}}{\partial x_m} - \sum_{\rho l} \begin{Bmatrix} im \\ \rho \end{Bmatrix} \begin{Bmatrix} \rho l \\ l \end{Bmatrix}. \tag{13a}$$

This tensor G_{im} is the only tensor available for the construction of generally covariant gravitational field equations.

If we posit that the gravitational field equations should be

$$G_{\mu\nu} = -\kappa T_{\mu\nu}. \tag{16b}$$

we have succeeded in finding generally covariant field equations. Together with the generally covariant laws for "material" processes, given by the absolute differential calculus, these field equations express the causal connections in nature in such a way that any special choice of coordinates, which *logically* has nothing to do with the laws of nature, is not used in their formulation either.

From this system one can easily get back to the system of equations in my last paper without actually changing anything in the physical laws by retroactively choosing specific coordinates. This is because it is clear that we can introduce a new coordinate system such that

$$\sqrt{-g} = 1$$

holds everywhere.[4] In that case, S_{im} vanishes and one get back to the system of field equations

---

[1] When I wrote my earlier paper I had not yet realized that the hypothesis $\sum T_\mu^\mu = 0$ is, in principle, admissible.

$$R_{\mu\nu} = -\kappa T_{\mu\nu} \tag{16}$$

of my last paper. The formulas of the absolute differential calculus degenerate exactly in the manner shown in this paper.[5] And with this choice of coordinates only transformations of determinant 1 are allowed.

The only difference between the field equations obtained from general covariance and those of my last paper is that the value of $\sqrt{-g}$ could not be prescribed in the latter. This value was instead determined by the equation

$$\sum_{\alpha\beta} \frac{\partial}{\partial x_\alpha}\left(g^{\alpha\beta}\frac{\partial \lg\sqrt{-g}}{\partial x_\beta}\right) = -\kappa \sum_\sigma T_\sigma^\sigma. \tag{21a}$$

According to this equation $\sqrt{-g}$ can only be constant if the scalar of the energy tensor vanishes.

In our present derivation $\sqrt{-g} = 1$ due to our arbitrary choice of coordinates. The vanishing of the scalar of the energy tensor of "matter" now follows from our field equations instead of from equation (21a). *Hence, the generally covariant field equations (16b), which form our starting point, do not lead to a contradiction only if the hypothesis given in the introduction is correct.* In that case, we are also allowed to add to our earlier field equation the restrictive condition:

$$\sqrt{-g} = 1. \tag{21b}$$

6.2.3 Commentary

This paper, a short addendum to Einstein (1915a, see Sec. 6.1), was submitted November 11, 1915 and published November 18, 1915. It is presented in facsimile as Doc. 22 in CPAE6.

1. This paper is Einstein's first and short-lived attempt to get around the condition (21a),
$$\left(g^{\alpha\beta}\left(\log\sqrt{-g}\right)_{,\beta}\right)_{,\alpha} = -\kappa T, \tag{21a}$$
in his first paper of November 1915, which prohibits the use of the natural unimodular coordinates in a theory covariant under unimodular transformations (cf. Sec. 6.1.3, note 18, and Pt. I, Sec. 5.4). Setting the trace (or "scalar") T of the energy-momentum tensor equal to zero, which allowed him to set $\sqrt{-g} = 1$, Einstein could now look upon the field equations of that first paper as generally covariant equations expressed in unimodular coordinates.
2. Such a non-linear generalization of Maxwell's theory had been proposed by Gustav Mie (1912; 1913, see Pt. I, Sec. 5.4, especially note 19, for discussion).
3. The order in which Einstein presents the two key steps in the argument in this paper is probably the opposite of the order in which he found them. Presumably, he first realized that setting $T = 0$ would allow him to look upon the field equations of his first November 1915 paper as generally covariant equations expressed in unimodular coordinates and only then realized that setting $T = 0$ could be justified on the basis of the electromagnetic worldview. Einstein did something similar in his cosmology paper of February 1917. In that case, he probably first realized that adding the

cosmological term to his field equations would allow a universe that is spatially closed and then thought of an argument to justify this extra term. In the published paper, however, Einstein (1917) gives this justification first and then shows that the field equations with the extra term allow his spatially closed cosmological model (see Pt. I, Sec. 7.1 and Janssen 2014, p. 201).

4. Before Einstein could set $\sqrt{-g} = 1$, he had to overcome his firm belief, in place since the early pages of the Zurich notebook, that the metric for a weak static field would only have one variable component (the 44 component), in which case its determinant cannot be a constant (cf. Sec. 1.3, note 16, and Pt. I, Ch. 3 and Sec. 5.3, especially 12). When he calculated the perihelion motion of Mercury on the basis of his 1915 theory, Einstein realized that, even in a first-order approximation, the metric for the weak static field of the sun has other variable components (see Sec. 6.3.3, note 8). This realization not only removed an obstacle to the use of unimodular coordinates, it also opened the door to a more elegant and convincing way to get around the puzzling condition (21a) (see note 1 above) that Einstein would present two weeks later in the fourth November 1915 paper (see Sec. 6.4.3, note 3).
5. Setting $\sqrt{-g} = 1$ simplifies the argument in Einstein (1915a, §§ 1–2) for replacing the definition of the covariant divergence in a general covariant setting by one appropriate in a unimodular setting. For instance, the energy-momentum balance equation in its generally covariant form immediately reduces to its counterpart in Einstein's unimodular theory once $\sqrt{-g}$ is set equal to 1 (cf. Sec. 6.1.3, note 10).

6.3 Explanation of the perihelion motion of Mercury on the basis of the general theory of relativity

6.3.1 Facsimile

Erklärung der Perihelbewegung des Merkur aus der allgemeinen Relativitätstheorie.

Von A. Einstein.

In einer jüngst in diesen Berichten erschienenen Arbeit, habe ich Feldgleichungen der Gravitation aufgestellt, welche bezüglich beliebiger Transformationen von der Determinante 1 kovariant sind. In einem Nachtrage habe ich gezeigt, daß jenen Feldgleichungen allgemein kovariante entsprechen, wenn der Skalar des Energietensors der »Materie« verschwindet, und ich habe dargetan, daß der Einführung dieser Hypothese, durch welche Zeit und Raum der letzten Spur objektiver Realität beraubt werden, keine prinzipiellen Bedenken entgegenstehen[1]. [1]

In der vorliegenden Arbeit finde ich eine wichtige Bestätigung dieser radikalsten Relativitätstheorie; es zeigt sich nämlich, daß sie die von Leverrier entdeckte säkulare Drehung der Merkurbahn im Sinne der Bahnbewegung, welche etwa 45″ im Jahrhundert beträgt qualitativ und quantitativ erklärt, ohne daß irgendwelche besondere Hypothese zugrunde gelegt werden müßte[2]. [2]

Es ergibt sich ferner, daß die Theorie eine stärkere (doppelt so starke) Lichtstrahlenkrümmung durch Gravitationsfelder zur Konsequenz hat als gemäß meinen früheren Untersuchungen.

---

[1] In einer bald folgenden Mitteilung wird gezeigt werden, daß jene Hypothese entbehrlich ist. Wesentlich ist nur, daß eine solche Wahl des Bezugssystems möglich ist, daß die Determinante $|g_{\mu\nu}|$ den Wert -1 annimmt. Die nachfolgende Untersuchung ist hiervon unabhängig.

[2] Über die Unmöglichkeit, die Anomalien der Merkurbewegung auf der Basis der Newtonschen Theorie befriedigend zu erklären, schrieb E. Freundlich jüngst einen beachtenswerten Aufsatz (Astr. Nachr. 4803, Bd. 201. Juni 1915).

§ 1. Das Gravitationsfeld.

Aus meinen letzten beiden Mitteilungen geht hervor, daß das Gravitationsfeld im Vakuum bei geeignet gewähltem Bezugssystem folgenden Gleichungen zu genügen hat

$$\sum_{\alpha} \frac{\partial \Gamma^{\alpha}_{\mu\nu}}{\partial x_{\alpha}} + \sum_{\alpha\beta} \Gamma^{\alpha}_{\mu\beta} \Gamma^{\beta}_{\nu\alpha} = 0, \qquad (1)$$

wobei die $\Gamma^{\alpha}_{\mu\nu}$ durch die Gleichung definiert sind

$$\Gamma^{\alpha}_{\mu\nu} = -\begin{Bmatrix} \mu\nu \\ \alpha \end{Bmatrix} = -\sum_{\beta} g^{\alpha\beta} \begin{bmatrix} \mu\nu \\ \beta \end{bmatrix} = -\frac{1}{2} \sum_{\beta} g^{\alpha\beta} \left(\frac{\partial g_{\mu\beta}}{\partial x_{\nu}} + \frac{\partial g_{\nu\beta}}{\partial x_{\mu}} - \frac{\partial g_{\mu\nu}}{\partial x_{\alpha}} \right). \qquad (2)$$

Machen wir außerdem die in der letzten Mitteilung begründete Hypothese, daß der Skalar des Energietensors der »Materie« stets verschwinde, so tritt hierzu die Determinantengleichung

$$|g_{\mu\nu}| = -1. \qquad (3)$$

Es befinde sich im Anfangspunkt des Koordinatensystems ein Massenpunkt (die Sonne). Das Gravitationsfeld, welches dieser Massenpunkt erzeugt, kann aus diesen Gleichungen durch sukzessive Approximation berechnet werden. [3]

Es ist indessen wohl zu bedenken, daß die $g_{\mu\nu}$ bei gegebener Sonnenmasse durch die Gleichungen (1) und (3) mathematisch noch nicht vollständig bestimmt sind. Es folgt dies daraus, daß diese Gleichungen bezüglich beliebiger Transformationen mit der Determinante 1 kovariant sind. Es dürfte indessen berechtigt sein, vorauszusetzen, daß alle diese Lösungen durch solche Transformationen aufeinander reduziert werden können, daß sie sich also (bei gegebenen Grenzbedingungen) nur formell, nicht aber physikalisch voneinander unterscheiden. Dieser Überzeugung folgend begnüge ich mich vorerst damit, hier eine Lösung abzuleiten, ohne mich auf die Frage einzulassen, ob es die einzig mögliche sei. [4]

Wir gehen nun in solcher Weise vor. Die $g_{\mu\nu}$ seien in »nullter Näherung« durch folgendes, der ursprünglichen Relativitätstheorie entsprechende Schema gegeben

$$\left.\begin{matrix} -1 & 0 & 0 & 0 \\ 0 & -1 & 0 & 0 \\ 0 & 0 & -1 & 0 \\ 0 & 0 & 0 & +1 \end{matrix}\right\}, \qquad (4)$$

oder kürzere

$$\left.\begin{aligned} g_{\rho\sigma} &= \delta_{\rho\sigma} \\ g_{\rho 4} &= g_{4\rho} = 0 \\ g_{44} &= 1 \end{aligned}\right\}. \qquad (4a)$$

Hierbei bedeuten ρ und σ die Indizes 1, 2, 3; $\delta_{\rho\sigma}$ ist gleich 1 oder 0, je nachdem $\rho = \sigma$ oder $\rho \neq \sigma$ ist.

Wir setzen nun im folgenden voraus, daß sich die $g_{\mu\nu}$ von den in (4a) angegebenen Werten nur um Größen unterscheiden, die klein sind gegenüber der Einheit. Diese Abweichungen behandeln wir als kleine Größen »erster Ordnung«, Funktionen $n$ten Grades dieser Abweichungen als »Größen $n$ter Ordnung«. Die Gleichungen (1) und (3) setzen uns in den Stand, von (4a) ausgehend, durch sukzessive Approximation das Gravitationsfeld bis auf Größen $n$ter Ordnung genau zu berechnen. Wir sprechen in diesem Sinne von der »$n$ten Approximation«; die Gleichungen (4a) bilden die »nullte Approximation«. [5]

Die im folgenden gegebene Lösung hat folgende, das Koordinatensystem festlegende Eigenschaften:

1. Alle Komponenten sind von x_4 unabhängig.
2. Die Lösung ist (räumlich) symmetrisch um den Anfangspunkt des Koordinatensystems, in dem Sinne, daß man wieder auf dieselbe Lösung stößt, wenn man sie einer linearen orthogonalen (räumlichen) Transformation unterwirft.
3. Die Gleichungen $g_{\varrho 4} = g_{4\varrho} = 0$ gelten exakt (für $\rho = 1$ bis 3).
4. Die $g_{\mu\nu}$ besitzen im Unendlichen die in (4a) gegebenen Werte.

Erste Approximation.

Es ist leicht zu verifizieren, daß in Größen erster Ordnung den Gleichungen (1) und (3) sowie den eben genannten 4 Bedingungen genügt wird durch den Ansatz [6]

$$\left.\begin{aligned} g_{\varrho\sigma} &= -\delta_{\varrho\sigma} + \alpha\left(\frac{\partial^2 r}{\partial x_\varrho \partial x_\sigma} - \frac{\delta_{\varrho\sigma}}{r}\right) = -\delta_{\varrho\sigma} - \alpha\frac{x_\varrho x_\sigma}{r^3} \\ g_{44} &= 1 - \frac{\alpha}{r} \end{aligned}\right\}. \quad (4\,\mathrm{b})$$

Die $g_{4\varrho}$ bzw. $g_{\varrho 4}$ sind dabei durch Bedingung 3 festgelegt. r bedeutet die Größe $+\sqrt{x_1^2 + x_2^2 + x_3^2}$, α eine durch die Sonnenmasse bestimmte Konstante.

Daß (3) in Gliedern erster Ordnung erfüllt ist, sieht man sogleich. Um in einfacher Weise einzusehen, daß auch die Feldgleichungen (1) in erster Näherung erfüllt sind, braucht man nur zu beachten, daß bei Vernachlässigung von Größen zweiter und höherer Ordnung die linke Seite der Gleichungen (1) sukzessive durch [7]

$$\sum_\alpha \frac{\partial \Gamma^\alpha_{\mu\nu}}{\partial x_\alpha}$$

$$\sum_\alpha \frac{\partial}{\partial x_\alpha}\begin{bmatrix}\mu\nu\\ \alpha\end{bmatrix}$$

versetzt werden kann, wobei α nur von 1—3 läuft.

Wie man aus (4b) ersieht, bringt es unsere Theorie mit sich, daß im Falle einer ruhenden Masse die Komponenten g_{11} bis g_{33} bereits in den Größen erster Ordnung von null verschieden sind. Wir werden [8] später sehen, daß hierdurch kein Widerspruch gegenüber Newtons Gesetz (in erster Näherung) entsteht. Wohl aber ergibt sich hieraus ein etwas anderer Einfluß des Gravitationsfeldes auf einen Lichtstrahl als nach meinen früheren Arbeiten; denn die Lichtgeschwindigkeit ist durch die Gleichung

$$\sum g_{\mu\nu} dx_\mu dx_\nu = 0 \tag{5}$$

bestimmt. Unter Anwendung von Huygens' Prinzip findet man aus (5) und (4b) durch eine einfache Rechnung, daß ein an der Sonne im Abstand Δ vorbeigehender Lichtstrahl eine Winkelablenkung von der Größe $\frac{2\alpha}{\Delta}$ erleidet, während die früheren Rechnungen, bei welchen die Hypothese $\sum T^\mu_\mu = 0$ nicht zugrunde gelegt war, den Wert $\frac{\alpha}{\Delta}$ ergeben hatten. Ein an der Oberfläche der Sonne vorbeigehender Lichtstrahl soll eine Ablenkung von 1.7″ (statt 0.85″) erleiden. Hin- [9] gegen bleibt das Resultat betreffend die Verschiebung der Spektrallinien durch das Gravitationspotential, welches durch Herrn Freundlich an den Fixsternen der Größenordnung nach bestätigt wurde, ungeändert bestehen, da dieses nur von g_{44} abhängt. [10]

Nachdem wir die $g_{\mu\nu}$ in erster Näherung erlangt haben, können wir auch die Komponenten $\Gamma^\alpha_{\mu\nu}$ des Gravitationsfeldes in erster Näherung berechnen. Aus (2) und (4b) ergibt sich

$$\Gamma^\tau_{\rho\sigma} = -\alpha\left(\delta_{\rho\sigma}\frac{x_\tau}{r^3} - \frac{3}{2}\frac{x_\rho x_\sigma x_\tau}{r^5}\right), \tag{6a}$$

wobei ρ, σ, τ irgendwelche der Indizes 1, 2, 3 bedeuten,

$$\Gamma^\sigma_{44} = \Gamma^4_{4\sigma} = -\frac{\alpha}{2}\frac{x_\sigma}{r^3}, \tag{6b}$$

wobei σ den Index 1, 2 oder 3 bedeutet. Diejenigen Komponenten, in welchen der Index 4 einmal oder dreimal auftritt, verschwinden.

Zweite Approximation.

Es wird sich nachher ergeben, daß wir nur die drei Komponenten Γ^σ_{44} in Größen zweiter Ordnung genau zu ermitteln brauchen, um die Planetenbahnen mit dem entsprechenden Genauigkeitsgrade ermitteln zu können. Hierfür genügt uns die letzte Feldgleichung zusammen mit den allgemeinen Bedingungen, welche wir unserer Lösung auferlegt haben. Die letzte Feldgleichung

$$\sum_{\sigma} \frac{\partial \Gamma^{\sigma}_{44}}{\partial x_{\sigma}} + \sum_{\sigma\tau} \Gamma^{\sigma}_{4\tau} \Gamma^{\tau}_{4\sigma} = 0$$

geht mit Rücksicht auf (6b) bei Vernachlässigung von Größen dritter
und höherer Ordnung über in [11]

$$\sum_{\sigma} \frac{\Gamma^{\sigma}_{44}}{\partial x_{\sigma}} = \frac{\alpha^2}{2 r^4}.$$

Hieraus folgern wir mit Rücksicht auf (6b) und die Symmetrieeigen-
schaften unserer Lösung [12]

$$\Gamma^{\sigma}_{44} = -\frac{\alpha}{2} \frac{x_{\sigma}}{r^3} \left(1 - \frac{\alpha}{r}\right). \tag{6c}$$

§ 2. Die Planetenbewegung.

Die von der allgemeinen Relativitätstheorie gelieferten Bewegungsgleichungen des materiellen Punktes im Schwerefelde lauten

$$\frac{d^2 x_{\nu}}{ds^2} = \sum_{\sigma\tau} \Gamma^{\nu}_{\sigma\tau} \frac{dx_{\sigma}}{ds} \frac{dx_{\tau}}{ds}. \tag{7}$$

Aus diesen Gleichungen folgern wir zunächst, daß sie die NEWTONschen Bewegungsgleichungen als erste Näherung enthalten. Wenn nämlich die Bewegung des Punktes mit gegen die Lichtgeschwindigkeit kleiner Geschwindigkeit stattfindet, so sind dx_1, dx_2, dx_3 klein gegen dx_4. Folglich bekommen wir eine erste Näherung, indem wir auf der rechten Seite jeweilen nur das Glied $\sigma = \tau = 4$ berücksichtigen. Man erhält dann mit Rücksicht auf (6b)

$$\left.\begin{aligned} \frac{d^2 x_{\nu}}{ds^2} &= \Gamma^{\nu}_{44} = -\frac{\alpha}{2} \frac{x_{\nu}}{r^3} \quad (\nu = 1, 2, 3) \\ \frac{d^2 x_4}{ds^2} &= 0 \end{aligned}\right\}. \tag{7a}$$

Diese Gleichungen zeigen, daß man für eine erste Näherung $s = x_4$ setzen kann. Dann sind die ersten drei Gleichungen genau die NEWTONschen. Führt man in der Bahnebene Polargleichungen r, ϕ ein, so
liefern der Energie- und der Flächensatz bekanntlich die Gleichungen [13]

$$\left.\begin{aligned} \frac{1}{2} u^2 + \Phi &= A \\ r^2 \frac{d\phi}{ds} &= B \end{aligned}\right\}, \tag{8}$$

wobei A und B die Konstanten des Energie- bzw. Flächensatzes bedeuten, wobei zur Abkürzung

$$\left.\begin{aligned} \Phi &= -\frac{\alpha}{2r} \\ u^2 &= \frac{dr^2 + r^2 d\phi^2}{ds^2} \end{aligned}\right\} \qquad (8a)$$

gesetzt ist.

Wir haben nun die Gleichungen (7) um eine Größenordnung genauer auszuwerten. Die letzte der Gleichungen (7) liefert dann zusammen mit (6b) [14]

$$\frac{d^2 x_4}{ds^2} = 2 \sum_\tau \Gamma^4_{\sigma 4} \frac{dx_\tau}{ds} \frac{dx_4}{ds} = -\frac{dg_{44}}{ds} \frac{dx_4}{ds}$$

oder in Größen erster Ordnung genau

$$\frac{dx_4}{ds} = 1 + \frac{\alpha}{r}. \qquad (9)$$

Wir wenden uns nun zu den ersten drei Gleichungen (7). Die rechte Seite liefert [15]

a) für die Indexkombination $\sigma = \tau = 4$

$$\Gamma^\nu_{44} \left(\frac{dx_4}{ds}\right)^2$$

oder mit Rücksicht auf (6c) und (9) in Größen zweiter Ordnung genau

$$-\frac{\alpha}{2} \frac{x_\nu}{r^3} \left(1 + \frac{\alpha}{r}\right),$$

b) für die Indexkombinationen $\sigma \neq 4$ $\tau \neq 4$ (welche allein noch in Betracht kommen) mit Rücksicht darauf, daß die Produkte $\frac{dx_\sigma}{ds} \frac{dx_\tau}{ds}$ mit Rücksicht auf (8) als Größen erster Ordnung anzusehen sind[1], ebenfalls auf Größen zweiter Ordnung genau

$$-\frac{\alpha x_\nu}{r^3} \sum_{\sigma\tau} \left(\delta_{\sigma\tau} - \frac{3}{2} \frac{x_\sigma x_\tau}{r^2}\right) \frac{dx_\sigma}{ds} \frac{dx_\tau}{ds}.$$

Die Summation ergibt

$$-\frac{\alpha x_\nu}{r^3} \left(u^2 - \frac{3}{2} \left(\frac{dr}{ds}\right)^2\right).$$

---

[1] Diesem Umstand entsprechend können wir uns bei den Feldkomponenten $\Gamma^\nu_{\sigma\tau}$ mit der in Gleichung (6a) gegebenen ersten Näherung begnügen.

Mit Rücksicht hierauf erhält man für die Bewegungsgleichungen die in Größen zweiter Ordnung genaue Form

$$\frac{d^2 x_\nu}{ds^2} = -\frac{\alpha}{2}\frac{x_\nu}{r^3}\left(1 + \frac{\alpha}{r} + 2u^2 - 3\left(\frac{dr}{ds}\right)^2\right), \tag{7b}$$

welche zusammen mit (9) die Bewegung des Massenpunktes bestimmt. Nebenbei sei bemerkt, daß (7b) und (9) für den Fall der Kreisbewegung keine Abweichungen vom dritten KEPLERschen Gesetze ergeben.

Aus (7b) folgt zunächst die exakte Gültigkeit der Gleichung

$$r^2 \frac{d\phi}{ds} = B, \tag{10}$$

wobei B eine Konstante bedeutet. Der Flächensatz gilt also in Größen zweiter Ordnung genau, wenn man die »Eigenzeit« des Planeten zur Zeitmessung verwendet. Um nun die säkulare Drehung der Bahnellipse aus (7b) zu ermitteln, ersetzt man die Glieder erster Ordnung in der Klammer der sechsten Seite am vorteilhaftesten vermittels (10) und der ersten der Gleichungen (8), durch welches Vorgehen die Glieder zweiter Ordnung auf der rechten Seite nicht geändert werden. Die Klammer nimmt dadurch die Form an [16]

$$\left(1 - 2A + \frac{3B^2}{r^2}\right).$$

Wählt man endlich $s\sqrt{1 - 2A}$ als Zeitvariable, und nennt man letztere wieder s, so hat man bei etwas geänderter Bedeutung der Konstanten B: [17]

$$\left.\begin{aligned} \frac{d^2 x_\nu}{ds^2} &= -\frac{\partial \Phi}{\partial x_\nu} \\ \Phi &= -\frac{\alpha}{2}\left[1 + \frac{B^2}{r^2}\right] \end{aligned}\right\}. \tag{7c}$$

Bei der Bestimmung der Bahnform geht man nun genau vor wie im NEWTONschen Falle. Aus (7c) erhält man zunächst [18]

$$\frac{dr^2 + r^2 d\phi^2}{ds^2} = 2A - 2\Phi.$$

Eliminiert man aus dieser Gleichung ds mit Hilfe von (10), so ergibt sich, indem man mit x die Größe $\frac{1}{r}$ bezeichnet:

$$\left(\frac{dx}{d\phi}\right)^2 = \frac{2A}{B^2} + \frac{\alpha}{B^2}x - x^2 + \alpha x^3, \tag{11}$$

welche Gleichung sich von der entsprechenden der NEWTONschen Theorie nur durch das letzte Glied der rechten Seite unterscheidet.

Der vom Radiusvektor zwischen dem Perihel und dem Aphel beschriebene Winkel wird demnach durch das elliptische Integral [19]

$$\phi = \int_{\alpha_1}^{\alpha_2} \frac{dx}{\sqrt{\frac{2A}{B^2} + \frac{\alpha}{B^2}x - x^2 + \alpha x^3}},$$

wobei α_1 und α_2 diejenigen Wurzeln der Gleichung

$$\frac{2A}{B^2} + \frac{\alpha}{B^2}x - x^2 + \alpha x^3 = 0$$

bedeuten, welchen sehr benachbarte Wurzeln derjenigen Gleichung entsprechen, die aus dieser durch Weglassen des letzten Gliedes entsteht.

Hierfür kann mit der von uns zu fordernden Genauigkeit gesetzt werden [20]

$$\phi = [1 + \alpha(\alpha_1 + \alpha_2)] \cdot \int_{\alpha_1}^{\alpha_2} \frac{dx}{\sqrt{-(x-\alpha_1)(x-\alpha_2)(1-\alpha x)}},$$

oder nach Entwicklung von $(1-\alpha x)^{-\frac{1}{2}}$

$$\phi = [1 + \alpha(\alpha_1 + \alpha_2)] \int_{\alpha_1}^{\alpha_2} \frac{\left(1 + \frac{\alpha}{2}x\right)dx}{\sqrt{-(x-\alpha_1)(x-\alpha_2)}}.$$

Die Integration liefert [21]

$$\phi = \pi\left[1 + \frac{3}{4}\alpha(\alpha_1 + \alpha_2)\right],$$

oder, wenn man bedenkt, daß α_1 und α_2 die reziproken Werte der maximalen bzw. minimalen Sonnendistanz bedeuten,

$$\phi = \pi\left(1 + \frac{3}{2}\frac{\alpha}{a(1-e^2)}\right). \qquad (12)$$

Bei einem ganzen Umlauf rückt also das Perihel um

$$\varepsilon = 3\pi \frac{\alpha}{a(1-e^2)} \qquad (13)$$

im Sinne der Bahnbewegung vor, wenn mit a die große Halbachse, mit e die Exzentrizität bezeichnet wird. Führt man die Umlaufszeit T

(in Sekunden) ein, so erhält man, wenn c die Lichtgeschwindigkeit in cm/sec. bedeutet: [22]

$$\varepsilon = 24\pi^3 \frac{a^2}{T^2 c^2 (1 - e^2)} . \tag{14}$$

Die Rechnung liefert für den Planeten Merkur ein Vorschreiten des Perihels um 43″ in hundert Jahren, während die Astronomen 45″ ± 5″ als unerklärten Rest zwischen Beobachtungen und NEWTONscher Theorie angeben. Dies bedeutet volle Übereinstimmung.

Für Erde und Mars geben die Astronomen eine Vorwärtsbewegung von 11″ bzw. 9″ in hundert Jahren an, während unsere Formel nur 4″ bzw. 1″ liefert. Es scheint jedoch diesen Angaben wegen der zu geringen Exzentrizität der Bahnen jener Planeten ein geringer Wert eigen zu sein. Maßgebend für die Sicherheit der Konstatierung der Perihelbewegung ist ihr Produkt mit der Exzentrizität $\left(e\frac{d\pi}{dt}\right)$. Betrachtet man die für diese Größe von NEWCOMB angegebenen Werte

	$e\frac{d\pi}{dt}$
Merkur	8.48″ ± 0.43
Venus	−0.05 ± 0.25
Erde	0.10 ± 0.13
Mars	0.75 ± 0.35,

welche ich Hrn. Dr. FREUNDLICH verdanke, so gewinnt man den Eindruck, daß ein Vorrücken des Perihels überhaupt nur für Merkur wirklich nachgewiesen ist. Ich will jedoch ein endgültiges Urteil hierüber gerne den Fachastronomen überlassen. [23]

6.3.2 Translation

Explanation of the Perihelion Motion of Mercury from the General Theory of Relativity

In a paper recently published in these proceedings, I introduced gravitational field equations that are covariant under arbitrary transformations with determinant 1. In a supplement I showed that these equations correspond to generally covariant ones if the scalar of the energy tensor of "matter" vanishes, and I showed that the introduction of this hypothesis, which robs time and space of the last vestige of objective reality, faces no fundamental objections.[1],[1]

In this paper I find an important corroboration of this most radical theory of relativity, for it turns out that it explains qualitatively and quantitatively the secular precession of the orbit of Mercury (in the direction of the orbital motion), which was discovered by LEVERRIER and which amounts to 45″ per century.[2],[2]

It also turns out that the theory entails a bending of light rays by gravitational fields that is twice the size of that found in my earlier investigations.

§ 1. The Gravitational Field

From my last two communications it follows that the gravitational field in a vacuum has to satisfy, given a properly chosen reference frame, the equations

$$\sum_{\alpha} \frac{\partial \Gamma^{\alpha}_{\mu\nu}}{\partial x_{\alpha}} + \sum_{\alpha\beta} \Gamma^{\alpha}_{\mu\beta} \Gamma^{\beta}_{\nu\alpha} = 0, \tag{1}$$

where $\Gamma^{\alpha}_{\mu\nu}$ is defined by the equations

$$\begin{aligned} \Gamma^{\alpha}_{\mu\nu} &= -\begin{Bmatrix} \mu\nu \\ \alpha \end{Bmatrix} = -\sum_{\beta} g^{\alpha\beta} \begin{bmatrix} \mu\nu \\ \beta \end{bmatrix} \\ &= -\frac{1}{2} \sum_{\beta} g^{\alpha\beta} \left(\frac{\partial g_{\mu\beta}}{\partial x_{\nu}} + \frac{\partial g_{\nu\beta}}{\partial x_{\mu}} - \frac{\partial g_{\mu\nu}}{\partial x_{\beta}} \right). \end{aligned} \tag{2}$$

If, moreover, we adopt the hypothesis argued for in my last communication, that the scalar of the energy tensor of "matter" always vanishes, we get an additional equation for the determinant:

---

[1] In a forthcoming communication it will be shown that one can dispense with this hypothesis. The only thing that matters is that it is possible to choose the reference frame in such a way that the determinant $|g_{\mu\nu}|$ takes on the value -1. The following investigation is independent of this.

[2] E. FREUNDLICH recently wrote a noteworthy article on the impossibility of satisfactorily explaining the anomalies in the motion of Mercury on the basis of the NEWTONian theory (*Astronomische Nachrichten 201*, 49 [1915]).

$$|g_{\mu\nu}| = -1. \tag{3}$$

Suppose a point mass (the sun) is located at the origin of the coordinate system. The gravitational field this point mass produces can be calculated from these equations through iterative approximation.[3]

One should keep in mind, however, that the $g_{\mu\nu}$ for a given solar mass are not fully determined mathematically by equations (1) and (3). This is because these equations are covariant under arbitrary transformations with determinant 1. Yet it might be justified to assume that all these solutions can be reduced to one another by such transformations and that (for given boundary conditions) they differ only formally, not physically. Following this conviction, I am satisfied for the time being to derive *one* solution, without entering into a discussion of whether it is the only possible one.[4]

We proceed as follows. Let the $g_{\mu\nu}$ in "zeroth approximation" be given by the following scheme corresponding to the original theory of relativity:

$$\left.\begin{matrix} -1 & 0 & 0 & 0 \\ 0 & -1 & 0 & 0 \\ 0 & 0 & -1 & 0 \\ 0 & 0 & 0 & +1 \end{matrix}\right\}, \tag{4}$$

or, abbreviated,

$$\left.\begin{aligned} g_{\rho\sigma} &= \delta_{\rho\sigma} \\ g_{\rho 4} = g_{4\rho} &= 0 \\ g_{44} &= 1 \end{aligned}\right\}. \tag{4a}$$

Here ρ and σ refer to the indices 1, 2, 3; $\delta_{\rho\sigma}$ is equal to 1 or to 0 depending on whether $\rho = \sigma$ or $\rho \neq \sigma$.

In the following, we assume that the $g_{\mu\nu}$ only differ from the values in equation (4a) by quantities that are small compared to unity. We will treat these deviations as small quantities of "first order" and functions of the nth degree in these deviations as "quantities of nth order". Equations (1) and (3) enable us, starting with equation (4a), to calculate through successive approximations the gravitational field up to quantities of nth order. The equation (4a) form the "zeroth approximation".[5]

The solution given below has the following properties, which fix the coordinate system:

1. All components are independent of x_4.
2. The solution is spatially symmetric around the origin of the coordinate system, in the sense that we get back to the same solution if we subject it to a linear orthogonal (spatial) transformation.
3. The equations $g_{\rho 4} = g_{4\rho} = 0$ hold exactly (for $\rho = 1,2,3$).
4. At infinity, $g_{\mu\nu}$ has the values in equation (4a).

First approximation

It is easy to verify that, up to quantities of first order, equations (1) and (3) as well as the four conditions just listed are satisfied if we posit[6]

$$\left.\begin{aligned} g_{\rho\sigma} &= -\delta_{\rho\sigma} + \alpha\left(\frac{\partial^2 r}{\partial x_\rho \partial x_\sigma} - \frac{\delta_{\rho\sigma}}{r}\right) = -\delta_{\rho\sigma} - \alpha\frac{x_\rho x_\sigma}{r^3} \\ g_{44} &= 1 - \frac{\alpha}{r} \end{aligned}\right\}. \tag{4b}$$

The $g_{4\rho}$ and $g_{\rho 4}$ are fixed by condition 3, r stands for the quantity $+\sqrt{x_1^2+x_2^2+x_3^2}$, and α is a constant determined by the mass of the sun.

One immediately sees that condition 3 is satisfied to first order. For a simple way to see that the field equations (1) are also satisfied in first approximation, one only needs to note that, if quantities of second and higher order are neglected, the left-hand side of equation (1) can be replaced by[7]

$$\sum_\alpha \frac{\partial \Gamma^\alpha_{\mu\nu}}{\partial x_\alpha}$$

$$\sum_\alpha \frac{\partial}{\partial x_\alpha}\begin{bmatrix}\mu\nu \\ \alpha\end{bmatrix},$$

where α only runs from 1 to 3.

As one sees from Eq. (4b), our theory implies that in the case of a mass at rest, the components g_{11} through g_{33} already differ from zero in terms of quantities of first order.[8] As we shall see later, no discrepancy with Newton's law (in first approximation) arises from this. The theory, however, does give a slightly different result for the effect of a gravitational field on a light ray than my earlier papers. The velocity of light is determined by the equation

$$\sum g_{\mu\nu} dx_\mu dx_\nu = 0. \tag{5}$$

Applying Huygens's principle, we find from equations (5) and (4b), after a simple calculation, that a light ray passing by the sun at a distance Δ experiences an angular deflection of magnitude $\frac{2\alpha}{\Delta}$, while my earlier calculation, which was not based on the hypothesis $\sum T^\mu_\mu = 0$, led to the value $\frac{\alpha}{\Delta}$. A light ray grazing the surface of the sun should experience a deflection of 1.7″ (instead of 0.85″).[9] By contrast, the result concerning the shift of spectral lines by the gravitational potential, the order of magnitude of which Mr. Freundlich has confirmed for the fixed stars, remains unaffected, because this depends only on g_{44}.[10]

Now that we have obtained the $g_{\mu\nu}$ in first approximation, we can also calculate the components $\Gamma^\alpha_{\mu\nu}$ of the gravitational field in first approximation. From equations (2) and (4b) we have

$$\Gamma^{\tau}_{\rho\sigma} = -\alpha\left(\delta_{\rho\sigma}\frac{x_\tau}{r^3} - \frac{3x_\rho x_\sigma x_\tau}{2r^5}\right) \tag{6a}$$

where ρ, σ, τ take on the values 1, 2, 3, and

$$\Gamma^{\sigma}_{44} = \Gamma^{4}_{4\sigma} = -\frac{\alpha x_\sigma}{2r^3}, \tag{6b}$$

where σ takes on the values 1, 2, 3. Those components in which the index 4 appears once or three times vanish.

SECOND APPROXIMATION

As we will see later, we only need to determine the three components Γ^{σ}_{44} to second order to determine the orbits of the planets to the same degree of accuracy. For this purpose, the last field equation suffices along with the general conditions we have imposed on our solution. If we take into account equation (6b) and neglect quantities of third and higher order, the last field equation,

$$\sum_{\sigma}\frac{\partial\Gamma^{\sigma}_{44}}{\partial x_\sigma} + \sum_{\sigma\tau}\Gamma^{\sigma}_{4\tau}\Gamma^{\tau}_{4\sigma} = 0,$$

turns into[11]

$$\sum_{\sigma}\frac{\partial\Gamma^{\sigma}_{44}}{\partial x_\sigma} = -\frac{\alpha^2}{2r^4}.$$

From this we deduce, using equation (6b) and the symmetry properties of our solution,[12]

$$\Gamma^{\sigma}_{44} = -\frac{\alpha}{2}\frac{x_\sigma}{r^3}\left(1 - \frac{\alpha}{r}\right). \tag{6c}$$

§ 2. The Motion of the Planets

The equations of motion of a point mass in a gravitational field given by the general theory of relativity is

$$\frac{d^2x_\nu}{ds^2} = \sum_{\sigma\tau}\Gamma^{\nu}_{\sigma\tau}\frac{dx_\sigma}{ds}\frac{dx_\tau}{ds}. \tag{7}$$

We first show that these equations contain the NEWTONian equations of motion in first approximation. If the motion of the planet is with a velocity that is small compared to the velocity of light, then dx_1, dx_2 and dx_3 are small compared to dx_4. We thus obtain a first approximation by keeping only the term with $\sigma = \tau = 4$ on the right-hand side. Using equation (6b), we then find:

$$\left.\begin{aligned} \frac{d^2x_\nu}{ds^2} &= \Gamma^{\nu}_{44} = -\frac{\alpha}{2}\frac{x_\nu}{r^3} \quad (\nu = 1,2,3) \\ \frac{d^2x_4}{ds^2} &= 0 \end{aligned}\right\}. \tag{7a}$$

These equations show that we can set $s = x_4$ in first approximation. The first three equations are thus exactly the NEWTONian equations. If we introduce polar coordinates r, φ in the orbital plane, then, as is well known, the energy law and the area law give the equations[13]

$$\left.\begin{aligned} \frac{1}{2}u^2 + \Phi &= A \\ r^2 \frac{d\varphi}{ds} &= B \end{aligned}\right\}, \tag{8}$$

where A and B are the constants of the energy law and the area law, respectively, and where we used the abbreviations

$$\left.\begin{aligned} \Phi &= -\frac{\alpha}{2r} \\ u^2 &= \frac{dr^2 + r^2 d\varphi^2}{ds^2} \end{aligned}\right\} \tag{8a}$$

We now have to evaluate the equations to the next order. The last of the equations (7) then yields, together with equation (6b),[14]

$$\frac{d^2 x_4}{ds^2} = 2\sum_\sigma \Gamma^4_{\sigma 4} \frac{dx_\sigma}{ds}\frac{dx_4}{ds} = -\frac{dg_{44}}{ds}\frac{dx_4}{ds},$$

or, correct to the first order,

$$\frac{dx_4}{ds} = 1 + \frac{a}{r}. \tag{9}$$

We now turn to the first of the three equations (7).[15] The right-hand side gives:

a) for the index combination $\sigma = \tau = 4$

$$\Gamma^\nu_{44}\left(\frac{dx_4}{ds}\right)^2,$$

or, with the help equations (6c) and (9), to second order:

$$-\frac{\alpha}{2}\frac{x_\nu}{r^3}\left(1 + \frac{\alpha}{r}\right);$$

b) for the index combination $\sigma \neq 4$, $\tau \neq 4$ (which are the only ones we still need to consider), given that the products $\frac{dx_\sigma}{ds}\frac{dx_\tau}{ds}$ should, on the basis of equation (8), be seen as quantities of first order,[1] likewise up to quantities

[1] Because of this we only need the field components $\Gamma^\nu_{\sigma\tau}$ to first order as given in equation (6a).

of second order:

$$-\frac{\alpha x_\nu}{r^3}\sum\left(\delta_{\sigma\tau}-\frac{3}{2}\frac{x_\sigma x_\tau}{r^2}\right)\frac{dx_\sigma}{ds}\frac{dx_\tau}{ds}.$$

The summation gives

$$-\frac{\alpha x_\nu}{r^3}\left(u^2-\frac{3}{2}\left(\frac{dr}{ds}\right)^2\right).$$

Using these results we obtain the equations of motion in a form correct to second order,

$$\frac{d^2x_\nu}{ds^2}=-\frac{\alpha}{2}\frac{x_\nu}{r^3}\left(1+\frac{\alpha}{r}+2u^2-3\left(\frac{dr}{ds}\right)^2\right), \tag{7b}$$

which together with equation (9) determine the motion of the mass point. As an aside, I note that equations (7b) and (9) do give deviations from KEPLER's third law in the case of circular motion.

From equation (7b) follows, first of all, the exact validity of the equation

$$r^2\frac{d\varphi}{ds}=B, \tag{10}$$

where B is a constant. The area law therefore remains valid to second order if we use the planet's "proper time" to measure time. To determine the secular rotation of the orbital ellipse from equation (7b), it is most convenient to replace the quantities of first order in the expression in parentheses on the right-hand side with the help of (10) and the first of equations (8), which does not affect quantities of second order on the right-hand side. The expression in parentheses then takes the form[16]

$$\left(1-2A+\frac{3B^2}{r^2}\right).$$

Finally, if we choose $s\sqrt{(1-2A)}$ as the time variable and call that variable s again, we have, with a slightly different definition of the constant B;[17]

$$\left.\begin{aligned}\frac{d^2x_\nu}{ds^2}&=-\frac{\partial\Phi}{\partial x_\nu}\\ \Phi&=-\frac{\alpha}{2r}\left[1+\frac{B^2}{r^2}\right]\end{aligned}\right\}. \tag{7c}$$

To determine the equation of the orbit, we now proceed exactly as in the NEWTONian case. From equation (7c) we obtain first[18]

$$\frac{dr^2+r^2d\varphi^2}{ds^2}=2A-2\Phi.$$

If we eliminate ds from this equation with the help of (10), we find

$$\left(\frac{dx}{d\varphi}\right)^2 = \frac{2A}{B^2} + \frac{\alpha}{B^2}x - x^2 + \alpha x^3, \tag{11}$$

where x is defined as $1/r$. The only difference between this equation differs and the corresponding one in NEWTONian theory is the last term on the right-hand side.

The angle traversed by the radius vector between perihelion and aphelion is therefore given by the integral for the ellipse[19]

$$\varphi = \int_{\alpha_1}^{\alpha_2} \frac{dx}{\sqrt{\frac{2A}{B^2} + \frac{\alpha}{B^2}x - x^2 + \alpha x^3}},$$

where α_1 and α_2 are the roots of the equation

$$\frac{2A}{B^2} + \frac{\alpha}{B^2}x - x^2 + \alpha x^3 = 0,$$

which will be very close to the roots of the equation obtained if the last term is omitted.

To the degree of accuracy required, this angle is given by[20]

$$\varphi = \left[1 + \frac{\alpha(\alpha_1 + \alpha_2)}{2}\right] \cdot \int_{\alpha_1}^{\alpha_2} \frac{dx}{\sqrt{-(x-\alpha_1)(x-\alpha_2)(1-\alpha x)}},$$

or, upon expansion of $(1-\alpha x)^{-1/2}$,

$$\varphi = \left[1 + \frac{\alpha(\alpha_1 + \alpha_2)}{2}\right] \cdot \int_{\alpha_1}^{\alpha_2} \frac{\left(1 + \frac{\alpha}{2}x\right) dx}{\sqrt{-(x-\alpha_1)(x-\alpha_2)}}.$$

The result of the integration is[21]

$$\varphi = \pi\left[1 + \frac{3}{4}\alpha(\alpha_1 + \alpha_2)\right],$$

or if we bear in mind that α_1 and α_2 are the reciprocals of the maximum and minimum distance from the sun, respectively,

$$\varphi = \pi\left(1 + \frac{3}{2}\frac{\alpha}{a(1-e^2)}\right). \tag{12}$$

In one complete orbit, the perihelion thus advances by

$$\varepsilon = 3\pi \frac{\alpha}{\alpha(1-e^2)} \tag{13}$$

in the direction of the orbital motion, where a is the semimajor axis and e is the eccentricity. If we introduce the orbital period T (in seconds) and the velocity of light in cm/sec., we obtain[22]

$$\varepsilon = 24\pi^3 \frac{\alpha^2}{T^2c^2(1-e^2)}. \tag{14}$$

Calculation gives a perihelion advance of 43″ per century for the planet Mercury, while astronomers find an unexplained difference between observations and NEWTONian theory of 45″ ± 5″ per century. Hence there is now complete agreement.

For Earth and Mars, astronomers have found advances of 11″ and 9″ per century, respectively, whereas our formula gives 4″ and 1″, respectively. However, it would seem that little weight is to be attached to these values because the eccentricity of the orbits of these planets is too small. What determines the reliability of the observation of the perihelion motion is the product of this motion and the eccentricity $\left(e\frac{d\pi}{dt}\right)$. Considering the values for these quantities given by NEWCOMB,

	$e\frac{d\pi}{dt}$
Mercury	8.48″ ± 0.43
Venus	−0.05 ± 0.25
Earth	0.10 ± 0.13
Mars	0.75 ± 0.35,

which I owe to Dr. FREUNDLICH,[23] one gets the impression that an advance of the perihelion has clearly been demonstrated only for Mercury. A definitive verdict on these matters, however, I am happy to leave to professional astronomers.

6.3.3 Commentary

This paper is the written version of a lecture delivered to the Prussian Academy of Sciences, Berlin, November 18, 1915. It was published November 25, 1915. It is presented in facsimile as Doc. 24 in CPAE6. Our commentary is based on Earman and Janssen (1993, secs. 5–7, pp. 138–158). In the translation, we corrected a number of typos in the German original. Some more important ones are noted in the annotation (see notes 7, 11, 17 and 20). Our translation follows the one by Brian Doyle used in the companion volume to CPAE6, originally published in Lang and Gingerich (1979).

Introduction

1. In his first November 1915 paper, Einstein (1915a) had found a condition that prohibited the use of unimodular coordinates even though these are the natural coordinates to use in the theory covariant under unimodular transformations that Einstein proposed in this paper. In the second and fourth paper, Einstein (1915b, 1915d) proposed ways around this prohibition, by setting $T = 0$ and adding a term with T to the field equations, respectively. This allowed him to look upon the (amended) field equations of the first paper as generally covariant equations expressed in unimodular coordinates. See Sec. 6.1.3, note 18, Sec. 6.2.3, note 1, Sec. 6.4.3, note 1 and Pt. I, Sec. 5.4 for further discussion.
2. In this paper, Einstein's protégé Freundlich (1915a) criticized a paper on the perihelion problem by a leading astronomer (Seeliger 1915, cf. Sec. 4.3, note 1). For the history of the perihelion problem, see Roseveare (1982).

§ 1

3. This iterative approximation procedure follows the one used in the Einstein-Besso manuscript (see Ch. 2, [pp. 1, 3, 4, 6 and 7]).
4. In a letter to Einstein of January 1, 1916 (CPAE8, Doc. 177a in CPAE13), Ehrenfest quoted the last two sentences of this paragraph, calling them "your incantation" (*Deine Beschwörungsformel*) and adding that "Leiden takes note of this with an icy-polite smile but does not endorse it." In the Einstein-Besso manuscript, Besso already raised the question whether the solution for the static field of the sun is unique (CPAE4, Doc. 14, [p. 16]). This question may have triggered the "hole argument" (see Pt. I, Sec. 4.1) in the original form in which it can be found in the Besso memo of August 28, 1913 (Janssen 2007, p. 820). Ehrenfest's skeptical comment was actually made in the context of a discussion of the hole argument. In a letter to Ehrenfest of January 16, 1916, also in the context of the hole argument (see Pt. I, Sec. 4.1, note 5), Lorentz also questioned the uniqueness of the solution of the field equations, both in the specific case of the field of the sun and in general (Kox 2018, Doc. 179). Birkhoff (1923) would show that the Schwarzschild solution, the exact solution corresponding to the approximate one Einstein gives here (see note 6) is the unique spherically symmetric vacuum solution of the Einstein field equations (for an elementary proof of this result, known as Birkhoff's theorem, see Carroll 2004, sec. 5.2, pp. 197–204).
5. The actual calculations make it clear that Einstein is assuming that $g_{\mu\nu}$ can be written as a power series

$$g_{\mu\nu} = \overset{(0)}{g}_{\mu\nu} + \overset{(1)}{g}_{\mu\nu} + \overset{(2)}{g}_{\mu\nu} + \ldots,$$

as he did in the Einstein-Besso manuscript (see our commentary on [p. 1] in Sec. 2.3 and Earman and Janssen 1993, pp. 142–148). In the Einstein-Besso manuscript, the components of the metric, the potential for the gravitational field, are calculated to second order; in this paper, the components of gravitational field itself, defined as minus the Christoffel symbols, are.

6. Derivations of the general form of a static spherically symmetric metric in Cartesian coordinates compatible with Newtonian theory for weak fields can be found, e.g., on [p. 7] of the Einstein-Besso manuscript (see Ch. 2) and in Droste (1915, pp. 999–1000). Here we follow Earman and Janssen (1993, p. 144, though our choice of the symbols R, T and Φ below follows the notation used in the Einstein-Besso manuscript; T is not to be confused with the trace of the energy-momentum tensor for matter). To find the metric at some arbitrary point P with spatial coordinates (x,y,z) in some arbitrary Cartesian coordinate system in a spherically symmetric space-time, we rotate the coordinate system such that the x-axis of the new coordinate system goes through P. The spatial coordinates of P in this new coordinate system are $(x',y',z') = (r,0,0)$, where $r \equiv \sqrt{x^2+y^2+z^2}$. Because of spherical symmetry, the metric in primed coordinates has the simple form $g'_{\mu\nu} = \mathrm{diag}(R,T,T,\Phi)$, where R, T and Φ are yet to be determined functions of the coordinates. Transforming back to the coordinate system we started from (cf. Sec. 2.3, commentary on [p. 7]), we find a metric of the form:

$$g_{\mu\nu} = T\,\delta_{ij} + \frac{x^i x^j}{r^2}(R-T), \quad g_{i4} = g_{4i} = 0, \quad g_{44} = \Phi$$

$(i,j = 1,2,3)$. Since the space-time is Minkowskian at spatial infinity, T must be equal to -1. To recover the Newtonian theory for weak static fields, Φ must be equal to $1 - \frac{\alpha}{r}$ with $\alpha \equiv \frac{2GM}{c^2}$ (where G is Newton's gravitational constant, M the mass of the sun and c the velocity of light). To ensure that $g = RT^2\Phi = -1$ to first order in $\frac{\alpha}{r}$, R must be equal to $-\left(1+\frac{\alpha}{r}\right)$, which means that $R - T = -\frac{\alpha}{r}$. We thus arrive at Eq. (4b):

$$g_{ij} = -\delta_{ij} - \alpha\frac{x^i x^j}{r^3}, \quad g_{44} = 1 - \frac{\alpha}{r}. \tag{4b}$$

The line element corresponding to this metric field is

$$ds^2 = \left(1-\frac{\alpha}{r}\right)c^2dt^2 - \sum_{i,j=1}^{3}\left(\delta_{ij} + \alpha\frac{x^i x^j}{r^3}\right)dx^i dx^j.$$

In spherical coordinates (r,ϑ,φ), which Einstein could not use because the transformation from Cartesian to spherical coordinates does not have determinant 1, this line element becomes

$$ds^2 = \left(1-\frac{\alpha}{r}\right)c^2dt^2 - \left(1+\frac{\alpha}{r}\right)dr^2 - r^2\left(d\vartheta^2 + \sin^2\vartheta\, d\varphi^2\right),$$

which, to order $\frac{\alpha}{r}$, is just the line element of the Schwarzschild solution,

$$ds^2 = \left(1-\frac{\alpha}{r}\right)c^2dt^2 - \frac{1}{1-\frac{\alpha}{r}}dr^2 - r^2\left(d\vartheta^2 + \sin^2\vartheta\, d\varphi^2\right).$$

7. To check that the metric field in Eq. (4b) is a solution of $\Gamma^\alpha_{\mu\nu,\alpha} = 0$, the vacuum field equations to first order in α/r, we first need to verify Eqs. (6a) and (6b) for the components of the gravitational field by inserting Eq. (4b) for $g_{\mu\nu}$ into the definition of $\Gamma^\alpha_{\mu\nu}$ in Eq. (2):

$$\Gamma^\alpha_{\mu\nu} = -\frac{1}{2}g^{\alpha\rho}\left(g_{\rho\mu,\nu} + g_{\rho\nu,\mu} - g_{\mu\nu,\rho}\right). \tag{2}$$

As Einstein notes in the two lines below Eqs. (6a) and (6b), $\Gamma^\alpha_{\mu\nu}$ is zero if either one of its indices is 4 or all three of them are. Since $g^{ij} \approx -\delta^{ij}$ and $g^{44} \approx 1$ and since $g_{i4} = 0$

and $g_{\mu\nu,4}=0$, we have

$$\Gamma^i_{j4}=\Gamma^i_{4j}=\frac{1}{2}\delta^{ik}\left(g_{kj,4}+g_{k4,j}-g_{j4,k}\right)=0,$$

$$\Gamma^4_{ij}=-\frac{1}{2}\left(g_{4i,j}+g_{4j,i}-g_{ij,4}\right)=0,\quad \Gamma^4_{44}=-\frac{1}{2}g_{44,4}=0.$$

If two of the three indices of $\Gamma^\alpha_{\mu\nu}$ are 4, we have $\Gamma^i_{44}=\frac{1}{2}\left(2g_{i4,4}-g_{44,i}\right)=-\frac{1}{2}g_{44,i}$ and $\Gamma^4_{4i}=-\frac{1}{2}g_{44,i}$. Inserting $g_{44}=1-\frac{\alpha}{r}$ on the right-hand sides of these equations, we arrive at Eq. (6b):

$$\Gamma^i_{44}=\Gamma^4_{4i}=\frac{1}{2}\frac{\partial}{\partial x^i}\left(\frac{\alpha}{r}\right)=-\frac{\alpha}{2r^2}\frac{\partial r}{\partial x^i}=-\frac{\alpha}{2}\frac{x^i}{r^3}. \tag{6b}$$

That leaves the case where none of the indices are 4. In that case,

$$\begin{aligned}\Gamma^k_{ij}&=\frac{1}{2}\delta^{kl}\left(g_{li,j}+g_{lj,i}-g_{ij,l}\right)\\&=-\frac{\alpha}{2}\left(\frac{\partial}{\partial x^j}\left(\frac{x^kx^i}{r^3}\right)+\frac{\partial}{\partial x^i}\left(\frac{x^kx^j}{r^3}\right)-\frac{\partial}{\partial x^k}\left(\frac{x^ix^j}{r^3}\right)\right).\end{aligned}$$

Evaluating the derivatives, we find:

$$\begin{aligned}\Gamma^k_{ij}=-\frac{\alpha}{2}&\left\{\left(\delta^k_jx^i+\delta^i_jx^k+\delta^k_ix^j+\delta^j_ix^k-\delta^i_kx^j-\delta^j_kx^i\right)\frac{1}{r^3}\right.\\&\left.+x^kx^i\frac{\partial}{\partial x^j}\left(\frac{1}{r^3}\right)+x^kx^j\frac{\partial}{\partial x^i}\left(\frac{1}{r^3}\right)-x^ix^j\frac{\partial}{\partial x^k}\left(\frac{1}{r^3}\right)\right\}.\end{aligned}$$

The six terms in parentheses on the first line add up to $2\delta_{ij}x^k$. Noting that

$$\frac{\partial}{\partial x^j}\left(\frac{1}{r^3}\right)=-\frac{3}{r^4}\frac{\partial r}{\partial x^j}=-\frac{3x^j}{r^5},$$

we see that the three terms on the second line are the same, except for the minus sign in front of the third. We thus arrive at Eq. (6a) (after correcting a typo: r^2 in the first term should be r^3):

$$\Gamma^k_{ij}=-\alpha\left(\delta_{ij}\frac{x^k}{r^3}-\frac{3}{2}\frac{x^ix^jx^k}{r^5}\right). \tag{6a}$$

Inserting Eqs. (6a) and (6b) into $\Gamma^\alpha_{\mu\nu,\alpha}=0$, we now verify that the metric in Eq. (4b) is indeed a solution of the vacuum field equations to first order in α/r. For the index combinations $\mu\nu=4i$ (or, equivalently, $i4$), we get $\Gamma^\alpha_{\mu\nu,\alpha}=\Gamma^4_{4i,4}+\Gamma^j_{4i,j}$, which vanishes, since $\Gamma^\alpha_{\mu\nu}$ is time-independent and $\Gamma^j_{4i}=0$. For the index combination $\mu\nu=44$, we find

$$\Gamma^i_{44,i}=-\frac{\alpha}{2}\frac{\partial}{\partial x^i}\left(\frac{x^i}{r^3}\right)=-\frac{\alpha}{2}\left(\frac{3}{r^3}+x^i\frac{\partial}{\partial x^i}\left(\frac{x^i}{r^3}\right)\right)=0,$$

where in the last step we used that $\frac{\partial}{\partial x^i}\left(\frac{x^i}{r^3}\right)=-\frac{3x^i}{r^5}$ and that $\sum_{i=1}^{3}(x^i)^2=r^2$. That leaves the index combinations $\mu\nu=ij$:

$$\Gamma^k_{ij,k} = -\alpha\left(\delta_{ij}\frac{\partial}{\partial x^k}\left(\frac{x^k}{r^3}\right) - \frac{3}{2}\frac{\partial}{\partial x^k}\left(\frac{x^i x^j x^k}{r^5}\right)\right).$$

The first term on the right-hand side vanishes, as we just saw. The derivative in the second term gives:

$$\frac{\partial}{\partial x^k}\left(\frac{x^i x^j x^k}{r^5}\right) = \left(\delta^{ik}x^j x^k + \delta^{jk}x^i x^k + 3x^i x^j\right)\frac{1}{r^5} - \frac{5x^i x^j x^k}{r^6}\frac{\partial r}{\partial x^k}.$$

The first term on the right-hand side gives $\dfrac{5x^i x^j}{r^5}$, which cancels against the last term, since $\dfrac{\partial r}{\partial x^k} = \dfrac{x^k}{r}$ and $\sum_k (x^k)^2 = r^2$. Hence, $\Gamma^\alpha_{\mu\nu,\alpha} = 0$ for all values of μ and ν. The metric in Eq. (4b) is a solution of the vacuum field equations to order α/r.

8. Einstein mentioned this feature in three of the four postcards to Besso in which he reported his success in accounting for the perihelion motion of Mercury after the failure of their efforts to do so on the basis of the *Entwurf* theory recorded in the Einstein-Besso manuscript (see Ch. 2 and Pt. I, Ch. 6). As he told Besso in the fourth of these, on January 3, 1916: "That the effect is so much larger than in our calculation is because in the new theory the g_{11}–g_{33} appear in first order and thus contribute to the perihelion motion" (CPAE8, Doc. 178). Ever since he had started working on his metric theory of gravity, Einstein had expected weak static fields to be represented by a spatially flat metric (cf. Sec. 1.3, note 16, and Pt. I, Ch. 3 and Sec. 5.3, especially note 12). His calculation of the perihelion motion of Mercury showed him that this need not be the case. This realization removed obstacles to (a) the use of unimodular coordinates (see Sec. 6.2.3, note 4) and (b) the addition of a term with the trace of the energy-momentum tensor to the field equations (see Sec. 6.4.3, note 3).
9. Einstein (1911) derived the original value for the bending of light in the field of the sun directly from the equivalence principle. In the review article on *Entwurf* theory the previous year, Einstein (1914c, p. 1084) indicated how this result is recovered in that theory. In the review article on the new theory the following year, Einstein (1916b, pp. 821–822) went through the derivation of the result of the new theory. At this point, Einstein assumed that the use of unimodular coordinates required that he set the trace T of the energy-momentum tensor for matter equal to zero. He would soon discover that he could use these coordinates without putting any restrictions on T. For accounts of the 1919 (and subsequent) eclipse expeditions in which Einstein's new prediction for light bending was tested, see Kennefick (2019) and Crelinsten (2006).
10. For discussion of Freundlich's controversial claim to have confirmed Einstein's prediction of a gravitational redshift, see Sec. 4.3, note 1, and Hentschel (1992, 1994).
11. The 44-component of the field equations to second order in $\dfrac{\alpha}{r}$ is

$$\overset{(2)}{\Gamma}{}^i_{44,i} + \overset{(1)}{\Gamma}{}^\alpha_{4\beta}\overset{(1)}{\Gamma}{}^\beta_{4\alpha} = 0,$$

where we have enhanced Einstein's notation to indicate to what order the different Γ's need to evaluated. Using the expressions we found in note 7 for $\overset{(1)}{\Gamma}{}^\alpha_{\mu\nu}$ (note that $\overset{(0)}{\Gamma}{}^\alpha_{\mu\nu} = 0$) and recalling, in particular, that this quantity vanishes if one or three of its indices are 4, we have:

$$\overset{(1)}{\Gamma}{}^\alpha_{4\beta}\overset{(1)}{\Gamma}{}^\beta_{4\alpha} = \overset{(1)}{\Gamma}{}^i_{44}\overset{(1)}{\Gamma}{}^4_{4i} + \overset{(1)}{\Gamma}{}^4_{4i}\overset{(1)}{\Gamma}{}^i_{44} = 2\sum_{i=1}^{3}\left(-\frac{\alpha}{2}\frac{x^i}{r^3}\right)\left(-\frac{\alpha}{2}\frac{x^i}{r^3}\right) = \frac{\alpha^2}{2r^4},$$

where we used Eq. (6b) for $\overset{(1)}{\Gamma}{}^4_{4i} = \overset{(1)}{\Gamma}{}^i_{44}$ and $\sum_i (x^i)^2 = r^2$. We thus arrive at the equation above Eq. (6c) (after correcting a typo: there is a minus sign missing on the right-hand side):

$$\Gamma^i_{44,i} = -\frac{\alpha^2}{2r^4}.$$

We already saw that $\overset{(1)}{\Gamma}{}^i_{44,i} = 0$ (see note 7) so Γ^i_{44} can be replaced by $\overset{(2)}{\Gamma}{}^i_{44}$ in this equation.

12. We check that

$$\Gamma^i_{44} = \overset{(1)}{\Gamma}{}^i_{44} + \overset{(2)}{\Gamma}{}^i_{44} = -\frac{\alpha}{2}\frac{x^i}{r^3}\left(1 - \frac{\alpha}{r}\right) \tag{6c}$$

is a solution of the 44-component of the field equations to second order in $\frac{\alpha}{r}$ (see note 11). Using that $\overset{(1)}{\Gamma}{}^i_{44,i} = 0$ (see note 7) and inserting $\overset{(2)}{\Gamma}{}^i_{44} = \frac{\alpha^2}{2}\frac{x^i}{r^4}$ on the left-hand side of this component of the field equations, we find:

$$\begin{aligned}
\overset{(2)}{\Gamma}{}^i_{44,i} &= \frac{\alpha^2}{2}\sum_{i=1}^{3}\frac{\partial}{\partial x^i}\left(\frac{x^i}{r^4}\right) = \frac{\alpha^2}{2}\left(\frac{3}{r^4} - \sum_{i=1}^{3}\frac{4x^i}{r^5}\frac{x^i}{r}\right) \\
&= \frac{\alpha^2}{2}\left(\frac{3}{r^4} - \frac{4}{r^4}\right) = -\frac{\alpha^2}{2}\frac{1}{r^4},
\end{aligned}$$

which is in accordance with the right-hand side of this component of the field equations (see note 11).

§ 2

13. Einstein's analysis of the slow motion of a planet in the spherically symmetric weak static field of the sun in this section closely follows the analysis of the same problem in the context of the *Entwurf* theory in the Einstein-Besso manuscript (cf. Ch. 2, [pp. 8–11, 14 and 26]). As we noted in our commentary in Sec. 2.3 on [p. 8] of the manuscript, where Besso derived the analogue of Eq. (8) in this paper for the conservation of energy and angular momentum (the area law), it follows from the virial theorem that potential and kinetic energy, and hence the quantities $\frac{\alpha}{r}$ and $\left(\frac{dx^i}{ds}\right)^2$, are of the same order of magnitude, as Einstein notes explicitly under point (b) on p. 836.

14. The only contribution to the $\nu = 4$ component of Eq. (7) of second order in $\frac{\alpha}{r}$ comes from the index combinations $\sigma\tau = i4$ and $\sigma\tau = 4i$. Using Eq. (6b) for $\overset{(1)}{\Gamma}{}^4_{i4}$, we find that

$$\frac{d^2x^4}{ds^2} = 2\overset{(1)}{\Gamma}{}^4_{i4}\frac{dx^i}{ds}\frac{dx^4}{ds} = -\alpha\sum_{i=1}^{3}\frac{x^i}{r^3}\frac{dx^i}{ds}\frac{dx^4}{ds}.$$

Since $\frac{dx^4}{ds} = 1$ to zeroth order and

$$\sum_{i=1}^{3} x^i\frac{dx^i}{ds} = \frac{1}{2}\frac{d}{ds}\left(\sum_{i=1}^{3}(x^i)^2\right) = \frac{1}{2}\frac{dr^2}{ds} = r\frac{dr}{ds},$$

this reduces, to first order in $\frac{\alpha}{r}$, to

$$\frac{d}{ds}\left(\frac{dx^4}{ds}\right) = -\frac{\alpha}{r^2}\frac{dr}{ds} = \frac{d}{ds}\left(\frac{\alpha}{r}\right),$$

from which, again to first order, Einstein's Eq. (9) follows

$$\frac{dx^4}{ds} = 1 + \frac{\alpha}{r}. \tag{9}$$

15. To second order in $\frac{\alpha}{r}$ the spatial components of the equations of motion in Eq. (7) are

$$\frac{d^2x^i}{ds^2} = \left(\overset{(1)}{\Gamma}{}^i_{44} + \overset{(2)}{\Gamma}{}^i_{44}\right)\frac{dx^4}{ds}\frac{dx^4}{ds} + 2\overset{(1)}{\Gamma}{}^i_{4j}\frac{dx^4}{ds}\frac{dx^j}{ds} + \overset{(1)}{\Gamma}{}^i_{jk}\frac{dx^j}{ds}\frac{dx^k}{ds}.$$

The second term on the right-hand side vanishes because $\overset{(1)}{\Gamma}{}^i_{4j} = 0$ (see note 7). Following Einstein, we evaluate the first term on the right-hand side under (a) and the third under (b).

(a) Using Eq. (6c) for $\overset{(1)}{\Gamma}{}^i_{44} + \overset{(2)}{\Gamma}{}^i_{44}$ (see note 12) and Eq. (9) for $\frac{dx^4}{ds}$ (see note 14), we find:

$$\left(\overset{(1)}{\Gamma}{}^i_{44} + \overset{(2)}{\Gamma}{}^i_{44}\right)\frac{dx^4}{ds}\frac{dx^4}{ds} = -\frac{\alpha}{2}\frac{x^i}{r^3}\left(1 - \frac{\alpha}{r}\right)\left(1 + 2\frac{\alpha}{r}\right) = -\frac{\alpha}{2}\frac{x^i}{r^3}\left(1 + \frac{\alpha}{r}\right).$$

(b) Using equation (6a) for $\overset{(1)}{\Gamma}{}^i_{jk}$ (see note 7) and restoring the summation signs, we have:

$$\sum_{j,k=1}^{3}\overset{(1)}{\Gamma}{}^i_{jk}\frac{dx^j}{ds}\frac{dx^k}{ds} = -\frac{\alpha x^i}{r^3}\sum_{j,k=1}^{3}\left(\delta_{jk} - \frac{3}{2}\frac{x^j x^k}{r^2}\right)\frac{dx^j}{ds}\frac{dx^k}{ds}.$$

Substituting u^2, the square of the velocity of the planet in its orbit (see Eq. (8a)), for $\sum_{i,j}\delta_{ij}\frac{dx^i}{ds}\frac{dx^j}{ds}$ and $r\frac{dr}{ds}$ for $\sum_i \frac{dx^i}{ds}$ (see note 14), we can rewrite this as

$$\sum_{j,k=1}^{3}\overset{(1)}{\Gamma}{}^i_{jk}\frac{dx^j}{ds}\frac{dx^k}{ds} = -\frac{\alpha x^i}{r^3}\left(u^2 - \frac{3}{2}\left(\frac{dr}{ds}\right)^2\right).$$

Combining the results found under (a) and (b), we arrive at Eq. (7b), the correction to the Newtonian Eq. (7a) once terms of order $\left(\frac{\alpha}{r}\right)^2$ are taken into account:

$$\frac{d^2x^i}{ds^2} = -\frac{\alpha}{2}\frac{x^i}{r^3}\left(1 + \frac{\alpha}{r} + 2u^2 - 3\left(\frac{dr}{ds}\right)^2\right). \tag{7b}$$

16. With the help of Eq. (8a) for u^2 in polar coordinates, the expression in parentheses on the right-hand side of Eq. (7b) can be rewritten as

$$1 + \frac{\alpha}{r} + 2u^2 - 3\left(u^2 - r^2\frac{d\varphi^2}{ds^2}\right).$$

If we substitute $\frac{B}{r^2}$ for $\frac{d\varphi}{ds}$ (see Eq. (10)), $2A - 2\Phi$ for u^2 (see Eq. (8)), and -2Φ for $\frac{\alpha}{r}$ (see Eq. (8a)), this turns into the expression above Eq. (7c):

$$1 - 2\Phi - 2A + 2\Phi + 3r^2\left(\frac{B^2}{r^4}\right) = 1 - 2A + 3\frac{B^2}{r^2}.$$

17. Using the result derived in note 16, we can rewrite Eq. (7b) as

$$\frac{d^2x^i}{ds^2} = -\frac{\alpha}{2}\frac{x^i}{r^3}\left(1-2A+3\frac{B^2}{r^2}\right).$$

If both sides are divided by $1-2A$, this turns into

$$\frac{1}{1-2A}\frac{d^2x^i}{ds^2} = -\frac{\alpha}{2}\frac{x^i}{r^3}\left(1+\frac{3B^2}{(1-2A)r^2}\right).$$

Introducing the new time variable $s' \equiv s\sqrt{1-2A}$ and the new area-law constant $B' \equiv \frac{B}{\sqrt{1-2A}}$, we can rewrite this as

$$\frac{d^2x^i}{ds'^2} = -\frac{\alpha}{2}\frac{x^i}{r^3}\left(1+3\frac{B'^2}{r^2}\right).$$

As we will check below, the right-hand side is equal to minus the gradient of the potential

$$\Phi \equiv -\frac{\alpha}{2r}\left(1+\frac{B'^2}{r^2}\right). \tag{7c}$$

The factor in parentheses is a correction to the Newtonian potential, also called Φ, in Eq. (8a) (there is a typo in Eq. (7c): a factor $1/r$ is missing on the right-hand side). Taking minus the gradient of this potential, we find:

$$-\frac{\partial\Phi}{\partial x^i} = \frac{\partial}{\partial x^i}\left(\frac{\alpha}{2r}+\frac{\alpha B'^2}{2r^3}\right) = -\frac{\alpha}{2}\frac{1}{r^2}\frac{\partial r}{\partial x^i} - \frac{\alpha B'^2}{2r}\frac{3}{r^4}\frac{\partial r}{\partial x^i}.$$

Substituting $\frac{x^i}{r}$ for $\frac{\partial r}{\partial x^i}$, we recover the right-hand side of the equations of motion above:

$$-\frac{\partial\Phi}{\partial x^i} = -\frac{\alpha}{2}\frac{x^i}{r^3}\left(1+3\frac{B'^2}{r^2}\right).$$

18. Energy conservation, $\frac{1}{2}u^2+\Phi = A$, allows us to write (see the equation above Eq. (11)):

$$\frac{dr^2+r^2d\varphi^2}{ds^2} = 2A-2\Phi = 2A+\frac{\alpha}{r}+\frac{\alpha B^2}{r^3},$$

where, following Einstein, we wrote u^2 in polar coordinates (see Eq. (8a)) and dropped the prime on the new area-law constant.

Using the area law $r^2\frac{d\varphi}{ds} = B$ (see Eq. (10)) to eliminate ds from the equation above, we obtain a differential equation for $\frac{d\varphi}{dr}$. Integrating this equation between the values of r for perihelion and aphelion, we expect to find a result slightly deviating from π, with the small deviation giving the motion of the perihelion per half a revolution. This is how the perihelion motion was calculated in the Einstein-Besso manuscript (see Ch. 2, [pp. 9–11, 14]) and it is how Einstein calculates it here, except that he uses $x = \frac{1}{r}$ and derives (and then integrates) a differential equation for $\frac{d\varphi}{dx}$.

If $(r^2/B)\,d\varphi$ is substituted for ds, the left-hand side of the equation above becomes:

$$\frac{dr^2+r^2d\varphi^2}{(r^4/B^2)d\varphi^2} = B^2\left(\frac{1}{r^4}\frac{dr^2}{d\varphi^2}+\frac{1}{r^2}\right).$$

If we switch from r to $\frac{1}{x}$, noting that $dr = -\frac{dx}{x^2}$, this expression reduces to

$$B^2\left(\frac{dx^2}{d\varphi^2}+x^2\right)$$

and the equation we started from turns into Einstein's Eq. (11)

$$\frac{dx^2}{d\varphi^2} = \frac{2A}{B^2} + \frac{\alpha}{B^2}x - x^2 + \alpha x^3. \tag{11}$$

19. On [pp. 10–11] of the Einstein-Besso manuscript (see Ch. 2), Einstein performed some contour integrations to evaluate similar integrals to find a formula for the perihelion motion of a planet in the field of the sun as given by the *Entwurf* theory. The switch from r to $x = 1/r$ simplifies these calculations considerably. For a modern version, see Møller (1972, pp. 495–497, Eqs. (12.56)–(12.65)).
20. We first consider the corresponding calculation in Newtonian theory. In that case, the term αx^3 in Eq. (11) is missing. The quadratic expression that remains can be written as $(x-\alpha_1)(\alpha_2 - x)$, where α_1 and α_2 are the values of $x = 1/r$ at aphelion and perihelion, respectively. So we have

$$\alpha_1 + \alpha_2 = \frac{1}{r_1} + \frac{1}{r_2} = \frac{1}{a(1+e)} + \frac{1}{a(1-e)} = \frac{2}{a(1-e^2)},$$

where a is the semi-major axis of the planet's elliptical orbit and e is the orbit's eccentricity. Integrating the resulting differential equation,

$$\frac{d\varphi}{dx} = \frac{1}{\sqrt{\frac{2A}{B^2} + \frac{\alpha}{B^2}x - x^2}} = \frac{1}{\sqrt{-(x-\alpha_1)(x-\alpha_2)}},$$

between aphelion (where x has its minimum value α_1) and perihelion (where x has its maximum value α_2) gives π (see Eq. 2.17 in Sec. 2.3). In Newtonian theory the perihelion of a single planet in the field of the sun will not show any movement.

If the term αx^3 is taken into account, the third-order polynomial on the right-hand side of Eq. (11) can be written as

$$\alpha(x-\alpha_1)(x-\alpha_2)(x-\alpha_3),$$

where two of the three roots, as Einstein notes, will only differ insignificantly from α_1 and α_2 in the Newtonian case. Eq. (11) tells us that the coefficient of x^2 should be -1. It follows that

$$\alpha(\alpha_1 + \alpha_2 + \alpha_3) = 1.$$

This relation allows us to replace the factor $(x - \alpha_3)$ by an expression depending only on the other two roots. Using that $\alpha\alpha_3 = 1 - \alpha(\alpha_1 + \alpha_2)$, we can write

$$\alpha(\alpha_3 - x) = \alpha\alpha_3 - \frac{\alpha x}{\alpha\alpha_3} = \left(1 - \alpha(\alpha_1+\alpha_2)\right)\left(1 - \frac{\alpha x}{1-\alpha(\alpha_1+\alpha_2)}\right).$$

To first order in α, the polynomial can thus be rewritten as

$$-(x-\alpha_1)(x-\alpha_2)(1-\alpha(\alpha_1+\alpha_2))(1-\alpha x).$$

Eq. (11), again to first order in α, then gives the following differential equation:

$$\frac{d\varphi}{dx} = \frac{1}{\sqrt{-(x-\alpha_1)(x-\alpha_2)(1-\alpha(\alpha_1+\alpha_2))(1-\alpha x)}}$$

$$= \frac{1+\dfrac{\alpha(\alpha_1+\alpha_2)}{2}+\dfrac{\alpha}{2}x}{\sqrt{-(x-\alpha_1)(x-\alpha_2)}},$$

which leads to the first two of the three equations for φ above Eq. (12) (after correction of a typo in both of them: there is a factor $\frac{1}{2}$ missing in the term $\alpha(\alpha_1+\alpha_2)$ in the factor in front of the integral).

21. To find an expression for the angle φ between aphelion and perihelion, we need the values of two integrals:

$$\int_{\alpha_1}^{\alpha_2} \frac{dx}{\sqrt{-(x-\alpha_1)(x-\alpha_2)}} = \pi$$

(see note 20 and Sec. 2.3, Eq. 2.17) and

$$\int_{\alpha_1}^{\alpha_2} \frac{x\,dx}{\sqrt{-(x-\alpha_1)(x-\alpha_2)}} = \frac{\alpha_1+\alpha_2}{2}\pi.$$

Splitting this second integral into two parts,

$$\int_{\alpha_1}^{\alpha_2}\left(\frac{\left(x-\dfrac{\alpha_1+\alpha_2}{2}\right)dx}{\sqrt{-(x-\alpha_1)(x-\alpha_2)}}+\frac{\dfrac{\alpha_1+\alpha_2}{2}dx}{\sqrt{-(x-\alpha_1)(x-\alpha_2)}}\right),$$

we see that the first part vanishes (the integrand is anti-symmetric around the middle of the interval $[\alpha_1,\alpha_2]$), while the second part is a constant times the first integral. To first order in α, φ is thus given by

$$\varphi = \left(1+\frac{\alpha(\alpha_1+\alpha_2)}{2}\right)\pi + \frac{\alpha}{2}\left(\frac{\alpha_1+\alpha_2}{2}\pi\right) = \pi\left(1+\frac{3}{4}\alpha(\alpha_1+\alpha_2)\right).$$

Substituting $\dfrac{2}{a(1-e^2)}$ for $\alpha_1+\alpha_2$ (see note 20), we arrive at Eq. (12).

22. Using Kepler's third law—which, as Einstein noted under Eq. (7b) on p. 837, continues to hold in his new theory—to substitute $\dfrac{8\pi^2a^3}{T^2c^2}$ for α (cf. Eq. (2.15) in our commentary on [p. 14] of the Einstein-Besso manuscript in Sec. 2.3), we get from Eq. (13) to Eq. (14):

$$\varepsilon = 3\pi\frac{\alpha}{a(1-e^2)} = 24\pi^3\frac{a^2}{T^2c^2(1-e^2)}. \tag{13, 14}$$

This is more than double (i.e., $\dfrac{12}{5}$ times) the size of the effect found in the *Entwurf* theory (see Eq. 2.16 in Sec. 2.3). See our commentary on [p. 28] of the Einstein-Besso manuscript in Sec. 2.3 for the conversion of the perihelion advance from radians per revolution to seconds of arc per century.

23. The values listed here are part of a table in Newcomb (1895, p. 119) and are reproduced in Freundlich (1915a, Col. 52). For discussion, see Roseveare (1982, pp. 51–52).

6.4 The field equations of gravitation

6.4.1 Facsimile

Die Feldgleichungen der Gravitation.

Von A. Einstein.

In zwei vor kurzem erschienenen Mitteilungen[1] habe ich gezeigt, wie man zu Feldgleichungen der Gravitation gelangen kann, die dem Postulat allgemeiner Relativität entsprechen, d. h. die in ihrer allgemeinen Fassung beliebigen Substitutionen der Raumzeitvariabeln gegenüber kovariant sind.

Der Entwicklungsgang war dabei folgender. Zunächst fand ich Gleichungen, welche die Newtonsche Theorie als Näherung enthalten und beliebigen Substitutionen von der Determinante 1 gegenüber kovariant waren. Hierauf fand ich, daß diesen Gleichungen allgemein kovariante entsprechen, falls der Skalar des Energietensors der »Materie« verschwindet. Das Koordinatensystem war dann nach der einfachen Regel zu spezialisieren, daß $\sqrt{-g}$ zu 1 gemacht wird, wodurch die Gleichungen der Theorie eine eminente Vereinfachung erfahren. Dabei mußte aber, wie erwähnt, die Hypothese eingeführt werden, daß der Skalar des Energietensors der Materie verschwinde.

Neuerdings finde ich nun, daß man ohne Hypothese über den Energietensor der Materie auskommen kann, wenn man den Energietensor der Materie in etwas anderer Weise in die Feldgleichungen einsetzt, als dies in meinen beiden früheren Mitteilungen geschehen ist. Die Feldgleichungen für das Vakuum, auf welche ich die Erklärung der Perihelbewegung des Merkur gegründet habe, bleiben von dieser Modifikation unberührt. Ich gebe hier nochmals die ganze Betrachtung, damit der Leser nicht genötigt ist, die früheren Mitteilungen unausgesetzt heranzuziehen. [1]

Aus der bekannten Riemannschen Kovariante vierten Ranges leitet man folgende Kovariante zweiten Ranges ab:

$$G_{im} = R_{im} + S_{im} \tag{1}$$

$$R_{im} = -\sum_{l} \frac{\partial \begin{Bmatrix} im \\ l \end{Bmatrix}}{\partial x_l} + \sum_{l\varrho} \begin{Bmatrix} il \\ \varrho \end{Bmatrix} \begin{Bmatrix} m\varrho \\ l \end{Bmatrix} \tag{1a}$$

$$S_{im} = \sum_{l} \frac{\partial \begin{Bmatrix} il \\ l \end{Bmatrix}}{\partial x_m} - \sum_{l\varrho} \begin{Bmatrix} im \\ \varrho \end{Bmatrix} \begin{Bmatrix} \varrho l \\ l \end{Bmatrix} \tag{1b}$$

[1] Sitzungsber. XLIV, S. 778 und XLVI, S. 799, 1915.

Die allgemein kovarianten zehn Gleichungen des Gravitationsfeldes in Räumen, in denen »Materie« fehlt, erhalten wir, indem wir ansetzen

$$G_{im} = 0 . \qquad (2)$$

Diese Gleichungen lassen sich einfacher gestalten, wenn man das Bezugssystem so wählt, daß $\sqrt{-g} = 1$ ist. Dann verschwindet S_{im} wegen (1b), so daß man statt (2) erhält

$$R_{im} = \sum_l \frac{\partial \Gamma^l_{im}}{\partial x_l} + \sum_{\varrho l} \Gamma^l_{i\varrho} \Gamma^\varrho_{ml} = 0 \qquad (3)$$

$$\sqrt{-g} = 1 . \qquad (3a)$$

Dabei ist

$$\Gamma^l_{im} = -\begin{Bmatrix} im \\ l \end{Bmatrix} \qquad (4)$$

gesetzt, welche Größen wir als die »Komponenten« des Gravitationsfeldes bezeichnen. [2]

Ist in dem betrachteten Raume »Materie« vorhanden, so tritt deren Energietensor auf der rechten Seite von (2) bzw. (3) auf. Wir setzen [3]

$$G_{im} = -\varkappa \left(T_{im} - \frac{1}{2} g_{im} T \right), \qquad (2a)$$

wobei

$$\sum_{\varrho\sigma} g^{\varrho\sigma} T_{\varrho\sigma} = \sum_\sigma T^\sigma_\sigma = T \qquad (5)$$

gesetzt ist; T ist der Skalar des Energietensors der »Materie«, die rechte Seite von (2a) ein Tensor. Spezialisieren wir wieder das Koordinatensystem in der gewohnten Weise, so erhalten wir an Stelle von (2a) die äquivalenten Gleichungen

$$R_{im} = \sum_l \frac{\partial \Gamma^l_{im}}{\partial x_l} + \sum_{\varrho l} \Gamma^l_{i\varrho} \Gamma^\varrho_{ml} = -\varkappa \left(T_{im} - \frac{1}{2} g_{im} T \right) \qquad (6)$$

$$\sqrt{-g} = 1 . \qquad (3a)$$

Wie stets nehmen wir an, daß die Divergenz des Energietensors der Materie im Sinne des allgemeinen Differentialkalkuls verschwinde (Impulsenergiesatz). Bei der Spezialisierung der Koordinatenwahl gemäß (3a) kommt dies darauf hinaus, daß die T_{im} die Bedingungen

$$\sum_\lambda \frac{\partial T^\lambda_\sigma}{\partial x_\lambda} = -\frac{1}{2} \sum_{\mu\nu} \frac{\partial g^{\mu\nu}}{\partial x_\sigma} T_{\mu\nu} \qquad (7)$$

oder

$$\sum_\lambda \frac{\partial T^\lambda_\sigma}{\partial x_\lambda} = -\sum_{\mu\nu} \Gamma^\mu_{\sigma\nu} T^\nu_\mu \qquad (7a)$$

erfüllen sollen. [4]

Multipliziert man (6) mit $\frac{\partial g^{im}}{\partial x_\sigma}$ und summiert über i und m, so erhält man[1] mit Rücksicht auf (7) und auf die aus (3a) folgende Relation [5]

$$\frac{1}{2}\sum_{im} g_{im}\frac{\partial g^{im}}{\partial x_\sigma} = -\frac{\partial lg\sqrt{-g}}{\partial x_\sigma} = 0$$

den Erhaltungssatz für Materie und Gravitationsfeld zusammen in der Form

$$\sum_\lambda \frac{\partial}{\partial x_\lambda}(T_\sigma^\lambda + t_\sigma^\lambda) = 0\,, \tag{8}$$

wobei t_σ^λ (der »Energietensor« des Gravitationsfeldes) gegeben ist durch

$$\varkappa t_\sigma^\lambda = \frac{1}{2}\delta_\sigma^\lambda \sum_{\mu\nu\alpha\beta} g^{\mu\nu}\Gamma_{\mu\beta}^\alpha\Gamma_{\nu\alpha}^\beta - \sum g^{\mu\nu}\Gamma_{\mu\sigma}^\alpha\Gamma_{\nu\alpha}^\lambda\,. \tag{8a}$$

Die Gründe, welche mich zur Einführung des zweiten Gliedes auf der rechten Seite von (2a) und (6) veranlaßt haben, erhellen erst aus den folgenden Überlegungen, welche den an der soeben angeführten Stelle (S. 785) gegebenen völlig analog sind. [6]

Multiplizieren wir (6) mit g^{im} und summieren wir über die Indizes i und m, so erhalten wir nach einfacher Rechnung

$$\sum_{\alpha\beta}\frac{\partial^2 g^{\alpha\beta}}{\partial x_\alpha \partial x_\beta} - \varkappa(T+t) = 0\,, \tag{9}$$

wobei entsprechend (5) zur Abkürzung gesetzt ist

$$\sum_{\varrho\sigma} g^{\varrho\sigma} t_{\varrho\sigma} = \sum_\sigma t_\sigma^\sigma = t\,. \tag{8b}$$

Man beachte, daß es unser Zusatzglied mit sich bringt, daß in (9) der Energietensor des Gravitationsfeldes neben dem der Materie in gleicher Weise auftritt, was in Gleichung (21) a. a. O. nicht der Fall ist. [7]

Ferner leitet man an Stelle der Gleichung (22) a. a. O. auf dem dort angegebenen Wege mit Hilfe der Energiegleichung die Relationen ab:

$$\frac{\partial}{\partial x_\mu}\left[\sum_{\alpha\beta}\frac{\partial^2 g^{\alpha\beta}}{\partial x_\alpha \partial x_\beta} - \varkappa(T+t)\right] = 0\,. \tag{10}$$

Unser Zusatzglied bringt es mit sich, daß diese Gleichungen gegenüber (9) keine neue Bedingung enthalten, so daß über den Energie-

[1] Über die Ableitung vgl. Sitzungsber. XLIV, 1915, S. 784/785. Ich ersuche den Leser, für das Folgende auch die dort auf S. 785 gegebenen Entwicklungen zum Vergleiche heranzuziehen.

tensor der Materie keine andere Voraussetzung gemacht werden muß als die, daß er dem Impulsenergiesatze entspricht. [8]

Damit ist endlich die allgemeine Relativitätstheorie als logisches Gebäude abgeschlossen. Das Relativitätspostulat in seiner allgemeinsten Fassung, welches die Raumzeitkoordinaten zu physikalisch bedeutungslosen Parametern macht, führt mit zwingender Notwendigkeit zu einer ganz bestimmten Theorie der Gravitation, welche die Perihelbewegung des Merkur erklärt. Dagegen vermag das allgemeine Relativitätspostulat uns nichts über das Wesen der übrigen Naturvorgänge zu offenbaren, was nicht schon die spezielle Relativitätstheorie gelehrt hätte. Meine in dieser Hinsicht neulich an dieser Stelle geäußerte Meinung war irrtümlich. Jede der speziellen Relativitätstheorie gemäße physikalische Theorie kann vermittels des absoluten Differentialkalkuls in das System der allgemeinen Relativitätstheorie eingereiht werden, ohne daß letztere irgendein Kriterium für die Zulässigkeit jener Theorie lieferte.

Ausgegeben am 2. Dezember.

6.4.2 Translation

The Field Equations of Gravitation

In two recently published papers[1] I have shown how to obtain field equations of gravitation that are in accordance with the postulate of general relativity, i.e., covariant, in their general form, under arbitrary transformations of the space-time coordinates.

The sequence of events was as follows. First, I found equations that give the NEWTONIAN theory in first approximation and that are covariant under arbitrary transformations with determinant 1. Then I found that these equations correspond to generally-covariant ones if the scalar of the energy tensor of "matter" vanishes. The coordinate system could then be specified by the simple rule that $\sqrt{-g}$ must be equal to 1, which leads to a remarkable simplification of the equations of the theory. As I mentioned, however, this did require the introduction of the hypothesis that the scalar of the energy tensor of matter vanishes.

Quite recently I found that one can manage without this hypothesis about the energy tensor of matter, if one inserts the energy tensor of matter into the field equations in a slightly different way. The vacuum field equations, on which I based the explanation of the Mercury perihelion, remain unaffected by this modification. I will repeat the entire argument here so that the reader does not constantly have to consult the earlier publications.[1]

From the well-known RIEMANN-covariant of rank four, one derives the following covariant of rank two:

$$G_{im} = R_{im} + S_{im} \tag{1}$$

$$R_{im} = -\sum_{l} \frac{\partial \begin{Bmatrix} im \\ l \end{Bmatrix}}{\partial x_l} + \sum_{l\rho} \begin{Bmatrix} il \\ \rho \end{Bmatrix} \begin{Bmatrix} m\rho \\ l \end{Bmatrix} \tag{1a}$$

$$S_{im} = \sum_{l} \frac{\partial \begin{Bmatrix} il \\ l \end{Bmatrix}}{\partial x_m} - \sum_{l\rho} \begin{Bmatrix} im \\ \rho \end{Bmatrix} \begin{Bmatrix} \rho l \\ l \end{Bmatrix}. \tag{1b}$$

We obtain the ten generally-covariant equations for the gravitational field in regions without "matter" by setting

$$G_{im} = 0. \tag{2}$$

These equations can be simplified by choosing a reference frame such that $\sqrt{-g} = 1$. S_{im} then vanishes because of (16) and, instead of (2), one gets

[1] *Proceedings* 44, p. 778, and 46, p. 799 (1915).

$$R_{im} = \sum_l \frac{\partial \Gamma^l_{im}}{\partial x_l} + \sum_{\rho l} \Gamma^l_{i\rho} \Gamma^\rho_{ml} = 0 \tag{3}$$

$$\sqrt{-g} = 1. \tag{3a}$$

Here we have set

$$\Gamma^l_{im} = -\begin{Bmatrix} im \\ l \end{Bmatrix}. \tag{4}$$

We call these quantities the "components" of the gravitational field.[2]

When there is "matter" present in the region under consideration, its energy tensor occurs on the right-hand sides of (2) and (3), respectively. We set[3]

$$G_{im} = -\kappa \left(T_{im} - \frac{1}{2} g_{im} T \right), \tag{2a}$$

where

$$\sum_{\rho\sigma} g^{\rho\sigma} T_{\rho\sigma} = \sum_\sigma T^\sigma_\sigma = T. \tag{5}$$

T is the scalar of the energy tensor of "matter," and the right-hand side of (2a) is a tensor. If once again we specify the coordinate system in the usual way, we get, instead of (2a), the equivalent equations

$$R_{im} = \sum_l \frac{\partial \Gamma^l_{im}}{\partial x_l} + \sum_{\rho l} \Gamma^l_{i\rho} \Gamma^\rho_{ml} = -\kappa \left(T_{im} - \frac{1}{2} g_{im} T \right) \tag{6}$$

$$\sqrt{-g} = 1. \tag{3a}$$

We assume, as always, that the energy tensor of matter has a vanishing divergence in the sense of the general differential calculus (energy-momentum theorem). If the coordinates are chosen according to (3a), this means that T_{im} should satisfy the conditions[4]

$$\sum_\lambda \frac{\partial T^\lambda_\sigma}{\partial x_\lambda} = -\frac{1}{2} \sum_{\mu\nu} \frac{\partial g^{\mu\nu}}{\partial x_\sigma} T_{\mu\nu} \tag{7}$$

or

$$\sum_\lambda \frac{\partial T^\lambda_\sigma}{\partial x_\lambda} = -\sum_{\mu\nu} \Gamma^\mu_{\sigma\nu} T^\nu_\mu. \tag{7a}$$

If one multiplies (6) by $\partial g^{im}/\partial x_\sigma$ and sums over i and m, one gets[1,][5] because of (7) and because of the relation

$$\frac{1}{2} \sum_{im} g_{im} \frac{\partial g^{im}}{\partial x_\sigma} = -\frac{\partial \lg \sqrt{-g}}{\partial x_\sigma} = 0,$$

[1] On the derivation see *Proceedings* 44 (1915), pp. 784–785. For the following I ask the reader also to consult, for a comparison, the deliberations given there on p. 785.

which follows from (3a), the conservation law for matter and gravitational field combined, in the form

$$\sum_{\lambda} \frac{\partial}{\partial x_{\lambda}} (T_{\sigma}^{\lambda} + t_{\sigma}^{\lambda}) = 0, \tag{8}$$

where t_{σ}^{λ} (the "energy tensor" of the gravitational field) is given by

$$-\kappa t_{\sigma}^{\lambda} = \frac{1}{2} \delta_{\sigma}^{\lambda} \sum_{\mu\nu\alpha\beta} g^{\mu\nu} \Gamma_{\mu\beta}^{\alpha} \Gamma_{\nu\alpha}^{\beta} - \sum_{\mu\nu\alpha} g^{\mu\nu} \Gamma_{\mu\sigma}^{\alpha} \Gamma_{\nu\alpha}^{\lambda}. \tag{8a}$$

The reasons that led me to introduce the second term on the right-hand sides of (2a) and (6) will only become clear from the following considerations, which are completely analogous to those in the passage I just referred to (p. 785).[6]

If we multiply (6) by g_{im} and sum over i and m, we obtain after a simple calculation

$$\sum_{\alpha\beta} \frac{\partial^2 g^{\alpha\beta}}{\partial x_{\alpha} \partial x_{\beta}} - \kappa(T + t) = 0, \tag{9}$$

where, in analogy with (5), we used the abbreviation

$$\sum_{\rho\sigma} g^{\rho\sigma} t_{\rho\sigma} = \sum_{\sigma} t_{\sigma}^{\sigma} = t. \tag{8b}$$

It should be noted that our additional term is such that the energy tensor of the gravitational field occurs in (9) in the same way as the one for matter, which was not the case in equation (21) l.c.[7]

Furthermore, one derives instead of equation (22) l.c. and along the same lines as there, with the help of the energy equation, the relations

$$\frac{\partial}{\partial x_{\mu}} \left[\sum_{\alpha\beta} \frac{\partial^2 g^{\alpha\beta}}{\partial x_{\alpha} \partial x_{\beta}} - \kappa(T + t) \right] = 0. \tag{10}$$

Our additional term ensures that these equations contain no additional conditions compared to (9); we thus need not make any hypotheses about the energy tensor of matter other than that it is compatible with energy-momentum conservation.[8]

This then finally completes the general theory of relativity as a logical structure. The postulate of relativity in its most general formulation (which turns space-time coordinates into physically meaningless parameters) leads with inescapable necessity to a specific theory of gravity, which explains the perihelion motion of Mercury. The postulate of general relativity, however, tell us nothing about the essence of the various processes in nature that the special theory of relativity has not already taught us. The opinion I recently expressed here in this regard was mistaken. Any physical theory compatible with the special theory of relativity can, through the methods of the absolute differential calculus, be absorbed into the framework of the general theory of relativity, without the latter providing any criteria for the admissibility of that theory.

6.4.3 Commentary

This paper, the sequel to Einstein (1915a; 1915b, see Secs. 6.1 and 6.2), was submitted November 25, 1915 and published December 2, 1915. It is presented in facsimile as Doc. 25 in CPAE6. Our commentary is based on sec. 7 of "Untying the knot" (Janssen and Renn 2007).

1. In this fourth and final paper of November 1915, Einstein presents a much more satisfactory way to get around the problematic condition (21a),

$$\left(g^{\alpha\beta}\left(\log\sqrt{-g}\right)_{,\beta}\right)_{,\alpha}=-\kappa T,$$

in his first paper than he did in the second, a short addendum to the first (cf. Sec. 6.1.3, note 18, Sec. 6.2.3, note 1, and and Pt. I, Secs. 5.3–5.4, especially notes 15, 16 and 21). In the second paper he appealed to the electromagnetic view of nature to set the trace (or "scalar") T of the energy-momentum tensor of matter equal to zero. In this fourth paper he adds a term with T to the field equations in such a way that the right-hand side of condition (21a) vanishes regardless of the value of T. It is not hard to see how the field equations of first November 1915 paper, covariant under unimodular transformations, would have to be modified to bring about this result (Janssen and Renn 2007, pp. 889–891).

Condition (21a) follows from the combination of two other conditions, which Einstein found by contracting the field equations of the first November 1915 paper in two different ways, These are the conditions we labeled (A) and (B) in note 17 in Sec. 6.1.3. Fully contracting these field equations (in the form of Eq. (16a)),

$$\Gamma^{\alpha}_{\mu\nu,\alpha}-\Gamma^{\alpha}_{\beta\mu}\Gamma^{\beta}_{\alpha\nu}=-\kappa T_{\mu\nu},$$

Einstein arrived at the first of these conditions:

$$g^{\alpha\beta}_{\alpha\beta}-\kappa t+\left(g^{\alpha\beta}\left(\log\sqrt{-g}\right)_{,\beta}\right)_{,\alpha}=-\kappa T, \tag{A}$$

where $\kappa t=g^{\mu\nu}\Gamma^{\alpha}_{\beta\mu}\Gamma^{\beta}_{\alpha\nu}$. Contracting the field equations with $g^{\nu\lambda}$, he found

$$(g^{\nu\lambda}\Gamma^{\alpha}_{\mu\nu})_{,\alpha}-\frac{1}{2}\delta^{\lambda}_{\mu}\kappa t=-\kappa(T^{\lambda}_{\mu}+t^{\lambda}_{\mu}).$$

Requiring the four-divergence of $T^{\lambda}_{\mu}+t^{\lambda}_{\mu}$ to vanish and using that the definition $\Gamma^{\alpha}_{\beta\mu}\equiv-\left\{\begin{matrix}\alpha\\ \beta\mu\end{matrix}\right\}$ of the gravitational field, he arrived at the condition

$$(g^{\alpha\beta}_{\alpha\beta}-\kappa t)_{,\mu}=0,$$

which is satisfied if the stronger condition,

$$g^{\alpha\beta}_{\alpha\beta}-\kappa t=0. \tag{B}$$

is imposed. Combining conditions (A) and (B), Einstein ended up with condition (21a).

It is easily seen that this unwelcome implication can be avoided if a term with the trace of the energy-momentum tensor is added to the field equations:

$$\Gamma^{\alpha}_{\mu\nu,\alpha} + \Gamma^{\alpha}_{\beta\mu}\Gamma^{\beta}_{\alpha\nu} = -\kappa\left(T_{\mu\nu} - \frac{1}{2}g_{\mu\nu}T\right).$$

This adds a term $\frac{1}{2}\kappa g^{\mu\nu}g_{\mu\nu}T = 2\kappa T$ to the right-hand side of (A), which then turns into:

$$g^{\alpha\beta}_{\alpha\beta} - \kappa(t+T) + \left(g^{\alpha\beta}\left(\log\sqrt{-g}\right)_{,\beta}\right)_{,\alpha} = 0. \tag{A$'$}$$

It likewise adds a term $\frac{1}{2}\kappa g^{\nu\lambda}g_{\mu\nu}T = \frac{1}{2}\delta^{\lambda}_{\mu}\kappa T$ to the right-hand side of the old field equations contracted with $g^{\nu\lambda}$, which for these new field equations becomes

$$(g^{\nu\lambda}\Gamma^{\alpha}_{\mu\nu})_{,\alpha} - \frac{1}{2}\delta^{\lambda}_{\mu}\kappa(t+T) = -\kappa(T^{\lambda}_{\mu} + t^{\lambda}_{\mu}).$$

Demanding that $\left(T^{\lambda}_{\mu} + t^{\lambda}_{\mu}\right)_{,\lambda} = 0$ in this case thus leads to the condition

$$\left(g^{\alpha\beta}_{\alpha\beta} - \kappa(t+T)\right)_{,\mu} = 0.$$

which is satisfied if the stronger condition

$$g^{\alpha\beta}_{\alpha\beta} - \kappa(t+T) = 0 \tag{B$'$}$$

is imposed. The addition of a trace term to the field equation thus turns conditions (A) and (B) into conditions (A′) and (B′) and changes condition (21a) into

$$\left(g^{\alpha\beta}\left(\log\sqrt{-g}\right)_{,\beta}\right)_{,\alpha} = 0.$$

This new condition allows for unimodular coordinates without requiring any assumptions about the energy-momentum tensor of matter. The new field equations can thus be seen as generally covariant field equations, $G_{\mu\nu} = -\kappa\left(T_{\mu\nu} - \frac{1}{2}g_{\mu\nu}T\right)$ (where $G_{\mu\nu}$ is the Ricci tensor), expressed in unimodular coordinates.

Note that the energy-momentum tensor for matter and the energy-momentum pseudo-tensor for the gravitational field enter conditions (A′) and (the equations leading up to) conditions (B′) in the exact same way. This is not true for the conditions (A) and (B) they replace (see Sec. 6.1.3, note 17, and note 7 below).

2. This new definition of the gravitational field played a key role on the transition from the *Entwurf* field equations to the field equations of the first November 1915 paper (cf. Sec. 6.1.3, note 11).
3. On p. 20L of the Zurich notebook, Einstein had already considered adding a trace term to the field equations in linear approximation but had decided against it, probably because a metric of the form Einstein expected in the case of weak static fields would not be a solution of these amended field equations (see Sec. 1.3, notes 13–16, and Pt. I, Ch. 3 and Sec. 5.3, especially note 12). The problem, in a nutshell (and in modern notation), is this. For weak static fields, the field equations with trace term reduce to:

$$\Delta g_{\mu\nu} = \kappa\left(T_{\mu\nu} - \frac{1}{2}\eta_{\mu\nu}T\right),$$

where $\eta_{\mu\nu} \equiv \mathrm{diag}(-1,-1,-1,1)$ is the standard diagonal Minkowski metric. For a static mass distribution, the energy-momentum tensor is $T_{\mu\nu} = \mathrm{diag}(0,0,0,\rho)$ with trace $T = \rho$. The only non-trivial components of the field equations in this case are therefore:

$$\Delta g_{11} = \Delta g_{22} = \Delta g_{33} = \Delta g_{44} = \frac{1}{2}\kappa\rho.$$

A metric of the form $\mathrm{diag}(-1,-1,-1,f(x,y,z))$, the spatially flat metric Einstein expected in this case, is not a solution of these equations. In his calculation of the perihelion motion of Mercury, however, Einstein had found that the weak static field of the sun is not spatially flat. This removed the objection to the addition of a trace term to the field equations as well as an objection against the use of unimodular coordinates (cf. Sec. 6.2.3, note 4, and Sec. 6.3.3, note 8).

4. In arbitrary coordinates, the vanishing of the covariant divergence of the energy-momentum tensor is given by:

$$T^{\nu}_{\sigma;\nu} = T^{\nu}_{\sigma,\nu} + \begin{Bmatrix} \nu \\ \tau\nu \end{Bmatrix} T^{\tau}_{\sigma} - \begin{Bmatrix} \tau \\ \sigma\nu \end{Bmatrix} T^{\nu}_{\tau} = 0.$$

In unimodular coordinates, this equation reduces to Einstein's Eqs. (7) and (7a) (cf. Sec. 6.1.3, note 10).

5. As Einstein notes in the footnote appended to this passage, the derivation of Eqs. (8) and (8a) only requires a slight modification of the derivation of Eqs. (20), (20a) and (20b) in Einstein (1915a, pp. 784–785; cf. Sec. 6.1.3, notes 15 and 16). We start from the energy-momentum balance law in unimodular coordinates in the form of Eq. (7):

$$T^{\lambda}_{\sigma,\lambda} + \frac{1}{2} g^{\mu\nu}_{\sigma} T_{\mu\nu} = 0 \tag{6.1}$$

Since

$$g^{\mu\nu}_{\sigma} g_{\mu\nu} = -(\log\sqrt{-g})_{,\sigma} = 0,$$

we can replace $T_{\mu\nu}$ by $T_{\mu\nu} - \frac{1}{2} g_{\mu\nu} T$ in the second term on the left-hand side:

$$\frac{1}{2} g^{\mu\nu}_{\sigma} \left(T_{\mu\nu} - \frac{1}{2} g_{\mu\nu} T \right).$$

As Einstein did in the first November 1915 paper, we can then substitute the left-hand side of the field equations in unimodular coordinates, given in Eq. (6):

$$-\frac{1}{2\kappa} g^{\mu\nu}_{\sigma} (\Gamma^{\alpha}_{\mu\nu,\alpha} + \Gamma^{\alpha}_{\beta\mu} \Gamma^{\beta}_{\alpha\nu}).$$

Einstein had already shown in this earlier paper that this expression can be rewritten as the divergence of the pseudo-tensor for gravitational energy-momentum, if the latter is defined as in Eq. (8a):

$$\kappa t^{\lambda}_{\sigma} = \frac{1}{2} \delta^{\lambda}_{\sigma} g^{\mu\nu} \Gamma^{\alpha}_{\beta\mu} \Gamma^{\beta}_{\alpha\nu} - g^{\mu\nu} \Gamma^{\alpha}_{\mu\sigma} \Gamma^{\lambda}_{\alpha\nu}$$

The energy-momentum balance law we started from thus takes the form of Eq. (8), the vanishing of the ordinary divergence of the sum $T^{\lambda}_{\sigma} + t^{\lambda}_{\sigma}$.

6. Eq. (9), obtained if we contract Eq. (6), the field equations in unimodular coordinates, with g^{im}, is the condition labeled (A′) in note 1. Eq. (10), obtained if we contract Eq. (6) with g^{ij} $(j \neq m)$, is the equation right above the condition labeled (B′) in note 1. Note that Einstein, contrary to what he promised earlier in the paper, refers to his first November 1915 paper for the derivations of these two equations and leaves

it to the reader to figure out how these earlier derivations need to be modified if a trace term is added to the right-hand side of the field equations. Einstein's argument proved hard to follow even for Lorentz and Ehrenfest in Leiden who had closely followed the development of the theory. In a letter to Ehrenfest early the following year, Einstein gave a self-contained presentation of the considerations that led him to these new field equations (CPAE8, Doc. 185, see Ch. 8). This letter helped shape the sections on the field equations in the review article published a few months later (Einstein 1916b, see Ch. 9; cf. Pt. I, Sec. 2.2, note 8).

7. An important argument in favor of adding a term with the trace of the energy-momentum tensor for matter to the field equations is that in this way all energy-momentum enters the field equations in the exact same way (cf. note 1).
8. To reiterate (see note 1): unlike the conditions we labeled (A) and (B) in the first November 1915 paper (see Sec. 6.1.3, note 17), the conditions (A′) and (B′) that take their place in this fourth and final one do not lead to any restrictions on the determinant of the metric over and above the one that follows directly from the restriction to unimodular transformations. This removed the odd prohibition Einstein had found against unimodular coordinates (in which $\sqrt{-g} = 1$) in that first paper even though the theory proposed in it was covariant under unimodular transformations (under which $\sqrt{-g}$ transforms as a scalar). That (A) and (B) got replaced by (A′) and (B′) is a direct result of the addition of a term with the trace T of the energy-momentum tensor $T_{\mu\nu}$ to the right-hand side of the field equations. The addition of this trace term thus allowed Einstein to look upon the amended version of the field equations of his first November 1915 paper as generally covariant equations expressed in unimodular coordinates *without* making any assumptions about $T_{\mu\nu}$ as he had done in the second November 1915 paper.

Chapter 7
Einstein to Arnold Sommerfeld, November 28, 1915

7.1 Transcription

Berlin, 28. XI. [1915][1]

Lieber Sommerfeld!

Sie dürfen mir nicht böse sein, dass ich erst heute auf Ihren freundlichen und interessanten Brief antworte. Aber ich hatte im letzten Monat eine der [a]ufregendsten, anstrengendsten Zeiten meines Lebens, allerdings auch der erfolgreichsten. Ans Schreiben konnte ich nicht denken.[2]

Ich erkannte nämlich dass meine bisherigen Feldgleichungen der Gravitation gänzlich haltlos waren! Dafür ergaben sich folgende Anhaltspunkte[3]

1) Ich bewies, dass das Gravitationsfeld auf einem gleichförmig rotierenden System den Feldgleichungen nicht genügt.[4]

2) Die Bewegung des Merkur-Perihels ergab sich zu 18″ statt 45″ pro Jahrhundert.[5]

3) Die Kovarianzbetrachtung in meiner Arbeit vom letzten Jahre liefert die Hamilton-Funktion H nicht. Sie lässt, wenn sie sachgemäss verallgemeinert wird, ein beliebiges H zu.[6]Daraus ergab sich, dass die Kovarianz bezüglich „angepasster" Koordinatensysteme ein Schlag ins Wasser war.[7]

Nachdem so jedes Vertrauen im Resultate und Methode der früheren Theorie gewichen war, sah ich klar, dass nur durch einen Anschluss an die allgemeine Kovariantentheorie, d. h. an Riemanns Kovariante, eine befriedigende Lösung gefunden werden konn[t]e.[8] Die letzten Irrtümer in diesem Kampfe habe ich leider in den Akademie-Arbeiten, die ich Ihnen bald senden kann, verevigt.[9] Das endgültige Ergebnis ist folgendes.

Die Gleichungen des Gravitationsfeldes sind allgemein kovariant. Ist

$$(ik, lm)$$

M. Janssen, J. Renn, *How Einstein Found His Field Equations*, Classic Texts in the Sciences, https://doi.org/10.1007/978-3-030-97955-3_14

der Christoffel'sche Tensor vierten Ranges, so ist $G_{im} = \sum_{kl} g^{kl}(ik, lm)$ ein symmetrischer Tensor zweiten Ranges.[10] Die Gleichungen lauten

$$G_{im} = -\kappa \left(T_{im} - \frac{1}{2} g_{im} \underbrace{\sum_{\alpha\beta} (g^{\alpha\beta} T_{\alpha\beta})} \right)$$

Skalar des Energietensors der „Materie", für den ich im Folgenten „T" schreibe.

Es ist natürlich leicht, diese allgemein kovarianten Gleichungen hinzusetzen, schwer aber, einzusehen, dass sie Verallgemeinerungen von Poissons Gleichungen sind, und nicht leicht, einzusehen, dass sie den Erhaltungssätzen Genüge leisten.[11]

Man kann nun die ganze Theorie eminent vereinfachen, indem man das Bezugssystem so wählt, dass $\sqrt{-g} = 1$ wird. Dann nehmen die Gleichungen die Form an,

$$-\sum_{l} \frac{\partial \left\{ {im \atop l} \right\}}{\partial x_l} + \sum_{\alpha\beta} \left\{ {i\alpha \atop \beta} \right\} \left\{ {m\beta \atop \alpha} \right\} = -\kappa (T_{im} - \frac{1}{2} g_{im} T)$$

Diese Gleichungen hatte ich schon vor 3 Jahren mit Grossmann erwogen/bis auf das zweite Glied der rechten Seite,[12] war aber damals zu dem Ergebnis gelangt, dass sie nicht Newtons Näherung liefere, was irrtümlich war. Den Schlüssel zu dieser Lösung lieferte mir die Erkenntnis, dass nicht

$$\sum g^{l\alpha} \frac{\partial g_{\alpha i}}{\partial x_m}$$

sondern die damit verwandten Christoffel'schen Symbole $\left\{ {im \atop l} \right\}$ als natürlichen Ausdruck für die „Komponente" des Gravitationsfeldes anzusehen ist.[13] Hat man dies gesehen, so ist die obige Gleichung denkbar einfach, weil man nicht in Versuchung kommt, sie behufs allgemeiner Interpretation umzuformen durch Ausrechnen der Symbole.

Das Herrliche, was ich erlebte, war nun, dass sich nicht nur Newtons Theorie als erste Näherung, sondern auch die Perihelbewegung des Merkur (43″ pro Jahrhundert) als zweite Näherung ergab.[14] Für die Li[ch]tablenkung an der Sonne ergab sich der doppelte Betrag wie früher.[15]

Freundlich hat eine Methode, um die Lichtablenkung an Jupiter zu messen.[16] Nur die Intriguen armseliger Menschen verhindern es, dass diese letzte wichtige Prüfung der Theorie ausgeführt wird.[17] Dies ist mir aber doch nicht so schmerzlich, weil mir die Theorie besonders auch mit Rücksicht auf die qualitative Bestätigung der Verschiebung der Spektr[a]llinien genügend gesichert erscheint.[18]

Ihre beiden Abhandlungen[19] werde ich jetzt studieren und Ihnen dann wieder zusenden. Herzliche Grüsse von Ihrem rabiaten

Einstein.

Die Akademie-Arbeiten sende ich dann alle auf einmal.

7.2 Translation

Berlin, 28 November 1915[1]

Dear Sommerfeld,

Please don't be upset that I am only answering your kind and interesting letter today. But this last month I had one of the most stimulating and exhausting times of my life, albeit one of the most successful. Writing was simply out of the question.[2]

I realized that my earlier gravitational field equations were completely untenable! The indications for this were the following:[3]

1) I proved that the gravitational field in a uniformly rotating system does not satisfy the field equations.[4]

2) The motion of Mercury's perihelion came to 18″ rather than 45″ per century.[5]

3) The covariance considerations in my paper of last year do not yield the Hamiltonian function H. When properly generalized, they allow an arbitrary H.[6] It followed that the idea of covariance with respect to "adapted" coordinate systems was completely misguided.[7]

After all confidence in the result and the method of the earlier theory had thus given way, I clearly saw that a satisfactory solution could only be found through a connection to the general theory of covariants, i.e., Riemann's covariant.[8] Unfortunately, I have immortalized the final errors in this struggle in the Academy papers, which I will soon be able to send you.[9] The final result is as follows.

The gravitational field equations are generally covariant. If

$$(ik, lm)$$

is the Christoffel tensor of the fourth order, then $G_{im} = \sum_{kl} g^{kl}(ik, lm)$ is a symmetrical tensor of the second order.[10] The equations are

$$G_{im} = -\kappa \left(T_{im} - \frac{1}{2} g_{im} \underbrace{\sum_{\alpha\beta} (g^{\alpha\beta} T_{\alpha\beta})} \right).$$

Scalar derived from the energy tensor of "matter," for which I write "T" in the following.

It is easy, of course, to write down these generally-covariant field equations but difficult to see that they are a generalization of Poisson's equations and not easy to see that they satisfy the conservation laws.[11]

The whole theory can be eminently simplified if one chooses the reference system in such a way that $\sqrt{-g} = 1$. In that case, the equations take the form

$$-\sum_l \frac{\partial \left\{ \begin{matrix} im \\ l \end{matrix} \right\}}{\partial x_l} + \sum_{\alpha\beta} \left\{ \begin{matrix} i\alpha \\ \beta \end{matrix} \right\} \left\{ \begin{matrix} m\beta \\ \alpha \end{matrix} \right\} = -\kappa (T_{im} - \frac{1}{2} g_{im} T).$$

I had already considered these equations with Grossmann 3 years ago, with the exception of the second term on the right-hand side.[12] At the time, however, I arrived at the result that they did not have the correct Newtonian limit, which was a mistake. The key to this solution was my realization that not

$$\sum g^{l\alpha} \frac{\partial g_{\alpha i}}{\partial x_m}$$

but the related Christoffel symbols $\left\{ \begin{matrix} im \\ l \end{matrix} \right\}$ are to be regarded as the natural expression for the "components" of the gravitational field.[13] Once this is recognized, the above equation is about the simplest one conceivable, because there is no temptation to rewrite it for the purpose of a general interpretation by computing the symbols.

The wonderful thing I experienced was that not only Newton's theory resulted in first approximation, but also Mercury's perihelion motion (43″ per century) in second approximation.[14] The deflection of light by the Sun came out twice as large as before.[15]

Freundlich has a method of measuring light deflection by Jupiter.[16] Only the intrigues of pathetic individuals prevent this last important test of the theory from being carried out.[17] But this is not too painful for me because the theory strikes me as sufficiently secure, especially in view of the qualitative verification of the shift in spectral lines.[18]

I am now going to study both of your articles[19] and I will then send them back to you. Cordial greetings from your raving

Einstein.

I will send you the Academy papers all at once.

7.3 Commentary

The original, a signed autograph letter, is in the Archive of the *Deutsches Museum* in Munich (*Sommerfeld-Nachlaß*, 1977-28/A, 78). There are copies in the Einstein Archive with the designations EA 21 382 and 21 382.1 (cf. Pt. I, Ch. 1, notes 2 and 15). The transcription presented here comes from CPAE8, Doc. 153. The letter is also presented in transcription in Sommerfeld (2000, Doc. 221).

1. The year was added by Sommerfeld.
2. In this letter, Einstein gives a blow-by-blow account of the transition from the *Entwurf* theory to the theory of November 1915. In Pt. I, Sec. 2.3, we conjectured that he sent this account to Sommerfeld in part to secure priority for the new theory. Einstein knew that Sommerfeld was in touch with Hilbert, with whom Einstein found himself in a race to find satisfactory field equations for his theory.
3. Einstein listed these same three problems in a letter to Hendrik A. Lorentz of January 1, 1916, in which he also specified the order in which he found them: "I had already found earlier that the perihelion motion of Mercury came out too small. In addition I found that the equations were not covariant under substitutions corresponding to uniform rotation of the (new) frame of reference. Finally, I found that the argument I gave last year to determine the Lagrangian H for the gravitational field was completely illusory" (CPAE8, Doc. 177).
4. In a letter to Erwin Freundlich of September 30, 1915 (Ch. 4), Einstein showed that his iterative approximation procedure for solving the *Entwurf* field equations fails to reproduce the metric of a Minkowski space-time in rotating coordinates. Einstein's formulation of the problem of rotation in the letter to Lorentz quoted in the preceding note suggests that the problem was that the *Entwurf* field equations are not invariant under ordinary (what Einstein called *autonomous*) transformations to rotating coordinate systems. As the statement in this letter to Sommerfeld makes clear, the real problem was that the *Entwurf* field equations are not even invariant under *non-autonomous* transformations to rotating coordinate system in the special case of a Minkowski metric (see Pt. I, Sec. 3.1 for the distinction between autonomous and non-autonomous transformations). This is also how Einstein presented the problem in a letter to Willem de Sitter of January 23, 1917: "I calculated directly that the field equations I then held are not satisfied in a system rotating in a Galilean space" (CPAE8, Doc. 290; see Janssen, 1999, pp. 129–131, for a comparison of these different formulations of the problem of rotation). In the first of his November 1915 papers, Einstein (1915a, p. 786; Sec. 6.1) explicitly pointed out that his new field equations are covariant under (autonomous) transformations to uniformly rotating frames.
5. In collaboration with Michele Besso, Einstein had found in 1913 that the *Entwurf* theory, predicts that the perihelion of Mercury should advance an additional 18″ per century compared to the Newtonian prediction (see Ch. 2 and Pt. I, Ch. 7). A value of 17″ per century is recorded in Einstein's so-called "Scratch Notebook" (CPAE3, Appendix A, [p. 61]; see the annotation of [p. 28] of the Einstein-Besso manuscript in Sec. 2.3). In December 1914, Johannes Droste (1915, p. 981; p. 1010 in the English translation) had published the value of 18″, attributing it to De Sitter.
6. In the lengthy exposition of the *Entwurf* theory published in November 1914, Einstein (1914c, p. 1074–1076; Ch. 3) had argued that covariance considerations uniquely lead to the Lagrangian for the *Entwurf* field equations. In a letter to Lorentz of October 12, 1915 (Ch. 5), Einstein showed that this argument is fallacious. He retracted it in print in the first of the November 1915 papers (Einstein 1915a, p. 778; Sec. 6.1).
7. For the definition of "adapted coordinates", see Einstein (1914c, p. 1070; Ch. 3); for discussion, see Pt. I, Secs. 3.1 and 4.2.
8. This passage strongly suggests that an eleventh-hour switch from physics to mathematics was responsible for the breakthrough in Einstein's search for satisfactory

gravitational field equations. However, the November 1915 papers rely heavily on the formalism of Einstein's November 1914 exposition of the *Entwurf* theory (especially in its handling of the compatibility of the field equations and energy-momentum conservation). This casts doubt on Einstein's claim that he had lost all confidence in the *method* of the earlier theory. Moreover, his observation that "a satisfactory solution could only be found through a connection to [Riemannian geometry]" would seem to be at odds with his observation further down that "the key to [the] solution" was his redefinition of the gravitational field. See Pt. I, Sec. 5.3 for discussion of the relative importance of mathematical and physical considerations for the transition from the *Entwurf* field equations to the field equations of November 1915.

9. The four November 1915 papers in Ch. 6 (Einstein 1915a, 1915b, 1915c, 1915d).
10. (ik, lm) is the Riemann tensor; G_{im} the Ricci tensor.
11. This acknowledgment of physical considerations (the Newtonian limit and the compatibility with energy-momentum conservation) is an important qualification of Einstein's suggestion earlier in the letter that the breakthrough of November 1915 was due to him putting his faith in mathematics rather than physics. Einstein, however, may well have had an ulterior motive in emphasizing the importance of these physical considerations. Hilbert had been relying solely on mathematical considerations in his search for gravitational field equations. The denigration of Hilbert's approach, implied by Einstein's observation that finding the equations was easy and that the hard part was to determine whether they pass muster from a physics point of view, may well have been deliberate.
12. These equations, except for the term with the trace $T = g^{\mu\nu}T_{\mu\nu}$ of the energy-momentum tensor, can already be found on p. 22R of the Zurich Notebook (see Ch. 1 and Pt. I, Sec. 1.2, Fig. 1.3). On p. 20L, moreover, Einstein had already considered weak-field equations with a trace term (see Sec. 1.3, notes 13–16, and Sec. 6.4.3, note 3).
13. In the first of his November 1915 papers, Einstein (1915a, p. 782; Sec. 6.1) called the choice of the former expression for the components of the gravitational field a "fatal prejudice" (for discussion, see Pt. I, Sec. 5.1).
14. Years later, Einstein told Lorentz's student, Adriaan D. Fokker, that the result had given him heart palpitations (Pais 1982, p. 253; cf. Pt. I, Ch. 6).
15. These results were published in Einstein (1915c, see Sec. 6.3).
16. A year earlier, in April 1914, Einstein had written to Paul Ehrenfest that "Freundlich has found a method to measure the refraction of light [*Lichtbrechung*] by the gravitational field of Jupiter" (CPAE8, Doc. 2). For discussion of Freundlich's efforts to test Einstein's theory of gravity, see Hentschel (1992, 1994).
17. Einstein is referring to Hermann Struve, Freundlich's superior at the Neubabelsberg Observatory outside Potsdam, and the Munich astronomer Hugo von Seeliger, whose wrath Freundlich had incurred with his papers on the astronomical consequences of Einstein's theory (Freundlich 1915a, 1915b). For further discussion of the tension between Freundlich, Struve, and von Seeliger, see Sec. 4.3, note 1. In the postscript to another letter to Sommerfeld a week and a half later, on December 9, 1915, Einstein wrote: "Tell your colleague Selinger [sic] that he has a horrible temperament. I got a taste of it recently from a reply to the astronomer Freundlich" (CPAE8, Doc. 161). In this reply, Seeliger (1915) severely criticized Freundlich (1915b). Einstein's complaints did not fall on deaf ears. Sommerfeld tried to intervene on Freundlich's behalf with Karl Schwarzschild. In a letter of December 28, 1915, after complimenting Schwarzschild on finding the exact solution for the field of the Sun, Sommerfeld wrote:

> Einstein complains that Freundlich is not allowed to observe. 'Pathetic intrigues of deplorable individuals' [*Armselige Intrigen trauriger Menschen*]. Seeliger is like poison for Freundlich. I appeal to you that you will allow Einstein and Freundlich to measure the light bending by [Jupiter] in your observatory ... Please make sure that German astronomy does not embarrass itself! It has

> not had an opportunity like this for decades to show that it is top-notch. In Göttingen I shook my head when I was told that you are also in the clique against Freundlich (??). So Fr. may not be a great mind. Einstein's, however, is all the greater—better yet: is the greatest we've had since Gauß and Newton (Sommerfeld 2000, Doc. 226).

In an undated letter of January 1916, Sommerfeld sent another letter to Schwarzschild apologizing for trying to win him over for Freundlich. After several conversations with Seeliger, Sommerfeld had "felt obligated to warn Einstein for this astronomical expert [*Gewährsmann*] (just as energetically as I wanted to recommend him to you)" (Doc. 231). Sommerfeld's letter to Einstein with this warning is no longer extant but Einstein's reply is. On February 2, 1915, Einstein wrote to Sommerfeld that he recognized Freundlich's weaknesses but that he nonetheless continued to support him as Freundlich, for all his foibles, had been the only astronomer who had taken the trouble to check some of the astronomical predictions of his theory (CPAE8, Doc. 186; Sommerfeld, 2000, Doc. 232).

18. Einstein's assessment that the gravitational redshift had at least qualitatively been established is based on the work of Freundlich (1915a), who examined the spectra of a large number of stars and found that on average they showed a non-zero red shift. Freundlich's analysis, however, was severely criticized by von Seeliger (1915, see note 17). For the history of attempts to measure the gravitational redshift predicted by general relativity in subsequent years, see Hentschel (1992, 1993, 1994) and Crelinsten (2006).
19. These are the papers elaborating Bohr's atomic model presented to the Bavarian Academy of Sciences in December 1915 and published in its Proceedings (Sommerfeld 1915a; 1915b, see Eckert, 2013, sec. 7.4).

Chapter 8
Einstein to Paul Ehrenfest, January, 1916

8.1 Facsimile

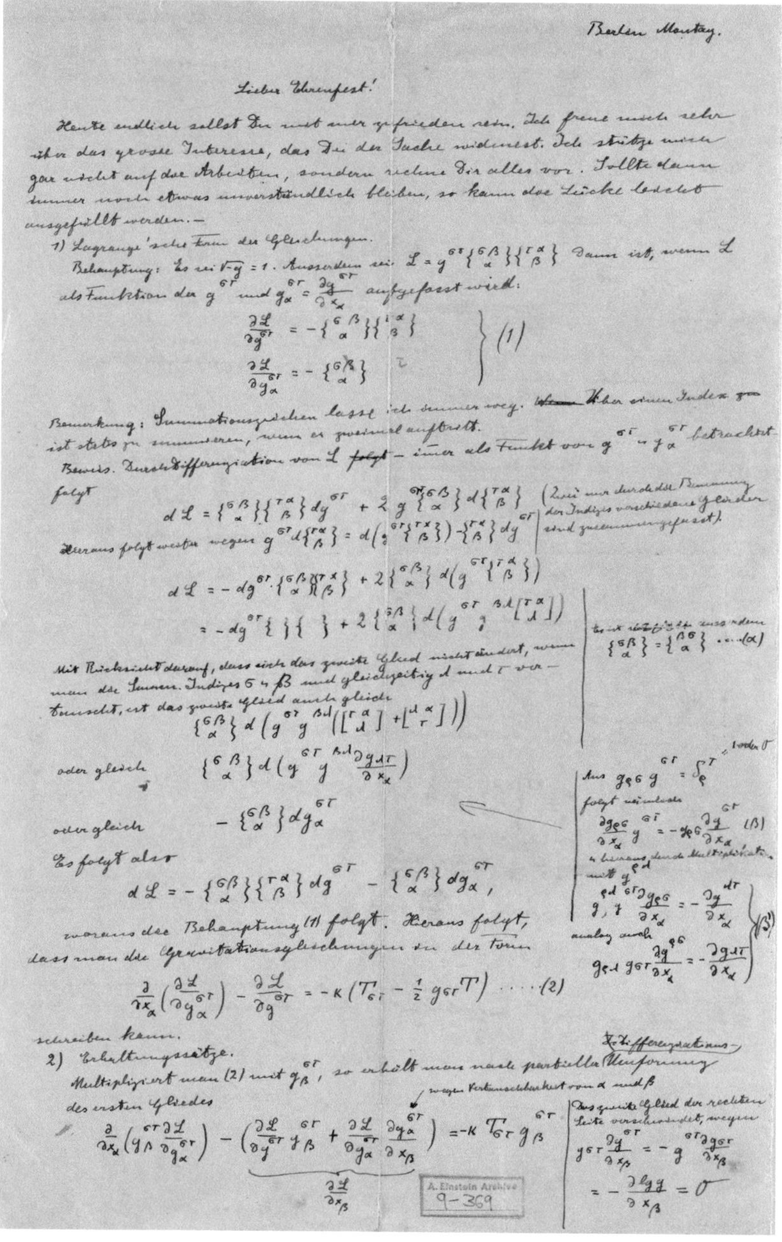

Berlin Montag.

Lieber Ehrenfest!

Heute endlich sollst Du mit mir zufrieden sein. Ich freue mich sehr über das grosse Interesse, das Du der Sache widmest. Ich stütze mich gar nicht auf die Arbeiten, sondern rechne Dir alles vor. Sollte dann immer noch etwas unverständlich bleiben, so kann die Lücke leicht ausgefüllt werden. —

1) Lagrange'sche Form der Gleichungen.

Behauptung: Es sei $\sqrt{-g} = 1$. Ausserdem sei $\mathfrak{L} = g^{\sigma\tau}\left\{{\sigma\beta \atop \alpha}\right\}\left\{{\tau\alpha \atop \beta}\right\}$ Dann ist, wenn $\mathfrak{L}$ als Funktion der $g^{\sigma\tau}$ und $g^{\sigma\tau}_\alpha = \frac{\partial g^{\sigma\tau}}{\partial x_\alpha}$ aufgefasst wird:

$$\left.\begin{aligned} \frac{\partial \mathfrak{L}}{\partial g^{\sigma\tau}} &= -\left\{{\sigma\beta \atop \alpha}\right\}\left\{{\tau\alpha \atop \beta}\right\} \\ \frac{\partial \mathfrak{L}}{\partial g^{\sigma\tau}_\alpha} &= -\left\{{\sigma\beta \atop \alpha}\right\} \end{aligned}\right\} (1)$$

Bemerkung: Summationszeichen lasse ich immer weg. Über einen Index ist stets zu summieren, wenn er zweimal auftritt.

Beweis. Durch Differenziation von $\mathfrak{L}$ folgt — immer als Funkt. von $g^{\sigma\tau}$ u. $g^{\sigma\tau}_\alpha$ betrachtet — folgt

$$d\mathfrak{L} = \left\{{\sigma\beta \atop \alpha}\right\}\left\{{\tau\alpha \atop \beta}\right\} dg^{\sigma\tau} + 2 g^{\sigma\tau}\left\{{\sigma\beta \atop \alpha}\right\} d\left\{{\tau\alpha \atop \beta}\right\}$$

(Zwei nur durch die Benennung der Indizes verschiedene Glieder sind zusammengefasst).

Hieraus folgt weiter wegen $g^{\sigma\tau} d\left\{{\tau\alpha \atop \beta}\right\} = d\left(g^{\sigma\tau}\left\{{\tau\alpha \atop \beta}\right\}\right) - \left\{{\tau\alpha \atop \beta}\right\} dg^{\sigma\tau}$

$$d\mathfrak{L} = -dg^{\sigma\tau}\left\{{\sigma\beta \atop \alpha}\right\}\left\{{\tau\alpha \atop \beta}\right\} + 2\left\{{\sigma\beta \atop \alpha}\right\} d\left(g^{\sigma\tau}\left\{{\tau\alpha \atop \beta}\right\}\right)$$

$$= -dg^{\sigma\tau}\{\ \}\{\ \} + 2\left\{{\sigma\beta \atop \alpha}\right\} d\left(g^{\sigma\tau} g^{\beta\lambda}\left[{\tau\alpha \atop \lambda}\right]\right)$$

Es ist $\left\{{\sigma\beta \atop \alpha}\right\} = \left\{{\beta\sigma \atop \alpha}\right\}$ ....(α)

Mit Rücksicht darauf, dass sich das zweite Glied nicht ändert, wenn man die Summen. Indizes σ u. β und gleichzeitig λ und τ vertauscht, ist das zweite Glied auch gleich

$$\left\{{\sigma\beta \atop \alpha}\right\} d\left(g^{\sigma\tau} g^{\beta\lambda}\left(\left[{\tau\alpha \atop \lambda}\right] + \left[{\lambda\alpha \atop \tau}\right]\right)\right)$$

oder gleich

$$\left\{{\sigma\beta \atop \alpha}\right\} d\left(g^{\sigma\tau} g^{\beta\lambda}\frac{\partial g_{\lambda\tau}}{\partial x_\alpha}\right)$$

oder gleich

$$-\left\{{\sigma\beta \atop \alpha}\right\} dg^{\sigma\tau}_\alpha$$

Aus $g_{\rho\sigma} g^{\sigma\tau} = \delta_\rho^\tau$ folgt nämlich

$$\frac{\partial g_{\rho\sigma}}{\partial x_\alpha} g^{\sigma\tau} = -g_{\rho\sigma}\frac{\partial g^{\sigma\tau}}{\partial x_\alpha} \quad (\beta)$$

u. hieraus durch Multiplikation mit $g^{\rho\lambda}$

$$g^{\rho\lambda} g^{\sigma\tau}\frac{\partial g_{\rho\sigma}}{\partial x_\alpha} = -\frac{\partial g^{\lambda\tau}}{\partial x_\alpha}$$

analog auch

$$g_{\rho\lambda} g_{\sigma\tau}\frac{\partial g^{\rho\sigma}}{\partial x_\alpha} = -\frac{\partial g_{\lambda\tau}}{\partial x_\alpha} \quad (\beta')$$

Es folgt also

$$d\mathfrak{L} = -\left\{{\sigma\beta \atop \alpha}\right\}\left\{{\tau\alpha \atop \beta}\right\} dg^{\sigma\tau} - \left\{{\sigma\beta \atop \alpha}\right\} dg^{\sigma\tau}_\alpha,$$

woraus die Behauptung (1) folgt. Hieraus folgt, dass man die Gravitationsgleichungen in der Form

$$\frac{\partial}{\partial x_\alpha}\left(\frac{\partial \mathfrak{L}}{\partial g^{\sigma\tau}_\alpha}\right) - \frac{\partial \mathfrak{L}}{\partial g^{\sigma\tau}} = -\kappa\left(T_{\sigma\tau} - \tfrac{1}{2} g_{\sigma\tau} T\right) \ \ldots (2)$$

schreiben kann.

2) Erhaltungssätze.

Multipliziert man (2) mit $g^{\sigma\tau}_\beta$, so erhält man nach partieller Differenziations-Umformung des ersten Gliedes

$$\frac{\partial}{\partial x_\alpha}\left(g^{\sigma\tau}_\beta\frac{\partial \mathfrak{L}}{\partial g^{\sigma\tau}_\alpha}\right) - \underbrace{\left(\frac{\partial \mathfrak{L}}{\partial g^{\sigma\tau}} g^{\sigma\tau}_\beta + \frac{\partial \mathfrak{L}}{\partial g^{\sigma\tau}_\alpha}\frac{\partial g^{\sigma\tau}_\alpha}{\partial x_\beta}\right)}_{\frac{\partial \mathfrak{L}}{\partial x_\beta}} = -\kappa\, T_{\sigma\tau} g^{\sigma\tau}_\beta$$

(wegen Vertauschbarkeit von α und β)

Das zweite Glied der rechten Seite verschwindet, wegen

$$g_{\sigma\tau}\frac{\partial g^{\sigma\tau}}{\partial x_\beta} = -g^{\sigma\tau}\frac{\partial g_{\sigma\tau}}{\partial x_\beta} = -\frac{\partial \lg g}{\partial x_\beta} = 0$$

A. Einstein Archive 9-369

M. Janssen, J. Renn, *How Einstein Found His Field Equations*, Classic Texts in the Sciences, https://doi.org/10.1007/978-3-030-97955-3_15

Nun schreibe ich den Erhaltungssatz für die Materie, indem ich ihn formal einführe, ohne seine Gültigkeit vorauszusetzen

$$\frac{\partial T_\mu^\sigma}{\partial x_\sigma} + \frac{1}{2} g^{\alpha\tau}_\mu T_{\alpha\tau} = A_\mu \quad \cdots\cdots (3)$$

(A_μ: unbekannte Raumfunktionen)

Das zweite Glied der linken Seite kann wegen (β) in die Form

$$-\frac{1}{2}\frac{\partial g_{\alpha\tau}}{\partial x_\mu} T^{\alpha\tau}$$

gebracht werden. Ebenso lässt sich die { } einführen, was ich aber hier nicht brauche.

Mit Hilfe davon lässt sich die rechte Seite der letzten Gleichung durch

$$2\kappa \frac{\partial T_\beta^\alpha}{\partial x_\alpha} - 2\kappa A_\beta$$

ersetzen, und man erhält:

$$\frac{\partial}{\partial x_\alpha}\Bigg(\underbrace{\frac{1}{2\kappa}\left[\mathfrak{L}\delta_\beta^\alpha - g^{\sigma\tau}_\beta \frac{\partial \mathfrak{L}}{\partial g^{\sigma\tau}_\alpha}\right]}_{t_\beta^\alpha} + T_\beta^\alpha\Bigg) = A_\beta \quad \cdots\cdots (4)$$

Würden die A_β verschwinden, so wäre dies die Erhaltungsgleichung für Materie und Gravitation zusammen. Einstweilen operieren wir mit (4). Mit Rücksicht auf (1) folgt aus der Definition der t_β^α

$$t_\alpha^\alpha = t = \frac{1}{\kappa}\mathfrak{L} \quad \cdots\cdots (5)$$

3) Gemischte Form der Gravitationsgleichungen. Wir schreiben letztere in der Form

$$-\frac{\partial \left\{ {\sigma\tau \atop \alpha} \right\}}{\partial x_\alpha} + \left\{ {\sigma\beta \atop \alpha} \right\}\left\{ {\tau\alpha \atop \beta} \right\} = -\kappa\left(T_{\sigma\tau} - \frac{1}{2} g_{\sigma\tau} T\right) \cdots (2a)$$

und multiplizieren mit $g^{\tau\nu}$. Das erste Glied links liefert durch partielle Diff. Umformung

$$-\frac{\partial}{\partial x_\alpha}\left(g^{\tau\nu}\left\{ {\sigma\tau \atop \alpha} \right\}\right) + \left\{ {\sigma\tau \atop \alpha} \right\}\frac{\partial g^{\tau\nu}}{\partial x_\alpha},$$

wovon das zweite Glied mit Hilfe der nebenstehend abgeleiteten Gleichung (γ) umgeformt werden kann, sodass man erhält

$$-\frac{\partial}{\partial x_\alpha}\left(g^{\tau\nu}\left\{ {\sigma\tau \atop \alpha} \right\}\right) - g^{\tau\varepsilon}\left\{ {\varepsilon\alpha \atop \nu} \right\}\left\{ {\tau\sigma \atop \alpha} \right\} - g^{\nu\varepsilon}\left\{ {\varepsilon\alpha \atop \tau} \right\}\left\{ {\sigma\tau \atop \alpha} \right\}$$

Hilfsrechnung:

$$\frac{\partial g^{\tau\nu}}{\partial x_\alpha} = -g^{\tau\xi} g^{\nu\zeta}\frac{\partial g_{\xi\zeta}}{\partial x_\alpha} \quad \text{(wegen } \beta'\text{)}$$

$$= -g^{\tau\xi} g^{\nu\zeta}\left(\left[{\xi\alpha \atop \zeta} \right] + \left[{\zeta\alpha \atop \xi} \right]\right)$$

also

$$\frac{\partial g^{\tau\nu}}{\partial x_\alpha} = -g^{\tau\varepsilon}\left\{ {\varepsilon\alpha \atop \nu} \right\} - g^{\nu\varepsilon}\left\{ {\varepsilon\alpha \atop \tau} \right\} \quad \cdots\cdots (\gamma)$$

Das dritte dieser Glieder hebt sich gegenüber dem aus dem zweiten von (2a) entstehenden weg. Man erhält also zunächst

$$\frac{\partial}{\partial x_\alpha}\left(g^{\tau\nu}\left\{ {\sigma\tau \atop \alpha} \right\}\right) + g^{\varepsilon\tau}\left\{ {\varepsilon\alpha \atop \nu} \right\}\left\{ {\tau\sigma \atop \alpha} \right\} = \kappa\left(T_\sigma^\nu - \frac{1}{2}\delta_\sigma^\nu T\right) \quad \cdots\cdots (6)$$

Aus der Definition von t_β^α und den Gleichungen (1) (β) erhält man

$$t_\beta^\alpha = \frac{1}{2} t\,\delta_\beta^\alpha + \frac{1}{2\kappa}\left\{ {\sigma\tau \atop \alpha} \right\}\frac{\partial g^{\sigma\tau}}{\partial x_\beta}$$

Nach Umformung des zweiten Gliedes gemäss (γ) und Vereinigung beider so entstehenden Glieder

$$t_\beta^\alpha - \frac{1}{2}\delta_\beta^\alpha t = -\frac{1}{\kappa} g^{\sigma\varepsilon}\left\{ {\sigma\tau \atop \alpha} \right\}\left\{ {\varepsilon\beta \atop \tau} \right\} \quad \cdots (7)$$

Abgesehen vom Faktor $-\frac{1}{\kappa}$ und der Bezeichnung der Indizes stimmt die rechte Seite von (7) mit dem zweiten Glied in (6) überein, sodass man schreiben kann

$$\frac{\partial}{\partial x_\alpha}\left(g^{\tau\nu}\left\{ {\sigma\tau \atop \alpha} \right\}\right) = \kappa\left((T_\sigma^\nu + t_\sigma^\nu) - \frac{1}{2}\delta_\sigma^\nu (T + t)\right) \quad \cdots (8)$$

Diese Gleichung ist interessant, weil sie zeigt, dass das Entspringen der Gravitationslinien allein durch die Summe $T_\sigma^\nu + t_\sigma^\nu$ bestimmt ist, wie man ja auch erwarten muss.

(Zweites Blatt)

4) Beweis, dass die A_μ verschwinden. Jetzt kommt noch die Hauptsache.

a) Multipliziert man (8) mit δ_ν^σ so erhält man die skalare Gleichung

$$\frac{\partial}{\partial x_\alpha}\left(g^{\nu\tau}\left\{ {\nu\tau \atop \alpha} \right\}\right) = -\kappa(T+A) \quad \cdots (9)$$

Die linke Seite lautet ausführlich

$$\frac{1}{2}\frac{\partial}{\partial x_\alpha}\left(g^{\nu\tau} g^{\alpha\beta}\left(\frac{\partial g_{\beta\nu}}{\partial x_\tau}+\frac{\partial g_{\beta\tau}}{\partial x_\nu}-\frac{\partial g_{\nu\tau}}{\partial x_\beta}\right)\right)$$

Das dritte Glied liefert nichts, weil $g^{\nu\tau}\frac{\partial g_{\nu\tau}}{\partial x_\beta} = \frac{\partial \lg g}{\partial x_\beta} = 0$. Die beiden ersten gehen durch Vertauschung von ν und τ in einander über, sodass sie sich vereinigen lassen. Durch Anwendung von (B') erhält man endlich

$$-\frac{\partial^2 g^{\alpha\tau}}{\partial x_\alpha \partial x_\tau}.$$

Aus (9) wird also

$$\frac{\partial^2 g^{\alpha\beta}}{\partial x_\alpha \partial x_\beta} - \kappa(T+A) = 0 \quad \cdots\cdots (9a)$$

Wir führen an (8) die Operation $\frac{\partial}{\partial x_\nu}$ aus und erhalten mit Rücksicht auf (4)

$$\frac{\partial}{\partial x_\alpha}\frac{\partial}{\partial x_\nu}\left(g^{\tau\nu}\left\{ {\sigma\tau \atop \alpha} \right\}\right) = -\frac{1}{2}\frac{\partial(T+A)}{\partial x_\sigma} + \kappa A_\sigma \quad \cdots (10)$$

Die linke Seite ist ausführlicher

$$\frac{1}{2}\frac{\partial}{\partial x_\alpha}\frac{\partial}{\partial x_\nu}\left(g^{\tau\nu} g^{\alpha\beta}\left(\frac{\partial g_{\sigma\beta}}{\partial x_\tau}+\frac{\partial g_{\tau\beta}}{\partial x_\sigma}-\frac{\partial g_{\sigma\tau}}{\partial x_\beta}\right)\right)$$

Vertauscht man im ersten Glied α und ν sowie β und τ, so geht es abgesehen vom Vorzeichen ins dritte über. Es bleibt nur das zweite, welches mit Rücksicht auf (B') in

$$-\frac{1}{2}\frac{\partial^3 g^{\alpha\nu}}{\partial x_\alpha \partial x_\nu \partial x_\sigma}$$

übergeht

Man erhält daher anstelle von (10)

$$\frac{\partial}{\partial x_\sigma}\left(\frac{\partial^2 g^{\alpha\nu}}{\partial x_\alpha \partial x_\nu} - \kappa(T+A)\right) = -2\kappa A_\sigma \quad \cdots (10a)$$

Aus (9a) und (10a) ergibt sich $A_\sigma = 0$, d. h. genau das (3) der Erhaltungssatz der Materie als Konsequenz der Feldgleichungen (2). —

Du wirst nun wohl keine Schwierigkeit mehr finden. Zeige die Sache auch Lorentz, der die Notwendigkeit der Struktur der rechten Seite der Feldgleichungen auch noch nicht empfindet. Es wäre mir lieb, wenn Du mir diese Blätter dann wieder zurückgäbest, weil ich die Sachen sonst nirgends so hübsch beisammen habe.

Mit besten Grüssen

Dein Einstein.

8.2 Transcription

Berlin Montag.

Lieber Ehrenfest!

Heute endlich sollst Du mit mir zufrieden sein. Ich freue mich sehr über das grosse Interesse, das Du der Sache widmest.[1] Ich stütze mich gar nicht auf die Arbeiten, sondern rechne Dir alles vor.[2] Sollte dann immer noch etwas unverständlich bleiben, so kann die Lücke leicht ausgefüllt werden.—

1) Lagrange'sche Form der Gleichungen.

Behauptung: Es sei $\sqrt{-g}=1$. Ausserdem sei $L=g^{\sigma\tau}\left\{ {\sigma\beta \atop \alpha} \right\}\left\{ {\tau\alpha \atop \beta} \right\}$. Dann ist, wenn L als Funktion der $g^{\sigma\tau}$ und $g^{\sigma\tau}_{\alpha}=\dfrac{\partial g^{\sigma\tau}}{\partial x_\alpha}$ aufgefasst wird:[3]

$$\left.\begin{aligned} \frac{\partial L}{\partial g^{\sigma\tau}} &= -\left\{ {\sigma\beta \atop \alpha} \right\}\left\{ {\tau\alpha \atop \beta} \right\} \\ \frac{\partial L}{\partial g^{\sigma\tau}_{\alpha}} &= -\left\{ {\sigma\beta \atop \alpha} \right\} \end{aligned}\right\} \tag{1}$$

Bemerkung: Summationszeichen lasse ich immer weg. Über einen Index ist stets zu summieren, wenn er zweimal auftritt.[4]

Beweis. Durch Differenziation von L—immer als Funkt von $g^{\sigma\tau}$ & $g^{\sigma\tau}_{\alpha}$ betrachtet—folgt

$$dL=\left\{ {\sigma\beta \atop \alpha} \right\}\left\{ {\tau\alpha \atop \beta} \right\}dg^{\sigma\tau}+2g^{\sigma\tau}\left\{ {\sigma\beta \atop \alpha} \right\}d\left\{ {\tau\alpha \atop \beta} \right\}.$$

(Zwei nur durch die Benennung der Indizes verschiedene Glieder sind zusammengefasst).

Hieraus folgt weiter wegen $g^{\sigma\tau}d\left\{ {\tau\alpha \atop \beta} \right\}=d\left(g^{\sigma\tau}\left\{ {\tau\alpha \atop \beta} \right\}\right)-\left\{ {\tau\alpha \atop \beta} \right\}dg^{\sigma\tau}$

$$\begin{aligned} dL &= -dg^{\sigma\tau}\left\{ {\sigma\beta \atop \alpha} \right\}\left\{ {\tau\alpha \atop \beta} \right\}+2\left\{ {\sigma\beta \atop \alpha} \right\}d\left(g^{\sigma\tau}\left\{ {\tau\alpha \atop \beta} \right\}\right) \\ &= -dg^{\sigma\tau}\left\{\ \right\}\left\{\ \right\}+2\left\{ {\sigma\beta \atop \alpha} \right\}d\left(g^{\sigma\tau}g^{\beta\lambda}\left[{\tau\alpha \atop \lambda} \right]\right). \end{aligned}$$

Es ist ausserdem

$$\left\{ {\sigma\beta \atop \alpha} \right\}=\left\{ {\beta\sigma \atop \alpha} \right\} \tag{α}$$

Mit Rücksicht darauf, dass sich das zweite Glied nicht ändert, wenn man die Summ. Indizes σ & β und gleichzeitig λ und τ vertauscht, ist das zweite

Glied auch gleich

$$\begin{Bmatrix}\sigma\beta\\ \alpha\end{Bmatrix} d\left(g^{\sigma\tau}g^{\beta\lambda}\left(\begin{bmatrix}\tau\alpha\\ \lambda\end{bmatrix}+\begin{bmatrix}\lambda\alpha\\ \tau\end{bmatrix}\right)\right)$$

oder gleich[5] $$\begin{Bmatrix}\sigma\beta\\ \alpha\end{Bmatrix} d\left(g^{\sigma\tau}g^{\beta\lambda}\frac{\partial g_{\lambda\tau}}{\partial x_\alpha}\right)$$

oder gleich $$-\begin{Bmatrix}\sigma\beta\\ \alpha\end{Bmatrix} dg_\alpha^{\sigma\tau}$$

Aus $g_{\rho\sigma}g^{\sigma\tau}=\delta_\rho^\tau=1$ oder 0 folgt nämlich

$$\frac{\partial g_{\rho\sigma}}{\partial x_\alpha}g^{\sigma\tau}=-g_{\rho\sigma}\frac{\partial g^{\sigma\tau}}{\partial x_\alpha}, \qquad (\beta)$$

& hieraus durch Multiplikation mit $g^{\rho\lambda}$

analog auch

$$\left.\begin{aligned} g^{\rho\lambda}g^{\sigma\tau}\frac{\partial g_{\rho\sigma}}{\partial x_\alpha} &= -\frac{\partial g^{\lambda\tau}}{\partial x_\alpha}\\ g_{\rho\lambda}g_{\sigma\tau}\frac{\partial g^{\rho\sigma}}{\partial x_\alpha} &= -\frac{\partial g_{\lambda\tau}}{\partial x_\alpha}\end{aligned}\right\}. \qquad (\beta')$$

Es folgt also[6]

$$dL=-\begin{Bmatrix}\sigma\beta\\ \alpha\end{Bmatrix}\begin{Bmatrix}\tau\alpha\\ \beta\end{Bmatrix}dg^{\sigma\tau}-\begin{Bmatrix}\sigma\beta\\ \alpha\end{Bmatrix}dg_\alpha^{\sigma\tau},$$

woraus die Behauptung (1) folgt. Hieraus folgt, dass man die Gravitationsgleichungen in der Form

$$\frac{\partial}{\partial x_\alpha}\left(\frac{\partial L}{\partial g_\alpha^{\sigma\tau}}\right)-\frac{\partial L}{\partial g^{\sigma\tau}}=-\kappa\left(T_{\sigma\tau}-\frac{1}{2}g_{\sigma\tau}T\right) \qquad (2)$$

schreiben kann.[7]

2) Erhaltungssätze.[8]

Multipliziert man (2) mit $g_\beta^{\sigma\tau}$, so erhält man nach partieller Differenziations-Umformung des ersten Gliedes (wegen Vertauschbarkeit von α und β),

$$\frac{\partial}{\partial x_\alpha}\left(g_\beta^{\sigma\tau}\frac{\partial L}{\partial g_\alpha^{\sigma\tau}}\right)-\underbrace{\left(\frac{\partial L}{\partial g^{\sigma\tau}}g_\beta^{\sigma\tau}+\frac{\partial L}{\partial g_\alpha^{\sigma\tau}}\frac{\partial g_\alpha^{\sigma\tau}}{\partial x_\beta}\right)}_{\frac{\partial L}{\partial x_\beta}}=-\kappa T_{\sigma\tau}g_\beta^{\sigma\tau}.$$

Das zweite Glied der rechten Seite verschwindet, wegen[9]

$$g^{\sigma\tau}\frac{\partial g^{\sigma\tau}}{\partial x_\beta} = -g^{\sigma\tau}\frac{\partial t g_{\sigma\tau}}{\partial x_\beta} = -\frac{\partial \lg g}{\partial x_\beta} = 0.$$

Nun schreibe ich den Erhaltungssatz für die Materie, indem ich ihn formal einführe, ohne seine Gültigkeit vorauszusetzen[10]

$$\frac{\partial T_\mu^\sigma}{\partial x_\sigma} + \frac{1}{2}g_\mu^{\alpha\tau}T_{\alpha\tau} = A_\mu \quad \begin{array}{c}\text{unbekannte}\\ \text{Raumfunktionen}\end{array} \tag{3}$$

Das zweite Glied der linken Seite kann wegen (β) in die Form $-\frac{1}{2}\frac{\partial g_{\alpha\tau}}{\partial x_\mu}$ gebracht werden. Ebenso lässt sich die $\left\{\ \right\}$ einführen, was ich aber hier nicht brauche.

Mit Hilfe davon lässt sich die rechte Seite der letzten Gleichung[11] durch

$$2\kappa\frac{\partial T_\beta^\alpha}{\partial x_\alpha} - 2\kappa A_\beta,$$

ersetzen, und man erhält:

$$\frac{\partial}{\partial x_\alpha}\Bigg(\underbrace{\frac{1}{2\kappa}\left[L\delta_\beta^\alpha - g_\beta^{\sigma\tau}\frac{\partial L}{\partial g_\alpha^{\sigma\tau}}\right]}_{t_\beta^\alpha} + T_\beta^\alpha\Bigg) = A_\beta. \tag{4}$$

Würde die A_β verschwinden, so wäre dies die Erhaltungsgleichung für Materie und Gravitation zusammen. Einstweilen operieren wir mit (4). Mit Rücksicht auf (1) folgt aus der Definition der t_β^α[12]

$$t_\alpha^\alpha = t = \frac{1}{\kappa}L \tag{5}$$

3) *Gemischte Form der Gravitationsgleichungen.*[13] Wir schreiben letztere in der Form

$$-\frac{\partial\begin{Bmatrix}\sigma\beta\\ \alpha\end{Bmatrix}}{\partial x_\alpha} + \begin{Bmatrix}\sigma\beta\\ \alpha\end{Bmatrix}\begin{Bmatrix}\tau\alpha\\ \beta\end{Bmatrix} = -\kappa\left(T_{\sigma\tau} - \frac{1}{2}g_{\sigma\tau}T\right) \tag{2a}$$

und multiplizieren mit $g^{\tau\nu}$. Das erste Glied links liefert durch partielle Diff. Umformung

$$-\frac{\partial}{\partial x_\alpha}\left(g^{\tau\nu}\begin{Bmatrix}\sigma\tau\\ \alpha\end{Bmatrix}\right) + \begin{Bmatrix}\sigma\tau\\ \alpha\end{Bmatrix}\frac{\partial g^{\tau\nu}}{\partial x_\alpha},$$

wovon das zweite Glied mit Hilfe der nebenstehend abgeleiteten Gleichung (γ) umgeformt werden kann,[14] sodass man erhält

$$-\frac{\partial}{\partial x_\alpha}\left(g^{\tau\nu}\begin{Bmatrix}\sigma\tau\\ \alpha\end{Bmatrix}\right)-g^{\tau\varepsilon}\begin{Bmatrix}\varepsilon\alpha\\ \nu\end{Bmatrix}\begin{Bmatrix}\tau\sigma\\ \alpha\end{Bmatrix}-g^{\nu\varepsilon}\begin{Bmatrix}\varepsilon\alpha\\ \tau\end{Bmatrix}\begin{Bmatrix}\sigma\tau\\ \alpha\end{Bmatrix}.$$

Hilfsrechung:

$$\begin{aligned}\frac{\partial g^{\tau\nu}}{\partial x_\alpha}&=g^{\tau\varepsilon}g^{\nu\xi}\frac{\partial g_{\varepsilon\xi}}{\partial x_\alpha}\qquad\text{(wegen }\beta')\\ &=g^{\tau\varepsilon}g^{\nu\xi}\left(\begin{bmatrix}\varepsilon\alpha\\ \xi\end{bmatrix}+\begin{bmatrix}\xi\alpha\\ \varepsilon\end{bmatrix}\right).\end{aligned}$$

also

$$\frac{\partial g^{\tau\nu}}{\partial x_\alpha}=-g^{\tau\varepsilon}\begin{Bmatrix}\varepsilon\alpha\\ \nu\end{Bmatrix}-g^{\tau\varepsilon}\begin{Bmatrix}\varepsilon\alpha\\ \tau\end{Bmatrix}\tag{γ}$$

Das dritte dieser Glieder hebt sich gegenüber dem aus dem zweiten von (2a) entstehenden weg. Man erhält also zunächst

$$\frac{\partial}{\partial x_\alpha}\left(g^{\tau\nu}\begin{Bmatrix}\sigma\tau\\ \alpha\end{Bmatrix}\right)+g^{\varepsilon\tau}\begin{Bmatrix}\varepsilon\alpha\\ \nu\end{Bmatrix}\begin{Bmatrix}\tau\sigma\\ \alpha\end{Bmatrix}=\kappa\left(T_\sigma^\nu-\frac{1}{2}\delta_\sigma^\nu T\right)\tag{6}$$

Aus der Definition von t_β^α und den Gleichungen (1) & (5) erhält man[15]

$$t_\beta^\alpha=\frac{1}{2}t\delta_\beta^\alpha+\frac{1}{2\kappa}\begin{Bmatrix}\sigma\tau\\ \alpha\end{Bmatrix}\frac{\partial g^{\sigma\tau}}{\partial x_\beta}.$$

Nach Umformung des zweiten Gliedes gemäss (γ) und Vereinigung beider so entstehender Glieder

$$t_\beta^\alpha-\frac{1}{2}\delta_\beta^\alpha t=-\frac{1}{\kappa}g^{\sigma\varepsilon}\begin{Bmatrix}\sigma\tau\\ \alpha\end{Bmatrix}\begin{Bmatrix}\varepsilon\beta\\ \tau\end{Bmatrix}\tag{7}$$

Abgesehen vom Faktor $-\frac{1}{\kappa}$ und der Bezeichnung der Indizes stimmt die rechte Seite von (7) mit dem zweiten Glied in (6) überein, sodass man schreiben kann

$$\frac{\partial}{\partial x_\alpha}\left(g^{\tau\nu}\begin{Bmatrix}\sigma\tau\\ \alpha\end{Bmatrix}\right)=\kappa\left((T_\sigma^\nu+t_\sigma^\nu)-\frac{1}{2}\delta_\sigma^\nu(T+t)\right)\tag{8}$$

Diese Gleichung ist interessant, weil sie zeigt, dass das Entspringen der Gravitationslinien allein durch die Summe $T_\sigma^\nu+t_\sigma^\nu$ bestimmt ist, wie man ja auch erwarten muss.[16]

(Zweites Blatt)

4) *Beweis, dass die A_μ verschwinden.*[17] Jetzt kommt noch die Hauptsache.

a) Multipliziert man (8) mit δ_ν^σ so erhält man die skalare Gleichung

$$\frac{\partial}{\partial x_\alpha}\left(g^{\tau\nu}\begin{Bmatrix}\nu\tau\\ \alpha\end{Bmatrix}\right) = -\kappa(T+t) \tag{9}$$

Die linke Seite lautet ausführlich

$$\frac{1}{2}\frac{\partial}{\partial x_\alpha}\left(g^{\nu\tau}g^{\alpha\beta}\left(\frac{\partial g_{\beta\nu}}{\partial x_\tau}+\frac{\partial g_{\beta\tau}}{\partial x_\nu}-\frac{\partial g_{\nu\tau}}{\partial x_\beta}\right)\right).$$

Das dritte Glied liefert nichts, weil $g^{\nu\tau}\frac{\partial g_{\nu\tau}}{\partial x_\beta} = \frac{\partial \lg g}{\partial x_\beta} = 0$. Die beiden ersten gehen durch Vertauschung von ν und τ in einander über, sodass sie sich vereinigen lassen. Durch Anwendung von (β') erhält man endlich

$$-\frac{\partial^2 g^{\alpha\tau}}{\partial x_\alpha \partial x_\tau}.$$

Aus (9) wird also

$$\frac{\partial^2 g^{\alpha\beta}}{\partial x_\alpha \partial x_\beta} - \kappa(T+t) = 0 \tag{9a}$$

Wir führen an (8) die Operation $\frac{\partial}{\partial x_\nu}$ aus und erhalten mit Rücksicht auf (4)[18]

$$\frac{\partial}{\partial x_\alpha}\frac{\partial}{\partial x_\nu}\left(g^{\tau\nu}\begin{Bmatrix}\sigma\tau\\ \alpha\end{Bmatrix}\right) = \frac{1}{2}\frac{\partial(T+t)}{\partial x_\sigma} + \kappa A_\sigma \tag{10}$$

Die linke Seite ist ausführlicher

$$\frac{1}{2}\frac{\partial}{\partial x_\alpha}\frac{\partial}{\partial x_\nu}\left(g^{\tau\nu}g^{\alpha\beta}\left(\frac{\partial g_{\sigma\beta}}{\partial x_\tau}+\frac{\partial g_{\tau\beta}}{\partial x_\sigma}-\frac{\partial g_{\sigma\tau}}{\partial x_\beta}\right)\right).$$

Vertauscht man im ersten Glied α und ν so wie β und τ, so geht es abgesehen vom Vorzeichen ins dritte über. Es bleibt nur das zweite, welches mit Rücksicht auf (β') in

$$-\frac{1}{2}\frac{\partial^3 g^{\alpha\nu}}{\partial x_\alpha \partial x_\nu \partial x_\sigma}$$

übergeht. Man erhält daher anstelle von (10)

$$\frac{\partial}{\partial x_\sigma}\left(\frac{\partial^2 g^{\alpha\nu}}{\partial x_\alpha \partial x_\nu} - \kappa(T+t)\right) = -2\kappa A_\sigma \tag{10a}$$

Aus (9a) und (10a) ergibt sich $A_\sigma = 0$, d. h. gemäss (3) der Erhaltungssatz der Materie als Konsequenz der Feldgleichungen (2).[19]

Du wirst nun wohl keine Schwierigkeit mehr finden. Zeige die Sache auch Lorentz, der die Notwendigkeit der Struktur der rechten Seite der Feldgleichungen auch noch nicht empfindet.[20] Es wäre mir lieb, wenn Du mir diese Blätter dann wieder zurückgäbest, weil ich die Sachen sonst nirgends so hübsch beisammen habe.[21]

Mit besten Grüssen Dein

Einstein.

8.3 Translation

Berlin, Monday

Dear Ehrenfest,

Today you should finally be satisfied. I am very happy about the strong interest you take in the matter.[1] I will not rely on the papers at all but will go through all calculations for you.[2] If some things then still remain unclear, the gap can be easily closed.

1) Lagrangian form of the equations.

Assertion: Let $\sqrt{-g} = 1$. In addition, let $L = g^{\sigma\tau} \left\{ {\sigma\beta \atop \alpha} \right\} \left\{ {\tau\alpha \atop \beta} \right\}$. Then it follows, if we conceive of L as a function of $g^{\sigma\tau}$ and $g_\alpha^{\sigma\tau} = \dfrac{\partial g^{\sigma\tau}}{\partial x_\alpha}$:[3]

$$\left.\begin{aligned} \frac{\partial L}{\partial g^{\sigma\tau}} &= -\left\{ {\sigma\beta \atop \alpha} \right\} \left\{ {\tau\alpha \atop \beta} \right\} \\ \frac{\partial L}{\partial g_\alpha^{\sigma\tau}} &= -\left\{ {\sigma\beta \atop \alpha} \right\} \end{aligned}\right\}. \qquad (1)$$

Comment: I omit all summation signs. An index must be summed over whenever it appears twice.[4]

Proof: From the differentiation of L—always considered as a function of $g^{\sigma\tau}$ & $g_\alpha^{\sigma\tau}$—it follows that

$$dL = \left\{ {\sigma\beta \atop \alpha} \right\} \left\{ {\tau\alpha \atop \beta} \right\} dg^{\sigma\tau} + 2g^{\sigma\tau} \left\{ {\sigma\beta \atop \alpha} \right\} d \left\{ {\tau\alpha \atop \beta} \right\}.$$

(Two terms that differ only in the labeling of the indices have been combined.)

Since $g^{\sigma\tau} d \left\{ {\tau\alpha \atop \beta} \right\} = d\left(g^{\sigma\tau} \left\{ {\tau\alpha \atop \beta} \right\} \right) - \left\{ {\tau\alpha \atop \beta} \right\} dg^{\sigma\tau}$, it furthermore follows that

$$dL = -dg^{\sigma\tau}\begin{Bmatrix}\sigma\beta\\ \alpha\end{Bmatrix}\begin{Bmatrix}\tau\alpha\\ \beta\end{Bmatrix} + 2\begin{Bmatrix}\sigma\beta\\ \alpha\end{Bmatrix} d\left(g^{\sigma\tau}\begin{Bmatrix}\tau\alpha\\ \beta\end{Bmatrix}\right)$$

$$= -dg^{\sigma\tau}\begin{Bmatrix} \\ \end{Bmatrix}\begin{Bmatrix} \\ \end{Bmatrix} + 2\begin{Bmatrix}\sigma\beta\\ \alpha\end{Bmatrix} d\left(g^{\sigma\tau}g^{\beta\lambda}\begin{bmatrix}\tau\alpha\\ \lambda\end{bmatrix}\right).$$

In addition,

$$\begin{Bmatrix}\sigma\beta\\ \alpha\end{Bmatrix} = \begin{Bmatrix}\beta\sigma\\ \alpha\end{Bmatrix}. \qquad (\alpha)$$

Since the second term does not change when the summation indices σ and β and, at the same time, λ and τ are interchanged, the second term is also equal to

$$\begin{Bmatrix}\sigma\beta\\ \alpha\end{Bmatrix} d\left(g^{\sigma\tau}g^{\beta\lambda}\left(\begin{bmatrix}\tau\alpha\\ \lambda\end{bmatrix} + \begin{bmatrix}\lambda\alpha\\ \tau\end{bmatrix}\right)\right)$$

or equal to[5]

$$\begin{Bmatrix}\sigma\beta\\ \alpha\end{Bmatrix} d\left(g^{\sigma\tau}g^{\beta\lambda}\frac{\partial g_{\lambda\tau}}{\partial x_\alpha}\right)$$

or equal to

$$-\begin{Bmatrix}\sigma\beta\\ \alpha\end{Bmatrix} dg^{\sigma\tau}_{\alpha}.$$

For it follows from $g_{\rho\sigma}g^{\sigma\tau} = \delta^{\tau}_{\rho} = 1$ or 0 that

$$\frac{\partial g_{\rho\sigma}}{\partial x_\alpha}g^{\sigma\tau} = -g_{\rho\sigma}\frac{\partial g^{\sigma\tau}}{\partial x_\alpha}, \qquad (\beta)$$

& from this through multiplication by $g^{\rho\lambda}$

$$\left.\begin{aligned} g^{\rho\lambda}g^{\sigma\tau}\frac{\partial g_{\rho\sigma}}{\partial x_\alpha} &= -\frac{\partial g^{\lambda\tau}}{\partial x_\alpha} \\ \text{analogously also} \qquad g_{\rho\lambda}g_{\sigma\tau}\frac{\partial g^{\rho\sigma}}{\partial x_\alpha} &= -\frac{\partial g_{\lambda\tau}}{\partial x_\alpha}\end{aligned}\right\}. \qquad (\beta')$$

It thus follows that[6]

$$dL = -\begin{Bmatrix}\sigma\beta\\ \alpha\end{Bmatrix}\begin{Bmatrix}\tau\alpha\\ \beta\end{Bmatrix} dg^{\sigma\tau} - \begin{Bmatrix}\sigma\beta\\ \alpha\end{Bmatrix} dg^{\sigma\tau}_{\alpha},$$

from which assertion (1) follows. From this it follows that the gravitation equations can be written in the form[7]

$$\frac{\partial}{\partial x_\alpha}\left(\frac{\partial L}{\partial g^{\sigma\tau}_{\alpha}}\right) - \frac{\partial L}{\partial g^{\sigma\tau}} = -\kappa\left(T_{\sigma\tau} - \frac{1}{2}g_{\sigma\tau}T\right). \qquad (2)$$

2) Conservation laws.[8]

If (2) is multiplied by $g_\beta^{\sigma\tau}$, one obtains, after rewriting the first term with the help of partial differentiation (because of the interchangeability of α and β),

$$\frac{\partial}{\partial x_\alpha}\left(g_\beta^{\sigma\tau}\frac{\partial L}{\partial g_\alpha^{\sigma\tau}}\right)-\underbrace{\left(\frac{\partial L}{\partial g^{\sigma\tau}}g_\beta^{\sigma\tau}+\frac{\partial L}{\partial g_\alpha^{\sigma\tau}}\frac{\partial g_\alpha^{\sigma\tau}}{\partial x_\beta}\right)}_{\frac{\partial L}{\partial x_\beta}}=-\kappa T_{\sigma\tau}g_\beta^{\sigma\tau}.$$

The second term on the right-hand side vanishes since[9]

$$g^{\sigma\tau}\frac{\partial g^{\sigma\tau}}{\partial x_\beta}=-g^{\sigma\tau}\frac{\partial t g_{\sigma\tau}}{\partial x_\beta}=-\frac{\partial \lg g}{\partial x_\beta}=0.$$

I now write down the conservation law of matter, introducing it formally and without assuming its validity[10]

$$\frac{\partial T_\mu^\sigma}{\partial x_\sigma}+\frac{1}{2}g_\mu^{\alpha\tau}T_{\alpha\tau}=A_\mu. \quad \text{unknown functions of the spatial coordinates} \tag{3}$$

The second term on the left-hand side can, on account of (β), be written in the form $-\frac{1}{2}\frac{\partial g_{\alpha\tau}}{\partial x_\mu}$. One could also introduce the $\left\{\right\}$ but I have no need for that at this point.
With the help of this, the right-hand side of the last equation[11] can be replaced by

$$2\kappa\frac{\partial T_\beta^\alpha}{\partial x_\alpha}-2\kappa A_\beta,$$

and one obtains:

$$\frac{\partial}{\partial x_\alpha}\Bigg(\underbrace{\frac{1}{2\kappa}\left[L\delta_\beta^\alpha-g_\beta^{\sigma\tau}\frac{\partial L}{\partial g_\alpha^{\sigma\tau}}\right]}_{t_\beta^\alpha}+T_\beta^\alpha\Bigg)=A_\beta. \tag{4}$$

Were A_β to vanish, this would be the conservation law for matter and gravity combined. For the time being we will use (4). If we take (1) into account, it follows from the definition of t_β^α that[12]

$$t_\alpha^\alpha=t=\frac{1}{\kappa}L. \tag{5}$$

3) *Mixed form of the gravitation equations.*[13] We write the field equations in the form

$$-\frac{\partial \left\{ \begin{matrix} \sigma\beta \\ \alpha \end{matrix} \right\}}{\partial x_\alpha} + \left\{ \begin{matrix} \sigma\beta \\ \alpha \end{matrix} \right\} \left\{ \begin{matrix} \tau\alpha \\ \beta \end{matrix} \right\} = -\kappa \left(T_{\sigma\tau} - \frac{1}{2} g_{\sigma\tau} T \right) \tag{2a}$$

and multiply by $g^{\tau\nu}$. With the help of partial differentiation, the first term yields

$$-\frac{\partial}{\partial x_\alpha} \left(g^{\tau\nu} \left\{ \begin{matrix} \sigma\tau \\ \alpha \end{matrix} \right\} \right) + \left\{ \begin{matrix} \sigma\tau \\ \alpha \end{matrix} \right\} \frac{\partial g^{\tau\nu}}{\partial x_\alpha},$$

The second term can be rewritten with the help of equation (γ) derived in the passage to the right[14] and we obtain

$$-\frac{\partial}{\partial x_\alpha} \left(g^{\tau\nu} \left\{ \begin{matrix} \sigma\tau \\ \alpha \end{matrix} \right\} \right) - g^{\tau\varepsilon} \left\{ \begin{matrix} \varepsilon\alpha \\ \nu \end{matrix} \right\} \left\{ \begin{matrix} \tau\sigma \\ \alpha \end{matrix} \right\} - g^{\nu\varepsilon} \left\{ \begin{matrix} \varepsilon\alpha \\ \tau \end{matrix} \right\} \left\{ \begin{matrix} \sigma\tau \\ \alpha \end{matrix} \right\}.$$

Auxiliary calculation:

$$\begin{aligned} \frac{\partial g^{\tau\nu}}{\partial x_\alpha} &= g^{\tau\varepsilon} g^{\nu\xi} \frac{\partial g_{\varepsilon\xi}}{\partial x_\alpha} \qquad \text{(because of } \beta'\text{)} \\ &= g^{\tau\varepsilon} g^{\nu\xi} \left(\left[\begin{matrix} \varepsilon\alpha \\ \xi \end{matrix} \right] + \left[\begin{matrix} \xi\alpha \\ \varepsilon \end{matrix} \right] \right). \end{aligned}$$

Hence

$$\frac{\partial g^{\tau\nu}}{\partial x_\alpha} = -g^{\tau\varepsilon} \left\{ \begin{matrix} \varepsilon\alpha \\ \nu \end{matrix} \right\} - g^{\tau\varepsilon} \left\{ \begin{matrix} \varepsilon\alpha \\ \tau \end{matrix} \right\}. \tag{γ}$$

The third of these terms cancels against the one arising from the second term in (2a). One thus obtains initially

$$\frac{\partial}{\partial x_\alpha} \left(g^{\tau\nu} \left\{ \begin{matrix} \sigma\tau \\ \alpha \end{matrix} \right\} \right) + g^{\varepsilon\tau} \left\{ \begin{matrix} \varepsilon\alpha \\ \nu \end{matrix} \right\} \left\{ \begin{matrix} \tau\sigma \\ \alpha \end{matrix} \right\} = \kappa \left(T_\sigma^\nu - \frac{1}{2} \delta_\sigma^\nu T \right). \tag{6}$$

From the definition of t_β^α and equations (1) & (5) one obtains[15]

$$t_\beta^\alpha = \frac{1}{2} t \delta_\beta^\alpha + \frac{1}{2\kappa} \left\{ \begin{matrix} \sigma\tau \\ \alpha \end{matrix} \right\} \frac{\partial g^{\sigma\tau}}{\partial x_\beta}.$$

When the second term is rewritten with the help of (γ) and the two resulting terms are combined, this gives:

$$t_\beta^\alpha - \frac{1}{2} \delta_\beta^\alpha t = -\frac{1}{\kappa} g^{\sigma\varepsilon} \left\{ \begin{matrix} \sigma\tau \\ \alpha \end{matrix} \right\} \left\{ \begin{matrix} \varepsilon\beta \\ \tau \end{matrix} \right\}. \tag{7}$$

Except for the factor $-\frac{1}{\kappa}$ and the labeling of the indices, the right-hand side of (7) matches the second term in (6), so that one can write

$$\frac{\partial}{\partial x_\alpha}\left(g^{\tau\nu}\left\{\begin{matrix}\sigma\tau\\ \alpha\end{matrix}\right\}\right)=\kappa\left(\left(T_\sigma^\nu+t_\sigma^\nu\right)-\frac{1}{2}\delta_\sigma^\nu(T+t)\right). \tag{8}$$

This equation is interesting because it shows that the source of the gravitational field lines is determined solely by the sum $T_\sigma^\nu+t_\sigma^\nu$, as one should expect.[16]

(Second sheet)

4) *Proof that A_μ vanishes.*[17] Now comes the main point.

a) Multiplying (8) by δ_ν^σ, one obtains the scalar equation

$$\frac{\partial}{\partial x_\alpha}\left(g^{\tau\nu}\left\{\begin{matrix}\nu\tau\\ \alpha\end{matrix}\right\}\right)=-\kappa(T+t). \tag{9}$$

The left-hand side expands to

$$\frac{1}{2}\frac{\partial}{\partial x_\alpha}\left(g^{\nu\tau}g^{\alpha\beta}\left(\frac{\partial g_{\beta\nu}}{\partial x_\tau}+\frac{\partial g_{\beta\tau}}{\partial x_\nu}-\frac{\partial g_{\nu\tau}}{\partial x_\beta}\right)\right).$$

The third term does not contribute anything, since $g^{\nu\tau}\frac{\partial g_{\nu\tau}}{\partial x_\beta}=\frac{\partial \lg g}{\partial x_\beta}=0$. The first two turn into each other if ν and τ are interchanged and can thus be combined. Applying (β') one then obtains

$$-\frac{\partial^2 g^{\alpha\tau}}{\partial x_\alpha \partial x_\tau}.$$

(9) thus turns into

$$\frac{\partial^2 g^{\alpha\beta}}{\partial x_\alpha \partial x_\beta}-\kappa(T+t)=0. \tag{9a}$$

We now perform the operation $\frac{\partial}{\partial x_\nu}$ on (8) and, with the help of (4), obtain[18]

$$\frac{\partial}{\partial x_\alpha}\frac{\partial}{\partial x_\nu}\left(g^{\tau\nu}\left\{\begin{matrix}\sigma\tau\\ \alpha\end{matrix}\right\}\right)=\frac{1}{2}\frac{\partial\kappa(T+t)}{\partial x_\sigma}+\kappa A_\sigma. \tag{10}$$

The left-hand side expands to

$$\frac{1}{2}\frac{\partial}{\partial x_\alpha}\frac{\partial}{\partial x_\nu}\left(g^{\tau\nu}g^{\alpha\beta}\left(\frac{\partial g_{\sigma\beta}}{\partial x_\tau}+\frac{\partial g_{\tau\beta}}{\partial x_\sigma}-\frac{\partial g_{\sigma\tau}}{\partial x_\beta}\right)\right).$$

If α and ν as well as β and τ are interchanged in the first term, it turns into the third except for the sign. Hence only the second term remains, which on account of (β') becomes

$$-\frac{1}{2}\frac{\partial^3 g^{\alpha\nu}}{\partial x_\alpha \partial x_\nu \partial x_\sigma}.$$

Instead of (10), one thus obtains

$$\frac{\partial}{\partial x_\sigma}\left(\frac{\partial^2 g^{\alpha\nu}}{\partial x_\alpha \partial x_\nu} - \kappa(T+t)\right) = -2\kappa A_\sigma. \tag{10a}$$

From (9a) and (10a) it follows that $A_\sigma = 0$, i.e., the conservation law of matter (3) holds as a consequence of the field equations (2).[19]

I trust you will not find any further difficulties now. Show the material to Lorentz as well, who also has not yet appreciated the necessity of the structure of the right-hand side of the field equations.[20] I would appreciate it if you could then return these pages to me, as I do not have these things together in such a nice form anywhere else.[21]

With best regards, yours,

Einstein.

8.4 Commentary

The original, a signed autograph letter, is in the Einstein Archive (designation: EA 9 369, cf. Pt. I, Ch. 1, notes 2 and 15). The letter is a follow-up to a letter to Ehrenfest (CPAE8, Doc. 182) of January 17, 1916, which was a Monday. Since the undated letter presented here says "Monday" at the top, its earliest possible date is January 24. The letter clearly predates the review article on the new theory (Einstein 1916b, see Ch. 9), which was received by *Annalen der Physik* on March 20, 1916. It is much more likely that the letter was written on or a week after January 24 than shortly before March 20.

The transcription presented here follows CPAE8, Doc. 185. This transcription does not reflect the disposition of some comments and short auxiliary calculations on the first two pages of this letter. In CPAE8, deviations are explained in the annotation. Since the letter is reproduced in facsimile in this volume, we only note one such deviation (see note 14).

1. The group in Leiden around Lorentz and Ehrenfest had taken an active interest in the development of Einstein's theory, leading to a lively three-way correspondence in late 1915 and early 1916 (discussed in Kox 1988). In the letter to Ehrenfest before this one, Einstein was even more effusive voicing his appreciation of the enthusiasm of Ehrenfest and Lorentz for his new theory: "you form a beautiful little corner on this barren planet. There is cleverness in many places but goodness and generosity are depressingly rare" (CPAE6, Doc. 182). Unfortunately, no letters from Lorentz to Einstein from this period seem to have survived but three letters from Einstein to Lorentz have (January 1, 17 and 19; CPAE8, Docs. 177, 183 and 184; see also Kox 2008, Docs. 304–306; in the second of these Einstein writes that he has three letters from Lorentz). Only one letter from Ehrenfest to Einstein is still extant (January 1, 1916, CPAE8, Doc. 177a, in CPAE13) but six from Einstein to Ehrenfest are (December 26 and 29, 1915 and January 3, 5, 17 and [on or after January 19], 1916; CPAE8, Docs. 173, 174, 179, 180, 182 and 185). In addition, fourteen (!) letters Lorentz and Ehrenfest exchanged between December 23, 1915 and January 28, 1916 survive, nine from Lorentz and five from Ehrenfest (see Kox 2018, Docs. 174–187, pp.

441–487, for transcriptions of the mostly Dutch originals, English translations (except for the few passages in German) and commentary). Lorentz and Ehrenfest also shared the letters they received from Einstein (Lorentz enclosed a letter from Einstein to Ehrenfest with two of his own letters to Ehrenfest [Docs. 182 and 187]). Much of this three-way correspondence (including the one extant letter from Ehrenfest to Einstein) deals with the "hole argument" and the "point-coincidence argument" (see Pt. I, Sec. 4.1, especially notes 5 and 6). The letter presented here, however, focuses on the field equations and their compatibility with energy-momentum conservation. The points labeled (1) through (4) in this letter all have to do with adjustments on this score in calculations in the first November 1915 paper (Einstein 1915a, sec. 3, pp. 783–785) necessitated by the addition of a trace term to the field equations in the fourth (Einstein 1915d).

2. In earlier letters to Ehrenfest, Einstein did not give self-contained derivations but relied heavily on his papers of November 1915 (Einstein 1915a; 1915b; 1915d, see Secs. 6.1, 6.2 and 6.4), which in turn rely heavily on his review article on the *Entwurf* (Einstein 1914c, see Ch. 3).

(1) Lagrangian form of the field equations

3. In the first paper of November 1915, Einstein (1915a, p. 784, Eqs. (17)–(19a)) had cast his new field equations in Lagrangian form, using the Lagrangian given here, $L = g^{\sigma\tau}\Gamma^{\alpha}_{\beta\sigma}\Gamma^{\beta}_{\alpha\tau}$ with $\Gamma^{\alpha}_{\sigma\tau} \equiv -\begin{Bmatrix}\alpha\\ \sigma\tau\end{Bmatrix}$. He had given the expressions for $\partial L/\partial g^{\sigma\tau}$ and $\partial L/\partial g^{\sigma\tau}_{\alpha}$ in Eqs. (19) and (19a) without deriving them. These derivations are supplied here. Note that β should be τ in the expression for $\partial L/\partial g^{\sigma\tau}_{\alpha}$. This error, which is the result of a minor slip in Einstein's derivation of an expression for dL below (see note 5), is corrected in red pencil, probably by Ehrenfest. The error does not affect any of the results in the letter.
4. This is the first extant document in which Einstein explicitly formulates this convention, now known as the Einstein summation convention. It first appeared in print in Einstein (1916b, p. 781, see Sec. 9.3, note 1).
5. Using the definition of the Christoffel symbols of the first kind to substitute

$$[\tau\alpha,\lambda]+[\lambda\alpha,\tau] = \frac{1}{2}\left(g_{\lambda\tau,\alpha}+g_{\lambda\alpha,\tau}-g_{\tau\alpha,\lambda}+g_{\tau\lambda,\alpha}+g_{\tau\alpha,\lambda}-g_{\lambda\alpha,\tau}\right) = g_{\lambda\tau,\alpha}$$

into

$$\begin{Bmatrix}\alpha\\ \sigma\beta\end{Bmatrix} d\left(g^{\sigma\tau}g^{\beta\lambda}\left([\tau\alpha,\lambda]+[\lambda\alpha,\tau]\right)\right)$$

and using that

$$g^{\sigma\tau}g^{\beta\lambda}g_{\lambda\tau,\alpha} = -g^{\sigma\tau}\left(g_{\lambda\tau}g^{\beta\lambda}_{\alpha}\right) = -\delta^{\sigma}_{\lambda}g^{\beta\lambda}_{\alpha} = -g^{\beta\sigma}_{\alpha}$$

(cf. the equations Einstein labeled (β) and (β')), we can rewrite this expression as

$$-\begin{Bmatrix}\alpha\\ \sigma\beta\end{Bmatrix} dg^{\sigma\beta}_{\alpha}.$$

Einstein erroneously wrote $dg^{\sigma\tau}_{\alpha}$ instead of $dg^{\sigma\beta}_{\alpha}$. This error carries over to the expression for dL below and is responsible for the error corrected in Eq. (1) (see note 3).
6. In the last term β should be τ (see notes 3 and 5).
7. In this letter, Einstein had, in fact, not yet introduced the field equations of his fourth November 1915 paper, which in unimodular coordinates are

$$\Gamma^{\alpha}_{\sigma\tau,\alpha} + \Gamma^{\alpha}_{\beta\sigma}\Gamma^{\beta}_{\alpha\tau} = -\kappa\left(T_{\sigma\tau} - \frac{1}{2}g_{\sigma\tau}T\right)$$

(see Eq. (2a) under point (3) and note 13 below), nor his new definition of the gravitational field as minus the Christoffel symbols of the second kind,

$$\Gamma^{\alpha}_{\sigma\tau} \equiv -\begin{Bmatrix}\alpha \\ \sigma\tau\end{Bmatrix} = -g^{\alpha\rho}[\sigma\tau,\rho].$$

(see Einstein 1915d, p. 845, Eqs. (6) and (4), respectively). The calculation above shows that

$$\frac{\partial L}{\partial g^{\sigma\tau}} = -\Gamma^{\alpha}_{\beta\sigma}\Gamma^{\beta}_{\alpha\tau}, \qquad \frac{\partial}{\partial x^{\alpha}}\left(\frac{\partial L}{\partial g^{\sigma\tau}_{\alpha}}\right) = \Gamma^{\alpha}_{\sigma\tau,\alpha}$$

(ibid., Eqs, (19) and (19a)). In other words, the vacuum field equations in unimodular coordinates are the Euler-Lagrange equations for the Lagrangian $L = g^{\sigma\tau}\Gamma^{\alpha}_{\beta\sigma}\Gamma^{\beta}_{\alpha\tau}$.

(2) Conservation laws

8. Just as he filled in some of the details behind Eqs. (19) and (19a) on p. 784 of Einstein (1915a) under point (1) in this letter, Einstein fills in some of the details behind Eqs. (20)–(20b) on pp. 784–785 under point (2). He shows that these equations remain unchanged if the trace term of Einstein (1915d) is added to the field equations (see note 9). See Sec. 6.1.3, note 15, for our reconstruction of this derivation.
9. As he had already noted in his fourth November 1915 paper (Einstein 1915d, p. 846),the contraction of $g^{\sigma\tau}_{\beta}$ with the trace term, $-\frac{1}{2}g_{\sigma\tau}T$ vanishes (see Sec. 6.4.3, note 5).
10. In his first November paper, Einstein (1915a, p. 782, Eq. (14)) had shown that the equation below can serve as the energy-momentum balance law in a theory covariant under unimodular transformations (see Sec. 6.1.3, note 10). The term $-g^{\tau\mu}g_{\mu\nu,\sigma}T^{\nu}_{\tau}$ in Eq. (14) can be rewritten as $g^{\tau\mu}_{\sigma}g_{\mu\nu}T^{\nu}_{\tau} = g^{\tau\mu}_{\sigma}T_{\tau\mu}$. In unimodular coordinates, the generally covariant version of the energy-momentum balance law, the vanishing of the covariant divergence of the energy-momentum tensor for matter ($T^{\mu\nu}{}_{;\nu} = 0$), reduces to this simpler equation (with $A_{\mu} = 0$).
11. The "last equation" here refers to the first equation under the heading (2). Using the energy-momentum balance law (3), Einstein substitutes $2\kappa(T^{\alpha}_{\beta,\alpha} - A_{\beta})$ for $-\kappa T_{\sigma\tau}g^{\sigma\tau}_{\beta}$ on the right-hand side of this equation.
12. Contrary to what he promised at the beginning of the letter, Einstein appears to leave it to Ehrenfest at this point to check that inserting the Lagrangian L (see note 3) into the definition of the gravitational energy-momentum density t^{α}_{β} (a definition indicated by the upwards curly bracket under the first two terms of the expression in parentheses in Eq. (4)), one finds that its trace, $t = t^{\alpha}_{\alpha}$, is equal to L/κ (Eq. (5)). See Sec. 6.1.3, note 16, for our derivation of Eq. (20b) in Einstein (1915a, p. 785) for $\kappa t^{\lambda}_{\sigma}$ in terms of $\Gamma^{\alpha}_{\sigma\tau} \equiv -\begin{Bmatrix}\alpha \\ \sigma\tau\end{Bmatrix}$:

$$\kappa t^{\lambda}_{\sigma} = \frac{1}{2}\delta^{\lambda}_{\sigma}g^{\mu\nu}\Gamma^{\alpha}_{\mu\beta}\Gamma^{\beta}_{\nu\alpha} - g^{\mu\nu}\Gamma^{\alpha}_{\mu\sigma}\Gamma^{\lambda}_{\alpha\nu}.$$

Relabeling the summation indices, we readily confirm that $\kappa t = g^{\sigma\tau}\Gamma^{\alpha}_{\sigma\beta}\Gamma^{\beta}_{\alpha\tau} = L$ (we already used t in Sec. 6.1.3, note 17, and Sec. 6.4.3, note 1). In the next section of his letter, under the header (3), Einstein derives an expression for t^{α}_{β} (Eq. (7)) equivalent to the one for $\kappa t^{\lambda}_{\sigma}$ above. He assumes Eq. (5) in this derivation but that can easily be avoided (see note 15).

(3) Mixed form of the gravitational equations

13. Under point (3) in this letter, Einstein contracts the covariant form of the vacuum field equations in unimodular coordinates with a trace term of his fourth November 1915 paper (Einstein 1915d, p. 845, Eq. (6)),

$$\Gamma^{\alpha}_{\mu\nu,\alpha} + \Gamma^{\alpha}_{\mu\beta}\Gamma^{\beta}_{\nu\alpha} = -\kappa\left(T_{\mu\nu} - \frac{1}{2}g_{\mu\nu}T\right) \tag{2a}$$

with $g^{\nu\sigma}$, just as he had done for the field equations without this trace term in his first November 1915 paper (Einstein 1915a, p. 785, Eq. (21) and the two equations below it), and shows that the equations with the trace term in their mixed form are (see note 15 below):

$$(g^{\nu\sigma}\Gamma^{\alpha}_{\mu\nu})_{,\alpha} = -\kappa\left(\left[t^{\sigma}_{\mu} + T^{\sigma}_{\mu}\right] - \frac{1}{2}\delta^{\sigma}_{\mu}\left[t + T\right]\right). \tag{8}$$

14. The "auxiliary calculation" (*Hilfsrechnung*) with the derivation of Eq. (γ) is given to the right of the two equations following Eq. (2a) (see the facsimile in Sec. 8.1). In the transcription, it is given following these two equations. The calculation shows that the covariant derivative—or "extension" (*Erweiterung*) as Einstein called it—of the metric vanishes:

$$g^{\sigma\tau}{}_{;\beta} = g^{\sigma\tau}_{\beta} + \left\{ \begin{matrix} \sigma \\ \alpha\beta \end{matrix} \right\} g^{\alpha\tau} + \left\{ \begin{matrix} \tau \\ \alpha\beta \end{matrix} \right\} g^{\sigma\alpha} = 0. \tag{γ}$$

In the mathematical part of the *Entwurf* paper, Grossmann already noted that this is easy to prove (Einstein and Grossmann 1913, p. 28, note). So did Einstein in the mathematical section B of his review article about the *Entwurf* theory (Einstein 1914c, p. 1052).

15. The gravitational energy-momentum density t^{α}_{β} is defined in Eq. (4). The expressions for t^{α}_{β} and $t^{\alpha}_{\beta} - \frac{1}{2}\delta^{\alpha}_{\beta}t$ below are equivalent to the expression for $\kappa t^{\lambda}_{\sigma}$ in note 12. As Einstein notes, inserting Eqs. (1) and (5),

$$\frac{\partial L}{\partial g^{\sigma\tau}_{\alpha}} = -\left\{ \begin{matrix} \alpha \\ \sigma\beta \end{matrix} \right\}, \qquad t^{\alpha}_{\alpha} = t = \frac{L}{\kappa},$$

into

$$t^{\alpha}_{\beta} = \frac{1}{2\kappa}\left(L\delta^{\alpha}_{\beta} - g^{\sigma\tau}_{\beta}\frac{\partial L}{\partial g^{\sigma\tau}_{\alpha}}\right), \tag{4}$$

we arrive at the equation above Eq. (7)

$$t^{\alpha}_{\beta} = \frac{1}{2\kappa}\left(\kappa t\,\delta^{\alpha}_{\beta} + g^{\sigma\tau}_{\beta}\left\{ \begin{matrix} \alpha \\ \sigma\tau \end{matrix} \right\}\right).$$

Using Eq. (γ) (see note 14), we can write the second term on the right-hand side as

$$-\frac{1}{2\kappa}\left(\left\{ \begin{matrix} \sigma \\ \varepsilon\beta \end{matrix} \right\} g^{\varepsilon\tau} + \left\{ \begin{matrix} \tau \\ \varepsilon\beta \end{matrix} \right\} g^{\sigma\varepsilon}\right)\left\{ \begin{matrix} \alpha \\ \sigma\tau \end{matrix} \right\} = -\frac{1}{\kappa}g^{\sigma\varepsilon}\left\{ \begin{matrix} \tau \\ \varepsilon\beta \end{matrix} \right\}\left\{ \begin{matrix} \alpha \\ \sigma\tau \end{matrix} \right\},$$

where in the last step we used that $\left\{ \begin{matrix} \alpha \\ \sigma\tau \end{matrix} \right\}$ is symmetric in σ and τ. Substituting this result into the equation above, we arrive at

$$t^{\alpha}_{\beta} - \frac{1}{2}\delta^{\alpha}_{\beta}t = -\frac{1}{\kappa}g^{\sigma\varepsilon}\left\{ \begin{matrix} \alpha \\ \sigma\tau \end{matrix} \right\}\left\{ \begin{matrix} \tau \\ \varepsilon\beta \end{matrix} \right\}. \tag{7}$$

Combining Eq. (6) with Eq. (7), Einstein casts the field equations in the form of Eq. (8). Contracting the covariant field equations in Eq. (16a) of the first November 1915 paper (without the trace term $-\frac{1}{2}g_{\mu\nu}T$) with $g^{\nu\sigma}$, Einstein (1915a, p. 785, the equation above Eq. (22)) had found (see Sec. 6.3.1, note 17):

$$(g^{\nu\sigma}\Gamma^{\alpha}_{\mu\nu})_{,\alpha} - \frac{1}{2}\delta^{\sigma}_{\mu}\,\kappa t = -\kappa\left(t^{\sigma}_{\mu} + T^{\sigma}_{\mu}\right),$$

where we substituted κt for $g^{\mu\nu}\Gamma^{\alpha}_{\beta\mu}\Gamma^{\beta}_{\alpha\nu}$ to facilitate comparison with Eq. (8) in this letter.

Einstein's reliance on Eq. (5) in his derivation of Eq. (7) (see note 12) can be avoided if instead of Eq. (5), we insert the expression for the Lagrangian

$$L = g^{\sigma\tau}\begin{Bmatrix}\mu\\ \sigma\nu\end{Bmatrix}\begin{Bmatrix}\nu\\ \tau\mu\end{Bmatrix},$$

in the first-term on the right-hand side of Eq. (4) and rewrite the second term the way we did above, following Einstein, using that $g^{\sigma\tau}{}_{;\beta} = 0$. We then find

$$t^{\alpha}_{\beta} = \frac{1}{\kappa}\left(\frac{1}{2}\delta^{\alpha}_{\beta}\,g^{\sigma\tau}\begin{Bmatrix}\mu\\ \sigma\nu\end{Bmatrix}\begin{Bmatrix}\nu\\ \tau\mu\end{Bmatrix} - g^{\sigma\varepsilon}\begin{Bmatrix}\alpha\\ \sigma\tau\end{Bmatrix}\begin{Bmatrix}\tau\\ \varepsilon\beta\end{Bmatrix}\right).$$

Eq. (5) follows directly from this expression:

$$t = t^{\alpha}_{\alpha} = \frac{1}{\kappa}g^{\sigma\tau}\begin{Bmatrix}\mu\\ \sigma\nu\end{Bmatrix}\begin{Bmatrix}\nu\\ \tau\mu\end{Bmatrix} = \frac{L}{\kappa}.$$

Moreover, with the substitution

$$\Gamma^{\alpha}_{\sigma\tau} \equiv -\begin{Bmatrix}\alpha\\ \sigma\tau\end{Bmatrix},$$

this expression for t^{α}_{β} turns into the one in note 12.

16. In his review article a couple of months later, Einstein used the requirement that $t_{\mu\nu}$ and $T_{\mu\nu}$ enter the field equations in the same way to generalize vacuum field equations in unimodular coordinates,

$$(g^{\nu\sigma}\Gamma^{\alpha}_{\mu\nu})_{,\alpha} = -\kappa\left(t^{\sigma}_{\mu} - \frac{1}{2}\delta^{\sigma}_{\mu}t\right).$$

to the field equations in Eq. (8) in the presence of matter, also in unimodular coordinates (Einstein 1916b, §16, p. 807; cf. Ch. 9).

(4) Proof that A_{μ} vanishes

17. In this final entry of the letter, Einstein goes over the adjustments he had to make to calculations in the last part of sec. 3 of his first November 1915 paper (Einstein 1915a, p. 785, Eqs. (22), (22a) and (21a)) as a result of adding the trace term to the field equations in the fourth (Einstein 1915d). See Sec. 6.1.3, note 17, for our reconstruction of this part of Einstein (1915a) and Sec. 6.4.3, note 1, for our reconstruction of its counterpart in Einstein (1915d).
18. In the transcription and in the translation, we corrected a typo in Eqs. (10) and (10a): $(T+t)$ should be $\kappa(T+t)$. Taking the divergence on both sides of Eq. (8), we find

$$\frac{\partial}{\partial x^\nu}\frac{\partial}{\partial x^\alpha}\left(g^{\tau\nu}\left\{\begin{matrix}\alpha\\ \sigma\tau\end{matrix}\right\}\right)=\kappa\frac{\partial}{\partial x^\nu}\left(\left[t_\sigma^\nu+T_\sigma^\nu\right]-\frac{1}{2}\delta_\sigma^\nu\left[t+T\right]\right).$$

If Eq. (4), $\frac{\partial}{\partial x^\nu}\left(t_\sigma^\nu+T_\sigma^\nu\right)=A_\sigma$ is used for the first term on the right-hand side, this can be rewritten as

$$\frac{\partial}{\partial x^\nu}\frac{\partial}{\partial x^\alpha}\left(g^{\tau\nu}\left\{\begin{matrix}\alpha\\ \sigma\tau\end{matrix}\right\}\right)=\kappa A_\sigma-\frac{1}{2}\frac{\partial\kappa(t+T)}{\partial x^\sigma}. \tag{10}$$

19. The addition of the trace term to the field equations ensures that the field equations guarantee energy-momentum conservation, i.e., that A_μ (introduced in Eq. (3)) has to vanish. The fully contracted field equations,

$$\frac{\partial^2 g^{\alpha\beta}}{\partial x^\alpha\partial x^\beta}-\kappa(T+t)=0, \tag{9a}$$

ensure the vanishing of the left-hand side of Eq. (10a), a rewritten version of Eq. (10),

$$\frac{\partial}{\partial x^\sigma}\left(\frac{\partial^2 g^{\alpha\beta}}{\partial x^\alpha\partial x^\beta}-\kappa(T+t)\right)=-2\kappa A_\sigma \tag{10a}$$

(note that Einstein deleted $\frac{\partial}{\partial x_\sigma}$ in the second term on the left-hand side).

20. The justification of the trace term plays an important role in the correspondence between Einstein, Ehrenfest and Lorentz in this period. In the letter to Ehrenfest of January 17, 1916, from which we already quoted in note 1, Einstein refers to "the warrant (*Gewähr*) demanded by you for the "inevitability" (*Zwangsläufigkeit*) of the additional term $-\frac{1}{2}g_{im}T$" (CPAE8, Doc. 182).
21. It can no longer be established whether the letter was returned to Einstein (see Pt. I, Ch. 2, note 7) but it reads like a bridge between the presentation of the same material in the first and fourth of the November 1915 papers (Einstein 1915a; 1915d, see Ch. 6) and the introduction of the field equations and the proof of their compatibility with energy-momentum conservation (all in unimodular coordinates) in the review article of March 1916 (Einstein 1916b, see Ch. 9).

Chapter 9
The Foundation of the General Theory of Relativity. Riemann-Christoffel tensor (§ 12) and Theory of the Gravitational Field (Part C, §§ 13–18)

9.1 Facsimile

§ 12. **Der Riemann-Christoffelsche Tensor.** [1]

Wir fragen nun nach denjenigen Tensoren, welche aus dem Fundamentaltensor der $g_{\mu\nu}$ *allein* durch Differentiation gewonnen werden können. Die Antwort scheint zunächst auf der Hand zu liegen. Man setzt in (27) statt des beliebig gegebenen Tensors $A_{\mu\nu}$ den Fundamentaltensor der $g_{\mu\nu}$ ein und erhält dadurch einen neuen Tensor, nämlich die Erweiterung des Fundamentaltensors. Man überzeugt sich jedoch leicht, [2]
daß diese letztere identisch verschwindet. Man gelangt jedoch [3]
auf folgendem Wege zum Ziel. Man setze in (27)

$$A_{\mu\nu} = \frac{\partial A_\mu}{\partial x_\nu} - \begin{Bmatrix} \mu\nu \\ \varrho \end{Bmatrix} A_\varrho ,$$

M. Janssen, J. Renn, *How Einstein Found His Field Equations*, Classic Texts in the Sciences, https://doi.org/10.1007/978-3-030-97955-3_16

d. h. die Erweiterung des Vierervektors A_ν ein. Dann erhält man (bei etwas geänderter Benennung der Indizes) den Tensor dritten Ranges

$$A_{\mu\sigma\tau} = \frac{\partial^2 A_\mu}{\partial x_\sigma \partial x_\tau} - \begin{Bmatrix} \mu\sigma \\ \varrho \end{Bmatrix} \frac{\partial A_\varrho}{\partial x_\tau} - \begin{Bmatrix} \mu\tau \\ \varrho \end{Bmatrix} \frac{\partial A_\varrho}{\partial x_\sigma} - \begin{Bmatrix} \sigma\tau \\ \varrho \end{Bmatrix} \frac{\partial A_\mu}{\partial x_\varrho} + \left[- \frac{\partial}{\partial x_\tau} \begin{Bmatrix} \mu\sigma \\ \varrho \end{Bmatrix} + \begin{Bmatrix} \mu\tau \\ \alpha \end{Bmatrix} \begin{Bmatrix} \alpha\sigma \\ \varrho \end{Bmatrix} + \begin{Bmatrix} \sigma\tau \\ \alpha \end{Bmatrix} \begin{Bmatrix} \alpha\mu \\ \varrho \end{Bmatrix} \right] A_\varrho .$$

Dieser Ausdruck ladet zur Bildung des Tensors $A_{\mu\sigma\tau} - A_{\mu\tau\sigma}$ ein. Denn dabei heben sich folgende Terme des Ausdruckes für $A_{\mu\sigma\tau}$ gegen solche von $A_{\mu\tau\sigma}$ weg: das erste Glied, das vierte Glied, sowie das dem letzten Term in der eckigen Klammer entsprechende Glied; denn alle diese sind in σ und τ symmetrisch. Gleiches gilt von der Summe des zweiten und dritten Gliedes. Wir erhalten also

$$A_{\mu\sigma\tau} - A_{\mu\tau\sigma} = B^\varrho_{\mu\sigma\tau} A_\varrho , \tag{42}$$

$$\left\{ \begin{aligned} B^\varrho_{\mu\sigma\tau} = & - \frac{\partial}{\partial x_\tau} \begin{Bmatrix} \mu\sigma \\ \varrho \end{Bmatrix} + \frac{\partial}{\partial x_\sigma} \begin{Bmatrix} \mu\tau \\ \varrho \end{Bmatrix} \\ & - \begin{Bmatrix} \mu\sigma \\ \alpha \end{Bmatrix} \begin{Bmatrix} \alpha\tau \\ \varrho \end{Bmatrix} + \begin{Bmatrix} \mu\tau \\ \alpha \end{Bmatrix} \begin{Bmatrix} \alpha\sigma \\ \varrho \end{Bmatrix} . \end{aligned} \right. \tag{43}$$

Wesentlich ist an diesem Resultat, daß auf der rechten Seite von (42) nur die A_ϱ, aber nicht mehr ihre Ableitungen auftreten. Aus dem Tensorcharakter von $A_{\mu\sigma\tau} - A_{\mu\tau\sigma}$ in Verbindung damit, daß A_ϱ ein frei wählbarer Vierervektor ist, folgt, vermöge der Resultate des § 7, daß $B^\varrho_{\mu\sigma\tau}$ ein Tensor ist (Riemann-Christoffelscher Tensor).

Die mathematische Bedeutung dieses Tensors liegt im folgenden. Wenn das Kontinuum so beschaffen ist, daß es ein Koordinatensystem gibt, bezüglich dessen die $g_{\mu\nu}$ Konstanten sind, so verschwinden alle $B^\varrho_{\mu\sigma\tau}$. Wählt man statt des ursprünglichen Koordinatensystems ein beliebiges neues, so werden die auf letzteres bezogenen $g_{\mu\nu}$ nicht Konstanten sein. Der Tensorcharakter von $B^\varrho_{\mu\sigma\tau}$ bringt es aber mit sich, daß diese Komponenten auch in dem beliebig gewählten Bezugssystem sämtlich verschwinden. Das Verschwinden des Riemannschen Tensors ist also eine notwendige Bedingung dafür, daß durch geeignete Wahl des Bezugssystems die Konstanz der $g_{\mu\nu}$ herbeigeführt werden kann.[1]) In unserem Problem entspricht dies dem Falle, daß bei passender Wahl des Koordinatensystems in endlichen Gebieten die spezielle Relativitätstheorie gilt.

Durch Verjüngung von (43) bezüglich der Indizes τ und ϱ erhält man den kovarianten Tensor zweiten Ranges [4]

$$(44)\quad \begin{cases} B_{\mu\nu} = R_{\mu\nu} + S_{\mu\nu} \\ R_{\mu\nu} = -\dfrac{\partial}{\partial x_\alpha}\begin{Bmatrix}\mu\nu\\ \alpha\end{Bmatrix} + \begin{Bmatrix}\mu\alpha\\ \beta\end{Bmatrix}\begin{Bmatrix}\nu\beta\\ \alpha\end{Bmatrix} \\ S_{\mu\nu} = \dfrac{\partial \lg\sqrt{-g}}{\partial x_\mu \partial x_\nu} - \begin{Bmatrix}\mu\nu\\ \alpha\end{Bmatrix}\dfrac{\partial \lg\sqrt{-g}}{\partial x_\alpha}. \end{cases}$$

Bemerkung über die Koordinatenwahl. Es ist schon in § 8 im Anschluß an Gleichung (18a) bemerkt worden, daß die Koordinatenwahl mit Vorteil so getroffen werden kann, daß $\sqrt{-g} = 1$ wird. Ein Blick auf die in den beiden letzten Paragraphen erlangten Gleichungen zeigt, daß durch eine solche Wahl die Bildungsgesetze der Tensoren eine bedeutende Vereinfachung erfahren. Besonders gilt dies für den soeben entwickelten Tensor $B_{\mu\nu}$, welcher in der darzulegenden Theorie eine fundamentale Rolle spielt. Die ins Auge gefaßte Spezialisierung der Koordinatenwahl bringt nämlich das Verschwinden von $S_{\mu\nu}$ mit sich, so daß sich der Tensor $B_{\mu\nu}$ auf $R_{\mu\nu}$ reduziert. [5]

Ich will deshalb im folgenden alle Beziehungen in der vereinfachten Form angeben, welche die genannte Spezialisierung der Koordinatenwahl mit sich bringt. Es ist dann ein Leichtes, auf die *allgemein* kovarianten Gleichungen zurückzugreifen, falls dies in einem speziellen Falle erwünscht erscheint.

C. Theorie des Gravitationsfeldes.

§ 13. Bewegungsgleichung des materiellen Punktes im Gravitationsfeld. Ausdruck für die Feldkomponenten der Gravitation.

Ein frei beweglicher, äußeren Kräften nicht unterworfener Körper bewegt sich nach der speziellen Relativitätstheorie geradlinig und gleichförmig. Dies gilt auch nach der allgemeinen

1) Die Mathematiker haben bewiesen, daß diese Bedingung auch eine *hinreichende* ist.

Relativitätstheorie für einen Teil des vierdimensionalen Raumes, in welchem das Koordinatensystem K_0 so wählbar und so gewählt ist, daß die $g_{\mu\nu}$ die in (4) gegebenen speziellen konstanten Werte haben. [6]

Betrachten wir eben diese Bewegung von einem beliebig gewählten Koordinatensystem K_1 aus, so bewegt er sich von K_1 aus, beurteilt nach den Überlegungen des § 2 in einem Gravitationsfelde. Das Bewegungsgesetz mit Bezug auf K_1 [7]
ergibt sich leicht aus folgender Überlegung. Mit Bezug auf K_0 ist das Bewegungsgesetz eine vierdimensionale Gerade, also eine geodätische Linie. Da nun die geodätische Linie unabhängig vom Bezugssystem definiert ist, wird ihre Gleichung auch die Bewegungsgleichung des materiellen Punktes in bezug auf K_1 sein. Setzen wir

$$\Gamma^{\tau}_{\mu\nu} = -\begin{Bmatrix}\mu\nu\\ \tau\end{Bmatrix}, \tag{45}$$

so lautet also die Gleichung der Punktbewegung inbezug auf K_1

$$\frac{d^2 x_\tau}{d s^2} = \Gamma^{\tau}_{\mu\nu}\frac{d x_\nu}{d s}\frac{d x_\nu}{d s}\ . \tag{46}$$

Wir machen nun die sehr naheliegende Annahme, daß dieses allgemein kovariante Gleichungssystem die Bewegung des Punktes im Gravitationsfeld auch in dem Falle bestimmt, daß kein Bezugssystem K_0 existiert, bezüglich dessen in endlichen Räumen die spezielle Relativitätstheorie gilt. Zu dieser Annahme sind wir um so berechtigter, als (46) nur *erste* Ableitungen der $g_{\mu\nu}$ enthält, zwischen denen auch im Spezialfalle der Existenz von K_0 keine Beziehungen bestehen.[1])

Verschwinden die $\Gamma^{\tau}_{\mu\nu}$, so bewegt sich der Punkt geradlinig und gleichförmig; diese Größen bedingen also die Abweichung der Bewegung von der Gleichförmigkeit. Sie sind die Komponenten des Gravitationsfeldes. [8]

§ 14. Die Feldgleichungen der Gravitation bei Abwesenheit von Materie.

Wir unterscheiden im folgenden zwischen „Gravitationsfeld" und „Materie", in dem Sinne, daß alles außer dem Gravitationsfeld als „Materie" bezeichnet wird, also nicht nur

1) Erst zwischen den zweiten (und ersten) Ableitungen bestehen gemäß § 12 die Beziehungen $B^{\varrho}_{\mu\sigma\tau} = 0$.

die „Materie“ im üblichen Sinne, sondern auch das elektromagnetische Feld.

Unsere nächste Aufgabe ist es, die Feldgleichungen der Gravitation bei Abwesenheit von Materie aufzusuchen. Dabei verwenden wir wieder dieselbe Methode wie im vorigen Paragraphen bei der Aufstellung der Bewegungsgleichung des materiellen Punktes. Ein Spezialfall, in welchem die gesuchten Feldgleichungen jedenfalls erfüllt sein müssen, ist der der ursprünglichen Relativitätstheorie, in dem die $g_{\mu\nu}$ gewisse konstante Werte haben. Dies sei der Fall in einem gewissen endlichen Gebiete in bezug auf ein bestimmtes Koordinatensystem K_0. In bezug auf dies System verschwinden sämtliche Komponenten $B^{\varrho}_{\mu\sigma\tau}$ des Riemannschen Tensors [Gleichung (43)]. Diese verschwinden dann für das betrachtete Gebiet auch bezüglich jedes anderen Koordinatensystems.

Die gesuchten Gleichungen des materiefreien Gravitationsfeldes müssen also jedenfalls erfüllt sein, wenn alle $B^{\varrho}_{\mu\sigma\tau}$ verschwinden. Aber diese Bedingung ist jedenfalls eine zu weitgehende. Denn es ist klar, daß z. B. das von einem Massenpunkte in seiner Umgebung erzeugte Gravitationsfeld sicherlich durch keine Wahl des Koordinatensystems „wegtransformiert“, d. h. auf den Fall konstanter $g_{\mu\nu}$ transformiert werden kann.

Deshalb liegt es nahe, für das materiefreie Gravitationsfeld das Verschwinden des aus dem Tensor $B^{\varrho}_{\mu\sigma\tau}$ abgeleiteten symmetrischen Tensors $B_{\mu\nu}$ zu verlangen. Man erhält so 10 Gleichungen für die 10 Größen $g_{\mu\nu}$, welche im speziellen erfüllt sind, wenn sämtliche $B^{\varrho}_{\mu\sigma\tau}$ verschwinden. Diese Gleichungen lauten mit Rücksicht auf (44) bei der von uns getroffenen Wahl für das Koordinatensystem für das materiefreie Feld

$$(47) \qquad \left\{ \begin{aligned} &\frac{\partial \Gamma^{\alpha}_{\mu\nu}}{\partial x_{\alpha}} + \Gamma^{\alpha}_{\mu\beta}\,\Gamma^{\beta}_{\nu\alpha} = 0 \\ &\sqrt{-g} = 1\,. \end{aligned} \right.$$

Es muß darauf hingewiesen werden, daß der Wahl dieser Gleichungen ein Minimum von Willkür anhaftet. Denn es gibt außer $B_{\mu\nu}$ keinen Tensor zweiten Ranges, der aus den

52*

$g_{\mu\nu}$ und deren Ableitungen gebildet ist, keine höheren als zweite Ableitungen enthält und in letzteren linear ist.[1])

Daß diese aus der Forderung der allgemeinen Relativität auf rein mathematischem Wege fließenden Gleichungen in [10]
Verbindung mit den Bewegungsgleichungen (46) in erster Näherung das Newtonsche Attraktionsgesetz, in zweiter Näherung die Erklärung der von Leverrier entdeckten (nach [11]
Anbringung der Störungskorrektionen übrigbleibenden) Perihelbewegung des Merkur liefern, muß nach meiner Ansicht von der physikalischen Richtigkeit der Theorie überzeugen.

§ 15. Hamiltonsche Funktion für das Gravitationsfeld, Impulsenergiesatz.

Um zu zeigen, daß die Feldgleichungen dem Impulsenergiesatz entsprechen, ist es am bequemsten, sie in folgender Hamiltonscher Form zu schreiben: [12]

$$(47\text{a})\quad \left\{\begin{aligned} &\delta\left\{\int H\,d\tau\right\} = 0 \\ &H = g^{\mu\nu}\,\Gamma^{\alpha}_{\mu\beta}\,\Gamma^{\beta}_{\nu\alpha} \\ &\sqrt{-g} = 1\,. \end{aligned}\right.$$

Dabei verschwinden die Variationen an den Grenzen des betrachteten begrenzten vierdimensionalen Integrationsraumes.

Es ist zunächst zu zeigen, daß die Form (47a) den Gleichungen (47) äquivalent ist. Zu diesem Zweck betrachten wir H als Funktion der $g^{\mu\nu}$ und der

$$g^{\mu\nu}_{\sigma}\left(=\frac{\partial g^{\mu\nu}}{\partial x_\sigma}\right).$$

Dann ist zunächst

$$\begin{aligned} \delta H &= \Gamma^{\alpha}_{\mu\beta}\,\Gamma^{\beta}_{\nu\alpha}\,\delta g^{\mu\nu} + 2 g^{\mu\nu}\,\Gamma^{\alpha}_{\mu\beta}\,\delta\,\Gamma^{\beta}_{\nu\alpha} \\ &= -\,\Gamma^{\alpha}_{\mu\beta}\,\Gamma^{\beta}_{\nu\alpha}\,\delta g^{\mu\nu} + 2\,\Gamma^{\alpha}_{\mu\beta}\,\delta\left(g^{\mu\nu}\,\Gamma^{\beta}_{\nu\alpha}\right). \end{aligned}$$

Nun ist aber

$$\delta\left(g^{\mu\nu}\,\Gamma^{\beta}_{\nu\alpha}\right) = -\tfrac{1}{2}\,\delta\left[g^{\mu\nu}\,g^{\beta\lambda}\left(\frac{\partial g_{\nu\lambda}}{\partial x_\alpha} + \frac{\partial g^{\alpha\lambda}}{\partial x_\nu} - \frac{\partial g_{\alpha\nu}}{\partial x_\lambda}\right)\right].$$

---

1) Eigentlich läßt sich dies nur von dem Tensor $B_{\mu\nu} + \lambda g_{\mu\nu}\,(g^{\alpha\beta}\,B_{\alpha\beta})$ behaupten, wobei λ eine Konstante ist. Setzt man jedoch diesen $= 0$, so kommt man wieder zu den Gleichungen $B_{\mu\nu} = 0$. [9]

Die aus den beiden letzten Termen der runden Klammer hervorgehenden Terme sind von verschiedenem Vorzeichen und gehen auseinander (da die Benennung der Summationsindizes belanglos ist) durch Vertauschung der Indizes μ und β hervor. Sie heben einander im Ausdruck für δH weg, weil sie mit der bezüglich der Indizes μ und β symmetrischen Größe $\Gamma^{\alpha}_{\mu\beta}$ multipliziert werden. Es bleibt also nur das erste Glied der runden Klammer zu berücksichtigen, so daß man mit Rücksicht auf (31) erhält [13]

$$\delta H = -\Gamma^{\alpha}_{\mu\beta}\Gamma^{\beta}_{\nu\alpha}\,\delta g^{\mu\nu} - \Gamma^{\alpha}_{\mu\beta}\,\delta g^{\mu\beta}_{\alpha}\,.$$

Es ist also

$$\text{(48)}\quad \left\{\begin{aligned} \frac{\partial H}{\partial g^{\mu\nu}} &= -\Gamma^{\alpha}_{\mu\beta}\Gamma^{\beta}_{\nu\alpha} \\ \frac{\partial H}{\partial g^{\mu\nu}_{\sigma}} &= \Gamma^{\sigma}_{\mu\nu}\,. \end{aligned}\right.$$

Die Ausführung der Variation in (47a) ergibt zunächst das Gleichungssystem

$$\text{(47b)}\qquad \frac{\partial}{\partial x_{\alpha}}\left(\frac{\partial H}{\partial g^{\mu\nu}_{\alpha}}\right) - \frac{\partial H}{\partial g^{\mu\nu}} = 0\,,$$

welches wegen (48) mit (47) übereinstimmt, was zu beweisen war. — Multipliziert man (47b) mit $g^{\mu\nu}_{\sigma}$, so erhält man, weil [14]

$$\frac{\partial g^{\mu\nu}_{\sigma}}{\partial x_{\alpha}} = \frac{\partial g^{\mu\nu}_{\alpha}}{\partial x_{\sigma}}$$

und folglich

$$g^{\mu\nu}_{\sigma}\frac{\partial}{\partial x_{\alpha}}\left(\frac{\partial H}{\partial g^{\mu\nu}_{\alpha}}\right) = \frac{\partial}{\partial x_{\alpha}}\left(g^{\mu\nu}_{\sigma}\frac{\partial H}{\partial g^{\mu\nu}_{\alpha}}\right) - \frac{\partial H}{\partial g^{\mu\nu}_{\alpha}}\frac{\partial g^{\mu\nu}_{\alpha}}{\partial x_{\sigma}}$$

die Gleichung

$$\frac{\partial}{\partial x_{\alpha}}\left(g^{\mu\nu}_{\sigma}\frac{\partial H}{\partial g^{\mu\nu}_{\alpha}}\right) - \frac{\partial H}{\partial x_{\sigma}} = 0$$

oder[1])

$$\text{(49)}\quad \left\{\begin{aligned} \frac{\partial t^{\alpha}_{\sigma}}{\partial x_{\alpha}} &= 0 \\ -2\varkappa t^{\alpha}_{\sigma} &= g^{\mu\nu}_{\sigma}\frac{\partial H}{\partial g^{\mu\nu}_{\alpha}} - \delta^{\alpha}_{\sigma} H\,, \end{aligned}\right.$$

1) Der Grund der Einführung des Faktors $-2\varkappa$ wird später deutlich werden.

oder, wegen (48), der zweiten Gleichung (47) und (34) [15]

$$\varkappa t_\sigma^\alpha = \tfrac{1}{2}\delta_\sigma^\alpha g^{\mu\nu}\Gamma_{\mu\beta}^\alpha \Gamma_{\nu\alpha}^\beta - g^{\mu\nu}\Gamma_{\mu\beta}^\alpha \Gamma_{\nu\sigma}^\beta . \tag{50}$$

Es ist zu beachten, daß t_σ^α kein Tensor ist; dagegen gilt (49) für alle Koordinatensysteme, für welche $\sqrt{-g} = 1$ ist. Diese Gleichung drückt den Erhaltungssatz des Impulses und der Energie für das Gravitationsfeld aus. In der Tat liefert die Integration dieser Gleichung über ein *dreidimensionales* Volumen V die vier Gleichungen [16]

$$\frac{d}{d x_4}\left\{\int t_\sigma^4 \, d V\right\} = \int (t_\sigma^1 \alpha_1 + t_\sigma^2 \alpha_2 + t_\sigma^3 \alpha_3)\, d S, \tag{49a}$$

wobei α_1, α_2, α_3 der Richtungskosinus der nach innen gerichteten Normale eines Flächenelementes der Begrenzung von der Größe $d S$ (im Sinne der euklidischen Geometrie) bedeuten. Man erkennt hierin den Ausdruck der Erhaltungssätze in üblicher Fassung. Die Größen t_σ^α bezeichnen wir als die „Energiekomponenten" des Gravitationsfeldes.

Ich will nun die Gleichungen (47) noch in einer dritten Form angeben, die einer lebendigen Erfassung unseres Gegenstandes besonders dienlich ist. Durch Multiplikation der [17] Feldgleichungen (47) mit $g^{\nu\sigma}$ ergeben sich diese in der „gemischten" Form. Beachtet man, daß

$$g^{\nu\sigma}\frac{\partial \Gamma_{\mu\nu}^\alpha}{\partial x_\alpha} = \frac{\partial}{\partial x_\alpha}\left(g^{\nu\sigma}\Gamma_{\mu\nu}^\alpha\right) - \frac{\partial g^{\nu\sigma}}{\partial x_\alpha}\Gamma_{\mu\nu}^\alpha ,$$

welche Größe wegen (34) gleich

$$\frac{\partial}{\partial x_\alpha}(g^{\nu\sigma}\Gamma_{\mu\nu}^\alpha) - g^{\nu\beta}\Gamma_{\alpha\beta}^\sigma \Gamma_{\mu\nu}^\alpha - g^{\sigma\beta}\Gamma_{\beta\alpha}^\nu \Gamma_{\mu\nu}^\alpha ,$$

oder (nach geänderter Benennung der Summationsindizes) gleich [18]

$$\frac{\partial}{\partial x_\alpha}(g^{\sigma\beta}\Gamma_{\mu\beta}^\alpha) - g^{mn}\Gamma_{m\beta}^\sigma \Gamma_{n\mu}^\beta - g^{\nu\sigma}\Gamma_{\mu\beta}^\alpha \Gamma_{\nu\alpha}^\beta .$$

Das dritte Glied dieses Ausdrucks hebt sich weg gegen das aus dem zweiten Glied der Feldgleichungen (47) entstehende; an Stelle des zweiten Gliedes dieses Ausdruckes läßt sich nach Beziehung (50)

$$\varkappa(t_\mu^\sigma - \tfrac{1}{2}\delta_\mu^\sigma t)$$

setzen ($t = t_\alpha^\alpha$). Man erhält also an Stelle der Gleichungen (47)

$$\left\{\begin{aligned} \frac{\partial}{\partial x_\alpha}(g^{\sigma\beta}\Gamma_{\mu\beta}^\alpha) &= -\varkappa(t_\mu^\sigma - \tfrac{1}{2}\delta_\mu^\sigma t) \\ \sqrt{-g} &= 1 . \end{aligned}\right. \tag{51}$$

§ 16. Allgemeine Fassung der Feldgleichungen der Gravitation.

Die im vorigen Paragraphen aufgestellten Feldgleichungen für materiefreie Räume sind mit der Feldgleichung

$$\Delta \varphi = 0$$

der Newtonschen Theorie zu vergleichen. Wir haben die Gleichungen aufzusuchen, welche der Poissonschen Gleichung

$$\Delta \varphi = 4 \pi \varkappa \varrho$$

entspricht, wobei ϱ die Dichte der Materie bedeutet.

Die spezielle Relativitätstheorie hat zu dem Ergebnis geführt, daß die träge Masse nichts anderes ist als Energie, welche ihren vollständigen mathematischen Ausdruck in einem symmetrischen Tensor zweiten Ranges, dem Energietensor, findet. Wir werden daher auch in der allgemeinen Relativitätstheorie einen Energietensor der Materie T_σ^α einzuführen haben, der wie die Energiekomponenten t_σ^α [Gleichungen (49) und (50)] des Gravitationsfeldes gemischten Charakter haben wird, aber zu einem symmetrischen kovarianten Tensor gehören wird[1]).

Wie dieser Energietensor (entsprechend der Dichte ϱ in der Poissonschen Gleichung) in die Feldgleichungen der Gravitation einzuführen ist, lehrt das Gleichungssystem (51). Betrachtet man nämlich ein vollständiges System (z. B. das Sonnensystem), so wird die Gesamtmasse des Systems, also auch seine gesamte gravitierende Wirkung, von der Gesamtenergie des Systems, also von der ponderablen und Gravitationsenergie zusammen, abhängen. Dies wird sich dadurch ausdrücken lassen, daß man in (51) an Stelle der Energiekomponenten t_μ^σ des Gravitationsfeldes allein die Summen $t_\mu^\sigma + T_\mu^\sigma$ der Energiekomponenten von Materie und Gravitationsfeld einführt. Man erhält so statt (51) die Tensorgleichung [19]

$$\text{(52)} \quad \left\{ \begin{aligned} &\frac{\partial}{\partial x_\alpha} \left(g^{\sigma\beta} \Gamma_{\mu\beta}^\alpha \right) = - \varkappa \left[\left(t_\mu^\sigma + T_\mu^\sigma \right) - \tfrac{1}{2} \delta_\mu^\sigma (t + T) \right] \\ &\sqrt{-g} = 1, \end{aligned} \right.$$

wobei $T = T_\mu^\mu$ gesetzt ist (Lauescher Skalar). Dies sind die gesuchten allgemeinen Feldgleichungen der Gravitation in ge- [20]

1) $g_{\sigma\tau} T_\sigma^\alpha = T_{\sigma\tau}$ und $g^{\sigma\beta} T_\sigma^\alpha = T^{\alpha\beta}$ sollen symmetrische Tensoren sein.

mischter Form. An Stelle von (47) ergibt sich daraus rückwärts das System [21]

$$(53)\quad \begin{cases} \dfrac{\partial \Gamma^{\alpha}_{\mu\nu}}{\partial x_\alpha} + \Gamma^{\alpha}_{\mu\beta}\,\Gamma^{\beta}_{\nu\alpha} = -\varkappa\left(T_{\mu\nu} - \frac{1}{2} g_{\mu\nu} T\right), \\ \sqrt{-g} = 1\,. \end{cases}$$

Es muß zugegeben werden, daß diese Einführung des Energietensors der Materie durch das Relativitätspostulat allein nicht gerechtfertigt wird; deshalb haben wir sie im vorigen aus der Forderung abgeleitet, daß die Energie des Gravitationsfeldes in gleicher Weise gravitierend wirken soll, wie jegliche Energie anderer Art. Der stärkste Grund für die Wahl der vorstehenden Gleichungen liegt aber darin, daß sie zur Folge haben, daß für die Komponenten der Totalenergie Erhaltungsgleichungen (des Impulses und der Energie) gelten, welche den Gleichungen (49) und (49a) genau entsprechen. Dies soll im folgenden dargetan werden. [22]

§ 17. Die Erhaltungssätze im allgemeinen Falle.

Die Gleichung (52) ist leicht so umzuformen, daß auf der rechten Seite das zweite Glied wegfällt. Man verjünge (52) nach den Indizes μ und σ und subtrahiere die so erhaltene, mit $\frac{1}{2}\delta_\mu^{\ \sigma}$ multiplizierte Gleichung von (52). Es ergibt sich [23]

$$(52\text{a})\quad \frac{\partial}{\partial x_\alpha}\left(g^{\sigma\beta}\,\Gamma^{\alpha}_{\mu\beta} - \tfrac{1}{2}\delta_\mu^{\ \sigma}\, g^{\lambda\beta}\,\Gamma^{\alpha}_{\lambda\beta}\right) = -\varkappa\left(t_\mu^{\ \sigma} + T_\mu^{\ \sigma}\right).$$

An dieser Gleichung bilden wir die Operation $\partial/\partial x_\sigma$. Es ist

$$\frac{\partial^2}{\partial x_\alpha\,\partial x_\sigma}\left(g^{\sigma\beta}\,\Gamma^{\alpha}_{\mu\beta}\right) = -\frac{1}{2}\,\frac{\partial^2}{\partial x_\alpha\,\partial x_\sigma}\left[g^{\sigma\beta} g^{\alpha\lambda}\left(\frac{\partial g_{\mu\lambda}}{\partial x_\beta} + \frac{\partial g_{\beta\lambda}}{\partial x_\mu} - \frac{\partial g_{\mu\beta}}{\partial x_\lambda}\right)\right].$$

Das erste und das dritte Glied der runden Klammer liefern Beiträge, die einander wegheben, wie man erkennt, wenn man im Beitrage des dritten Gliedes die Summationsindizes α und σ einerseits, β und λ andererseits vertauscht. Das zweite Glied läßt sich nach (31) umformen, so daß man erhält

$$(54)\quad \frac{\partial^2}{\partial x_\alpha\,\partial x_\sigma}\left(g^{\sigma\beta}\,\Gamma^{\alpha}_{\mu\beta}\right) = \frac{1}{2}\,\frac{\partial^3 g^{\alpha\beta}}{\partial x_\alpha\,\partial x_\beta\,\partial x_\mu}\,.$$

Das zweite Glied der linken Seite von (52a) liefert zunächst

$$-\frac{1}{2}\,\frac{\partial^2}{\partial x_\alpha\,\partial x_\mu}\left(g^{\lambda\beta}\,\Gamma^{\alpha}_{\lambda\beta}\right)$$

oder

$$\frac{1}{4}\,\frac{\partial^2}{\partial x_\alpha\,\partial x_\mu}\left[g^{\lambda\beta}\,g^{\alpha\delta}\left(\frac{\partial g_{\delta\lambda}}{\partial x_\beta}+\frac{\partial g_{\delta\beta}}{\partial x_\lambda}-\frac{\partial g_{\lambda\beta}}{\partial x_\delta}\right)\right].$$

Das vom letzten Glied der runden Klammer herrührende Glied verschwindet wegen (29) bei der von uns getroffenen Koordinatenwahl. Die beiden anderen lassen sich zusammenfassen und liefern wegen (31) zusammen [24]

$$-\frac{1}{2}\,\frac{\partial^3 g^{\alpha\beta}}{\partial x_\alpha\,\partial x_\beta\,\partial x_\mu},$$

so daß mit Rücksicht auf (54) die Identität

$$\frac{\partial^2}{\partial x_\alpha\,\partial x_\sigma}\left(g^{\sigma\beta}\,\Gamma^\alpha_{\mu\beta}-\tfrac{1}{2}\,\delta_\mu^\sigma\,g^{\lambda\beta}\,\Gamma^\alpha_{\lambda\beta}\right)\equiv 0 \tag{55}$$

besteht. Aus (55) und (52a) folgt

$$\frac{\partial\,(t_\mu^\sigma+T_\mu^\sigma)}{\partial x_\sigma}=0\,. \tag{56}$$

Aus unseren Feldgleichungen der Gravitation geht also hervor, daß den Erhaltungssätzen des Impulses und der Energie Genüge geleistet ist. Man sieht dies am einfachsten nach [25] der Betrachtung ein, die zu Gleichung (49a) führt; nur hat man hier an Stelle der Energiekomponenten t_μ^σ des Gravitationsfeldes die Gesamtenergiekomponenten von Materie und Gravitationsfeld einzuführen.

§ 18. Der Impulsenergiesatz für die Materie als Folge der Feldgleichungen.

Multipliziert man (53) mit $\partial g^{\mu\nu}/\partial x_\sigma$, so erhält man auf dem in § 15 eingeschlagenen Wege mit Rücksicht auf das Verschwinden von

$$g_{\mu\nu}\frac{\partial g^{\mu\nu}}{\partial x_\sigma}$$

die Gleichung [26]

$$\frac{\partial t_\sigma^\alpha}{\partial x_\alpha}+\frac{1}{2}\,\frac{\partial g^{\mu\nu}}{\partial x_\sigma}\,T_{\mu\nu}=0,$$

oder mit Rücksicht auf (56)

$$\frac{\partial T_\sigma^\alpha}{\partial x_\alpha}+\frac{1}{2}\,\frac{\partial g^{\mu\nu}}{\partial x_\sigma}\,T_{\mu\nu}=0\,. \tag{57}$$

Ein Vergleich mit (41b) zeigt, daß diese Gleichung bei der getroffenen Wahl für das Koordinatensystem nichts anderes aussagt als das Verschwinden der Divergenz des Tensors der Energiekomponenten der Materie. Physikalisch zeigt das Auftreten des zweiten Gliedes der linken Seite, daß für die Materie allein Erhaltungssätze des Impulses und der Energie im eigentlichen Sinne nicht, bzw. nur dann gelten, wenn die $g^{\mu\nu}$ konstant sind, d. h. wenn die Feldstärken der Gravitation verschwinden. Dies zweite Glied ist ein Ausdruck für Impuls bzw. Energie, welche pro Volumen und Zeiteinheit vom Gravitationsfelde auf die Materie übertragen werden. Dies tritt noch klarer hervor, wenn man statt (57) im Sinne von (41) schreibt [27]

$$\frac{\partial T_\sigma^{\alpha}}{\partial x_\alpha} = - \Gamma^{\alpha}_{\sigma\beta} T_\alpha^{\beta}. \tag{57a}$$

Die rechte Seite drückt die energetische Einwirkung des Gravitationsfeldes auf die Materie aus.

Die Feldgleichungen der Gravitation enthalten also gleichzeitig vier Bedingungen, welchen der materielle Vorgang zu genügen hat. Sie liefern die Gleichungen des materiellen Vorganges vollständig, wenn letzterer durch vier voneinander unabhängige Differentialgleichungen charakterisierbar ist.[1])

D. Die „materiellen" Vorgänge.

Die unter B entwickelten mathematischen Hilfsmittel setzen uns ohne weiteres in den Stand, die physikalischen Gesetze der Materie (Hydrodynamik, Maxwellsche Elektrodynamik), wie sie in der speziellen Relativitätstheorie formuliert vorliegen, so zu verallgemeinern, daß sie in die allgemeine Relativitätstheorie hineinpassen. Dabei ergibt das allgemeine Relativitätsprinzip zwar keine weitere Einschränkung der Möglichkeiten; aber es lehrt den Einfluß des Gravitationsfeldes auf alle Prozesse exakt kennen, ohne daß irgendwelche neue Hypothese eingeführt werden müßte.

Diese Sachlage bringt es mit sich, daß über die physikalische Natur der Materie (im engeren Sinne) nicht notwendig bestimmte Voraussetzungen eingeführt werden müssen. Insbesondere kann die Frage offen bleiben, ob die Theorie des elektromagnetischen Feldes und des Gravitationsfeldes zu-

---

1) Vgl. hierüber D. Hilbert, Nachr. d. K. Gesellsch. d. Wiss. zu Göttingen, Math.-phys. Klasse. p. 3. 1915. [28]

9.2 Translation

The Foundation of the General Theory of Relativity. Riemann-Christoffel tensor (§ 12) and Theory of the Gravitational Field (Part C, §§ 13–18)

§ 12. The Riemann-Christoffel Tensor[1]

We now examine which tensors can be formed from the fundamental tensor $g_{\mu\nu}$ through differentiation *alone*. At first sight the solution seems obvious. Instead of an arbitrary tensor $A_{\mu\nu}$, we insert the fundamental tensor $g_{\mu\nu}$ in (27) and thus obtain a new tensor, namely, the extension of the fundamental tensor.[2] We easily convince ourselves, however, that this extension vanishes identically.[3] We reach our goal, however, in the following way. In (27) we insert

$$A_{\mu\nu} = \frac{\partial A_\mu}{\partial x_\nu} - \begin{Bmatrix} \mu\nu \\ \rho \end{Bmatrix} A_\rho,$$

i.e., the extension of the four-vector A_μ. Then (with a somewhat different labeling of the indices) we obtain the third-rank tensor

$$\begin{aligned} A_{\mu\sigma\tau} = & \frac{\partial^2 A_\mu}{\partial x_\sigma \partial x_\tau} \\ & - \begin{Bmatrix} \mu\sigma \\ \rho \end{Bmatrix} \frac{\partial A_\rho}{\partial x_\tau} - \begin{Bmatrix} \mu\tau \\ \rho \end{Bmatrix} \frac{\partial A_\rho}{\partial x_\sigma} - \begin{Bmatrix} \sigma\tau \\ \rho \end{Bmatrix} \frac{\partial A_\mu}{\partial x_\rho} \\ & + \left[-\frac{\partial}{\partial x_\tau} \begin{Bmatrix} \mu\sigma \\ \rho \end{Bmatrix} + \begin{Bmatrix} \mu\tau \\ \alpha \end{Bmatrix} \begin{Bmatrix} \alpha\sigma \\ \rho \end{Bmatrix} + \begin{Bmatrix} \sigma\tau \\ \alpha \end{Bmatrix} \begin{Bmatrix} \alpha\mu \\ \rho \end{Bmatrix} \right] A_\rho . \end{aligned}$$

This expression suggests forming the tensor $A_{\mu\sigma\tau} - A_{\mu\tau\sigma}$. For, if we do, the following terms in the expression for $A_{\mu\sigma\tau}$ cancel against those in $A_{\mu\tau\sigma}$: the first, the fourth, and the term corresponding to the last term in square brackets. This is because they all are symmetrical in σ and τ. The same is true for the sum of the second and the third term. We thus obtain

$$A_{\mu\sigma\tau} - A_{\mu\tau\sigma} = B^\rho_{\mu\sigma\tau} A_\rho \tag{42}$$

with

$$\left. \begin{aligned} B^\rho_{\mu\sigma\tau} = & -\frac{\partial}{\partial x_\tau} \begin{Bmatrix} \mu\sigma \\ \rho \end{Bmatrix} + \frac{\partial}{\partial x_\sigma} \begin{Bmatrix} \mu\tau \\ \rho \end{Bmatrix} \\ & - \begin{Bmatrix} \mu\sigma \\ \alpha \end{Bmatrix} \begin{Bmatrix} \alpha\tau \\ \rho \end{Bmatrix} + \begin{Bmatrix} \mu\tau \\ \alpha \end{Bmatrix} \begin{Bmatrix} \alpha\sigma \\ \rho \end{Bmatrix} \end{aligned} \right\} . \tag{43}$$

The essential feature of this result is that on the right-hand side of (42) only A_ρ occurs, not its derivatives. From the tensor character of $A_{\mu\sigma\tau} - A_{\mu\tau\sigma}$ and the fact that A_ρ is an arbitrary vector, it follows, given the results of § 7, that $B^\rho_{\mu\sigma\tau}$ is a tensor (the RIEMANN-CHRISTOFFEL tensor).

The mathematical meaning of this tensor is as follows: If the continuum is such that there is a coordinate system in which the $g_{\mu\nu}$ are constants, then the $B^\rho_{\mu\sigma\tau}$ all vanish. If instead of the original coordinate system we choose a new one, the $g_{\mu\nu}$ in the new system will not be constants. Yet, because of their tensor character, the transformed components of $B^\rho_{\mu\sigma\tau}$ will still vanish. Thus the vanishing of the RIEMANN tensor is a necessary condition for there to be a coordinate system in which the $g_{\mu\nu}$ are constant.[1] In our case this corresponds to the situation in which, with a suitable choice of coordinate system, the special theory of relativity holds good in finite regions of the continuum.

Contracting (43) over τ and ρ, we obtain the second-rank covariant tensor[4]

$$\left.\begin{aligned} B_{\mu\nu} &= R_{\mu\nu} + S_{\mu\nu} \\ R_{\mu\nu} &= -\frac{\partial}{\partial x_\alpha}\begin{Bmatrix}\mu\nu\\ \alpha\end{Bmatrix} + \begin{Bmatrix}\mu\alpha\\ \beta\end{Bmatrix}\begin{Bmatrix}\nu\beta\\ \alpha\end{Bmatrix} \\ S_{\mu\nu} &= \frac{\partial^2 \log\sqrt{-g}}{\partial x_\mu \partial x_\nu} - \begin{Bmatrix}\mu\nu\\ \alpha\end{Bmatrix}\frac{\partial \log\sqrt{-g}}{\partial x_\alpha} \end{aligned}\right\}. \tag{44}$$

Note on the Choice of Coordinates. I already noted in § 8, in connexion with equation (18a),[5] that it is advantageous to choose coordinates in such a way that $\sqrt{-g} = 1$. A glance at the equations obtained in the last two sections shows that this choice leads to a considerable simplification of the laws for the formation of tensors. This is particularly true for $B_{\mu\nu}$, the tensor we just introduced and which plays a fundamental role in the theory to be developed here. For this special choice of coordinates entails that $S_{\mu\nu}$ vanishes, so that the tensor $B_{\mu\nu}$ reduces to $R_{\mu\nu}$.

In what follows, I will therefore give all relations in the simplified form which this special choice of coordinates entails. It will be easy to revert to the *generally* covariant equations, if this is deemed desirable in some special case.

C. Theory of the Gravitational Field

§ 13. Equations of Motion of a Material Point in a Gravitational Field. Expression for the Components of the Gravitational Field

A freely movable body not subjected to external forces moves, according to the special theory of relativity, in a straight line and uniformly. This is also the case, according to the general theory of relativity, for parts of four-

[1] Mathematicians have shown that this is also a *sufficient* condition.

dimensional space in which the coordinate system K_0 can be and is chosen such that the $g_{\mu\nu}$ have the special constant values given in (4).[6]

If we consider such motion in an arbitrarily chosen coordinate system K_1, the body, observed from K_1, moves, according to the considerations in § 2, in a gravitational field.[7] The law of motion in K_1 results without difficulty from the following consideration. In K_0 the law of motion corresponds to a four-dimensional straight line, i.e., to a geodesic. But since a geodesic is defined independently of the reference system, its equation will also be the equation of motion of the material point in K_1. If we set

$$\Gamma^{\tau}_{\mu\nu} = -\left\{ \begin{matrix} \mu\nu \\ \tau \end{matrix} \right\}, \tag{45}$$

the equation of the motion of the point with respect to K_1, becomes

$$\frac{d^2 x_\tau}{ds^2} = \Gamma^{\tau}_{\mu\nu} \frac{dx_\mu}{ds} \frac{dx_\nu}{ds}. \tag{46}$$

We now make the plausible assumption that this generally covariant system of equations also defines the motion of a point in a gravitational field in the case when there is no reference system K_0 in which the special theory of relativity holds good in a finite region. We are all the more justified in making this assumption as (46) contains only first derivatives of $g_{\mu\nu}$, which are independent of one another, even in the special case that some such K_0 exists.[1]

If the $\Gamma^{\tau}_{\mu\nu}$ vanish, then the point moves in a straight line and uniformly. These quantities therefore determine the deviation of the motion from uniformity. They are the components of the gravitational field.[8]

§ 14. The Field Equations of Gravitation in the Absence of Matter

We make a distinction hereafter between "gravitational field" and "matter" in the sense that we call everything but the gravitational field "matter." Our use of the term therefore includes not only matter in the ordinary sense but also the electromagnetic field.

Our next task is to find the field equations of gravitation in the absence of matter. We use the same method we used in the preceding section to formulate the equations of motion of a material point. A special case in which the equations we are looking for must definitely be satisfied is that of the original relativity theory, in which the $g_{\mu\nu}$ have certain constant values. Let this be the case in some finite region in some coordinate system K_0. In this coordinate system all components of the RIEMANN tensor $B^{\rho}_{\mu\sigma\tau}$ [equation (43)] vanish. For the region under consideration they will then also vanish in any other coordinate system.

[1] In that case, it is only the second (and first) derivatives that, according to § 12, are related through $B^{\rho}_{\mu\sigma\tau} = 0$.

Hence the equations for the matter-free gravitational field we are looking for must definitely be satisfied, if all components $B^{\rho}_{\mu\sigma\tau}$ vanish. But this requirement is too strong. For it is clear, e.g., that the gravitational field generated in the vicinity of a material point surely cannot be “transformed away” through any choice of coordinate system, i.e., it cannot be transformed to the case of constant $g_{\mu\nu}$.

This then suggests that, for the matter-free gravitational field, we require the vanishing of the symmetrical tensor $B_{\mu\nu}$, derived from the tensor $B^{\rho}_{\mu\nu\tau}$. Thus we obtain ten equations for the ten quantities $g_{\mu\nu}$, which are satisfied in the special case of vanishing $B^{\rho}_{\mu\nu\tau}$.

Given (44) and our choice of coordinates, the equations for the matter-free field are

$$\left.\begin{array}{l} \dfrac{\partial \Gamma^{\alpha}_{\mu\nu}}{\partial x_{\alpha}} + \Gamma^{\alpha}_{\mu\beta}\Gamma^{\beta}_{\nu\alpha} = 0 \\ \sqrt{-g} = 1 \end{array}\right\}. \tag{47}$$

It should be pointed out that there is only a minimum of arbitrariness in the choice of these equations. For other than $B_{\mu\nu}$ there is no second-rank tensor constructed out of $g_{\mu\nu}$ and its derivatives that contains no derivatives higher than of second order and is linear in those.[1]

That these equations, to which the requirement of general relativity led us via a purely mathematical route,[10] combined with the equations of motion (46), give us NEWTON’s law of attraction in first approximation and the perihelion motion discovered by LEVERRIER (after correction for perturbations) in second approximation[11] must, in my opinion, be taken as convincing evidence that the theory is correct from a physics point of view.

§ 15. The Hamiltonian Function for the Gravitational Field. Laws of Momentum and Energy

The most convenient way to show that the field equations are in accordance with the law of energy-momentum conservation is to write them in the following HAMILTONian form:[12]

$$\left.\begin{array}{l} \delta \int H d\tau = 0 \\ H = g^{\mu\nu}\Gamma^{\alpha}_{\mu\beta}\Gamma^{\beta}_{\nu\alpha} \\ \sqrt{-g} = 1 \end{array}\right\}. \tag{47a}$$

where the variations vanish on the boundary of the finite four-dimensional region over which we integrate.

[1] Strictly speaking, this is true only for the tensor $B_{\mu\nu} + \lambda g_{\mu\nu}(g^{\alpha\beta}B_{\alpha\beta})$, where λ is a constant. If, however, we set this constant $= 0$, we once again have the equations $B_{\mu\nu} = 0$.[9]

We first have to show that the form (47a) is equivalent to the equations (47). To this end we regard H as a function of $g^{\mu\nu}$ and

$$g_\sigma^{\mu\nu}\left(=\frac{\partial g^{\mu\nu}}{\partial x_\sigma}\right).$$

First note that

$$\begin{aligned}\delta H &= \Gamma^\alpha_{\mu\beta}\Gamma^\beta_{\nu\alpha}\,\delta g^{\mu\nu} + 2g^{\mu\nu}\Gamma^\alpha_{\mu\beta}\Gamma^\beta_{\nu\alpha}\\ &= -\Gamma^\alpha_{\mu\beta}\Gamma^\beta_{\nu\alpha}\,\delta g^{\mu\nu} + 2\Gamma^\alpha_{\mu\beta}\,\delta(g^{\mu\nu}\Gamma^\beta_{\nu\alpha}).\end{aligned}$$

Now use that

$$\delta(g^{\mu\nu}\Gamma^\beta_{\nu\alpha}) = -\frac{1}{2}\delta\left[g^{\mu\nu}g^{\beta\lambda}\left(\frac{\partial g_{\nu\lambda}}{\partial x_\alpha}+\frac{\partial g_{\alpha\lambda}}{\partial x_\nu}-\frac{\partial g_{\alpha\nu}}{\partial x_\lambda}\right)\right].$$

The terms arising from the last two terms in round brackets have opposite signs and turn into each other (since the labeling of the summation indices does not matter) if the indices μ and β are switched. They cancel each other in the expression for δH, because they are multiplied by the quantity $\Gamma^\alpha_{\mu\beta}$ which is symmetrical in the indices μ and β. Thus only the first term in round brackets remains to be considered and, taking into account (31),[13] we obtain

$$\delta H = -\Gamma^\alpha_{\mu\beta}\Gamma^\beta_{\nu\alpha}\,\delta g^{\mu\nu} + \Gamma^\alpha_{\mu\beta}\,\delta g^{\mu\beta}_\alpha.$$

Hence

$$\left.\begin{aligned}\frac{\partial H}{\partial g^{\mu\nu}} &= -\Gamma^\alpha_{\mu\beta}\Gamma^\beta_{\nu\alpha}\\ \frac{\partial H}{\partial g^{\mu\nu}_\sigma} &= \Gamma^\sigma_{\mu\nu}\end{aligned}\right\}. \tag{48}$$

Carrying out the variation in (47a), we find

$$\frac{\partial}{\partial x_\alpha}\left(\frac{\partial H}{\partial g^{\mu\nu}_\alpha}\right) - \frac{\partial H}{\partial g^{\mu\nu}} = 0, \tag{47b}$$

which, on account of (48), agrees with (47), as was to be shown.

If we multiply (47b) by $g^{\mu\nu}_\sigma$,[14] then, since

$$\frac{\partial g^{\mu\nu}_\sigma}{\partial x_\alpha} = \frac{\partial g^{\mu\nu}_\alpha}{\partial x_\sigma}$$

and therefore

$$g^{\mu\nu}_\sigma\frac{\partial}{\partial x_\alpha}\left(\frac{\partial H}{\partial g^{\mu\nu}_\alpha}\right) = \frac{\partial}{\partial x_\alpha}\left(g^{\mu\nu}_\sigma\frac{\partial H}{\partial g^{\mu\nu}_\alpha}\right) - \frac{\partial H}{\partial g^{\mu\nu}_\alpha}\frac{\partial g^{\mu\nu}_\alpha}{\partial x_\sigma},$$

we obtain the equation

$$\frac{\partial}{\partial x_\alpha}\left(g_\sigma^{\mu\nu}\frac{\partial H}{\partial g_\alpha^{\mu\nu}}\right)-\frac{\partial H}{\partial x_\sigma}=0$$

or[1]

$$\left.\begin{aligned}\frac{\partial t_\sigma^\alpha}{\partial x_\alpha}&=0\\ -2\kappa t_\sigma^\alpha&=g_\sigma^{\mu\nu}\frac{\partial H}{\partial g_\alpha^{\mu\nu}}-\delta_\sigma^\alpha H\end{aligned}\right\}, \tag{49}$$

where, on account of (48), the second equation of (47) and (34)[15]

$$\kappa t_\sigma^\alpha=\frac{1}{2}\delta_\sigma^\alpha g^{\mu\nu}\Gamma_{\mu\beta}^\lambda\Gamma_{\nu\lambda}^\beta-g^{\mu\nu}\Gamma_{\mu\beta}^\alpha\Gamma_{\nu\sigma}^\beta. \tag{50}$$

One should keep in mind that t_σ^α is not a tensor; (49), however, holds in all coordinate systems for which $\sqrt{-g}=1$. This equation expresses the conservation law for momentum and of energy of the gravitational field. Integration of this equation over a *three-dimensional* volume V yields the four equations[16]

$$\frac{d}{dx_4}\left\{\int t_\sigma^4\,dV\right\}=\int\left(t_\sigma^1\alpha_1+t_\sigma^2\alpha_2+t_\sigma^3\alpha_3\right)dS, \tag{49a}$$

where $\alpha_1,\alpha_2,\alpha_3$ are the components of an inward-pointing unit vector perpendicular to the boundary-surface element of size dS (in the sense of Euclidean geometry). In this equation we recognize the usual form of the conservation laws. We will call the quantities t_σ^α the "energy components" of the gravitational field.

I will now give equations (47) in a third form, which is particularly useful for a vivid grasp of our subject.[17] Multiplication of the field equations (47) by $g^{\nu\sigma}$ gives them in the "mixed" form. Note that

$$g^{\nu\sigma}\frac{\partial\Gamma_{\mu\nu}^\alpha}{\partial x_\alpha}=\frac{\partial}{\partial x_\alpha}\left(g^{\nu\sigma}\Gamma_{\mu\nu}^\alpha\right)-\frac{\partial g^{\nu\sigma}}{\partial x_\alpha}\Gamma_{\mu\nu}^\alpha,$$

which, because of (34), is equal to

$$\frac{\partial}{\partial x_\alpha}\left(g^{\nu\sigma}\Gamma_{\mu\nu}^\alpha\right)-g^{\nu\beta}\Gamma_{\alpha\beta}^\sigma\Gamma_{\mu\nu}^\alpha-g^{\sigma\beta}\Gamma_{\beta\alpha}^\nu\Gamma_{\mu\nu}^\alpha,$$

or (with different labels for the summation indices)[18]

$$\frac{\partial}{\partial x_\alpha}\left(g^{\sigma\beta}\Gamma_{\mu\beta}^\alpha\right)-g^{mn}\Gamma_{m\beta}^\sigma\Gamma_{n\mu}^\beta-g^{\nu\sigma}\Gamma_{\mu\beta}^\alpha\Gamma_{\nu\alpha}^\beta.$$

[1] The reason for the introduction of the factor -2κ will later become clear.

The third term of this expression cancels against the one that comes from the second term in the field equations (47). Using relation (50), we can write the second term as

$$\kappa\left(t_\mu^\sigma - \frac{1}{2}\delta_\mu^\sigma t\right),$$

where $t = t_\alpha^\alpha$. Thus instead of equations (47) we obtain

$$\left.\begin{array}{c}\dfrac{\partial}{\partial x_\alpha}\left(g^{\sigma\beta}\Gamma_{\mu\beta}^\alpha\right) = -\kappa\left(t_\mu^\sigma - \dfrac{1}{2}\delta_\mu^\sigma t\right) \\ \sqrt{-g} = 1\end{array}\right\}. \tag{51}$$

§ 16. The General Form of the Field Equations of Gravitation.

The field equations for matter-free regions formulated in § 15 should be compared with the field equation

$$\nabla^2\phi = 0$$

of NEWTON's theory. We need to find the equation corresponding to POISSON's equation

$$\nabla^2\phi = 4\pi\kappa\rho,$$

where ρ is the density of matter.

The special theory of relativity has led to the conclusion that inertial mass is no different from energy, which finds its complete mathematical expression in a symmetrical second-rank tensor, the energy-tensor. Thus in the general theory of relativity we must introduce a corresponding energy-tensor T_σ^α for matter, which, like the energy-components t_σ^α [equations (49) and (50)] of the gravitational field, will have mixed character, but corresponds to a symmetrical covariant tensor.[1]

The system of equations (51) tells us how this energy-tensor (the analogue of the density ρ in POISSON's equation) must enter the gravitational field equations. For if we consider a complete system (e.g., the solar system), the total mass of the system, and thus its total gravitating action, will depend on the total energy of the system, and therefore on the ponderable and gravitational energy taken together. The way to implement this is to insert in (51), instead of the energy-components t_μ^σ of the gravitational field alone, the sum $t_\mu^\sigma + T_\mu^\sigma$ of the energy-components of matter and gravitational field.[19] In this way, we obtain, instead of (51), the equation

$$\left.\begin{array}{c}\dfrac{\partial}{\partial x_\alpha}\left(g^{\sigma\beta}\Gamma_{\mu\beta}^\alpha\right) = -\kappa\left[\left(t_\mu^\sigma + T_\mu^\sigma\right) - \dfrac{1}{2}\delta_\mu^\sigma\left(t + T\right)\right] \\ \sqrt{-g} = 1\end{array}\right\}. \tag{52}$$

---

[1] $g_{\alpha\tau}T_\sigma^\alpha = T_{\sigma\tau}$ and $g^{\sigma\beta}T_\sigma^\alpha = T^{\alpha\beta}$ should be symmetrical tensors.

where we have set $T = T_\mu^\mu$ (LAUE's scalar).[20] These are the general field equations of gravitation we were looking for in mixed form. Working back from these, we get instead of (47)[21]

$$\left.\begin{aligned} \frac{\partial \Gamma_{\mu\nu}^\alpha}{\partial x_\alpha} + \Gamma_{\mu\beta}^\alpha \Gamma_{\nu\alpha}^\beta &= -\kappa\left(T_{\mu\nu} - \frac{1}{2} g_{\mu\nu} T\right) \\ \sqrt{-g} &= 1 \end{aligned}\right\}. \tag{53}$$

It must be admitted that this introduction of the energy-tensor of matter is not justified by the relativity postulate alone.[22] For this reason we have deduced it here from the requirement that the energy of the gravitational field has the same gravitating effect as any other kind of energy. The strongest reason for the choice of these equations, however, is that they entail that the components of the total energy satisfy conservation laws that correspond exactly to equations (49) and (49a). This will be shown below.

§ 17. The Laws of Conservation in the General Case.

Equation (52) can easily be rewritten in such a way that the second term on the right-hand side is eliminated. Contract (52) over the indices μ and σ, multiply the resulting equation by $\frac{1}{2}\delta_\mu^\sigma$ and subtract it from equation (52). This gives[23]

$$\frac{\partial}{\partial x_\alpha}\left(g^{\sigma\beta}\Gamma_{\mu\beta}^\alpha - \frac{1}{2}\delta_\mu^\sigma g^{\lambda\beta}\Gamma_{\lambda\beta}^\alpha\right) = -\kappa\left(t_\mu^\sigma + T_\mu^\sigma\right). \tag{52a}$$

We perform the operation $\partial/\partial x_\sigma$ on this equation. We have

$$\frac{\partial^2}{\partial x_\alpha \partial x_\sigma}\left(g^\sigma \Gamma_{\beta\mu}^\alpha\right) = -\frac{1}{2}\frac{\partial^2}{\partial x_\alpha \partial x_\sigma}\left[g^{\sigma\beta} g^{\alpha\lambda}\left(\frac{\partial g_{\mu\lambda}}{\partial x_\beta} + \frac{\partial g_{\beta\lambda}}{\partial x_\mu} - \frac{\partial g_{\mu\beta}}{\partial x_\lambda}\right)\right].$$

The first and third terms of the expression in round brackets give contributions that cancel each other, as one sees by switching the summation indices α and σ as well as β and λ in the contribution coming from the third term. The second term can be rewritten with the help of (31), so that we have

$$\frac{\partial^2}{\partial x_\alpha \partial x_\sigma}\left(g^{\sigma\beta}\Gamma_{\mu\beta}^\alpha\right) = \frac{1}{2}\frac{\partial^3 g^{\alpha\beta}}{\partial x_\alpha \partial x_\beta \partial x_\mu}. \tag{54}$$

The second term on the left-hand side of (52a) yields

$$-\frac{1}{2}\frac{\partial^2}{\partial x_\alpha \partial x_\mu}\left(g^{\lambda\beta}\Gamma_{\lambda\beta}^\alpha\right)$$

or

$$\frac{1}{4}\frac{\partial^2}{\partial x_\alpha \partial x_\mu}\left[g^{\lambda\beta} g^{\alpha\delta}\left(\frac{\partial g_{\delta\lambda}}{\partial x_\beta} + \frac{\partial g_{\delta\beta}}{\partial x_\lambda} - \frac{\partial g_{\lambda\beta}}{\partial x_\delta}\right)\right].$$

Given our choice of coordinates, the contribution coming from the last term in the expression in round brackets vanishes on account of (29).[24] The other two can be combined and together, because of (31), they give

$$-\frac{1}{2}\frac{\partial^3 g^{\alpha\beta}}{\partial x_\alpha \partial x_\beta \partial x_\mu},$$

so that in view of (54), we have the identity

$$\frac{\partial^2}{\partial x_\alpha \partial x_\sigma}\left(g^{\rho\beta}\Gamma_{\mu\beta} - \frac{1}{2}\delta_\mu^\sigma g^{\lambda\beta}\Gamma_{\lambda\beta}^\alpha\right) \equiv 0. \tag{55}$$

From (55) and (52a), it follows that

$$\frac{\partial\left(t_\mu^\sigma + T_\mu^\sigma\right)}{\partial x_\sigma} = 0. \tag{56}$$

Hence it follows from our gravitational field equations that the conservation laws of momentum and energy are satisfied.[25] This is most easily seen from the consideration leading to equation (49a); except that, instead of the energy components t_μ^σ of the gravitational field, we now have to insert the total energy components of matter and gravitational field.

§ 18. The Laws of Momentum and Energy for Matter as a Consequence of the Field Equations

Multiplying (53) by $\partial g^{\mu\nu}/\partial x_\sigma$, we obtain, following the path we embarked on in § 15 and in view of the vanishing of

$$g_{\mu\nu}\frac{\partial g^{\mu\nu}}{\partial x_\sigma},$$

the equation[26]

$$-\frac{\partial t_\sigma^\alpha}{\partial x_\alpha} + \frac{1}{2}\frac{\partial g^{\mu\nu}}{\partial x_\sigma}T_{\mu\nu} = 0,$$

or, in view of (56),

$$\frac{\partial T_\sigma^\alpha}{\partial x_\alpha} + \frac{1}{2}\frac{\partial g^{\mu\nu}}{\partial x_\sigma}T_{\mu\nu} = 0. \tag{57}$$

Comparison with (41b) shows that in the coordinates we have chosen this equation simply expresses the vanishing of divergence of the energy tensor for matter.[27] Physically, the occurrence of the second term on the left-hand side shows that the conservation laws of momentum and energy strictly hold for matter alone only if the $g^{\mu\nu}$ are constant, i.e., when the gravitational field vanishes. This second term gives the transfer of momentum and energy per unit volume and per unit time from the gravitational field to matter. This is brought out even more clearly if we rewrite (57) in the form of (41):

$$\frac{\partial T_\sigma^\alpha}{\partial x_\alpha} = -\Gamma_{\alpha\sigma}^\beta T_\beta^\alpha. \tag{57a}$$

The right-hand side gives the energetic effect of the gravitational field on matter.

The gravitational field equations thus contain four conditions which govern the course of material phenomena. They give the equations of material phenomena completely, if the latter can be characterized by four independent differential equations.[1]

9.3 Commentary

Presented in this chapter is the part dealing with the field equations and their compatibility with energy-momentum conservation in the review article "The Foundation of the General Theory of Relativity" (Einstein 1916b, §§ 12–18, pp. 799–810). This part of the paper corresponds to [pp. 26–36] of the handwritten manuscript for it, which is still extant (see below). The paper can be seen as replacing the review article on the *Entwurf* theory of November 1914 (Einstein 1914c, see Ch. 3). As can be inferred from a note in pencil on the last page of the handwritten manuscript, *Annalen der Physik* received the article March 20, 1916. It was published May 11, 1916.

The handwritten manuscript of the article (EA 120 788)—consisting of 46 pages numbered 1 through 45 (including a page numbered '40a' but not counting part of a page pasted at the bottom of [p. 23] and numbered '23a')—is presented in facsimile in Gutfreund and Renn (2015, pp. 38–128) along with detailed commentary, written for a broad audience and illustrated by Laurent Taudin. For commentary written for a more specialized audience, see Janssen (2005) and Sauer (2005b).

Gutfreund and Renn (2015, pp. 130–138) also present in facsimile a five-page manuscript (EA 2 077) with an unpublished appendix (*Anhang*) to this review article, entitled "Formulation of the Theory on the Basis of a Variational Principle." At the top of the first page of this manuscript, Einstein wrote and then deleted "§ 14," which suggests that it is the original version of the part of the published paper dealing with the field equations, which starts in earnest in § 14. In the published version, Einstein introduces the field equations and analyzes their compatibility with energy-momentum conservation in unimodular coordinates. This is in keeping with his November 1915 papers (Einstein 1915a; 1915b; 1915c; 1915d, see Ch. 6) and his correspondence with Lorentz and Ehrenfest of December 1915 and January 1916 (see Ch. 8). In the unpublished appendix, he uses arbitrary coordinates, relying on the formalism developed in the review article on the *Entwurf* theory (Einstein 1914c, see Ch. 3). The manuscript for this appendix would serve as the basis for a paper published in November 1916, in which the theory is discussed in arbitrary rather than in unimodular coordinates (Einstein 1916d, see Ch. 10). In § 8 of the review article, Einstein defends his use of unimodular coordinates:

> One would be mistaken to think that [the choice of unimodular coordinates] amounts to abandoning the postulate of general relativity to some extent. We do not ask: "What are the laws of nature that are covariant under all transformation with determinant 1?" Our question is: "What are the *generally* covariant laws of nature?" It is only after we have formulated those that we simplify their expression by choosing special coordinates (Einstein 1916b, p. 789).

[1] On this question cf. H. Hilbert, Nachr. d. K. Gesellsch. d. Wiss. zu Göttingen, Math.-phys. Klasse, 1915, p. 3.[28]

The published paper is presented in facsimile as Doc. 30 in CPAE6; a transcription of the unpublished appendix as Doc. 31. The companion volume to CPAE6 uses the translation from the anthology *The Principle of Relativity* (Einstein et al. 1952, pp. 111–164), augmented by the translation of the first page of the article, which was omitted in this anthology. This first page has a generous acknowledgment of Marcel Grossmann (see also Einstein 1915a, p. 778, see Sec. 6.1):

> Finally, I want to acknowledge gratefully my friend, the mathematician Grossmann, whose help not only saved me the effort of studying the pertinent mathematical literature, but who also helped me in my search for the field equations of gravitation (Einstein 1916b, p. 769).

Gutfreund and Renn (2015, pp. 183–226) use this same translation. We will also follow it here. Our commentary is based, in part, on sec. 8 of "Untying the knot" (Janssen and Renn 2007).

§ 12

1. This is the final section of part B of the paper, "Mathematical tools (*Hilfsmittel*) for the formulation of generally-covariant equations" (Einstein 1916b, §§ 5–12, pp. 779–801). This part is very similar to the corresponding part B, "From the theory of covariants," in the review article on the *Entwurf* paper (Einstein 1914c, §§ 3–8, pp. 1034–1054). The introduction and discussion of the Riemann curvature tensor in § 12 of the new review article is a slightly expanded version of its introduction and discussion on p. 1053 of the old one.
   A "Note on a simplified way of writing the expressions" at the end of § 5 of part B marks the first time that Einstein mentioned what is now known as the Einstein summation convention in print (he had stated it before in a letter to Ehrenfest; see Sec. 8.4, note 4): "If an index occurs twice in one term of an expression, it is always to be summed over unless the contrary is explicitly stated" (Einstein 1916b, p. 781). This allowed Einstein to omit the summation signs ubiquitous in his earlier publications (cf. Chs. 3 and 6). The syntax of the equations, however, remained awkward as Einstein still wrote the index on the coordinates and the contravariant index of the Christoffel symbols "downstairs" and the covariant indices of the latter "upstairs".
2. Eq. (27) in the mathematical part B of the paper is the definition of the covariant derivative—or, as Einstein calls it, the "extension" (*Erweiterung*)—of the second-rank tensor $A_{\mu\nu}$:
   $$A_{\mu\nu\sigma} = \frac{\partial A_{\mu\nu}}{\partial x^{\sigma}} - \left\{ \begin{matrix} \tau \\ \sigma\mu \end{matrix} \right\} A_{\tau\nu} - \left\{ \begin{matrix} \tau \\ \sigma\nu \end{matrix} \right\} A_{\mu\tau} \qquad (27)$$
   (Einstein 1916b, p. 795).
3. Einstein included a simple proof that $g^{\mu\nu}{}_{;\nu} = 0$ in the letter to Ehrenfest presented in Ch. 8 (see Sec. 8.3, note 14).
4. This way of splitting the Ricci tensor $B_{\mu\nu} \equiv B^{\rho}_{\mu\nu\rho}$ into two parts, $R_{\mu\nu}$ and $S_{\mu\nu}$, goes back to Einstein's attempt in the Zurich notebook to extract field equations from the Ricci tensor covariant under unimodular *transformations* (see Ch. 1, [p. 22R] and Pt. I, Fig. 1.3). In the notebook, he already noted that the second part, $S_{\mu\nu}$, is equal to the covariant derivative of the vector $\left(\sqrt{-g}\right)_{,\mu}$ (cf. Pt. I, Ch. 3, note 7). We argued that Einstein found his way back to these field equations in 1915 by keeping the Lagrangian for the vacuum field equations of the *Entwurf* theory but changing the definition of the gravitational field (see Pt. I, Sec. 5.1 and note 8 below). These are the equations he published in the first paper of November 1915 (Einstein 1915a, p. 782, Eqs. (13), (13a) and (13b) where G_{im} is used instead of $B_{\mu\nu}$; see Sec. 6.1 and Pt. I, 1.1). In two short addenda to this paper, Einstein (1915c; 1915d, see Secs. 6.2 and 6.4) tweaked

these equations to get around the injunction against unimodular *coordinates* in the first, thereby arriving at generally covariant field equations expressed in unimodular coordinates (see, in particular, Sec. 6.4.3, note 1).

5. Eq. (18a) in the mathematical part B of the paper is $d\tau_0 = \sqrt{-g}\,d\tau$, where $d\tau \equiv dx^1 dx^2 dx^3 dx^4$ and $d\tau_0$ is the "natural volume element" (Einstein 1916b, p. 789).

§ 13

6. Eq. (4) in the mathematical part B of the paper gives the metric of Minkowski space-time in its standard form, $g_{\mu\nu} = \mathrm{diag}(-1,-1,-1,1)$ (Einstein 1916b, p. 778).
7. Einstein used a quantity that does not transform as a tensor to represent the gravitational field (see note 8). So there can be a gravitational field in flat Minkowski space-time and there need not be a gravitational field at points with a non-vanishing Riemann curvature tensor. For further discussion of Einstein's by modern lights rather idiosyncratic view of the gravitational field, see Janssen (2012).
8. In the review article on the *Entwurf* theory, Einstein (1914c, p. 1058, Eq. (46)) defined the components of the gravitational field as $\Gamma^{\tau}_{\mu\nu} = \dfrac{1}{2} g^{\tau\rho} g_{\rho\mu,\nu}$ (see Sec. 3.3, note 21). In the first November 1915 paper, Einstein (1915a, pp. 782–783, see Sec. 6.1.3, notes 11–13 and Pt. I, Ch. 3, note 15) called this definition a "fateful prejudice" and replaced it by the one given here

$$\Gamma^{\tau}_{\mu\nu} = -\left\{ \begin{matrix} \tau \\ \mu\nu \end{matrix} \right\}, \tag{45}$$

which Einstein, in a letter to Sommerfeld of November 28, 1915, called "the key to the solution" (see Ch. 7, especially Sec. 7.3, note 13).

§ 14

9. For $\lambda = -\frac{1}{2}$ this tensor (in modern notation: $R_{\mu\nu} + \lambda g_{\mu\nu} R$) is the Einstein tensor. Recall that Einstein (1915d, p. 845, Eq. (6)) arrived at the field equations of his fourth November 1915 paper by adding a term $\frac{1}{2}\kappa g_{\mu\nu} T$ on the right-hand side of the field equations $R_{\mu\nu} = -\kappa T_{\mu\nu}$ proposed in his first November paper (Einstein 1915a, p. 783, Eq. (16)), where $R_{\mu\nu}$ is the first half of the Ricci tensor written as $R_{\mu\nu} + S_{\mu\nu}$. It is unclear when Einstein came to realize that this is equivalent to adding a term $-\frac{1}{2} g_{\mu\nu} R$ on the left-hand side.

   In his cosmology paper published in February the following year, Einstein (1917) realized that one could still add a term $\lambda g_{\mu\nu}$ to the Einstein field equations (see Pt. I, Sec. 7.1 for discussion).
10. In the first November 1915 paper, Einstein (1915a, p. 783) suggested that such purely mathematical considerations had led him to the field equations he proposed in that paper. We argued that physical considerations played a more prominent role (see Pt. I, Sec. 5.3). For Einstein's conflation of general relativity and general covariance, see Pt. I, Ch. 1, note 3.
11. In his third November 1915 paper, Einstein (1915c, see Sec. 6.3) showed, using unimodular coordinates, that his theory reduces to the Newtonian theory in first-order approximation and that it gives the 43″ missing in the Newtonian account of the perihelion motion of Mercury in second-order approximation. (Whether the relativistic contribution to the perihelion motion comes out as a first-order or a second-order effect depends on the coordinates chosen; Earman and Janssen 1993, p. 146.)

§ 15

12. The Lagrangian for the *Entwurf* theory (Einstein 1914c, Eq. (78) on p. 1076 with $\Gamma^{\alpha}_{\mu\nu}$ defined in Eq. (46) on p. 1058) has the same form as the Lagrangian given here for the theory of November 1915 (originally given in Einstein 1915a, Eq. (17) on p. 784 with

$\Gamma^{\alpha}_{\mu\nu}$ defined in Eq. (15) on p. 783). The difference is in the definition of $\Gamma^{\alpha}_{\mu\nu}$ (see note 8 above). The proof in the first part of § 15 (Eqs. (47a)–(47b)) that the vacuum field equations in Eq. (47) are the Euler-Lagrange equations for this Lagrangian is given in more detail under point (1) in the letter from Einstein to Ehrenfest sometime after January 24 presented in Ch. 8 (see Sec. 8.4, notes 3–7, for detailed commentary).

13. Eq. (31) in the mathematical part B of the paper is (Einstein 1916b, p. 796):

$$dg^{\mu\nu} = -g^{\mu\alpha}g^{\nu\beta}dg_{\alpha\beta}, \quad \frac{\partial g^{\mu\nu}}{\partial x^{\sigma}} = -g^{\mu\alpha}g^{\nu\beta}\frac{\partial g_{\alpha\beta}}{\partial x^{\sigma}}. \tag{31}$$

14. The rationale behind the next part of § 15—from contracting the Euler-Lagrange equations in Eq. (47b) with $g^{\mu\nu}_{\sigma}$ to the expression for the gravitational energy-momentum pseudo-tensor in Eq. (50)—becomes clear if we compare it to the calculations under point (2) of the letter to Ehrenfest mentioned in note 12 (see Sec. 8.4, notes 8–12, for detailed commentary).

In unimodular coordinates, the vanishing of the covariant divergence of the energy-momentum tensor for matter, $T^{\mu\nu}{}_{;\nu} = 0$, giving the energy-momentum balance between matter and gravitational field, can be written as

$$T^{\mu}_{\sigma,\mu} + \frac{1}{2}g^{\mu\nu}_{\sigma}T_{\mu\nu} = 0$$

(see Eq. (3) in the letter to Ehrenfest presented in Ch. 8 and Einstein 1915a, p. 782, Eq. (14), discussed in Sec. 6.1.3, note 10). In this review article, this equation is not introduced until § 18 (Einstein 1916b, p. 809, Eq. (57)). It is only at that point that Einstein explains that the term with $g^{\mu\nu}_{\sigma}T_{\mu\nu}$ represents the energy-momentum transfer between matter and gravitational field. He had used similar expressions for this energy-momentum transfer ever since the Zurich notebook (see Pt. I, Eqs. (3.5)–(3.9)). Using the field equations to substitute an expression depending only on $g_{\mu\nu}$ and its derivatives for $T_{\mu\nu}$ in $\frac{1}{2}g^{\mu\nu}_{\sigma}T_{\mu\nu}$ and rewriting the resulting equation as the divergence of a quantity t^{μ}_{σ} representing gravitational energy-momentum, Einstein could then cast the equation above in the form $(T^{\mu}_{\sigma} + t^{\mu}_{\sigma})_{,\mu} = 0$ (see p. 809, Eq. (56)).

However, at this point in the review article, Einstein has not yet introduced the field equations *in the presence of matter*! In fact, as we will see in § 16, taking advantage of a result in the letter to Ehrenfest, Einstein arrived at these equations by writing the vacuum field equations in unimodular coordinates in terms of $t_{\mu\nu}$ and demanding that $T_{\mu\nu}$ enter the field equations in the exact same way. In unimodular coordinates, it turns out, the resulting field equations in the presence of matter can be written as

$$\frac{\partial}{\partial x_{\alpha}}\left(\frac{\partial H}{\partial g^{\mu\nu}_{\alpha}}\right) - \frac{\partial H}{\partial g^{\mu\nu}} = -\kappa\left(T_{\mu\nu} - \frac{1}{2}g_{\mu\nu}T\right).$$

As Einstein showed, the left-hand side is equal to $\Gamma^{\alpha}_{\mu\nu,\alpha} + \Gamma^{\alpha}_{\mu\beta}\Gamma^{\beta}_{\nu\alpha}$, so these are the Einstein field equations in unimodular coordinates (see p. 808, Eq. (53)).

With malice aforethought (as indicated by the footnote about the factor -2κ that seems to come out of nowhere at this point), Einstein could thus substitute minus the left-hand side divided by κ for $T_{\mu\nu}$ in $\frac{1}{2}g^{\mu\nu}_{\sigma}T_{\mu\nu}$ (in the letter to Ehrenfest he had shown that the contraction of $g^{\mu\nu}_{\sigma}$ with $g_{\mu\nu}T$ vanishes; see Sec. 8.4, note 9). This gives (cf. p. 807, Eq. (52))

$$-\frac{1}{2\kappa}g^{\mu\nu}_{\sigma}\left(\frac{\partial}{\partial x_{\alpha}}\left(\frac{\partial H}{\partial g^{\mu\nu}_{\alpha}}\right) - \frac{\partial H}{\partial g^{\mu\nu}}\right).$$

Rewriting this expression the way he did in the letter to Ehrenfest (see Sec. 8.4, notes 8–12), Einstein arrived at Eqs. (49) and (50).

15. Eq. (34) in the mathematical part B of the paper is (in our notation and after correction of a typo)

$$\frac{\partial g^{\mu\nu}}{\partial x^{\sigma}} = -g^{\mu\tau}\left\{\begin{matrix}\nu\\ \tau\sigma\end{matrix}\right\} - g^{\nu\tau}\left\{\begin{matrix}\mu\\ \tau\sigma\end{matrix}\right\} \tag{34}$$

and expresses the vanishing of the covariant derivative of the metric (Einstein 1916b, p. 796, cf. note 3 above).

16. Two years later, Einstein further explored the relation between differential and integral forms of the law of energy-momentum conservation, both in a paper (Einstein 1918d) and in correspondence with Felix Klein (see CPAE8, in particular Docs. 480, 487, 492), who published several papers on the subject (Klein 1917, 1918a, 1918b).

17. The final part of of § 15, the derivation of the mixed form of the vacuum field equations in Eq. (51), follows the derivation under point (3) of the letter to Ehrenfest mentioned in notes 12 and 14 (see Sec. 8.4, notes 13–16, for detailed commentary). The calculation in that letter, however, is not for the vacuum field equations but for the field equations in the presence of matter (cf. note 14). Instead of Eq. (51), Einstein thus arrived at

$$\frac{\partial}{\partial x_{\alpha}}\left(g^{\tau\nu}\left\{\begin{matrix}\alpha\\ \sigma\tau\end{matrix}\right\}\right) = \kappa\left((T_{\sigma}^{\nu}+t_{\sigma}^{\nu}) - \frac{1}{2}\delta_{\sigma}^{\nu}(T+t)\right)$$

and added the comment: "This equation is interesting because it shows that the source of the gravitational field lines is determined solely by the sum $T_{\sigma}^{\nu}+t_{\sigma}^{\nu}$, as one should expect" (see Sec. 8.3 and note 16 in Sec. 8.4). As we will see in the next section (§ 16), Einstein now used this observation to turn the vacuum field equations introduced in § 14 into field equations in the presence of matter.

18. Like the Greek indices, the Latin indices m and n take on the values 1 through 4.

§ 16

19. Einstein thus used the result under point (3) of his letter to Ehrenfest (see note 14 above) to arrive at gravitational field equations in the presence of matter. He already used the requirement that all energy (later: energy-momentum) enters the field equations in the same way in the second of two papers he wrote in Prague on a theory for static gravitational fields (Einstein 1912b, pp. 457–458; cf. Sec. 3.3, note 22, and Pt. I, Sec. 3.2).

20. The energy-momentum tensor plays a central role in Max Laue's (1911) textbook on special relativity. For the representation of matter in his new theory of gravity, Einstein relied heavily on the relativistic continuum mechanics that takes center stage in Laue's textbook (see Pt. I, Sec. 3.2, especially note 11).

21. Einstein arrived at Eq. (51) by contracting Eq. (47) with $g^{\nu\sigma}$. Hence,

$$g^{\nu\sigma}\left(\frac{\partial\Gamma_{\mu\nu}^{\alpha}}{\partial x_{\alpha}} + \Gamma_{\mu\beta}^{\alpha}\Gamma_{\nu\alpha}^{\beta}\right) = \frac{\partial}{\partial x_{\alpha}}\left(g^{\sigma\beta}\Gamma_{\mu\beta}^{\alpha}\right) + \kappa\left(t_{\mu}^{\sigma} - \frac{1}{2}\delta_{\mu}^{\sigma}t\right).$$

Eq. (52) can thus be rewritten as

$$g^{\nu\sigma}\left(\frac{\partial\Gamma_{\mu\nu}^{\alpha}}{\partial x_{\alpha}} + \Gamma_{\mu\beta}^{\alpha}\Gamma_{\nu\alpha}^{\beta}\right) = -\kappa\left(T_{\mu}^{\sigma} - \frac{1}{2}\delta_{\mu}^{\sigma}T\right) = -\kappa g^{\nu\sigma}\left(T_{\mu\nu} - \frac{1}{2}g_{\mu\nu}T\right),$$

from which Eq. (53) immediately follows. The right-hand side of Eq. (53) is a generally covariant tensor; the left-hand side is the Ricci tensor in unimodular coordinates (see § 12, Eq. (44)). Eq. (53) thus gives the generally covariant Einstein field equations in unimodular coordinates.

22. In § 14, Einstein (1916b, p. 804) argued that the principle of relativity does uniquely determine the *vacuum* field equations.

§ 17

23. Fully contracting Eq. (52), we find that

$$\frac{\partial}{\partial x_\alpha}\left(g^{\lambda\beta}\Gamma^\alpha_{\lambda\beta}\right) = -\kappa\left[\left(t^\lambda_\lambda + T^\lambda_\lambda\right) - \frac{1}{2}\delta^\lambda_\lambda\left(t+T\right)\right] = \kappa(T+t),$$

where in the last step we used that $\delta^\lambda_\lambda = 4$. Substituting this expression for $\kappa(T+t)$ into the second term on the right-hand side of Eq. (52) and moving this term to the left-hand side we arrive at Eq. (52a).

24. Eq. (29) in the mathematical part B of the paper is

$$\frac{1}{\sqrt{-g}}\frac{\partial\sqrt{-g}}{\partial x^\sigma} = \frac{1}{2}\frac{\partial \ln(-g)}{\partial x^\sigma} = \frac{1}{2}g^{\mu\nu}\frac{\partial g_{\mu\nu}}{\partial x^\sigma} = -\frac{1}{2}g_{\mu\nu}\frac{\partial g^{\mu\nu}}{\partial x^\sigma} \tag{29}$$

(Einstein 1916b, p. 796). In unimodular coordinates, this is equal to zero.

25. This relation between field equations and energy-momentum conservation is both similar to and interestingly different from its counterpart in the *Entwurf* theory (for discussion, see Pt. I, Secs. 4.2 and 5.2). The *Entwurf* field equations can be written as (Einstein 1914c, p. 1077, Eq. (81); see Ch. 3)

$$\frac{\partial}{\partial x^\alpha}\left(\sqrt{-g}g^{\alpha\beta}\Gamma^\nu_{\sigma\beta}\right) = -\kappa(\mathfrak{T}^\nu_\sigma + \mathfrak{t}^\nu_\sigma),$$

where $\Gamma^\nu_{\sigma\beta} \equiv \frac{1}{2}g^{\nu\tau}g_{\sigma\tau,\beta}$, $\mathfrak{T}^\nu_\sigma \equiv \sqrt{-g}T^\nu_\sigma$ and $\mathfrak{t}^\nu_\sigma \equiv \sqrt{-g}t^\nu_\sigma$. To ensure energy-momentum conservation, in the guise of the vanishing the four-divergence of the right-hand side, Einstein imposed the condition:

$$\frac{\partial^2}{\partial x^\mu \partial x^\alpha}\left(\sqrt{-g}g^{\alpha\beta}\Gamma^\nu_{\sigma\beta}\right) = 0.$$

This same condition, abbreviated $B_\mu = 0$, doubles as the condition for "adapted coordinates" in the *Entwurf* theory (Einstein 1914c, p. 1070, Eq. (67); see Sec. 3.3, note 9). Einstein thus came to expect that energy-momentum conservation gives a restriction on the covariance of the field equations.

However, the covariance of the field equations with which Einstein (1915a) replaced the *Entwurf* field equations in his first November 1915 paper appeared to be much broader than would seem to be allowed by the successor to the conditions $B_\mu = 0$ coming from energy-momentum conservation. We argued that this discrepancy is what motivated Einstein to replace the four components of this condition by one condition on the determinant of the metric and, subsequently, to replace that condition by one that is automatically satisfied in unimodular coordinates (for detailed discussion, see Sec. 6.1.3, note 17 and 18, and Sec. 6.4.3, note 1).

In this section of the review article, Einstein put together the relations he had found in his November 1915 papers in a new way to derive the identity (55) for the components of the metric tensor and their first- and second-order derivatives, which in conjunction with the field equations ensure energy-momentum conservation, at least in unimodular coordinates (see Janssen and Renn 2007, sec. 8, for further discussion and, in notes 123 and 124, a more detailed version of Einstein's derivation of the identity (55)).

In a paper published in November 1916, Einstein (1916d, see Ch. 10) showed—using the formalism developed in the review article on the *Entwurf* theory published two years earlier (Einstein 1914c, see Ch. 3)—that the field equations guarantee energy-momentum conservation in arbitrary coordinates with the help of an identity that follows from the covariance of the theory's Lagrangian under arbitrary transformations. This four-component identity, which we now recognize as the contracted Bianchi identities, is the direct analogue in the new theory of the condition $B_\mu = 0$ for "adapted coordinates" in the *Entwurf* theory.

§ 18

26. In § 15, Einstein had obtained Eq. (49) for the gravitational energy-momentum pseudo-tensor t_σ^α and the vanishing of its four-divergence by rewriting the contraction of the vacuum field equations and $g_\sigma^{\mu\nu}$. In doing so, we argued (see note 14), he implicitly made use of the field equations in the presence of matter he had yet to introduce: the interpretation of the divergence he found as that of the gravitational energy-momentum pseudo-tensor hinges on the interpretation of his starting point as $\frac{1}{2}g_\sigma^{\mu\nu}T_{\mu\nu}$. This earlier calculation in § 15 makes this one in § 18 easy to follow. Since $g_\sigma^{\mu\nu}g_{\mu\nu} = 0$, we can replace $\frac{1}{2}g_\sigma^{\mu\nu}T_{\mu\nu}$ by $\frac{1}{2}g_\sigma^{\mu\nu}\left(T_{\mu\nu} - \frac{1}{2}g_{\mu\nu}T\right)$. Setting $\frac{\partial}{\partial x_\alpha}\left(\frac{\partial H}{\partial g_\alpha^{\mu\nu}}\right) - \frac{\partial H}{\partial g^{\mu\nu}}$ equal to $-\kappa\left(T_{\mu\nu} - \frac{1}{2}g_{\mu\nu}T\right)$ and repeating the calculation in § 15, we find that

$$\frac{1}{2}g_\sigma^{\mu\nu}T_{\mu\nu} = t_{\sigma,\alpha}^\alpha,$$

which is equivalent to the equation above Eq. (57) (after correction of a sign error). Using Eq. (56) to replace $t_{\sigma,\alpha}^\alpha$ by $-T_{\sigma,\alpha}^\alpha$ we arrive at Eq. (57),

$$t_{\sigma,\alpha}^\alpha + \frac{1}{2}g_\sigma^{\mu\nu}T_{\mu\nu} = 0,$$

which expresses the vanishing of the covariant divergence of the energy-momentum tensor for matter in unimodular coordinates.

At this point, Einstein could only prove *in unimodular coordinates* that his field equations imply energy-momentum conservation. Lorentz already noted this when he read the November 1915 papers shortly after they appeared. On December 23, 1915, he wrote to Ehrenfest: "hypothesis (3a) [i.e., $\sqrt{-g} = 1$ in Einstein (1915d, p. 845)] is indeed essential for the final form of the theory, since the proof of the "conservation law," which is always regarded as something necessary, rests on the hypothesis" (Kox 2018, Doc. 174). In a paper published in November 1916, Einstein would prove *in arbitrary coordinates* that the field equations imply energy-momentum conservation (Einstein 1916d, see Ch. 10).

27. Eq. (41) and (41b) in the mathematical part B of the paper define the "[covariant] divergence of a mixed second-rank tensor

$$\sqrt{-g}A_\mu = \frac{\partial\left(\sqrt{-g}A_\mu^\sigma\right)}{\partial x^\sigma} - \begin{Bmatrix}\tau \\ \sigma\mu\end{Bmatrix}\sqrt{-g}A_\tau^\sigma, \tag{41}$$

$$\sqrt{-g}A_\mu = \frac{\partial\left(\sqrt{-g}A_\mu^\sigma\right)}{\partial x^\sigma} + \frac{1}{2}\frac{\partial g^{\rho\sigma}}{\partial x^\mu}\sqrt{-g}A_{\sigma\rho} \tag{41b}$$

(Einstein 1916b, p. 799).

28. The reference is to Hilbert (1915). For discussion, see Pt. I, Secs. 2.3 and 4.3, Sauer (1999, 2005a) and Renn and Stachel (2007).

Chapter 10
Hamilton's Principle and the General Theory of Relativity

M. Janssen, J. Renn, *How Einstein Found His Field Equations*, Classic Texts in the Sciences, https://doi.org/10.1007/978-3-030-97955-3_17

10.1 Facsimile

Hamiltonsches Prinzip und allgemeine Relativitätstheorie.

Von A. Einstein.

In letzter Zeit ist es H. A. Lorentz und D. Hilbert gelungen[1], der allgemeinen Relativitätstheorie dadurch eine besonders übersichtliche Gestalt zu geben, daß sie deren Gleichungen aus einem einzigen Variationsprinzipe ableiteten. Dies soll auch in der nachfolgenden Abhandlung geschehen. Dabei ist es mein Ziel, die fundamentalen Zusammenhänge möglichst durchsichtig und so allgemein darzustellen, als es der Gesichtspunkt der allgemeinen Relativität zuläßt. Insbesondere sollen über die Konstitution der Materie möglichst wenig spezialisierende Annahmen gemacht werden, im Gegensatz besonders zur
Hilbertschen Darstellung. Anderseits soll im Gegensatz zu meiner [2]
eigenen letzten Behandlung des Gegenstandes die Wahl des Koordinatensystems vollkommen freibleiben. [3]

§ 1. Das Variationsprinzip und die Feldgleichungen der Gravitation und der Materie.

Das Gravitationsfeld werde wie üblich durch den Tensor[2] der $g_{\mu\nu}$ (bzw. $g^{\mu\nu}$) beschrieben, die Materie (inklusive elektromagnetisches Feld) durch eine beliebige Zahl von Raum-Zeitfunktionen $q_{(\varrho)}$, deren invariantentheoretischer Charakter für uns gleichgültig ist. Es sei ferner $\mathfrak{H}$ eine Funktion der

$$g^{\mu\nu},\ g^{\mu\nu}_{\sigma}\left(=\frac{\partial g^{\mu\nu}}{\partial x_\sigma}\right) \text{ und } g^{\mu\nu}_{\sigma\tau}\left(=\frac{\partial^2 g^{\mu\nu}}{\partial x_\sigma \partial x_\tau}\right), \text{ der } q_{(\varrho)} \text{ und } q_{(\varrho)\alpha}\left(=\frac{\partial q_{(\varrho)}}{\partial x_\alpha}\right).$$

Dann liefert uns das Variationsprinzip

$$\delta\left\{\int \mathfrak{H}\, d\tau\right\} = 0 \qquad (1)$$

[1] Vier Abhandlungen von H. A. Lorentz in den Jahrgängen 1915 und 1916 d. Publikationer d. Koninkl. Akad. van Wetensch. te Amsterdam; D. Hilbert, Gött. Nachr.
1915. Heft 3. [1]

[2] Von dem Tensorcharakter der $g_{\mu\nu}$ wird vorläufig kein Gebrauch gemacht.

so viele Differentialgleichungen, wie zu bestimmende Funktionen $g_{\mu\nu}$ und $q_{(\varrho)}$ vorhanden sind, wenn wir festsetzen, daß die $g^{\mu\nu}$ und $q_{(\varrho)}$ abhängig voneinander zu variieren sind, und zwar derart, daß an den Integrationsgrenzen die $\delta q_{(\varrho)}$, $\delta g^{\mu\nu}$ und $\frac{\partial \delta g_{\mu\nu}}{\partial x_\sigma}$ alle verschwinden.

Wir wollen nun annehmen, daß $\mathfrak{H}$ in den $g^{\mu\nu}_{\sigma\tau}$ linear sei, und zwar derart, daß die Koeffizienten der $g^{\mu\nu}_{\sigma\tau}$ nur von den $g^{\mu\nu}$ abhängen. Dann kann man das Variationsprinzip (1) durch ein für uns bequemeres ersetzen. Durch geeignete partielle Integration erhält man nämlich

$$\int \mathfrak{H} d\tau = \int \mathfrak{H}^* d\tau + F, \tag{2}$$

wobei F ein Integral über die Begrenzung des betrachteten Gebietes bedeutet, die Größe $\mathfrak{H}^*$ aber nur mehr von den $g^{\mu\nu}$, $g^{\mu\nu}_\sigma$, $q_{(\varrho)}$, $q_{(\varrho)\alpha}$, aber nicht mehr von den $g^{\mu\nu}_{\sigma\tau}$ abhängt. Aus (2) ergibt sich für solche Variationen, wie sie uns interessieren

$$\delta\left\{\int \mathfrak{H} d\tau\right\} = \delta\left\{\int \mathfrak{H}^* d\tau\right\}, \tag{3}$$

so daß wir unser Variationsprinzip (1) ersetzen dürfen durch das bequemere

$$\delta\left\{\int \mathfrak{H}^* d\tau\right\} = 0. \tag{1a}$$

Durch Ausführung der Variation nach den $g^{\mu\nu}$ und nach den $q_{(\varrho)}$ erhält man als die Feldgleichungen der Gravitation und der Materie die Gleichungen[1]

$$\frac{\partial}{\partial x_\alpha}\left(\frac{\partial \mathfrak{H}^*}{\partial g^{\mu\nu}_\alpha}\right) - \frac{\partial \mathfrak{H}^*}{\partial g^{\mu\nu}} = 0 \tag{4}$$

$$\frac{\partial}{\partial x_\alpha}\left(\frac{\partial \mathfrak{H}^*}{\partial q_{(\varrho)\alpha}}\right) - \frac{\partial \mathfrak{H}^*}{\partial q_{(\varrho)}} = 0. \tag{5}$$

§ 2. Sonderexistenz des Gravitationsfeldes.

Wenn man über die Art und Weise, wie $\mathfrak{H}$ von den $g^{\mu\nu}$, $g^{\mu\nu}_\sigma$, $g^{\mu\nu}_{\sigma\tau}$, $q_{(\varrho)}$, $q_{(\varrho)\alpha}$ abhängt, keine spezialisierende Voraussetzung macht, können die Energiekomponenten nicht in zwei Teile gespalten werden, von denen der eine zum Gravitationsfelde, der andere zu der Materie gehört. Um diese Eigenschaft der Theorie herbeizuführen, machen wir folgende Annahme

$$\mathfrak{H} = \mathfrak{G} + \mathfrak{M}, \tag{6}$$

[1] Zur Abkürzung sind in den Formeln die Summenzeichen weggelassen. Es ist über diejenigen Indizes stets summiert zu denken, welche in einem Gliede zweimal vorkommen. In (4) bedeutet also z. B. $\frac{\partial}{\partial x_\alpha}\left(\frac{\partial \mathfrak{H}^*}{\partial g^{\mu\nu}_\alpha}\right)$ den Term $\sum_\alpha \frac{\partial}{\partial x_\alpha}\left(\frac{\partial \mathfrak{H}^*}{\partial g^{\mu\nu}_\alpha}\right)$. [4]

wobei $\mathfrak{G}$ nur von den $g^{\mu\nu}, g^{\mu\nu}_{\sigma}, g^{\mu\nu}_{\sigma\tau}$, $\mathfrak{M}$ nur von $g^{\mu\nu}, q_{(\varrho)}, q_{(\varrho)\alpha}$ abhänge. Die Gleichungen (4), (4a) nehmen dann die Form an

$$\frac{\partial}{\partial x_\alpha}\left(\frac{\partial \mathfrak{G}^*}{\partial g^{\mu\nu}_\alpha}\right) - \frac{\partial \mathfrak{G}^*}{\partial g^{\mu\nu}} = \frac{\partial \mathfrak{M}}{\partial g^{\mu\nu}} \tag{7}$$

$$\frac{\partial}{\partial x_\alpha}\left(\frac{\partial \mathfrak{M}}{\partial q_{(\varrho)\alpha}}\right) - \frac{\partial \mathfrak{M}}{\partial q_{(\varrho)}} = 0. \tag{8}$$

Dabei steht $\mathfrak{G}^*$ zu $\mathfrak{G}$ in derselben Beziehung wie $\mathfrak{H}^*$ zu $\mathfrak{H}$.

Es ist wohl zu beachten, daß die Gleichungen (8) bzw. (5) durch andere zu ersetzen wären, wenn wir annehmen würden, daß $\mathfrak{M}$ bzw. $\mathfrak{H}$ noch von höheren als den ersten Ableitungen der $q_{(\varrho)}$ abhängig wären. Ebenso wäre es denkbar, daß die $q_{(\varrho)}$ nicht als voneinander unabhängig, sondern als durch Bedingungsgleichungen miteinander verknüpft aufzufassen wären. All dies ist für die folgenden Entwicklungen ohne Bedeutung, da letztere allein auf die Gleichungen (7) gegründet sind, welche durch Variieren unseres Integrals nach den $g^{\mu\nu}$ gewonnen sind.

§ 3. Invariantentheoretische bedingte Eigenschaften der Feldgleichungen der Gravitation.

Wir führen nun die Voraussetzung ein, daß

$$ds^2 = g_{\mu\nu}dx_\mu dx_\nu \tag{9}$$

eine Invariante sei. Damit ist der Transformationscharakter der $g_{\mu\nu}$ festgelegt. Über den Transformationscharakter der die Materie beschreibenden $q_{(\varrho)}$ machen wir keine Voraussetzung. Hingegen seien die Funktionen $H = \frac{\mathfrak{H}}{\sqrt{-g}}$ sowie $G = \frac{\mathfrak{G}}{\sqrt{-g}}$ und $M = \frac{\mathfrak{M}}{\sqrt{-g}}$ Invarianten bezüglich beliebiger Substitutionen der Raum-Zeitkoordinaten. Aus diesen Voraussetzungen folgt die allgemeine Kovarianz der aus (1) gefolgerten Gleichungen (7) und (8). Ferner folgt, daß G (bis auf einen konstanten Faktor) gleich dem Skalar des Riemannschen Tensors der Krümmung sein muß; denn es gibt keine andere Invariante von den für G geforderten Eigenschaften[1]. Damit ist auch $\mathfrak{G}^*$ und damit die linke Seite der Feldgleichung (7) vollkommen festgelegt[2].

Aus dem allgemeinen Relativitätspostulat folgen gewisse Eigenschaften der Funktion $\mathfrak{G}^*$, die wir nun ableiten wollen. Zu diesem [6]

[1] Hierin liegt es begründet, daß die allgemeine Relativitätsforderung zu einer ganz bestimmten Gravitationstheorie führt.

[2] Man erhält durch Ausführung der partiellen Integration

$$\mathfrak{G}^* = \sqrt{-g}\, g^{\mu\nu}\left[\begin{Bmatrix}\mu\alpha\\ \beta\end{Bmatrix}\begin{Bmatrix}\nu\beta\\ \alpha\end{Bmatrix} - \begin{Bmatrix}\mu\nu\\ \alpha\end{Bmatrix}\begin{Bmatrix}\alpha\beta\\ \beta\end{Bmatrix}\right].$$ [5]

Zweck führen wir eine infinitesimale Transformation der Koordinaten durch, indem wir setzen

$$x'_\nu = x_\nu + \Delta x_\nu \, ; \qquad (10)$$

die Δx_ν sind beliebig wählbare, unendlich kleine Funktionen der Koordinaten. x'_ν sind die Koordinaten des Weltpunktes im neuen System, dessen Koordinaten im ursprünglichen System x_ν sind. Wie für die Koordinaten gilt für jede andere Größe ψ ein Transformationsgesetz vom Typus

$$\psi' = \psi + \Delta\psi \, ,$$

wobei sich $\Delta\psi$ stets durch die Δx_ν ausdrücken lassen muß. Aus der Kovarianteneigenschaft der $g^{\mu\nu}$ leitet man leicht für die $g^{\mu\nu}$ und $g^{\mu\nu}_\sigma$ die Transformationsgesetze ab: [7]

$$\Delta g^{\mu\nu} = g^{\mu\alpha}\frac{\partial \Delta x_\nu}{\partial x_\alpha} + g^{\nu\alpha}\frac{\partial \Delta x_\mu}{\partial x_\alpha} \qquad (11)$$

$$\Delta g^{\mu\nu}_\sigma = \frac{\partial(\Delta g^{\mu\nu})}{\partial x_\sigma} - g^{\mu\nu}_\alpha \frac{\partial \Delta x_\alpha}{\partial x_\sigma} \, . \qquad (12)$$

Da $\mathfrak{G}^*$ nur von den $g^{\mu\nu}$ und $g^{\mu\nu}_\sigma$ abhängt, ist es mit Hilfe von (13) und (14) möglich, $\Delta\mathfrak{G}^*$ zu berechnen. Man erhält so die Gleichung [8]

$$\sqrt{-g}\,\Delta\left(\frac{\mathfrak{G}^*}{\sqrt{-g}}\right) = S^\nu_\sigma \frac{\partial \Delta x_\sigma}{\partial x_\nu} + 2\frac{\partial \mathfrak{G}^*}{\partial g^{\mu\sigma}_\alpha} g^{\mu\nu} \frac{\partial^2 \Delta x_\sigma}{\partial x_\nu \partial x_\alpha} \, , \qquad (13)$$

wobei zur Abkürzung gesetzt ist

$$S^\nu_\sigma = 2\frac{\partial \mathfrak{G}^*}{\partial g^{\mu\sigma}} g^{\mu\nu} + 2\frac{\partial \mathfrak{G}^*}{\partial g^{\mu\sigma}_\alpha} g^{\mu\nu}_\alpha + \mathfrak{G}^*\delta^\nu_\sigma - \frac{\partial \mathfrak{G}^*}{\partial g^{\mu\alpha}_\nu} g^{\mu\alpha}_\sigma \, . \qquad (14)$$

Aus diesen beiden Gleichungen ziehen wir zwei für das Folgende wichtige Folgerungen. Wir wissen, daß $\frac{\mathfrak{G}}{\sqrt{-g}}$ eine Invariante ist bezüglich beliebiger Substitutionen, nicht aber $\frac{\mathfrak{G}^*}{\sqrt{-g}}$. Wohl aber ist es leicht, von letzerer Größe zu beweisen, daß sie bezüglich linearer Substitutionen der Koordinaten eine Invariante ist. Hieraus folgt, daß die rechte Seite von (13) stets verschwinden muß, wenn sämtliche $\frac{\partial^2 \Delta x_\sigma}{\partial x_\nu \partial x_\alpha}$ verschwinden. Es folgt daraus, daß $\mathfrak{G}^*$ der Identität

$$S^\nu_\sigma \equiv 0 \qquad (15)$$

genügen muß.

Wählen wir ferner die Δx_ν so, daß sie nur im Innern eines betrachteten Gebietes von null verschieden sind, in infinitesimaler Nähe der Begrenzung aber verschwinden, so ändert sich der Wert des in Gleichung (2) auftretenden, über die Begrenzung erstreckten Integrales nicht bei der ins Auge gefaßten Transformation; es ist also

$$\Delta(F) = 0$$

und somit[1]

$$\Delta\left\{\int \mathfrak{G} d\tau\right\} = \Delta\left\{\int \mathfrak{G}^* d\tau\right\}.$$

Die linke Seite der Gleichung muß aber verschwinden, da sowohl $\frac{\mathfrak{G}}{\sqrt{-g}}$ wie $\sqrt{-g}\,d\tau$ Invarianten sind. Folglich verschwindet auch die rechte Seite. Wir erhalten also mit Rücksicht auf (14), (15) und (16) zunächst die Gleichung

$$\int \frac{\partial \mathfrak{G}^*}{\partial g^{\mu\sigma}_{\alpha}} g^{\mu\nu} \frac{\partial^2 \Delta x_\sigma}{\partial x_\nu \partial x_\alpha} d\tau = 0. \qquad (16)$$

Formt man diese durch zweimalige partielle Integration um, so erhält man mit Rücksicht auf die freie Wählbarkeit der Δx_σ die Identität [9]

$$\frac{\partial^2}{\partial x_\nu \partial x_\alpha}\left(\frac{\partial \mathfrak{G}^*}{\partial g^{\mu\sigma}_{\alpha}} g^{\mu\nu}\right) \equiv 0. \qquad (17)$$

Aus den beiden Identitäten (16) und (17), welche aus der Invarianz von $\frac{\mathfrak{G}}{\sqrt{-g}}$, also aus dem Postulat der allgemeinen Relativität hervorgehen, haben wir nun Folgerungen zu ziehen.

Die Feldgleichungen (7) der Gravitation formen wir zunächst durch gemischte Multiplikation mit $g^{\mu\sigma}$ um. Man erhält dann (unter Vertauschung der Indizes σ und ν) die den Feldgleichungen (7) äquivalenten Gleichungen [10]

$$\frac{\partial}{\partial x_\alpha}\left(\frac{\partial \mathfrak{G}^*}{\partial g^{\mu\sigma}_{\alpha}} g^{\mu\nu}\right) = -(\mathfrak{T}^\nu_\sigma + \mathfrak{t}^\nu_\sigma), \qquad (18)$$

wobei gesetzt ist [11]

$$\mathfrak{T}^\nu_\sigma = -\frac{\partial \mathfrak{M}}{\partial g^{\mu\sigma}} g^{\mu\nu} \qquad (19)$$

$$\mathfrak{t}^\nu_\sigma = -\left(\frac{\partial \mathfrak{G}^*}{\partial g^{\mu\sigma}_{\alpha}} g^{\mu\nu}_{\alpha} + \frac{\partial \mathfrak{G}^*}{\partial g^{\mu\sigma}} g^{\mu\nu}\right) = \frac{1}{2}\left(\mathfrak{G}^* \delta^\nu_\sigma - \frac{\partial \mathfrak{G}^*}{\partial g^{\mu\alpha}_{\nu}} g^{\mu\alpha}_{\sigma}\right). \qquad (20)$$

Der letzte Ausdruck für $\mathfrak{t}^\nu_\sigma$ rechtfertigt sich aus (14) und (15). Durch Differenzieren von (18) nach x_ν und Summation über ν folgt mit Rücksicht auf (17)

$$\frac{\partial}{\partial x_\nu}(\mathfrak{T}^\nu_\sigma + \mathfrak{t}^\nu_\sigma) = 0. \qquad (21)$$

[1] Indem wir statt $\mathfrak{H}$ und $\mathfrak{H}^*$ die Größen $\mathfrak{G}$ und $\mathfrak{G}^*$ einführen.

Die Gleichung (21) drückt die Erhaltung des Impulses und der Energie aus. Wir nennen $\mathfrak{T}_\sigma^\nu$ die Komponenten der Energie der Materie, t_σ^ν die Komponenten der Energie des Gravitationsfeldes.

Aus den Feldgleichungen (7) der Gravitation folgt durch Multiplizieren mit $g_\sigma^{\mu\nu}$ und Summieren über μ und ν mit Rücksicht auf (20) [12]

$$\frac{\partial t_\sigma^\nu}{\partial x_\nu} + \frac{1}{2} g_\sigma^{\mu\nu} \frac{\partial \mathfrak{M}}{\partial g^{\mu\nu}} = 0$$

oder mit Rücksicht auf (19) und (21)

$$\frac{\partial \mathfrak{T}_\sigma^\nu}{\partial x_\nu} - \frac{1}{2} g_\sigma^{\mu\nu} \mathfrak{T}_{\mu\nu} = 0, \qquad (22)$$

wobei $\mathfrak{T}_{\mu\nu}$ die Größen $g_{\nu\sigma} \mathfrak{T}_\mu^\sigma$ bedeuten. Es sind dies 4 Gleichungen, welchen die Energie-Komponenten der Materie zu genügen haben.

Es ist hervorzuheben, daß die (allgemein kovarianten) Erhaltungssätze (21) und (22) aus den Feldgleichungen (7) der Gravitation in Verbindung mit dem Postulat der allgemeinen Kovarianz (Relativität) allein gefolgert sind, ohne Benutzung der Feldgleichungen (8) für die materiellen Vorgänge. [13]

Ausgegeben am 2. November.

10.2 Translation

Hamilton's Principle and the General Theory of Relativity

H. A. LORENTZ and D. HILBERT have recently succeeded[1] in presenting the general theory of relativity in a particularly clear form by deriving its equations from a single variational principle. The same will be done in this paper. My aim is to present the fundamental connections in as transparent and general a form as allowed from the point of view of general relativity. In particular I will make as few special assumptions as possible about the constitution of matter, in marked contrast to HILBERT's treatment of the subject.[2] And in contrast to my own most recent treatment of the subject, the choice of a coordinate system will be left completely open.[3]

§1. The Principle of Variation and the Field Equations of Gravitation and Matter

Let the gravitational field be described as usual by the tensor[2] $g_{\mu\nu}$ (or $g^{\mu\nu}$); matter (including the electromagnetic field) by an arbitrary number of space-time functions $q_{(\rho)}$, the covariance properties of which do not concern us here. Moreover, let $\mathfrak{H}$ be a function of

$$g^{\mu\nu},\ g^{\mu\nu}_{\sigma}\left(=\frac{\partial g^{\mu\nu}}{\partial x_\sigma}\right) \text{ and } g^{\mu\nu}_{\sigma\tau}\left(=\frac{\partial^2 g^{\mu\nu}}{\partial x_\sigma \partial x_\tau}\right),\ q_{(\rho)} \text{ and } q_{(\rho)\alpha}\left(=\frac{\partial q_{(\rho)}}{\partial x_\alpha}\right).$$

The variational principle

$$\delta\left\{\int \mathfrak{H}\, d\tau\right\} = 0 \tag{1}$$

then gives us as many differential equations as there are functions $g_{\mu\nu}$ and $q_{(\rho)}$ to be determined, provided we stipulate that $g^{\mu\nu}$ and $q_{(\rho)}$ are to be varied independently of one another and in such a way that $\delta q_{(\rho)}$, $\delta g^{\mu\nu}$ and $\dfrac{\partial \delta g_{\mu\nu}}{\partial x_\sigma}$ all vanish at the boundaries.

We will assume $\mathfrak{H}$ to be linear in $g^{\mu\nu}_{\sigma\tau}$ such that the coefficients of $g^{\mu\nu}_{\sigma\tau}$ depend only on $g^{\mu\nu}$. The variational principle (1) can then be replaced by one more convenient for us. Through suitable partial integration one gets

$$\int \mathfrak{H}\, d\tau = \int \mathfrak{H}^* d\tau + F, \tag{2}$$

[1] Four papers by H. A. LORENTZ in the Publications of the Royal Academy of Sciences in Amsterdam, 1915 and 1916; D. HILBERT, Göttingen Notices, 1915, Part 3.[1]

[2] For now, we make no use of the tensor character of $g_{\mu\nu}$.

where F denotes an integral over the boundary of the domain under consideration, while the quantity $\mathfrak{H}^*$ depends only on $g^{\mu\nu}$, $g_\sigma^{\mu\nu}$, $q_{(\rho)}$ and $q_{(\rho)\alpha}$ and no longer on $g_{\sigma\tau}^{\mu\nu}$. From (2) we obtain, for the variations of interest to us,

$$\delta\left\{\int \mathfrak{H}\, d\tau\right\} = \delta\left\{\int \mathfrak{H}^*\, d\tau\right\}, \tag{3}$$

so that we can replace the variational principle (1) by the more convenient one

$$\delta\left\{\int \mathfrak{H}^*\, d\tau\right\} = 0. \tag{1a}$$

By carrying out the variation of $g^{\mu\nu}$ and $q_{(\rho)}$ we obtain, as field-equations of gravitation and matter, the equations[1]

$$\frac{\partial}{\partial x_\alpha}\left(\frac{\partial \mathfrak{H}^*}{\partial g_\alpha^{\mu\nu}}\right) - \frac{\partial \mathfrak{H}^*}{\partial g^{\mu\nu}} = 0, \tag{4}$$

$$\frac{\partial}{\partial x_\alpha}\left(\frac{\partial \mathfrak{H}^*}{\partial q_{(\rho)\alpha}}\right) - \frac{\partial \mathfrak{H}^*}{\partial q_{(\rho)}} = 0. \tag{5}$$

§2. Separate Existence of the Gravitational Field

As long as one makes no special assumptions about how $\mathfrak{H}$ depends on $g^{\mu\nu}$, $g_\sigma^{\mu\nu}$, $g_{\sigma\tau}^{\mu\nu}$ and $q_{(\rho)}$, $q_{(\rho)\alpha}$, the energy components cannot be split into two parts, one pertaining to the gravitational field, the other to matter. To ensure that the theory does have this property, we make the following assumption:

$$\mathfrak{H} = \mathfrak{G} + \mathfrak{M}, \tag{6}$$

where $\mathfrak{G}$ depends only on $g^{\mu\nu}$, $g_\sigma^{\mu\nu}$ and $g_{\sigma\tau}^{\mu\nu}$, and $\mathfrak{M}$ only on $g^{\mu\nu}$, $q_{(\rho)}$ and $q_{(\rho)\alpha}$. Equations (4) and (5) then take the form

$$\frac{\partial}{\partial x_\alpha}\left(\frac{\partial \mathfrak{G}^*}{\partial g_\alpha^{\mu\nu}}\right) - \frac{\partial \mathfrak{G}^*}{\partial g^{\mu\nu}} = \frac{\partial \mathfrak{M}}{\partial g^{\mu\nu}}, \tag{7}$$

$$\frac{\partial}{\partial x_\alpha}\left(\frac{\partial \mathfrak{M}}{\partial q_{(\rho)\alpha}}\right) - \frac{\partial \mathfrak{M}}{\partial q_{(\rho)}} = 0. \tag{8}$$

Here $\mathfrak{G}^*$ stands in the same relation to $\mathfrak{G}$ as $\mathfrak{H}^*$ to $\mathfrak{H}$.

It must be noted that equations (8) or (5) would have to be replaced by others, if we were to assume that $\mathfrak{M}$ or $\mathfrak{H}$ also depend on derivatives of the $q_{(\rho)}$ higher than of first order. Similarly, one could imagine that the

---

[1] As an abbreviation, summation signs are omitted. Indices occurring twice in one term are to be summed over. Thus in (4), e.g., $\frac{\partial}{\partial x_\alpha}\left(\frac{\partial \mathfrak{H}^*}{\partial g_\alpha^{\mu\nu}}\right)$ stands for $\sum_\alpha \frac{\partial}{\partial x_\alpha}\left(\frac{\partial \mathfrak{H}^*}{\partial g_\alpha^{\mu\nu}}\right)$.[4]

$q_{(\rho)}$'s are not independent but connected to each other by further conditional equations. All this is of no importance for the following developments, which are based solely on the equations (7), found through variation of our integral with respect to $g^{\mu\nu}$.

§3. Properties of the Field Equations of Gravitation Required by the Theory of Invariants

We now introduce the assumption that

$$ds^2 = g_{\mu\nu} dx_\mu dx_\nu \tag{9}$$

is an invariant. This fixes the transformational character of $g_{\mu\nu}$. As to the transformational character of the $q_{(\rho)}$, which describe matter, we make no presuppositions. However, we will take the functions $H = \frac{\mathfrak{H}}{\sqrt{-g}}$, as well as $G = \frac{\mathfrak{G}}{\sqrt{-g}}$ and $M = \frac{\mathfrak{M}}{\sqrt{-g}}$ to be invariants under arbitrary transformations of the space-time coordinates. The general covariance of equations (7) and (8), derived from (1), follows from these assumptions. Moreover, it follows that G (apart from a constant factor) must be equal to the scalar of RIEMANN's tensor of curvature; because there is no other invariant with the properties required for G.[1] With this, $\mathfrak{G}^*$, and hence the left-hand side of field equation (7) is completely determined.[2]

The postulate of general relativity entails certain properties of the function $\mathfrak{G}^*$ which we will now derive.[6] For this purpose we carry out an infinitesimal transformation of the coordinates, by setting

$$x'_\nu = x_\nu + \Delta x_\nu, \tag{10}$$

where the Δx_ν are arbitrarily chosen, infinitely small functions of the coordinates; x'_ν are the coordinates of the world-point in the new system with coordinates x_ν in the original system. Just as for the coordinates, there is a transformation law for any other quantity ψ, of the type

$$\psi' = \psi + \Delta\psi,$$

where it must possible to express $\Delta\psi$ in terms of Δx_ν. From the covariant property of $g^{\mu\nu}$ one easily derives the transformation laws for $g^{\mu\nu}$ and $g_\sigma^{\mu\nu}$:[7]

---

[1] Herein lies the reason that requirement of general relativity leads to a very definite theory of gravitation.

[2] Through partial integration we obtain[5]

$$\mathfrak{G}^* = \sqrt{-g}\, g^{\mu\nu} \left[\begin{Bmatrix} \mu\alpha \\ \beta \end{Bmatrix} \begin{Bmatrix} \nu\beta \\ \alpha \end{Bmatrix} - \begin{Bmatrix} \mu\nu \\ \alpha \end{Bmatrix} \begin{Bmatrix} \alpha\beta \\ \beta \end{Bmatrix} \right].$$

$$\Delta g^{\mu\nu} = g^{\mu\alpha}\frac{\partial \Delta x_\nu}{\partial x_\alpha} + g^{\nu\alpha}\frac{\partial \Delta x_\mu}{\partial x_\alpha}, \tag{11}$$

$$\Delta g^{\mu\nu}_\sigma = \frac{\partial(\Delta g^{\mu\nu})}{\partial x_\sigma} - g^{\mu\nu}_\alpha \frac{\partial \Delta x_\alpha}{\partial x_\sigma}. \tag{12}$$

Since $\mathfrak{G}^*$ depends only on $g^{\mu\nu}$ and $g^{\mu\nu}_\sigma$, it is possible, with the help of (11) and (12), to calculate $\Delta\mathfrak{G}^*$. In this way we obtain the equation[8]

$$\sqrt{-g}\Delta\left(\frac{\mathfrak{G}^*}{\sqrt{-g}}\right) = S^\nu_\sigma \frac{\partial \Delta x_\sigma}{\partial x_\nu} + 2\frac{\partial \mathfrak{G}^*}{\partial g^{\mu\sigma}_\alpha} g^{\mu\nu}\frac{\partial^2 \Delta x_\sigma}{\partial x_\nu \partial x_\alpha}, \tag{13}$$

where we used the abbreviation

$$S^\nu_\sigma = 2\frac{\partial \mathfrak{G}^*}{\partial g^{\mu\sigma}} g^{\mu\nu} + 2\frac{\partial \mathfrak{G}^*}{\partial g^{\mu\sigma}_\alpha} g^{\mu\nu}_\alpha + \mathfrak{G}^*\delta^\nu_\sigma - \frac{\partial \mathfrak{G}^*}{\partial g^{\mu\alpha}_\nu} g^{\mu\alpha}_\sigma. \tag{14}$$

From these two equations we draw two conclusions that are important for what follows. We know that $\frac{\mathfrak{G}}{\sqrt{-g}}$ is an invariant under arbitrary transformations, but we do not know this about $\frac{\mathfrak{G}^*}{\sqrt{-g}}$. It is easy to show, however, that the latter quantity is an invariant under *linear* coordinate transformation. It follows that the right-hand side of (13) must vanish whenever all $\frac{\partial^2 \Delta x_\sigma}{\partial x_\nu \partial x_\alpha}$ do. Hence, $\mathfrak{G}^*$ must satisfy the identity

$$S^\nu_\sigma \equiv 0. \tag{15}$$

If, furthermore, we choose Δx_ν such that they differ from zero only inside the domain considered but vanish in a infinitesimal neighborhood of the boundary, then the value of the integral in equation (2) extended over the boundary is not changed by this transformation. Therefore

$$\Delta(F) = 0$$

and thus[1]

$$\Delta\left\{\int \mathfrak{G}\, d\tau\right\} = \Delta\left\{\int \mathfrak{G}^* d\tau\right\}.$$

The left-hand side of the equation must vanish, since both $\frac{\mathfrak{G}}{\sqrt{-g}}$ and $\sqrt{-g}\,d\tau$ are invariants. Consequently the right-hand side also vanishes. Thus, taking (13), (14) and (15) into consideration, we obtain

[1] If we introduce $\mathfrak{G}$ and $\mathfrak{G}^*$ instead of $\mathfrak{H}$ and $\mathfrak{H}^*$.

$$\int \frac{\partial \mathfrak{G}^*}{\partial g_\alpha^{\mu\sigma}} g^{\mu\nu} \frac{\partial^2 \Delta x_\sigma}{\partial x_\nu \partial x_\alpha} d\tau = 0. \tag{16}$$

Performing two partial integrations and using that Δx_σ can be arbitrarily chosen, we arrive at the identity[9]

$$\frac{\partial^2}{\partial x_\nu \partial x_\alpha} \left(g^{\mu\nu} \frac{\partial \mathfrak{G}^*}{\partial g_\alpha^{\mu\sigma}} \right) \equiv 0. \tag{17}$$

We now have to draw conclusions from the two identities (16) and (17), which follow from the invariance of $\frac{\mathfrak{G}}{\sqrt{-g}}$ and thus from the postulate of general relativity.

First, we rewrite the gravitational field equations (7) through mixed multiplication by $g^{\mu\sigma}$. We then obtain (switching the indices σ and ν) the following equations, equivalent to (7),[10]

$$\frac{\partial}{\partial x_\alpha} \left(g^{\mu\nu} \frac{\partial \mathfrak{G}^*}{\partial g_\alpha^{\mu\sigma}} \right) = -\left(\mathfrak{T}_\sigma^\nu + \mathfrak{t}_\sigma^\nu \right), \tag{18}$$

where we set[11]

$$\mathfrak{T}_\sigma^\nu = -\frac{\partial \mathfrak{M}}{\partial g^{\mu\sigma}} g^{\mu\nu}, \tag{19}$$

$$\mathfrak{t}_\sigma^\nu = -\left(\frac{\partial \mathfrak{G}^*}{\partial g_\alpha^{\mu\sigma}} g_\alpha^{\mu\nu} + \frac{\partial \mathfrak{G}^*}{\partial g^{\mu\sigma}} g^{\mu\nu} \right) = \frac{1}{2} \left(\mathfrak{G}^* \delta_\sigma^\nu - \frac{\partial \mathfrak{G}^*}{\partial g_\nu^{\mu\alpha}} g_\sigma^{\mu\alpha} \right). \tag{20}$$

The second expression for $\mathfrak{t}_\sigma^\nu$ is justified by (14) and (15). Differentiation of (18) with respect to x_ν and summation for ν gives, in view of (17):

$$\frac{\partial}{\partial x_\nu} \left(\mathfrak{T}_\sigma^\nu + \mathfrak{t}_\sigma^\nu \right) = 0. \tag{21}$$

Equation (21) expresses the conservation of momentum and energy. We call $\mathfrak{T}_\sigma^\nu$ the energy components of matter and $\mathfrak{t}_\sigma^\nu$ the energy components of the gravitational field.

From the gravitational field equations (7) it follows, after multiplication by $g_\sigma^{\mu\nu}$, summation over μ and ν, and on account of (20) that[12]

$$\frac{\partial \mathfrak{t}_\sigma^\nu}{\partial x_\nu} + \frac{1}{2} g_\sigma^{\mu\nu} \frac{\partial \mathfrak{M}}{\partial g^{\mu\nu}} = 0$$

or, in view of (19) and (21),

$$\frac{\partial \mathfrak{T}_\sigma^\nu}{\partial x_\nu} + \frac{1}{2} g_\sigma^{\mu\nu} \mathfrak{T}_{\mu\nu} = 0, \tag{22}$$

where $\mathfrak{T}_{\mu\nu}$ are the quantities $g_{\nu\sigma}\mathfrak{T}_{\mu}^{\sigma}$. These are four equations the energy components of matter have to satisfy.

It should be emphasized that the (generally covariant) conservation laws (21) and (22) have been derived from the gravitational field equations (7) *alone*, in conjunction with the postulate of general covariance (relativity) but without the use of the field equations (8) for material processes.[13]

10.3 Commentary

Presented in this chapter is the short paper, "Hamilton's Principle and the General Theory of Relativity" (Einstein 1916d), in which Einstein introduced the gravitational field equations of his theory via a variational (or, as he calls it, Hamiltonian) principle, showed that they imply energy-momentum conservation and, most importantly, did so in *arbitrary* coordinates rather than in the *unimodular* coordinates he had used both in the November 1915 papers (Einstein 1915a; 1915b; 1915c; 1915d, see Ch. 6) and in the review article written earlier in 1916 (Einstein 1916b, see Ch. 9).

"Hamilton's Principle ..." was submitted to the Berlin Academy October 26, 1916 and published in its Proceedings November 2, 1916. The paper, especially the crucial § 3, closely follows the corresponding sections of the review article on the *Entwurf* theory (Einstein 1914c, §§ 13–15, see Ch. 3), submitted to the same academy almost exactly two years earlier, on October 29, 1914.

The new paper can be found in facsimile as Doc. 41 in CPAE6. As the editors of CPAE6 write in the descriptive note for this document, "a three-page manuscript version of all but the last two paragraphs of § 3 of the paper and an additonal page of related calculations [5 034] are preserved" in the Einstein Archive. The paper grew out of a discarded appendix to the 1916 review article (Gutfreund and Renn 2015, pp. 130–138; cf. the introduction of Sec. 9.3). Results that would eventually find their way into this paper can be found in Einstein's correspondence of 1916, especially in a letter from Einstein to Lorentz, January 17, 1916 (CPAE8, Doc. 183; cf. note 5 below) and a letter from Einstein to Théophile de Donder, July 23, 1916 (CPAE8, Doc. 240; cf. note 11 below) (see Janssen and Renn 2007, pp. 900–903, for discussion).

An English translation of the paper can be found in the anthology *The Principle of Relativity* (Einstein et al. 1952, pp. 167–173; cf. Sec. 9.3). A new translation was prepared for the companion volume to CPAE6. In both translations, a number of obvious typos in the German original were corrected. Most of these are noted in the annotation of this document in CPAE6. Notes added to the translation in the companion volume draw attention to a few more. Our translation mostly follows the CPAE6 translation. We silently corrected the typos found by editors and translators.

Our commentary is based on sec. 9 of "Untying the knot" (Janssen and Renn 2007). Just as Einstein in this paper relied heavily on the sections dealing with the field equations and energy-momentum conservation in his review articles of 1914 and 1916 (Einstein 1914c; 1916b, see Chs. 3 and 9), we will in our commentary on this paper rely heavily on our commentary on these sections of his earlier papers in Secs. 3.3 and 9.3. We invite the reader to review these sections and our commentary on them before reading this paper, which can be seen as bringing Einstein's struggle to find the field equations of general relativity to a close.

Introduction

1. The references are to Lorentz (1914–15) with a variational formalism for the *Entwurf* theory; to the first three installments of Lorentz (1916–17), with a geometrical formulation of a variational formalism for its successor of November 1915 (for dis-

cussion, see Pt. I, Ch. 4, notes 5 and 6); and to Hilbert (1915, for discussion, see Pt. I, Secs. 2.3 and 4.3). Einstein followed Lorentz and Hilbert in picking the Riemann curvature scalar as the Lagrangian for the gravitational field but otherwise replicated some of the steps in his own review articles on the *Entwurf* theory and the theory of November 1915 (Einstein 1914c; 1916b, see notes 6 and 12 below).

2. Hilbert (1915) had endorsed the electromagnetic worldview of Gustav Mie (1912; 1913, see Pt. I, Sec. 5.4, especially note 19, for discussion). Einstein (1915b) had flirted briefly with the electromagnetic worldview, using it to set the trace T of the energy-momentum tensor for matter equal to zero in his second paper of November 1915 (see Sec. 6.2). In a footnote to the discarded appendix to his review article (Gutfreund and Renn 2015, p. 130), he characterized Hilbert's approach as "not very promising" (*wenig aussichtsvoll*). He was more dismissive in private correspondence. In a letter to Ehrenfest of May 1924, 1916, he wrote: "I don't care for Hilbert's presentation. It is unnecessarily specialized about 'matter', unnecessarily complicated, dishonest (= Gaussian) in its construction" (CPAE8, Doc. 220). In parentheses he added the unflattering description of Hilbert we already quoted in Pt. I, Ch. 7: "creating the impression of being superhuman by obfuscating one's methods." "Gaussian" probably refers to a comment attributed to Gauss that "a good building should not show its scaffolding when completed" and to his reputation of resembling "a fox erasing its tracks in the sand with its tail" (Janssen 2019, p. 103). Einstein was even more blunt in a letter to Hermann Weyl of November 23, 1916: "Hilbert's assumption about matter strikes me as infantile, in the sense of a child innocent of the tricks of the real world" (CPAE8, Doc. 278).
3. Both in the November 1915 papers (Einstein 1915a; 1915b; 1915c; 1915d, see Ch. 6) and in the review article of early 1916 (Einstein 1916b, see Ch. 9), Einstein had discussed the field equations and energy-momentum conservation in unimodular coordinates. In fact, he had only been able to prove in unimodular coordinates that the field equations entail energy-momentum conservation (see Sec. 9.3, note 26). In two letters to Lorentz on January 17 and 19, 1916 (CPAE8, Docs. 183 and 184), Einstein already acknowledged the desirability of deriving the field equations from a variational principle in arbitrary coordinates along the lines of Lorentz (1914–15) and made a modest start with this endeavor. He pursued this further in the discarded appendix to the review article (see the introduction to Sec. 9.3). On May 24, 1916, two weeks after this review article was published, in the same letter to Ehrenfest from which we quoted in note 2, Einstein wrote somewhat defensively: "My specialization of the coordinate system is not *just* out of laziness [*beruht nicht* nur *auf Faulheit*]. Perhaps I will at some point present the matter without such specialization, the way Lorentz does in his paper" (CPAE8, Doc. 220).
4. The Einstein summation convention was first introduced in print in the 1916 review article (Einstein 1916b, p. 781; cf. Sec. 9.3, note 1).

§ 3

5. A first attempt by Einstein to find the effective Lagrangian depending only on $g_{\mu\nu}$ and its first-order derivatives can be found in a letter from Einstein to Lorentz of January 17, 1916 (CPAE8, Doc. 183). In the letter to Weyl quoted in note 2, Einstein explicitly wrote that this is the Lagrangian in arbitrary coordinates which reduces to the one he had used earlier in expositions of the theory in unimodular coordinates (CPAE8, Doc. 278).
6. The calculations that follow—from Eq. (10) through Eq. (21)—closely follow those in §§ 13–15 of the review article on the *Entwurf* theory (Einstein 1914c, pp. 1067–1077). See Sec. 3.3 for detailed commentary on these earlier calculations (see also Janssen and Renn 2007, pp. 904–907, on their reprise in 1916).
7. Eq. (11) and (12) are the same as Eqs. (63) and (63a) in Einstein (1914c, § 13, p. 1069; cf. Sec. 3.3, note 7).

8. Eq. (13) for $\sqrt{-g}\Delta\left(\frac{\mathfrak{G}^*}{\sqrt{-g}}\right)$ is equivalent to Eq. (64) for ΔH in Einstein (1914c, § 13, p. 1069). $\frac{\mathfrak{G}^*}{\sqrt{-g}}$ in 1916 and H in 1914 could both assumed to be scalars under *linear* transformations. As he did for $\frac{\mathfrak{G}^*}{\sqrt{-g}}$ in 1916—see the paragraph below Eq. (14) arguing for Eq. (15)—Einstein used this property of H in 1914 to argue that the coefficients of the first-order derivatives of Δx_σ in ΔH vanish and did not bother to find an explicit expression for them, as he did in 1916. In 1914, he only evaluated the coefficients of the second-order derivatives of Δx_σ in ΔH. In our commentary on § 13 of 1914 review article, we did derive an expression for the coefficients of the first-order derivatives, which is exactly the same as the one Einstein found in 1916 (see Sec. 3.3, note 7). The expression for S_σ^ν in Eq. (14) in this 1916 paper is twice the expression we derived because Einstein considered $\frac{1}{2}\sqrt{-g}\Delta H$ in 1914 and $\sqrt{-g}\Delta\left(\frac{\mathfrak{G}^*}{\sqrt{-g}}\right)$ in 1916.

   In the 1914 review article, Einstein had already introduced the quantity S_σ^ν in Eq. (14) in this 1916 paper (modulo a factor 2) and set it equal to zero but *only in the context of his discussion of energy-momentum conservation* (in which it also plays a role in this 1916 paper: see Eq. (20) and note 11 below), *not* in the context of his examination of the covariance properties of the Lagrangian (§ 15, p. 1075, Eqs. (76a) and (77); cf. Sec. 3.3, note 15). Eq. (77) in the 1914 review article formed the basis for Einstein's ill-fated uniqueness argument for the *Entwurf* field equations (see Pt. I, Ch. 4.2 and the introduction of Ch. 5, for discussion). He mistakenly thought that the Lagrangian for the *Entwurf* field equations was the only one for which $S_\sigma^\nu = 0$. He only discovered a year later that $S_\sigma^\nu = 0$ for *any* Lagrangian covariant under linear transformations and explained his error in a letter to Lorentz of October 12, 1915 (see Ch. 5).

9. Substituting $\sqrt{-g}H$ for $\mathfrak{G}^*$ in Eq. (17), we recover the condition $B_\mu = 0$ in the review article on the *Entwurf* theory (Einstein 1914c, p. 1070, Eqs. (65a) and (67)). This condition did double duty in the *Entwurf* theory: it served as the condition for "adapted coordinates" and it ensured that the field equations implied energy-momentum conservation (ibid., p. 1077). Energy-momentum conservation thus *restricted* the covariance of field equations. In the new theory, the roles are reversed: the covariance of the field equations *guarantees* energy-momentum conservation (see Pt. I, Sec. 4.2 and 5.2, for discussion). With $\mathfrak{G}^*$ for $\sqrt{-g}H$, the condition $B_\mu = 0$ turns into the identity in Eq. (17), which, as Einstein shows below (see Eqs. (18) and (21)), ensures that the gravitational field equations entail energy-momentum conservation.

   In his review article, Einstein (1916b, § 17, p. 809, Eqs. (55)–(56)) had already shown that an identity ensured that the gravitational field equations entail energy-momentum conservation but both the identity and the conclusion drawn from it only hold in unimodular coordinates and the identity was not derived from the invariance of the action under unimodular transformations (see Sec. 9.3, note 25). As Einstein explained to Ehrenfest in a letter of November 7, 1916: "In my earlier presentation [in Einstein (1916b)] with $\sqrt{-g} = 1$, direct calculation establishes the identity that is here [in Einstein (1916d)] presented as a consequence of the invariance [of the action]" (CPAE8, Doc. 275; see Janssen and Renn 2007, p. 909, for discussion).

   As its role in guaranteeing energy-momentum conservation makes clear, Eq. (17) is a version of the contracted Bianchi identities (see the preface). Einstein seems to have been unaware of these identities at this point and when several correspondents (Tullio Levi-Civita, Rudolf Förster and Friedrich Kottler) alerted him to the (contracted) Bianchi identities in 1917-18 (CPAE8, Docs. 375, 463 and 495), he showed no interest in them: Einstein was interested in how the invariance of the action gave rise to identities guaranteeing energy-momentum conservation (see note 13 below and CPAE8, li).

10. If the first expression for $\mathfrak{t}_\sigma^\nu$ in Eq. (20) is substituted into Eq. (18), the latter turns into Eq. (80a) of the review article on the *Entwurf* theory (Einstein 1914c, p. 1076; see Ch. 3). Eq. (18) in the 1916 paper is derived the same way as Eq. (80a) in the 1914 one. Contracting Eq. (7) with $g^{\mu\sigma}$ and using Eq. (19) for the right-hand side, we find

$$g^{\mu\sigma}\left(\frac{\partial}{\partial x^\alpha}\left(\frac{\partial\mathfrak{G}^*}{\partial g_\alpha^{\mu\nu}}\right)-\frac{\partial\mathfrak{G}^*}{\partial g^{\mu\nu}}\right)=-\mathfrak{T}_\sigma^\nu,$$

which can be rewritten as

$$\frac{\partial}{\partial x^\alpha}\left(g^{\mu\sigma}\frac{\partial\mathfrak{G}^*}{\partial g_\alpha^{\mu\nu}}\right)-g_\alpha^{\mu\sigma}\frac{\partial\mathfrak{G}^*}{\partial g_\alpha^{\mu\nu}}-\frac{\partial\mathfrak{G}^*}{\partial g^{\mu\nu}}=-\mathfrak{T}_\sigma^\nu.$$

Setting the last two terms on the left-hand side equal to $\mathfrak{t}_\sigma^\nu$ as is done in Eq. (20) and moving $\mathfrak{t}_\sigma^\nu$ to the right-hand side, we arrive at Eq. (18). Although Einstein does not comment on this, Eq. (18) suggests that the quantity in parentheses on the left-hand side represents the gravitational field in the new theory (he also does not insert the expression for $\mathfrak{G}^*$ in note 2 on p. 1113 into this expression in Eq. (18) to see how this new definition is related to the one used in November 1915, i.e., minus the Christoffel symbols [cf. Sec. 6.1.3, note 11]). Finally, Eq. (18) shows that the theory in its generally covariant form meets the requirement that all energy-momentum enter the field equations in the same way, which Einstein had insisted on since 1912 (cf., e.g., Sec. 9.3, note 19).

11. The second expression in Eq. (20) for $\mathfrak{t}_\sigma^\nu$ can already be found in a letter from Einstein to Théophile de Donder of July 23, 1916 (CPAE8, Doc. 240).

12. The derivation of the equation above Eq. (22) is the analogue in arbitrary coordinates of a derivation in unimodular coordinates in the first paper of November 1915 (Einstein 1915a, p. 784; see Sec. 6.1.3, note 15). Contracting Eq. (7) with $g_\sigma^{\mu\nu}$, we find:

$$g_\sigma^{\mu\nu}\left(\frac{\partial}{\partial x^\alpha}\left(\frac{\partial\mathfrak{G}^*}{\partial g_\alpha^{\mu\nu}}\right)-\frac{\partial\mathfrak{G}^*}{\partial g^{\mu\nu}}\right)=g_\sigma^{\mu\nu}\frac{\partial\mathfrak{M}}{\partial g^{\mu\nu}}.$$

The left-hand side can be rewritten as

$$\frac{\partial}{\partial x^\alpha}\left(g_\sigma^{\mu\nu}\frac{\partial\mathfrak{G}^*}{\partial g_\alpha^{\mu\nu}}\right)-g_{\sigma\alpha}^{\mu\nu}\frac{\partial\mathfrak{G}^*}{\partial g_\alpha^{\mu\nu}}-g_\sigma^{\mu\nu}\frac{\partial\mathfrak{G}^*}{\partial g^{\mu\nu}}.$$

The last two terms combine to $-\dfrac{\partial\mathfrak{G}^*}{\partial x^\sigma}$. In the resulting expression

$$\frac{\partial}{\partial x^\alpha}\left(g_\sigma^{\mu\nu}\frac{\partial\mathfrak{G}^*}{\partial g_\alpha^{\mu\nu}}\right)-\frac{\partial\mathfrak{G}^*}{\partial x^\sigma}=\frac{\partial}{\partial x^\alpha}\left(g_\sigma^{\mu\nu}\frac{\partial\mathfrak{G}^*}{\partial g_\alpha^{\mu\nu}}-\delta_\sigma^\alpha\,\mathfrak{G}^*\right)$$

we recognize minus the divergence of twice the second expression for $\mathfrak{t}_\sigma^\nu$ in Eq. (20). The equation we started from thus turns into

$$-2\mathfrak{t}_{\sigma,\alpha}^\alpha=g_\sigma^{\mu\nu}\frac{\partial\mathfrak{M}}{\partial g^{\mu\nu}},$$

from which the equation above Eq. (22) directly follows. Eq. (22) itself is equivalent to $T^{\mu\nu}{}_{;\nu}=0$, the vanishing of the covariant divergence of the energy-momentum tensor (see Pt. I, Sec. 3.2, note 15).

13. It is clear from what Einstein wrote to various correspondents around this time that this was the central point of the paper for him (Janssen and Renn 2007, p. 908). In a letter to Ehrenfest of October 29, 1916, three days after he had submitted the paper,

he summarized its contents as follows: "I have now given a Hamiltonian treatment of the essential points of general relativity *to bring out the connection between relativity and the energy principle*" (CPAE8, Doc. 269, our emphasis). Two days later, on October 31, 1916, he wrote to Michele Besso: "You will soon receive a short paper of mine about the foundations of general relativity, in which it is shown how the requirement of relativity is connected with the energy principle. It is very amusing [*Es ist sehr amusant*]" (CPAE8, Doc. 270). Similarly, in a letter to Willem de Sitter of November 4, 1916, he wrote: "Take a look at the page proofs I sent to Ehrenfest. There the connection between relativity postulate and energy law is brought out very clearly" (CPAE8, Doc. 273). A little over a week later, on November 13, 1916, he sent Lorentz an offprint, describing it in the accompanying letter as "a short paper, in which I explained how in my opinion the relation of the conservation laws to the relativity postulate is to be understood" (CPAE8, Doc. 276). He emphasized that the conservation laws are satisfied regardless of how the Lagrangian for matter is chosen, adding: "So the choice [of $\mathfrak{M}$] made by Hilbert appears to have no justification" (cf. note 2 above). On November 23, 1916, in the letter to Weyl from which we already quoted in note 2, he made the same point and once more reiterated that the key point of the paper was that "[t]he connection between the requirement of general covariance and the conservation laws is made clearer" (CPAE8, Doc. 278).

References

Barbour, Julian B., and Herbert Pfister. 1995. *Mach's Principle. From Newton's Bucket to Quantum Gravity.* Boston: Birkhäuser. *Einstein Studies,* Vol. 6.

Bergmann, Peter, and Arthur Komar. 1972. "The Coordinate Group Symmetries of General Relativity." *International Journal of Theoretical Physics* 5:15–28.

Bianchi, Luigi. 1910. *Vorlesungen über Differentialgeometrie.* Leipzig: Teubner. 2nd expanded and rev. ed. Max Lukat (transl.)

Birkhoff, Garrett. 1923. *Relativity and Modern Physics.* Cambridge, MA: Harvard University Press.

Blum, Alexander S., Roberto Lalli, and Jürgen Renn. 2015. "The Reinvention of General Relativity: A Historiographical Framework for Assessing One Hundred Years of Curved Space-Time." *Isis* 106:598–620.

———, eds. 2020. *The Renaissance of General Relativity in Context.* Boston: Birkhäuser. *Einstein Studies,* Vol. 16.

Bohr, Niels, Hendrik A. Kramers, and John C Slater. 1924. "The Quantum Theory of Radiation." *Philosophical Magazine* 47:785–822.

Born, Max. 1914. "Der Impuls-Energie-Satz in der Elektrodynamik von Gustav Mie." *Königliche Gesellschaft der Wissenschaften zu Göttingen. Mathematisch-physikalische Klasse. Nachrichten* 1:23–36. English translation in Renn (2007a, Vol. 4, pp. 745–756).

Brading, Katherine. 2002. "Which Symmetry? Noether, Weyl, and the Conservation of Electric Charge." *Studies in History and Philosophy of Modern Physics* 33:3–22.

Bradonjić, Kaća. 2014. "Unimodular Conformal and Projective Relativity: An Illustrated Introduction." In *Frontiers of Fundamental Physics and Physics Education Research,* edited by Burra G. Sidharth, Marisa Michelini, and Lorenzo Santi, 197–203. Springer.

Brown, Harvey R., and Katherine Brading. 2002. "General Covariance from the Perspective of Noether's Theorems." *Diálogos* 79:59–86.

Bueno, Otávio, and Mark Colyvan. 2011. "An Inferential Conception of the Application of Mathematics." *Noûs* 45:345–374.

Carroll, Sean. 2004. *Spacetime and Geometry. An Introduction to General Relativity.* San Francisco: Addison Wesley.

M. Janssen, J. Renn, *How Einstein Found His Field Equations*, Classic Texts in the Sciences, https://doi.org/10.1007/978-3-030-97955-3

Castagnetti, Giuseppe, Peter Damerow, Werner Heinrich, Jürgen Renn, and Tilman Sauer. 1993. *Wissenschaft zwischen Grundlagenkrise und Politik. Einstein in Berlin.* Berlin: Max-Planck-Institut für Bildungsforschung.

Cattani, Carlo, and Michelangelo De Maria. 1989. "The 1915 Epistolary Controversy between Einstein and Tullio Levi-Civita." In Howard and Stachel (1989, pp. 175–200).

Clark, Ronald W. 1971. *Einstein: The Life and Time.* New York: Knopf.

Coover, Christopher. 1996. *The Einstein-Besso Working Manuscript.* New York: Catalog for Auction at Christie's, November 25, 1996.

———. 2002. *The History of Quantum Mechanics and the Theory of Relativity: The Harvey Plotnick Library.* New York: Catalog for Auction at Christie's, October 4, 2002.

Corry, Leo, Jürgen Renn, and John Stachel. 1997. "Belated Decision in the Hilbert-Einstein Priority Dispute." *Science* 278:1270–1273.

CPAE. 1987–. *The Collected Papers of Albert Einstein.* 15 Vols. Edited by John Stachel, Martin J. Klein, Robert Schulmann, Diana Kormos Buchwald, et al. Princeton: Princeton University Press. English translation companion volumes: Anna Beck, Alfred Engel, Ann Hentschel (translators); Don Howard, Engelbert Schücking, Klaus Hentschel (consultants).

Crelinsten, Jeffrey. 2006. *Einstein's Jury: The Race to Test Relativity.* Princeton: Princeton University Press.

De Sitter, Willem. 1916a. "De planetenbeweging en de beweging van de maan volgens de theorie van Einstein." *Koninklijke Akademie van Wetenschappen te Amsterdam. Wis- en Natuurkundige Afdeeling. Verslagen van de Gewone Vergaderingen* 25:232–245. Reprinted in translation as "Planetary Motion and the Motion of the Moon According to Einstein's Theory." *Koninklijke Akademie van Wetenschappen te Amsterdam. Section of Sciences. Proceedings* 19 (1916–17): 367–381.

———. 1916b. "De relativiteit der rotatie in de theorie van Einstein." *Koninklijke Akademie van Wetenschappen te Amsterdam. Wis- en Natuurkundige Afdeeling. Verslagen van de Gewone Vergaderingen* 25:499–504. Page reference to English translation, "On the Relativity of Rotation in Einstein's Theory" *Koninklijke Akademie van Wetenschappen te Amsterdam. Section of Sciences. Proceedings* 19 (1916–17): 527-532.

———. 1916c. "On Einstein's Theory of Gravitation, and its Astronomical Consequences. Second Paper." *Royal Astronomical Society. Monthly Notices* 77:155–184.

De Sitter, Willem. 1917a. "Over de relativiteit der traagheid: Beschouwingen naar aanleiding van Einstein's laatste hypothese." *Koninklijke Akademie van Wetenschappen te Amsterdam. Wis- en Natuurkundige Afdeeling. Verslagen van de Gewone Vergaderingen* 25:1268–1276. Reprinted in translation as "On the Relativity of Inertia: Remarks Concerning Einstein's Latest Hypothesis." *Koninklijke Akademie van Wetenschappen te Amsterdam. Section of Sciences. Proceedings* 19 (1917): 1217–1225.

———. 1917b. "Over de kromming der ruimte." *Koninklijke Akademie van Wetenschappen te Amsterdam. Wis- en Natuurkundige Afdeeling. Verslagen van de Gewone Vergaderingen* 26:222–236. Reprinted in translation as "On the Curvature of Space." *Koninklijke Akademie van Wetenschappen te Amsterdam. Section of Sciences. Proceedings* 20 (1917): 229–243.

———. 1917c. "On Einstein's Theory of Gravitation, and its Astronomical Consequences. Third Paper." *Royal Astronomical Society. Monthly Notices* 78:3–28.

Droste, Johannes. 1915. "Over het veld van een enkel centrum in Einstein's theorie der zwaartekracht." *Koninklijke Akademie van Wetenschappen te Amsterdam. Wis- en Natuurkundige Afdeeling. Verslagen van de Gewone Vergaderingen* 23:968–981. Page references to English translation: "On the Field of a Single Centre in Einstein's Theory of Gravitation." *Koninklijke Akademie van Wetenschappen te Amsterdam. Section of Sciences. Proceedings* 17 (1915): 998–1011.

Earman, John. 2003. "The Cosmological Constant, the Fate of the Universe, Unimodular Gravity, and all that." *Studies in History and Philosophy of Modern Physics* 34:559–577.

Earman, John, and Clark Glymour. 1978. "Lost in the Tensors: Einstein's Struggle with Covariance Principles, 1912–1916." *Studies in History and Philosophy of Science* 9:251–278.

Earman, John, and Michel Janssen. 1993. "Einstein's Explanation of the Motion of Mercury's Perihelion." In Earman *et al.* (1993, pp. 129–172).

Earman, John, Michel Janssen, and John D. Norton, eds. 1993. *The Attraction of Gravitation. New Studies in the History of General Relativity.* Boston: Birkhäuser. *Einstein Studies*, Vol. 5.

Eckert, Michael. 2013. *Arnold Sommerfeld. Atomphysiker und Kulturbote 1868–1951. Eine Biographie.* Göttingen: Wallstein. English translation: *Arnold Sommerfeld. Science, Life and Turbulent Times.* New York: Springer, 2013.

Eddington, Arthur Stanley. 1930. "On the Instability of Einstein's Spherical World." *Royal Astronomical Society. Monthly Notices* 90:668–678.

Einstein, Albert. 1911. "Über den Einfluß der Schwerkraft auf die Ausbreitung des Lichtes." *Annalen der Physik* 35:898–908. (CPAE3, Doc. 23).

———. 1912a. "Lichtgeschwindigkeit und Statik des Gravitationsfeldes." *Annalen der Physik* 38:355–369. (CPAE4, Doc. 3).

———. 1912b. "Zur Theorie des statischen Gravitationsfeldes." *Annalen der Physik* 38:443–458. (CPAE4, Doc. 4).

———. 1913. "Zum gegenwärtigen Stande des Gravitationsproblems." *Physikalische Zeitschrift* 14:1249–1262. (CPAE4, Doc. 17).

———. 1914a. "Bemerkungen." *Zeitschrift für Mathematik und Physik* 62:260–261. (CPAE4, Doc. 26).

———. 1914b. "Prinzipielles zur verallgemeinerten Relativitätstheorie." *Physikalische Zeitschrift* 15:176–180. (CPAE4, Doc. 25).

———. 1914c. "Die formale Grundlage der allgemeinen Relativitätstheorie." *Königlich Preußische Akademie der Wissenschaften* (Berlin). *Sitzungsberichte:* 1030–1085. (CPAE6, Doc. 9).

———. 1915a. "Zur allgemeinen Relativitätstheorie." *Königlich Preußische Akademie der Wissenschaften* (Berlin). *Sitzungsberichte:* 778–786. (CPAE6, Doc. 21).

———. 1915b. "Zur allgemeinen Relativitätstheorie. (Nachtrag)." *Königlich Preußische Akademie der Wissenschaften* (Berlin). *Sitzungsberichte:* 799–801. (CPAE6, Doc. 22).

———. 1915c. "Erklärung der Perihelbewegung des Merkur aus der allgemeinen Relativitätstheorie." *Königlich Preußische Akademie der Wissenschaften* (Berlin). *Sitzungsberichte:* 831–839. (CPAE6, Doc. 23).

———. 1915d. "Die Feldgleichungen der Gravitation." *Königlich Preußische Akademie der Wissenschaften* (Berlin). *Sitzungsberichte:* 844–847. (CPAE6, Doc. 25).

———. 1916a. "Eine neue formale Deutung der Maxwellschen Feldgleichungen der Elektrodynamik." *Königlich Preußische Akademie der Wissenschaften* (Berlin). *Sitzungsberichte:* 184–188.

———. 1916b. "Die Grundlage der allgemeinen Relativitätstheorie." *Annalen der Physik* 49:769–822. (CPAE6, Doc. 30). English translation in Einstein *et al.* (1952, 111–164).

———. 1916c. "Näherungsweise Integration der Feldgleichungen der Gravitation." *Königlich Preußische Akademie der Wissenschaften* (Berlin). *Sitzungsberichte:* 688–696. (CPAE6, Doc. 32).

Einstein, Albert. 1916d. "Hamiltonsches Prinzip und allgemeine Relativitätstheorie." *Königlich Preußische Akademie der Wissenschaften* (Berlin). *Sitzungsberichte:* 1111–1116. (CPAE6, Doc. 41). English translation in Einstein *et al.* (1952, 167–173).

———. 1917. "Kosmologische Betrachtungen zur allgemeinen Relativitätstheorie." *Königlich Preußische Akademie der Wissenschaften* (Berlin). *Sitzungsberichte:* 142–152. (CPAE 6, Doc. 43). English translation in Einstein *et al.* (1952, pp. 177–188).

———. 1918a. "Über Gravitationswellen." *Königlich Preußische Akademie der Wissenschaften* (Berlin). *Sitzungsberichte:* 154–167. (CPAE7, Doc. 1).

———. 1918b. "Prinzipielles zur allgemeinen Relativitätstheorie." *Annalen der Physik* 55:241–244. (CPAE7, Doc. 4).

———. 1918c. "Kritisches zu einer von Hrn. De Sitter gegebenen Lösung der Gravitationsgleichungen." *Königlich Preußische Akademie der Wissenschaften* (Berlin). *Sitzungsberichte:* 270–272. (CPAE7, Doc. 5).

———. 1918d. "Der Energiesatz in der allgemeinen Relativitätstheorie." *Königlich Preußische Akademie der Wissenschaften* (Berlin). *Sitzungsberichte:* 448–459. (CPAE7, Doc. 9).

———. 1919. "Spielen Gravitationsfelder im Aufbau der materiellen Elementarteilchen eine wesentliche Rolle?" *Königlich Preußische Akademie der Wissenschaften* (Berlin). *Sitzungsberichte:* 349–356. (CPAE7, Doc. 17). English translation in Einstein *et al.* (1952, 191–198).

Einstein, Albert, and Michele Besso. 1972. *Correspondance 1903–1955.* Edited by Pierre Speziali. Paris: Hermann. Translated by Speziali.

———. 2003. *The Einstein-Besso Manuscript: From Special Relativity to General Relativity/Le manuscrit Einstein-Besso: De la Relativité Restreinte é la Relativité Générale.* Paris: Scriptura / Aristophile.

Einstein, Albert, and Max Born. 1969. *Briefwechsel 1916–1955.* München: Nymphenburger Verlagshandlung. Page reference to English translation: Max Born, ed., *The Born-Einstein Letters. Correspondence Between Albert Einstein and Max and Hedwig Born from 1916 to 1955.* London: MacMillan, 1971.

Einstein, Albert, and Adriaan D Fokker. 1914. "Die Nordströmsche Gravitationstheorie vom Standpunkt des absoluten Differentialkalküls." *Annalen der Physik* 44:321–328. (CPAE4, Doc. 28).

Einstein, Albert, and Marcel Grossmann. 1913. *Entwurf einer verallgemeinerten Relativitätstheorie und einer Theorie der Gravitation.* Leipzig: Teubner. (CPAE4, Doc. 13).

———. 1914. "Kovarianzeigenschaften der Feldgleichungen der auf die verallgemeinerte Relativitätstheorie gegründeten Gravitationstheorie." *Zeitschrift für Mathematik und Physik* 63:215–225. (CPAE6, Doc. 2).

Einstein, Albert, Hendrik Antoon Lorentz, Hermann Minkowski, and Hermann Weyl. 1952. *The Principle of Relativity.* New York: Dover.

Einstein, Albert, Gustav Mie, Eduard Riecke, Hans Reißner, et al. 1913. "Discussion following lecture 'Zum gegenwärtigen Stande des Gravitationsproblems'." *Physikalische Zeitschrift* 14:1262–1266.(CPAE4, Doc. 18)

Einstein, Albert, and Nathan Rosen. 1937. "On Gravitational Waves." *Journal of the Franklin Institute* 223:43–54.

Eisenstaedt, Jean, and A. J. Kox. 1992. *Studies in the History of General Relativity.* Boston: Birkhäuser. *Einstein Studies*, Vol. 3.

Engler, Fynn Ole, and Jürgen Renn. 2013. "Hume, Einstein und Schlick über die Objektivität der Wissenschaft." In *Moritz Schlick: Die Rostocker Jahre und ihr Einfluss auf die Wiener Zeit,* edited by Fynn Ole Engler and Mathias Iven, 123–156. Leipzig: Universitätsverlag.

Freundlich, Erwin. 1915a. "Über die Erklärung der Anomalien im Planeten-System durch die Gravitationswirkung interplanetaren Massen." *Astronomische Nachrichten* 201 (4803): cols. 49–56.

———. 1915b. "Über die Gravitationsverschiebung der Spektrallinien bei Fixsternen." *Physikalische Zeitschrift* 16:115–117.

Gerber, Paul. 1917. "Die Fortpflanzungsgeschwindigheit der Gravitation." *Annalen der Physik* 52:415–441. Originally published in 1902 in *Programmabhanlung des städtischen Realgymnasiums zu Stargard in Pomerania.*

Goenner, Hubert, Jürgen Renn, Jim Ritter, and Tilman Sauer. 1999. *The Expanding Worlds of General Relativity.* Boston: Birkhäuser. *Einstein Studies,* Vol. 7.

Gordin, Michael D. 2020. *Einstein in Bohemia.* Princeton, Oxford: Princeton University Press.

Gutfreund, Hanoch, and Jürgen Renn. 2015. *The Road to Relativity: The History and Meaning of Einstein's "The Foundation of General Relativity".* Princeton: Princeton University Press.

———. 2017. *The Formative Years of General Relativity.* Princeton: Princeton University Press.

Harper, William L. 2011. *Isaac Newton's Scientific Method: Turning Data into Evidence about Gravity and Cosmology.* Oxford: Oxford University Press.

Hentschel, Klaus. 1992. *Der Einstein Turm: Erwin F. Freundlich und die Relativitätstheorie. Ansätze zu einer "dichten Beschreibung" von institutionellen, biographischen und theoriengeschichtlichen Aspekten.* Heidelberg, Berlin, and New York: Spektrum Akademie Verlag. English translation: *The Einstein Tower. An Intertexture of Dynamic Construction, Relativity Theory, and Astronomy.* Stanford: Stanford University Press, 1997.

———. 1993. "The Conversion of St. John. A Case Study on the Interplay of Theory and Experiment." *Science in Context* 6:137–194.

———. 1994. "Erwin Finlay Freundlich and Testing Einstein's Theory of Relativity." *Archive for History of Exact Sciences* 47:143–201.

Herglotz, Gustav. 1911. "Über die Mechanik des deformierbaren Körpers vom Standpunkte der Relativitätstheorie." *Annalen der Physik* 36:493–533.

Hilbert, David. 1915. "Die Grundlagen der Physik. (Erste Mitteilung)." *Königliche Gesellschaft der Wissenschaften zu Göttingen. Mathematisch-physikalische Klasse. Nachrichten:* 395–407. English translation in Renn (2007a, Vol. 4, pp. 1003–1015).

Hoefer, Carl. 1994. "Einstein's Struggle for a Machian Gravitation Theory." *Studies in History and Philosophy of Science* 25:287–335.

Howard, Don. 1999. "Point Coincidences and Pointer Coincidences: Einstein on the Invariant Content of Space-Time Theories." In Goenner et al. (1999, pp. 463–500).

Howard, Don, and John D Norton. 1993. "Out of the Labyrinth? Einstein, Hertz, and the Göttingen Answer to the Hole Argument." In Earman et al. (1993, pp. 30–62).

Howard, Don, and John Stachel, eds. 1989. *Einstein and the History of General Relativity.* Boston: Birkhäuser. *Einstein Studies*, Vol. 1.

Janssen, Michel. 1991. *Gravitational theory in late 1913: Einstein's Vienna lecture.* Unpublished manuscript.

———. 1992. "H. A. Lorentz's Attempt to Give a Coordinate-Free Formulation of the General Theory of Relativity." In Eisenstaedt and Kox (1992, pp. 344–363.

———. 1996. "The Einstein-Besso Manuscript, the Perihelion Motion of Mercury, and the Genesis of the General Theory of Relativity." In Coover (1996, pp. 15–21). Reprinted in Coover (2002, pp. 100–113) and in Einstein and Besso (2003).

———. 1999. "Rotation as the Nemesis of Einstein's *Entwurf* Theory." In Goenner *et al.* (1999, pp. 127–157).

———. 2004. "Relativity." In *New Dictionary of the History of Ideas.* Edited by Maryanne Cline Horowitz, 2039–2047. New York: Charles Scribner's Sons.

———. 2005. "Of Pots and Holes: Einstein's Bumpy Road to General Relativity." *Annalen der Physik* 14 (Supplement):58–85.

———. 2007. "What Did Einstein Know and When Did He Know It? A Besso Memo Dated August 1913." In Renn (2007a, Vol. 2, pp. 785–837).

———. 2012. "The Twins and the Bucket: How Einstein Made Gravity Rather Than Motion Relative in General Relativity." *Studies in History and Philosophy of Modern Physics* 43:159–175.

———. 2014. "'No Success Like Failure ...': Einstein's Quest for General Relativity, 1907–1920." In Janssen and Lehner (2014, pp. 167–227).

———. 2019. "Arches and Scaffolds: Bridging Continuity and Discontinuity in Theory Change." In *Beyond the Meme: Development and Structure in Cultural Evolution,* edited by Alan C. Love and William C. Wimsatt, 95–199. Minneapolis: University of Minnesota Press.

Janssen, Michel, and Christoph Lehner, eds. 2014. *The Cambridge Companion to Einstein.* Cambridge: Cambridge University Press.

Janssen, Michel, and Matthew Mecklenburg. 2007. "From Classical to Relativistic Mechanics: Electromagnetic Models of the Electron." In *Interactions: Mathematics, Physics and Philosophy, 1860–1930,* edited by Vincent F. Hendricks, Klaus Frovin Jørgensen, Jesper Lützen, and Stig Andur Pedersen, 65–134. Berlin: Springer.

Janssen, Michel, and Jürgen Renn. 2007. "Untying the Knot: How Einstein Found His Way Back to Field Equations Discarded in the Zurich Notebook." In Renn (2007a, Vol. 2, pp. 839–925).

———. 2015a. "Arch and Scaffold: How Einstein Found His Field Equations." *Physics Today* (November): 30–36.

———. 2015b. "Einstein was no lone genius." *Nature* 527 (November): 298–300.

———. 2015c. "Einsteins Weg zur allgemeinen Relativitätstheorie." *Spektrum der Wissenschaften* (October): 48–55.

———. 2015d. "Von verbogenen Räumen und krummen Zeiten." *Kultur und Technik,* no. 4: 10–15.

———. 2015e. "Ce que l'on doit aux mathématiques." *La Recherche,* no. 16: 18–21.

Janssen, Michel, and Robert Schulmann. 1998. "On the Dating of a Recently Published Einstein Manuscript: Could These Be the Calculations that Gave Einstein Heart Palpitations?" *Foundations of Physics Letters* 11:379–389.

Kennefick, Daniel. 2005. "Einstein versus the Physical Review." *Physics Today* 58, no. 9 (September): 43–48.

———. 2007. *Traveling at the Speed of Thought: Einstein and the Quest for Gravitational Waves.* Princeton: Princeton University Press.

———. 2014. "Einstein, Gravitational Waves, and the Theoretician's Regress." In Janssen and Lehner (2014, pp. 270–280).

———. 2019. *No Shadow of a Doubt. The 1919 Eclipse That Confirmed Einstein's Theory of Relativity.* Princeton: Princeton University Press.

Klein, Felix. 1917. "Zur Hilberts erster Note über die Grundlagen der Physik." *Königliche Gesellschaft der Wissenschaften zu Göttingen. Mathematisch-physikalische Klasse. Nachrichten:* 469–482.

———. 1918a. "Über die Differentialgesetze für die Erhaltung von Impuls und Energie in der Einsteinschen Gravitationstheorie." *Königliche Gesellschaft der Wissenschaften zu Göttingen. Mathematisch-physikalische Klasse. Nachrichten:* 171–189.

———. 1918b. "Über die Integralform der Erhaltungssätze und die Theorie der räumlichgeschlossenen Welt." *Königliche Gesellschaft der Wissenschaften zu Göttingen. Mathematisch-physikalische Klasse. Nachrichten:* 393–423.

Kossmann-Schwarzbach, Yvette. 2011. *The Noether Theorems: Invariance and Conservation Laws in the Twentieth Century.* New York: Springer. Translation (by Bertram E. Schwarzbach) of *Les Théorèmes de Noether: Invariance et lois de conservation au XXe siècle* (2nd. ed. Palaiseau: Éditions de l'École Polytechnique, 2006).

Kottler, Friedrich. 1912. "Über die Raumzeitlinien der Minkowski'schen Welt." *Kaiserliche Akademie der Wissenschaften* (Vienna). *Mathematisch-naturwissenschaftliche Klasse. Abteilung IIa. Sitzungsberichte* 121:1659–1759.

Kox, A. J. 1988. "Hendrik Antoon Lorentz, the Ether, and the General Theory of Relativity." *Archive for History of Exact Sciences* 38:67–78. Reprinted in Howard and Stachel (1989, pp. 201–212).

———, ed. 2008. *The Scientific Correspondence of H. A. Lorentz.* Vol. 1. New York: Springer.

———, ed. 2018. *The Scientific Correspondence of H. A. Lorentz.* Vol. 2. *The Dutch Correspondents.* New York: Springer.

Kox, A. J., and Jean Eisenstaedt, eds. 2005. *The Universe of General Relativity.* Boston: Birkhäuser. *Einstein Studies*, Vol. 11.

Kretschmann, Erich. 1915. "Über die prinzipielle Bestimmbarkeit der berechtigten Bezugssysteme beliebiger Relativitätstheorien (I)." *Annalen der Physik* 48:907–942.

———. 1917. "Über den physikalischen Sinn der Relativitätspostulate: A. Einsteins neue und seine ursprüngliche Relativitätstheorie." *Annalen der Physik* 48:907–942.

Landsman, Klaas. 2021. *Foundations of General Relativity: From Einstein to Black Holes.* Nijmegen: Radboud University Press.

Lang, Kenneth, and Owen Gingerich, eds. 1979. *Source Book in Astronomy and Astrophysics, 1900–1975.* Cambridge, MA: Harvard University Press.

Laue, Max. 1911. *Das Relativitätsprinzip.* Braunschweig: Vieweg.

Lehmkuhl, Dennis. 2014. "Why Einstein did not believe that general relativity geometrizes gravity." *Studies in History and Philosophy of Modern Physics* 46:316–326.

———. 2022. *Einstein's Principles. On the Interpretation of Gravity.* Oxford: Oxford University Press.

Lehner, Christoph, Jürgen Renn, and Matthias Schemmel, eds. 2012. *Einstein and the Changing Worldviews of Physics.* Boston: Birkhäuser. *Einstein Studies*, Vol. 12.

Lorentz, Hendrik Antoon. 1914–15. "Het beginsel van Hamilton in Einstein's theorie der zwaartekracht." *Koninklijke Akademie van Wetenschappen te Amsterdam. Wis- en Natuurkundige Afdeeling. Verslagen van de Gewone Vergaderingen* 23:1073–1089. English translation, "On Hamilton's Principle in Einstein's Theory of Gravitation," in *Koninklijke Akademie van Wetenschappen te Amsterdam, Section of Sciences, Proceedings* 19 (1916–17): 751–765. English translation reprinted in Lorentz (1935–39, Vol. 5, pp. 229–245).

———. 1916–17. "Over Einstein's theorie der zwaartekracht." *Koninklijke Akademie van Wetenschappen te Amsterdam. Wis- en Natuurkundige Afdeeling. Verslagen van de Gewone Vergaderingen.* 4 Pts. 24 (1915–16): 1389–1402 (I); 1759–1774 (II); 25 (1916–17): 468–486 (III), 1380–1396 (IV). English translation, "On Einstein's Theory of Gravitation," in *Konink- lijke Akademie van Wetenschappen te Amsterdam, Section of Sciences, Proceedings* 19 (1916-17): 1341–1354 (I), 1354–1369 (II); 20 (1917–18): 2–19 (III), 20–34 (IV). English translation reprinted in Lorentz (1935–39, Vol. 5, pp. 246–313).

Lorentz, Hendrik Antoon. 1935–39. *Collected Papers. 9 Vols.* Edited by Pieter Zeeman and Adriaan D Fokker. The Hague: Nijhoff.

Maudlin, Tim. 1990. "Substances and Space-Time: What Aristotle Would Have Said to Einstein." *Studies in History and Philosophy of Science* 21:531–561.

Mie, Gustav. 1912. "Grundlagen einer Theorie der Materie. Erste Mitteilung." *Annalen der Physik* 37:511–534. English translation in Renn (2007a, Vol. 4, pp. 633–651).

———. 1913. "Grundlagen einer Theorie der Materie. Dritte Mitteilung." *Annalen der Physik* 40:1–65. English translation of pp. 1–13 and pp. 25–65 in Renn (2007a, Vol. 4, pp. 663–697).

———. 1914. "Bemerkungen zu der Einsteinschen Gravitationstheorie. 2 Parts." *Physikalische Zeitschrift* 14:115–122 (I), 169–176 (II). English translation in Renn (2007a, Vol. 4, pp. 699–728).

———. 1917. "Die Einsteinsche Gravitationstheorie und das Problem der Materie. 3 Parts." *Physikalische Zeitschrift* 18:551–556 (I), 574–580 (II), 596–602 (III).

Misner, Charles W., Kip S. Thorne, and John Archibald Wheeler. 1973. *Gravitation.* New York: Freeman.

Møller, Christian. 1972. *The Theory of Relativity.* Oxford: Clarendon Press. 2nd ed.

Newcomb, Simon. 1895. *The Elements of the Four Inner Planets and the Fundamental Constants of Astronomy.* Washington, D.C.: Government Printing Office.

Noether, Emmy. 1918. "Invariante Variationsprobleme." *Königliche Gesellschaft der Wissenschaften zu Göttingen. Mathematisch-physikalische Klasse. Nachrichten:* 235–257.

Nordström, Gunnar. 1912. "Relativitätsprinzip und Gravitation." *Physikalische Zeitschrift* 13:1126–1129. English translation in Renn (2007a, Vol. 3, pp. 489–497).

———. 1913a. "Träge und schwere Masse in der Relativitätsmechanik." *Annalen der Physik* 40:856–878. English translation in Renn (2007a, Vol. 3, pp. 499–521).

———. 1913b. "Zur Theorie der Gravitation vom Standpunkt des Relativitätsprinzips." *Annalen der Physik* 42:533–554. English translation in Renn (2007a, Vol. 3, pp. 523–542).

Norton, John D. 1984. "How Einstein Found his Field Equations, 1912–1915." *Historical Studies in the Physical Sciences* 14:253–316. Page references to reprint in Howard and Stachel (1989, pp. 101–159).

———. 1992. "Einstein, Nordström and the Early Demise of Scalar Lorentz Covariant Theories of Gravitation." *Archive for the History of Exact Sciences* 45:17–94. Reprinted in Renn (2007a, Vol. 3, pp. 413–487).

———. 1999. "Geometries in Collision: Einstein, Klein, and Riemann." In *The Symbolic Universe: Geometry and Physics, 1890–1930,* edited by Jeremy Gray, 128–144. Oxford: Oxford University Press.

———. 2000. "'Nature is the Realisation of the Simplest Conceivable Mathematical Ideas': Einstein and the Canon of Mathematical Simplicity." *Studies in History and Philosophy of Modern Physics* 31:135–170.

———. 2005. "A Conjecture on Einstein, the Independent Reality of Spacetime Coordinate Systems and the Disaster of 1913." In *The Universe of General Relativity,* edited by A. J. Kox and Jean Eisenstaedt, 67–102. Boston: Birkhäuser.

———. 2007. "What was Einstein's "Fateful Prejudice"?" In Renn (2007a, Vol. 2, pp. 715–783).

———. 2018. "Einstein's Conflicting Heuristics: The Discovery of General Relativity." In *Proceedings of conference Einstein and General Relativity in Bern.*

Pais, Abraham. 1982. *'Subtle is the Lord ...' The Science and Life of Albert Einstein.* Oxford: Oxford University Press.

Pitts, J. Brian. 2016. "Einstein's physical strategy, energy conservation, symmetries, and stability: 'But Grossmann & I believed that the conservation laws were not satisfied'." *Studies in History and Philosophy of Modern Physics* 54:52–72.

Popper, Karl R. 1959. *The Logic of Scientific Discovery.* London: Hutchinson.

Räz, Tim. 2015. "Gone Till November. A Disagreement in Einstein Scholarship." In *The Philosophy of Historical Case Studies,* edited by Tilman Sauer and Raphael Scholl, 181–202. Boston: Springer.

Räz, Tim, and Tilman Sauer. 2015. "Outline of a Dynamical Inferential Conception of the Application of Mathematics." *Studies in History and Philosophy of Modern Physics* 49:57–72.

Renn, Jürgen. 2006. *Auf den Schultern von Riesen und Zwergen: Einsteins unvollendete Revolution.* New York: Wiley.

Renn, Jürgen, ed. 2007a. *The Genesis of General Relativity.* 4 Vols. New York, Berlin: Springer. Vols. 1–2, *Einstein's Zurich Notebook* (Michel Janssen, John D. Norton, Jürgen Renn, Tilman Sauer and John Stachel). Vols. 3–4, *Gravitation in the Twilight of Classical Physics* (Jürgen Renn and Matthias Schemmel, eds.)

———. 2007b. "Classical Physics in Disarray." In Renn (2007a, Vol. 1, pp. 21–80).

———. 2007c. "The Third Way to General Relativity." In Renn (2007a, Vol. 3, pp. 21–75).

———. 2020a. *The Evolution of Knowledge. Rethinking Science for the Anthropocene.* Princeton, Oxford: Princeton University Press.

———. 2020b. "The Genesis and Transformation of General Relativity." In *Einstein Was Right. The Science and History of Gravitational Waves,* edited by Jed Z. Buchwald, 76–110. Princeton: Princeton University Press.

Renn, Jürgen, and Tilman Sauer. 1999. "Heuristics and mathematical representation in Einstein's search for a gravitational field equation." In Goenner *et al.* (1999, pp. 87–125).

———. 2003. "Eclipses of the Stars: Mandl, Einstein, and the Early History of Gravitational Lensing." In *Revisiting the Foundations of Relativistic Physics,* edited by Abhay Ashtekar, Robert S. Cohen, Don Howard, Jürgen Renn, Sahotra Sarkar, and Abner Shimony, 69–92. Dordrecht: Kluwer.

———. 2007. "Pathways out of Classical Physics: Einstein's Double Strategy in Searching for the Gravitational Field Equation." In Renn (2007a, Vol. 1, pp. 113–312).

Renn, Jürgen, Tilman Sauer, and John Stachel. 1997. "The Origin of Gravitational Lensing: A Postscript to Einstein's 1936 *Science* Paper." *Science* 275:184–186. Reprinted in Stachel (2002a, pp. 347–352).

Renn, Jürgen, and John Stachel. 2007. "Hilbert's Foundation of Physics: From a Theory of Everything to a Constituent of General Relativity." In Renn (2007a, Vol. 4, pp. 857–973).

Röhle, Stefan. 2007. "Willem de Sitter in Leiden: ein Kapitel in der Entwicklung der relativistischen Kosmologie." PhD diss., Johannes Gutenberg Universität, Mainz.

Roseveare, N. T. 1982. *Mercury's Perihelion from Le Verrier to Einstein.* Oxford: Clarendon Press.

Rowe, David E. 1999. "The Göttingen Response to General Relativity and Emmy Noether's Theorems." In *The Symbolic Universe: Geometry and Physics, 1890–1930,* edited by Jeremy Gray, 189–234. Oxford: Oxford University Press.

———. 2001. "Einstein Meets Hilbert: At the Crossroads of Physics and Mathematics." *Physics in Perspective* 3:379–424.

———. 2006. "Review of Daniela Wuensch, *"Zwei wirkliche Kerle": Neues zur Entdeckung der Gravitationsgleichungen der Allgemeinen Relativitätstheorie durch Albert Einstein und David Hilbert* (Göttingen: Termessos, 2005)." *Historia Mathematica* 33:491–508.

———. 2021. "Felix Klein and Emmy Noether on Invariant Theory and Variational Principles." In *The Philosophy and Physics of Noether's Theorems,* edited by Nicholas Teh, James Read, and Bryan Roberts. Cambridge University Press.

Rowe, David E., Tilman Sauer, and Scott Walter, eds. 2018. *Beyond Einstein. Perspectives on Geometry, Gravitation, and Cosmology in the Twentieth Century.* Boston: Birkhäuser. *Einstein Studies*, Vol. 14.

Sauer, Tilman. 1999. "The Relativity of Discovery: Hilbert's First Note on the Foundations of Physics." *Archive for History of Exact Sciences* 53:529–575.

———. 2005a. "Einstein's Equations and Hilbert's Action: What Is Missing on Page 8 of the Proofs for Hilbert's First Communication on the Foundations of Physics." *Archive for History of Exact Sciences* 59:577–590. Reprinted in Renn (2007a, Vol. 4, pp. 975–988).

———. 2005b. "Einstein's Review Paper on General Relativity Theory." In *Landmark Writings in Western Mathematics, 1640–1940,* edited by I. Grattan-Guiness, 802–822. Amsterdam: Elsevier.

———. 2015. "Marcel Grossmann and His Contribution to the General Theory of Relativity." In *Proceedings of the 13th Marcel Grossmann Meeting on Recent Developments in Theoretical and Experimental General Relativity, Gravitation, and Relativistic Field Theory,* edited by Robert T. Jantzen, Kjell Rosquist, and Remo Ruffini, 456–503. Singapore: World Scientific.

Schulmann, Robert. 2012. *Seelenverwandte. Der Briefwechsel zwischen Albert Einstein und Heinrich Zangger (1910–1947).* Zurich: Verlag Neue Zürcher Zeitung.

Schwarzschild, Karl. 1916. "Über das Gravitationsfeld eines Massenpunktes nach der Einsteinschen Theorie." *Königlich Preußische Akademie der Wissenschaften* (Berlin). *Sitzungsberichte:* 189–196.

Seeliger, Hugo von. 1915. "Über die Anomalien in der Bewegung der inneren Planeten." *Astronomische Nachrichten* 201:cols. 273–280.

Smeenk, Christopher. 2014. "Einstein's Role in the Creation of Relativistic Cosmology." In Janssen and Lehner (2014, pp. 228–269).

Smeenk, Christopher, and Christopher Martin. 2007. "Mie's Theories of Matter and Gravitation." In Renn (2007a, Vol. 4, pp. 623–632).

Smith, George E. 2014. "Closing the Loop: Testing Newtonian Gravity, Then and Now." In *Newton and Empiricism,* edited by Zvi Biener and Eric Schliesser, 262–351. Oxford: Oxford University Press.

Sommerfeld, Arnold. 1915a. "Zur Theorie der Balmerschen Serie." *Königlich Bayerische Akademie der Wissenschaften zu München. Mathematisch-physikalische Klasse. Sitzungsberichte:* 425–458.

———. 1915b. "Die Feinstruktur der Wasserstoff- und der Wasserstoff-ähnlichen Linien." *Königlich Bayerische Akademie der Wissenschaften zu München. Mathematisch-physikalische Klasse. Sitzungsberichte:* 459–500.

———. 2000. *Wissenschaftlicher Briefwechsel. Band 1: 1892–1918.* Edited by Michael Eckert and Karl Märker. Berlin, Diepholz, München: Deutsches Museum. Verlag für die Geschichte der Naturwissenschaften und der Technik.

Stachel, John. 1989. "Einstein's Search for General Covariance, 1912–1915." In Howard and Stachel (1989, pp. 62–100). Reprinted in Stachel (2002a, pp. 301–337).

———. 2002a. *Einstein from 'B' to 'Z'.* Boston: Birkhäuser. *Einstein Studies,* Vol. 9.

———. 2002b. *The First Two Acts.* In Stachel (2002a, pp. 261–292). Reprinted in Renn (2007a, Vol. 1, pp. 81–111).

———. 2007. "The Story of Newstein or: Is Gravity just another Pretty Force?" In Renn (2007a, Vol. 4, pp. 1039–1976).

Stanley, Matthew. 2019. *Einstein's War. How Relativity Triumphed Amid the Vicious Nationalism of World War I.* New York: Dutton.

Van Dongen, Jeroen. 2010. *Einstein's Unification.* Cambridge: Cambridge University Press.

Walter, Scott. 1999. "Minkowski, Mathematicians, and the Mathematical Theory of Relativity." In Goenner et al. (1999, pp. 45–86).

———. 2007. "Breaking in the 4-Vectors: The Four-Dimensional Movement in Gravitation, 1905–1910." In Renn (2007a, Vol. 3, pp. 193–252).

Wazeck, Milena. 2009. *Einsteins Gegner. Die öffentliche Kontroverse um die Relativitätstheorie in den 1920er Jahren.* Frankfurt am Main: Campus. Translation: *Einstein's Opponents: The Public Controversy About the Theory of Relativity in the 1920s.* Cambridge: Cambridge University Press, 2014.

Weinberg, Steven. 1972. *Gravitation and Cosmology. Principles and Applications of the General Theory of Relativity.* New York: John Wiley & Sons.

Weinstein, Galina. 2018. "Why Did Einstein Reject the November Tensor in 1912–1913, Only to Come Back to It in November 1915?" *Studies in History and Philosophy of Modern Physics* 62:98–122.

Weyl, Hermann. 1918. "Reine Infinitesimalgeometrie." *Mathematische Annalen* 2:384–411. Translation of pp. 384–401 in Renn (2007a, Vol. 4, pp. 1089–1105).

Wiechert, Emil. 1916. "Perihelbewegung des Merkur und die allgemeine Mechanik." *Physikalische Zeitschrift* 17:442–448.

Wright, Joseph Edmund. 1908. *Invariants of Quadratic Differential Forms.* Cambridge: Cambridge University Press.

Printed in the United States
by Baker & Taylor Publisher Services